Y0-AEU-255

Recent Advances in Mechanistic and Synthetic Aspects of Polymerization

NATO ASI Series

Advanced Science Institutes Series

A Series presenting the results of activities sponsored by the NATO Science Committee, which aims at the dissemination of advanced scientific and technological knowledge, with a view to strengthening links between scientific communities.

The series is published by an international board of publishers in conjunction with the NATO Scientific Affairs Division

A Life Sciences B Physics	Plenum Publishing Corporation London and New York
C Mathematical and Physical Sciences	D. Reidel Publishing Company Dordrecht, Boston, Lancaster and Tokyo
D Behavioural and Social Sciences E Applied Sciences	Martinus Nijhoff Publishers Dordrecht, Boston and Lancaster
F Computer and Systems Sciences G Ecological Sciences H Cell Biology	Springer-Verlag Berlin, Heidelberg, New York, London, Paris, and Tokyo

Series C: Mathematical and Physical Sciences Vol. 215

Recent Advances in Mechanistic and Synthetic Aspects of Polymerization

edited by

M. Fontanille

University of Bordeaux I, Talence, France

and

A. Guyot

C.N.R.S. - Laboratory of Organic Materials,
Lyon-Vernaison, France

D. Reidel Publishing Company

Dordrecht / Boston / Lancaster / Tokyo

Published in cooperation with NATO Scientific Affairs Division

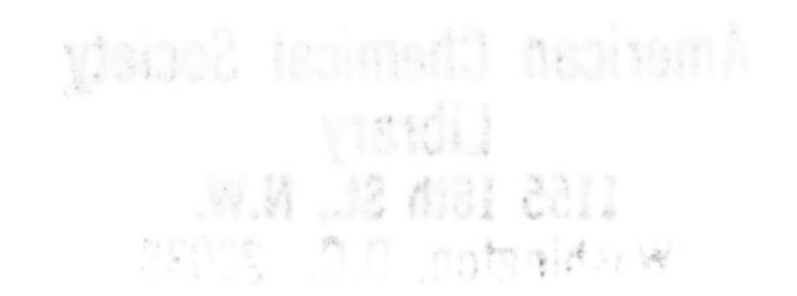

Proceedings of the NATO Advanced Research Workshop on
Frontiers in Polymerization Catalysis and Polymer Synthesis
Bandol, France
February 2-6, 1987

Library of Congress Cataloging in Publication Data

NATO Advanced Research Workshop on Frontiers in Polymerization Catalysis and Polymer Synthesis (1987: Bandol, France)
Recent advances in mechanistic and synthetic aspects of polymerization / edited by M. Fontanille and A. Guyot.
p. cm.—(NATO ASI Series. Series C, Mathematical and physical sciences; vol. 215)
"Proceedings of the NATO Advanced Research Workshop on Frontiers in Polymerization Catalysis and Polymer Synthesis, Bandol, France, February 2–6, 1987"—P. preceding CIP t.p.
"Published in cooperation with NATO Scientific Affairs Division."
Includes index.
ISBN 90–277–2602–7
1. Polymers and polymerization—Congresses. 2. Catalysis—Congresses. I. Fontanille, M. (Michel), 1936– . II. Guyot, A. (Alain), 1931– . III. North Atlantic Treaty Organization. Scientific Affairs Division. IV. Title. V. Series: NATO ASI series. Series C, Mathematical and physical sciences; no. 215.
QD380.N34 1987
541.3'93—dc 19
87–19383
CIP

Published by D. Reidel Publishing Company
P.O. Box 17, 3300 AA Dordrecht, Holland

Sold and distributed in the U.S.A. and Canada
by Kluwer Academic Publishers,
101 Philip Drive, Assinippi Park, Norwell, MA 02061, U.S.A.

In all other countries, sold and distributed
by Kluwer Academic Publishers Group,
P.O. Box 322, 3300 AH Dordrecht, Holland

D. Reidel Publishing Company is a member of the Kluwer Academic Publishers Group

Printed in The Netherlands

TABLE OF CONTENTS

IV - **POLYMERIZATIONS IN ORGANIZED MEDIUM**
Chairmen : K.H. REICHERT and S.L. REGEN

V - **CARBENES AND POLYMERIZATION**
Chairman : K.J. IVIN

VI - **CHIRAL POLYMERS**
Chairman : F. CIARDELLI

VII - **POLYMERIZATIONS BY COORDINATION CATALYSTS**
Chairman : A. GUYOT

LIST OF ALL PARTICIPANTS IN NATO ADVANCED RESEARCH WORKSHOP ON "FRONTIERS IN POLYMERIZATION CATALYSIS AND POLYMER SYNTHESIS"

BALLARD D.G.H., I.C.I. Ldt. Corporate Laboratory - Runcorn - CHESHIRE (U.K.)

BANDERMANN F., Universität Gesamthochschule Essen - ESSEN (F.R.G.)

BOILEAU S., Collège de France - PARIS (France)

CHIELLINI R., University of Pisa - PISA (Italy)

CIARDELLI F., University of Pisa - PISA (Italy)

CORRIU R., Université des Sciences et Techniques du Languedoc - MONTPELLIER (France)

CUZIN D., ATOCHEM - Centre d'Etude de Recherche et de Développement - SERQUIGNY (France)

DHAL K., Universität Düsseldorf - DUSSELDORF (F.R.G.)

FARINA M., Università Degli Studi di Milano - MILANO (Italy)

FAUVARQUE J.F., Laboratoire de Marcoussis - MARCOUSSIS (France)

FIJIKAWA N., Sophia University - TOKYO (Japan)

FINK G., Max Planck Institut für Kohlenforschung - MULHEIM a.d. RUHR (F.R.G.)

FONTANILLE M., Université de Bordeaux I - TALENCE (France)

GIANNINI U., Montedison Research Center - NOVARA (Italy)

GRUBBS R.H., California Institute of Technology - PASADENA, California (U.S.A.)

GUYOT A., C.N.R.S. - Laboratoire des Matériaux Organiques - VERNAISON (France)

INOUE S., University of Tokyo - TOKYO (Japan)

IVIN K.J., Queen's College - BELFAST (U.K.)

JEROME R., Institut de Chimie de l'Université de Liège - LIEGE (Belgique)

KAMINSKY W., Universität Hamburg - HAMBURG (F.R.G.)

KENNEDY J., University of Akron - Institute of Polymer Science - AKRON, Ohio (U.S.A.)

KLABUNDE U., Experimental Station - E.I. Du Pont de Nemours Inc. - WILMINGTON, Delaware (U.S.A.)

KRAUSS H.L., Universität Bayreuth - BAYREUTH (F.R.G.)

LEISING F., Rhône-Poulenc - Centre de Recherche d'Aubervilliers - AUBERVILLIERS (France)

McGRATH J., Virginia Polytechnic Institute - BLACKSBURG, Virginia (U.S.A.)

MACHON J.P., C.d.F. Chimie - Centre de Recherche Nord - MAZINGARBE (France)

MULLER A., Johannes Gutenberg Universität - Institut für Physikalische Chemie - MAINZ (F.R.G.)

NOLTE J.M., Organisch Chemisch Laboratorium, Afd. Fysisch Organische Chemie - UTRECHT (The Netherlands)

OGATA N., Sofia University - TOKYO (Japan)

REGEN S., Lehigh University - BETHLEHEM, Pensylvania (U.S.A.)

REICHERT K.H., Technische Universität Berlin - BERLIN (F.R.G.)

REVILLON A., C.N.R.S. - Laboratoire des Matériaux Organiques - VERNAISON (France)

ROSE J.B., University of Surrey - GUILDFORD, Surrey (U.K.)

SIGWALT P., Université Pierre et Marie Curie - PARIS (France)

SILLON B., CEMOTA - VERNAISON (France)

di SILVESTRO, Università Degli Studi di Milano - MILANO (Italy)

SIOVE A., Université de Paris XIII - VILLETANEUSE (France)

SOGAH D.Y., E.I. Du Pont de Nemours Inc. - WILMINGTON, Delaware (U.S.A.)

SOUM A., Université de Bordeaux I - TALENCE (France)

SPITZ R., C.N.R.S. - Laboratoire des Matériaux Organiques - VERNAISON (France)

TAN Y.Y., University of Groningen - GRONINGEN (The Netherlands)

TEYSSIE Ph., Institut de Chimie de l'Université de Liège - LIEGE (Belgium)

VAN DER LINDE R., D.S.M. Resins BV - ZWOLLE (The Netherlands)

WAGNER D., B.A.S.F. Aktiengesellschaft - LUDWIGSHAFFEN (F.R.G.)

WEBSTER O.W., E.I. Du Pont de Nemours Inc. - Central Research Department - WILMINGTON, Delaware (U.S.A.)

WITTE M., BAYER A.G. - LEVERKUSEN (F.R.G.)

WULFF G., Universität Dusseldorf - DUSSELDORF (F.R.G.)

PREFACE

Due to their specific properties, polymers with well-defined structures have been receiving increasing attention over the last several years. Owing to the wide variability of their properties, these specialty polymers have been used in various areas from biomedical engineering to electronics or energy applications. The synthesis of such polymers necessitates the use of new methods of polymerization which derived from an insight into the mechanism of polymerization reactions.

A NATO Advanced Research Workshop on "Frontiers in Polymerization Catalysis and Polymer Synthesis" was held in BANDOL (FRANCE) in February 1987. Its aim was to assess the new polymerization methods, as well as the latest advances in the mechanisms of conventional polymerization reactions together with their applications to the synthesis of new macromolecular structures.

The financial support from the NATO Scientific Affairs Division which covered the`lecturers' accomodation and travel expenses as well as the organization charges of this event gave it international scope. Several industrial companies participate at the meeting and contributed to it success. The organizors who are also editors of these proceedings, want to express their thanks to both NATO Scientific Affairs Division and the companies present at the meeting.

Michel FONTANILLE
Alain GUYOT

SCIENTIFIC CONTENT OF THE ARW

Although very significant progress has been obtained for the last decade in the field of polymeric materials, simply by blending conventional and engineering plastics in the presence of suitable additives, it is clear that most of the progress is to be expected from new paths in polymer synthesis in two main directions : for the materials to be produced on a large scale (commodity polymers), simpler polymerization processes with tighter control of the polymer structure are needed ; in the case of specialty polymers, which need a variety of very specific properties desired for very specific uses, a variety of synthetic methods allowing exact control of the structures as well as adequate processing conditions, are to be developed. In both cases, we have to start from simple and cheap small molecules in combination with suitable methods of activation.

The purpose of this workshop was to discuss the possibilities opened by the latest polymerization methods (group transfer polymerization for instance) as well as by the latest progress in polymerization catalysis (metallo-carbene, coordination chemistry) or also through the use of unusual and well-organized media (bacteria, chlatrates).

Much attention was paid to the methods allowing to get living polymers, i.e. systems in which initiation and propagation reactions are only observed without termination and transfer reactions ; it was the case of anionic polymerization ; it is also the case now for cationic polymerization, group transfer polymerization, metallo-carbene or transition metal alkyl compound-initiated polymerization, both for functional and hydrocarbon monomers. Such polymerizations allow to get polymers with well defined molecular weights and narrow molecular weight distribution (MWD). Even in polycondensation, narrow MWD were obtained upon phase transfer conditions. One case of "immortal" polymerization was described in which the molecular weight is controlled by a transfer reaction without consumption of the catalyst and, of course, without termination. It is important to note that, besides their interest for the synthesis of well-defined structures (narrow MWD, block copolymers) these living systems with stable active centers offer better possibilities for understanding the intimate nature of the polymerization mechanism.

A second general point of interest was the stereoregularity of the polymer structure. This can be achieved now by a variety of methods and is no longer restricted to heteregeneous coordination catalysis. 99 % isotactic polypropylene, optically active, was obtained through the use of homogeneous bimetallic Ziegler-type catalysts, but highly isotactic polypropylene may also be prepared by radical polymerization in chlatrates channels. Stereoregular advances were obtained in the methods of synthesis of chiral structures, the chirality being in the backbone and no longer only in side-chains ; it is worth noting here the use of precursors of chiral structures, as well as the induction of chirality in a growing chain.

Advances in more specific fields are also to be noted ; thus thermostable and processable oligomers obtained by polycondensation can be thermoset upon successive polymerization of telechelic functional groups ; further new polyhydrocarbon structures are obtained through a new type of coordination catalysts which allow the migration of the active centers along the chain ; other catalysts allow to introduce functional groups in polyolefins.

Despite such progress, a lot of efforts were devoted to understanding the mechanism of the actual polymerization catalysis, even if some industrial catalysts have now reached satisfactory activity and selectivity. It would also be important to develop new methods to produce high molecular weight polycondensates at lower temperatures possibly through biomimetic mechanisms. Some directions towards these and other objectives were identified during the final discussion and were included into the recommendation for future research which must be more and more active considering the very broad range of research possibilities.

I - GROUP TRANSFER POLYMERIZATION

Chairman : O.W. WEBSTER

RECENT ADVANCES IN THE CONTROLLED SYNTHESIS OF ACRYLIC POLYMERS BY GROUP TRANSFER POLYMERIZATION

O. W. Webster and D. Y. Sogah
E. I. Du Pont de Nemours and Company, Inc.
Central Research and Development Department
Experimental Station
Wilmington, Delaware 19898

ABSTRACT. Group transfer polymerization (GTP) is a new type of "living" polymerization for acrylic monomers. The polymerization proceeds by repeated addition of monomer to a growing chain end carrying a silyl ketene acetal group. During addition, the silyl group transfers to the incoming monomer regenerating a new ketene acetal function (Equation).

$$\mathrm{Me_2C{=}C(OSiMe_3)(OMe)} + x\,\mathrm{CH_2{=}C(Me)CO_2Me} \xrightarrow{\mathrm{HF_2^-}} \mathrm{Me}\left(\mathrm{C(CO_2Me)(Me){-}CH_2}\right)_x\mathrm{{-}C(Me){=}C(OSiMe_3)(OMe)}$$

A catalyst is required for GTP to operate. Bifluoride salts, certain Lewis acids and a variety of oxyanions work best. A key advantage of GTP is that it is a living system, which operates above room temperature. This property enables one to control molecular weight and polydispersity, as well as monomer sequences under industrially significant conditions. The synthesis of block, star, graft, chelic and telechelic polymers, their properties and their uses is described.

1. INTRODUCTION

Group Transfer Polymerization (GTP) is a fundamentally new method for the living synthesis of acrylic polymers[1-12]. It involves the repeated addition of monomer to a growing polymer chain end bearing a reactive silicon group. During the addition, the silyl group transfers to incoming monomer, hence the name group transfer polymerization. Scheme 1 illustrates the method for polymerization of methyl methacrylate in which the reactive end group is a silyl ketene acetal.

M. Fontanille and A. Guyot (eds.), Recent Advances in Mechanistic and Synthetic Aspects of Polymerization, 3–21.

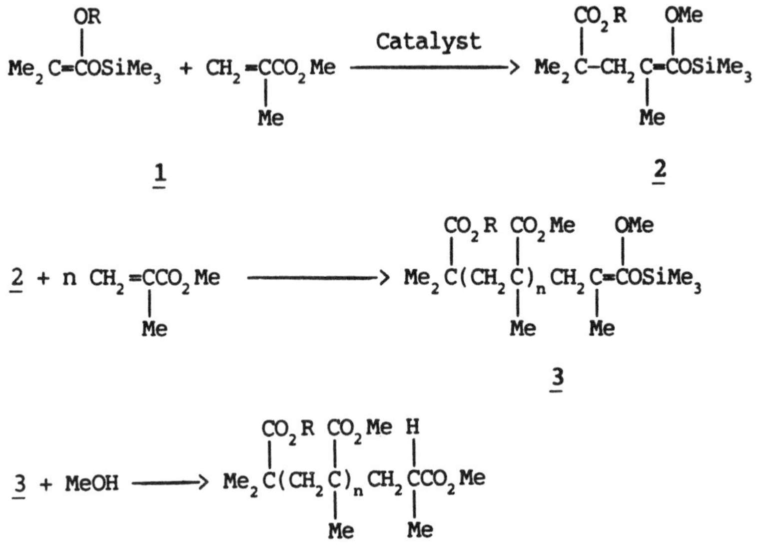

$$\underset{\mathbf{1}}{Me_2C{=}C(OR)OSiMe_3} + CH_2{=}C(Me)CO_2Me \xrightarrow{\text{Catalyst}} \underset{\mathbf{2}}{Me_2C(CO_2R){-}CH_2C(Me){=}C(OMe)OSiMe_3}$$

$$\mathbf{2} + n\,CH_2{=}C(Me)CO_2Me \longrightarrow \underset{\mathbf{3}}{Me_2C(CO_2R)(CH_2C(Me)(CO_2Me))_n CH_2C(Me){=}C(OMe)OSiMe_3}$$

$$\mathbf{3} + MeOH \longrightarrow Me_2C(CO_2R)(CH_2C(Me)(CO_2Me))_n CH_2CH(Me)CO_2Me$$

Scheme 1

After polymerization, the silicon function is removed from the polymer by methanol or other proton sources. A catalyst, for example a soluble bifuoride, is necessary for GTP to proceed. The chief characteristic of GTP is that it is a "living" polymerization and thus, molecular weight and polymer architecture can be controlled.

2. THE GTP PROCESS

2.1. Catalysts

Catalysts for GTP fall into two classes, anionic and Lewis acid. Among the anionic catalysts are fluoride[1], $Me_3SiF_2^{-}$[1], azide[1], cyanide[1], carboxylates[12], phenolates[12], sulfinates[12], phosphinates[12], nitrite[12] and cyanate[12]. Even though GTP does not proceed when protonic solvents or monomers are used, some of the better anionic catalysts are derived from the combination of an anion with its conjugate acid e.g. bifluoride[1] or biacetate[12]. The anionic catalysts are usually used as their tetraalkyl ammonium or tris(dimethylamino)sulfonium (TAS) salts. They catalyze GTP by coordination with the silicon atom[6,9]. GTP works best with low levels of catalyst[13,14] as little as 0.1% based on initiator.

Examples of Lewis acid catalysts for GTP are zinc halides, diakylaluminum halides and tetraalkyl aluminoxanes. These catalysts operate by coordination to monomer[3]. Relatively large amounts of catalyst are required, 10% or more based on initator. Anionic catalysts work best for methacrylates, Lewis acid catalysts for acrylates.

2.2. Initiators

To control molecular weight in a living polymerization, it is essential that the rate of initation be higher or the same as the rate of propagation. Therefore, the best initiators will be similar in structure to the living end of the polymer. Thus for methacrylate GTP, a silyl ketene actal with an alkyl group is the 2 position is ideal (1). Large alkyl groups on the silicon reduce the polymerization rate[10]. α-Trimethylsilyl, -stanyl and -germyl esters operate as GTP initiators, possible by first rearranging to the ketene acetals[10] (Scheme 2); but they tend to give poorer molecular weight control than do the silyl ketene acetals, since their rate of the rearrangement to the ketene acetal form is slower than the propagation rate.

$$Me_3SiR_2CO_2Me \rightleftharpoons R_2C{=}C(OMe)OSiMe_3$$

Scheme 2

The titanium enolate 4 initiates GTP at low temperatures[15].

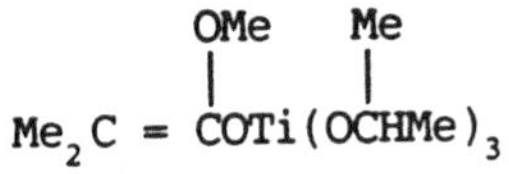

$$Me_2C = C(OMe)OTi(OCHMe)_3$$

4

Silyl derivatives that add to methacrylates to produce ketene acetals, for example, Me_3SiCN[1,13,14] (Scheme 3), Me_3SiSMe[10,11,16], Me_3SiSPh[10], Me_3SiCR_2CN[10], or $R_2PO\ SiMe_3$[8,10] also initiate GTP.

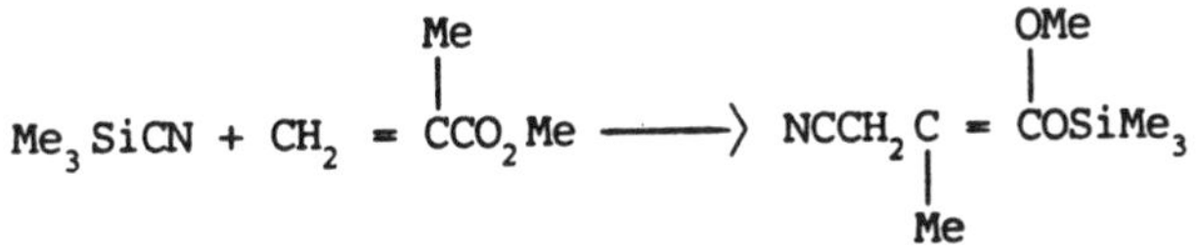

$$Me_3SiCN + CH_2 = C(Me)CO_2Me \longrightarrow NCCH_2C(Me) = C(OMe)OSiMe_3$$

Scheme 3

2.3. Reaction Conditions

All reagents and solvents must be dry and pure and the polymerization must be conducted in a dry atmosphere. Oxygen, however, does not interfere. Typical solvents include toluene, tetrahydrofuran, 1,2-dimethoxyethane, chlorobenzene, and N,N-dimethylformamide for nucleophilic catalysis and toluene, dichloromethane, and 1,2-dichloroethane for Lewis acid catalysis. Acetonitrile is silated by the initiator under some conditions and is not recommended for GTP[8,13,14].

Temperatures as low as -100°C and as high as about 120°C can be used but 0 to 50°C are preferred. For polymerization of acrylates, temperatures at or below 0°C work best. Polymerization is generally very fast and complete and can be controlled by the rate monomer is added during the course of the run. With certain initiators and catalysts, induction periods are noted, for example, with trimethylsilyl cyanide.

2.4. Molecular Weight Control

As in any other "living" polymerization, the molecular weight of polymer obtained by GTP is determined by the molar ratio of initiator to monomer. Although this property of GTP allows one to control molecular weight accurately in the 1,000 to 20,000 range, obtaining high molecular weight polymer is more difficult since the amount of initiator needed is so low that it begins to match interfering impurity levels. By using very pure monomers, solvents, catalysts, and initiators, however, one can make polymer in the 100,000 to 200,000 molecular weight range.

The molecular weight distribution, Mw/Mn, for GTP polymers under the best conditions is close to one, as it should be for a living polymerization. Dispersities higher than one could be due to chain termination by cyclization of the end group[17] (Scheme 4), reaction of the end group with impurities, an initiation rate which is slower than the propagation rate or a catalyst transfer rate which is slower than the propagation rate.

Polymer—C(CO_2Me)(Me)—CH_2—C(Me)(CO_2Me)—CH_2—C(Me)=C(OMe)$OSiMe_3$

↓

Polymer–(cyclohexanone ring bearing three CO_2Me groups) + $(Me)_3SiOMe$

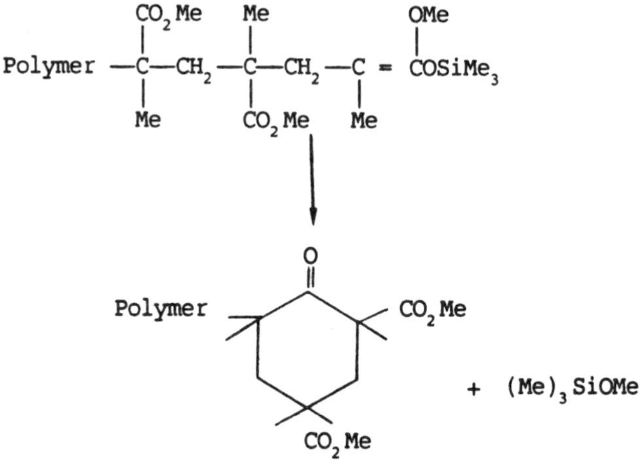

Scheme 4

2.5. Stereochemistry

The principle parameters influencing tacticity of GTP PMMA are catalyst and temperature. The precise variation of tactic composition of GTP PMMA with polymerization temperature was determined by performing isothermal polymerizations of MMA with $TASHF_2$ as catalyst and THF as solvent[7,10]. The triad composition of PMMA was determined by ^1H nmr at 360 mHz with polymerization temperatures varied from 60 to

-90°C. The isotactic component remains small under all conditions, while the syndiotatic component increases from 50 to 80% as the temperature is lowered. The dyad statistics are essentially Bernouillian. Therefore, an Arrhenius plot of 1000/polymerization temperature vs. ln(m/r) provides estimates of the enthalpy ($\Delta\Delta H^{\neq}$ = 0.97 kcal/mol) and entropy ($\Delta\Delta S^{\neq}$ = 0.98 e.u.) differences between m and r dyad formation. Most interestingly, these activation parameter differences are identical, within experimental error, to those reported by Fox and Schnecko[18] ($\Delta\Delta H^{\neq}$ = 1.07 Kcal/mol, $\Delta\Delta S^{\neq}$ = 0.99 e.u.) and by Otsu, Yamada, and Imoto[19] ($\Delta\Delta S^{\neq}$ = 0.96 Kcal/mol, $\Delta\Delta S^{\neq}$ = 0.94 e.u.) for free radical polymerization of MMA. There does not appear to be a detailed study of PMMA tacticity and polymerization temperature for anionic polymerization of MMA for comparison with GTP. Other work[20,21] on tacticity of GTP agrees with this data for poly(methyl methacrylate). However poly(t-butyl methacrylate) is anomalous[20].

Lewis acid catalysis of GTP of MMA generally provides a much more syndiotactic PPMA than do anion catalysts, but detailed temperature studies have not been carried out.

3. MECHANISM OF GTP

The principle mechanistic question to be answered is whether GTP occurs by dissociation of the silyl group and addition of monomer to the resulting anion or is the transfer taking place by some kind of bond rearrangement so that each silicon remains with the living chain it started with.

Two key experiments show that in GTP the identity of the silicon atom of the initiator remains invariant throughout the growth of the polymer chain. In bifluoride catalyzed GTP, there is no exchange of the silyl group on the chain end with added trialkylsilyl fluoride. Therefore, if the function of the catalyst is to generate a small amount of enolate anion for anionic polymerization, these chain ends are not recapped by silyl fluoride[6,9] (Scheme 5).

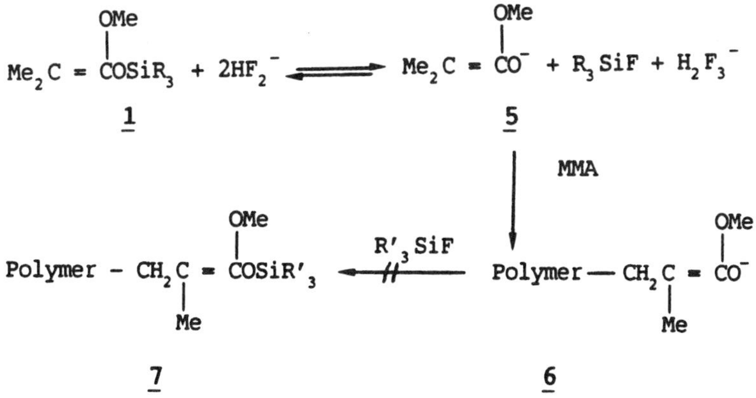

$$\underset{\mathbf{1}}{Me_2C=C(OMe)OSiR_3} + 2HF_2^- \rightleftharpoons \underset{\mathbf{5}}{Me_2C=C(OMe)O^-} + R_3SiF + H_2F_3^-$$

$$\mathbf{5} \xrightarrow{MMA} \underset{\mathbf{6}}{Polymer-CH_2C(Me)=C(OMe)O^-}$$

$$\underset{\mathbf{7}}{Polymer-CH_2C(Me)=C(OMe)OSiR'_3} \overset{R'_3SiF}{\not\longleftarrow} \underset{\mathbf{6}}{Polymer-CH_2C(Me)=C(OMe)O^-}$$

Scheme 5

In addition, in a double labelling experiment, it was shown that exchange does not occur between living poly(buty methacrylate) chain ends and living poly(methyl methacrylate) chain ends.[6,9] This results rules out the dissociative route depicted in Scheme 6 where at any one time there would be only trace amounts of 6 present.

$$\underline{1} + 2HF_2^- \longrightarrow \underline{5} + R_3SIF + H_3F_2^-$$

$$\underline{5} + MMA \longrightarrow \underline{6}$$

$$\underline{6} + \underline{7} \rightleftharpoons \underline{6}' + \underline{7}'$$

Scheme 6

Having dispensed with the two most likely dissociative processes, we proposed a mechanism wherein the nucleophilic catalyst first coordinates with the silicon atom, then monomer adds through a transition state in which it is also coordinated to the silicon[1,6,9] (see structure 8).

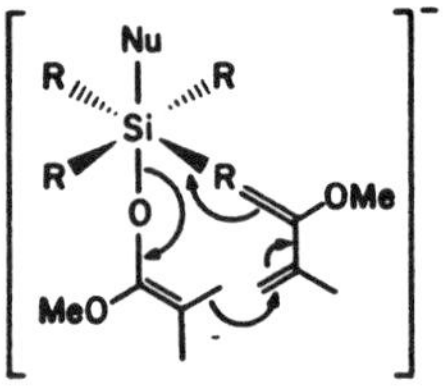

Nu
R
R
Si
R
R
O
OMe
MeO

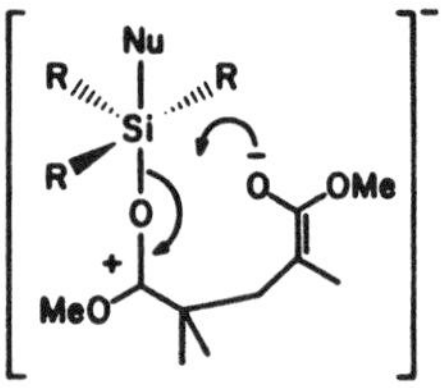

8

Nu
R
R
Si
R
O
O
OMe
+
MeO

9

Based on ab initio calculations[9,22], it appears likely that carbon-carbon bond formation is the first event after coordination of catalyst to silicon (see structure 9). In a more recent study[20], Wnek also came to similar conclusions. While we continue with the search for plausible mechanisms, we bring up the possibility that intermediate 10 or 11 intercedes before silyl transfer. This additional step gives one a pathway with six and four or four and six membered transition states (Scheme 7). In addition, these intermediates would unify Lewis Acid catalyzed, with nucleophile catalyzed GTP. We must emphasize that this is purely speculative in absence of any real evidence.

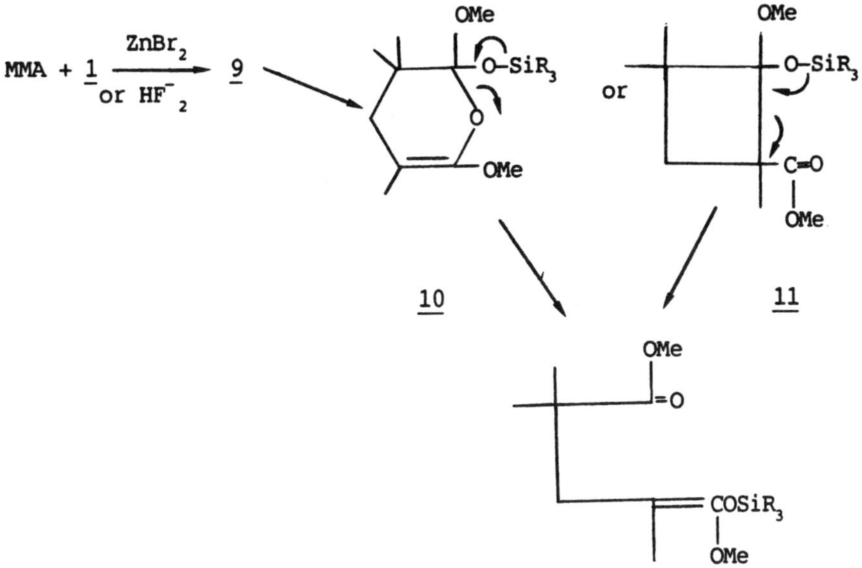

MMA + 1 $\xrightarrow[\text{or } HF_2^-]{ZnBr_2}$ 9 → 10 or 11 →

Scheme 7

Preliminary kinetic results indicate that the initiation rate is faster than propagation rate and that the rate is second order with respect to bifluoride ion concentration[17]. Several additional questions, as well as those mentioned above, need to be answered. How do other monmers, such as acrylonitrile, acrylates, and pentadienoate fit into the mechanistic picture? Why is a low level of catalyst the most effective[14]? How does chain termination occur? What is the cause of the induction period observed in a number of cases? Work directed to answering these questions is in progress in our laboratories and others.

With respect to the induction periods observed in GTP initiated with trimethylsilyl cyanide, the cause has been determined to be complexation of the catalyst with initiator[8,10,13,14]. Thus polymerization is slow until all of the trimethylsilyl cyanide is consumed by addition to monomer (Scheme 8).

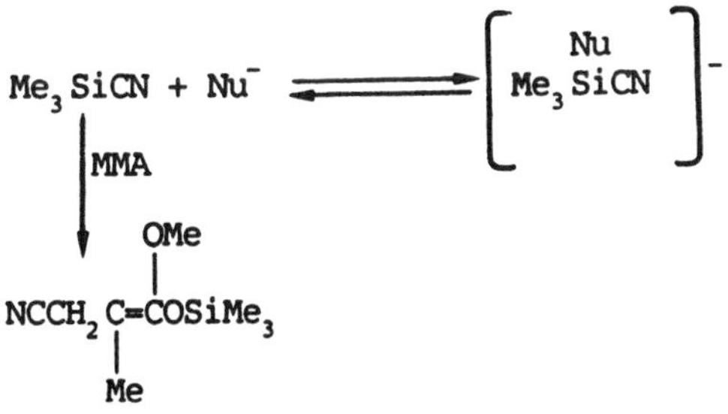

$$Me_3SiCN + Nu^- \rightleftharpoons [Me_3SiCN \cdot Nu]^-$$

$$Me_3SiCN \xrightarrow{MMA} NCCH_2C(Me){=}C(OMe)OSiMe_3$$

Scheme 8

4. GTP MONOMERS

4.1 Methacrylates

Group-Transfer polymerization works best with methacrylates, to which most of the published data refer[1,2,10,11]. Functional groups sensitive to free radicals, such as 4-vinylbenzyl[24] allyl and sorbyl, can be present on the ester groups of the monomer and remain unreacted in the polymer. This type of polymer is useful in air-drying finishes. Attempts to prepare polymers with pendent allyl or sorbyl groups by free-radical initiation give insoluble, cross-linked products.

Active hydrogen compounds interfere with GTP and stop chain growth if present in amounts greater than the initiator concentration. Thus, the polymerization of methacrylic acid or hydroxyethyl methacrylate requires the use of protective groups that can be subsequently removed. Carboxy groups can be introduced by using trimethylsilyl methacrylate <u>12</u> as the monomer for GTP[11,20] followed by mild hydrolysis. Pendent hydroxy groups can be introduced by using 2-(trimethylsilyloxy)ethyl methacrylate, <u>13</u> as the monomer[4,10]. The trimethyl silyl group is removed from this polymer with methanolic tetrabutylammonium fluoride. Glycidyl methacrylate can be polymerized without involving the epoxy groups provided the temperature is held below 0°C.

$$CH_2=C(Me)CO_2SiMe_3 \qquad\qquad CH_2=C(Me)CO_2CH_2CH_2OSiMe_3$$

12 **13**

Methacrylates bearing liquid crystalline ester functions[24,25] and other large groups such as carbazol[24] polymerize easily. Methacrylate ended macromonomers from polystyrene[26] and poly(2,6-dimethyl-1,4-phenylene oxide)[24] have been polymerized. Copolymers with these macromonomers and methyl methacrylate form more readily.

4.2 Acrylates

Although acrylates polymerize much faster than methacrylates, it is difficult to obtain high molecular weight polymers. Lewis acid catalysis is better than anion catalysis. GTP of simple acrylates using bifluoride catalysis generally leads to polymers with a broader molecular weight distribution than is observed in GTP of the corresponding methacrylate or GTP of acrylates using Lewis acid catalysts. Thus, polymerization of butyl acrylate in THF at 0°C gave poly(butyl acrylate) with M_n 27,200, M_w 59,400, D 2.16 (Theor. M_n 26,100).

To obtain insight into the causes of the molecular weight broadening, living ethyl acrylate oligomer (degree of polymerization, 4) was treated with p-nitrobenzyl bromide at -78°C. Chromatographic purification gave a 60% yield of the benzylated product containing both <u>internal</u> and <u>terminal</u> p-nitrobenzyl groups in the ratio of 9:1 (Scheme 9). These results suggest that the trimethylsilyl group is capable of isomerizing to an internal position of the polyacrylate chain. However, ^{13}C NMR studies of GTP poly(ethyl acrylate) gave no evidence of branching, suggesting that the internal silyl ketene acetal, while capable of reacting with a benzyl halide, is too sterically hindered to initiate a branch point. The isomerization is presumable slower than chain propagation so that polyacrylates remain living; but the decrease in the concentration of "useful" living ends caused by the isomerization may be accountable for the observed broadening of molecular weight distribution. The problems encountered in the anion catalyzed-GTP of acrylates may be avoided by using Lewis acid catalysts (*vide supra*).

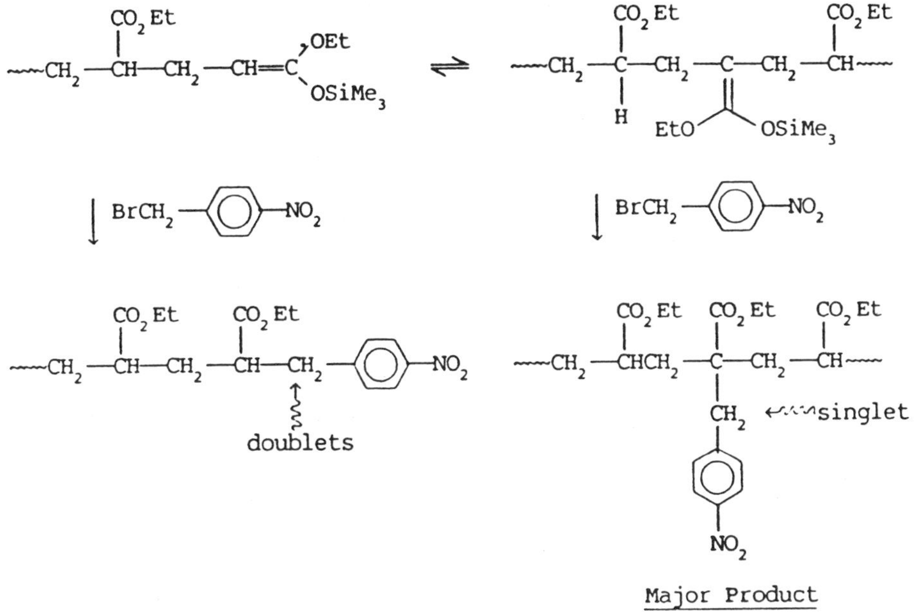

$$\sim CH_2-CH(CO_2Et)-CH_2-CH{=}C(OEt)(OSiMe_3) \rightleftharpoons \sim CH_2-CH(H)(CO_2Et)-CH_2-C(=C(OEt)(OSiMe_3))-CH_2-CH(CO_2Et)\sim$$

$\downarrow BrCH_2-C_6H_4-NO_2$ (both)

$$\sim CH_2-CH(CO_2Et)-CH_2-CH(CO_2Et)-CH_2-C_6H_4-NO_2$$

doublets

$$\sim CH_2-CH(CO_2Et)CH_2-C(CO_2Et)(CH_2-C_6H_4-NO_2)-CH_2-CH(CO_2Et)\sim$$

singlet

Major Product

Scheme 9

The difference in reactivity of acrylates and methacrylates can be exploited to make a polyacrylate with pendant methacrylate groups[3] (Scheme 10).

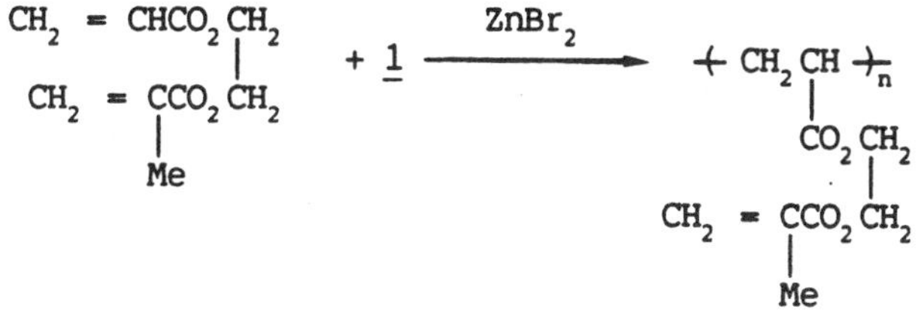

$$CH_2{=}CHCO_2CH_2CH_2O_2C(Me)C{=}CH_2 + \underline{1} \xrightarrow{ZnBr_2} -\!\!\left(CH_2CH(CO_2CH_2CH_2O_2CC(Me){=}CH_2)\right)\!\!-_n$$

Scheme 10

4.3. Other Acrylic Monomers

In addition to methacrylates and acrylates, GTP was used for polymerization of N,N-dimethylacrylamide, acrylonitrile, and methacrylonitrile.[10] Because of the solubility characteristics of poly(acrylonitrile), DMF is the solvent of choice for polymerization of acrylonitrile (AN). Since the polymerization rate of AN is very high compared with that of MMA, the normal technique of feeding the monomer resulted in uncontrolled localized polymerization of the monomer before complete mixing could be achieved. In a batch polymerization in which the TAS bifluoride was added to a solution of AN and 1 in DMF, uncontrolled polymerization occurred in the vicinity of the catalyst before mixing could be achieved. Best results were achieved with a batch polymerization in which the initiator, 1, was added to a DMF soltuion of AN and $TASHF_2$ cooled to -50°C. Although this procedure avoided gel formation, the molecular weight distribution is broad (D 3.79) because the rate of initiation by 1 is slower than the rate of propagation.

GTP of metharylonitrile in N,N-dimethylacetamide solution using 2-trimethylsilylisobutyronitrile[10] as initiator gave an 87% conversion to poly(methacrylonitrile) with M_n 5800, D 2.04 (Theor. M_n 3570). ^{13}C NMR studies of the resulting polymer showed nearly random tacticity.

GTP of α-methylene-γ-butyrolactone by a batch process in propylene carbonate at 20-50° gave poly(α-methylene-γ-butyrolactone) in 67% yield[10]. The polymer has an inherent viscosity of 0.072 g/dL, indicative of a substantially lower molecular weight than the polymers obtained by Akkapeddi[27] from α-methylene-γbutyrolactone by anionic- and by free radical-polymerization.

Stille and co-workers[28] have reported GTP of both racemic and chiral α-methylene-γ-methyl-γ-butyrolactone at -78°C in THF. The chiral polymer was much less soluble.

Pentadieneoates polymerize 1,4-by GTP[11,29] (Scheme 11).

$$CH_2 = CH - CH = CHCO_2Me \longrightarrow -\!\!\left(\!- CH_2CH = \underset{\displaystyle CO_2Me}{\underset{|}{CHCH}} -\!\right)_n$$

Scheme 11

5. CONTROL OF POLYMER STRUCTURE BY GTP

5.1. Homopolymer

GTP produces polymethacrylates with linear backbones unlike free radical polymerization where considerable branching occurs. As mentioned previously, molecular weights are controlled by the initator monomer ratio and have narrow dispersity. This narrow dispersity can be used to advantage in the preparation of high solids resin solutions for coatings. Solutions of narrow dispersity GTP polymer are less viscous and more easily applied than polymer solutions of the corresponding molecular weight prepared by free-radical polymerization.

5.1.1. Chelic Polymers. Polymers with reactive functionality on one end (chelic polymers) are ueful tools for synthesis of graft polymers, dispersing agents and ABA block polymers. They can be readily made by GTP by merely starting with a functionalized initiator. Polymers with terminal hydroxyl or carboxyl groups are readily prepared by using the functionalized initiators 14a or 14b, respectively (Scheme 12), and then hydrolyzing the resulting polymer with refluxing methanolic tetrabutylammonium fluoride or with dilute methanolic HCl at ambient temperature to give PMMA—OH (16a) or PMMA—COOH (16b), respectively.

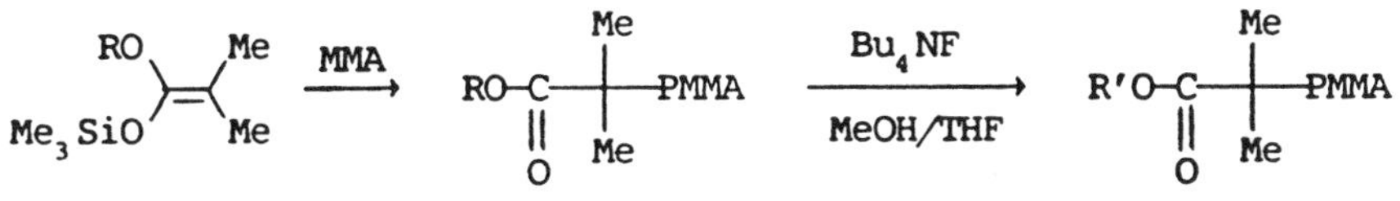

$$(RO)(Me_3SiO)C{=}CMe_2 \xrightarrow{MMA} RO{-}C({=}O){-}CMe_2{-}PMMA \xrightarrow[MeOH/THF]{Bu_4NF} R'O{-}C({=}O){-}CMe_2{-}PMMA$$

14 15 16

a. $R = CH_2CH_2OSiMe_3$ a. $R' = CH_2CH_2OH$

b. $R = SiMe_3$ b. $R' = H$

Scheme 12

Asami used the initator 17 to prepare a macromonomer[31] from methyl methacrylate (Scheme 13).

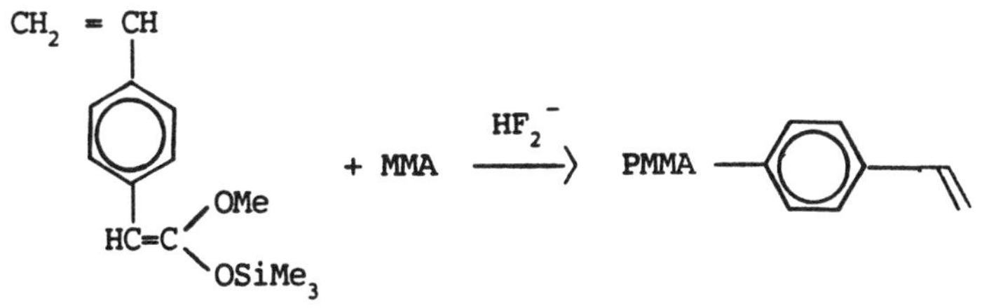

$$CH_2{=}CH{-}C_6H_4{-}HC{=}C(OMe)(OSiMe_3) + MMA \xrightarrow{HF_2^-} PMMA{-}C_6H_4{-}CH{=}CH_2$$

17

Scheme 13

A variety of phosporus-terminated polymers were synthesized by the use of phosphorus-containing silyl ketene acetals 18, prepared by the thermal addition of silyl phosphites, to acrylates and methacrylates in the absence of solvent (Scheme 14).

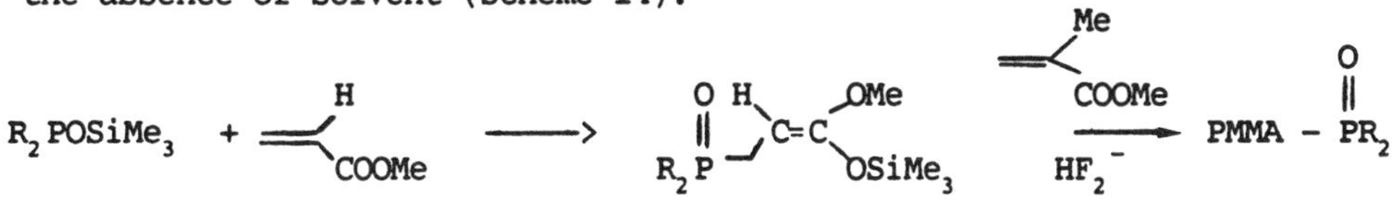

$$R_2POSiMe_3 + CH_2{=}CH{-}COOMe \longrightarrow R_2P({=}O){-}CH_2{-}CH{=}C(OMe)(OSiMe_3) \xrightarrow[HF_2^-]{CH_2{=}C(Me)COOMe} PMMA - P({=}O)R_2$$

18

a. R = EtO b. $R = Me_3SiO$ c. $R = Me_2N$

Scheme 14

Polymers containing a terminal silyl phosphonate group are easily hydrolyzed to the corresponding polymer with a terminal phosphonic acid group. (Scheme 15).

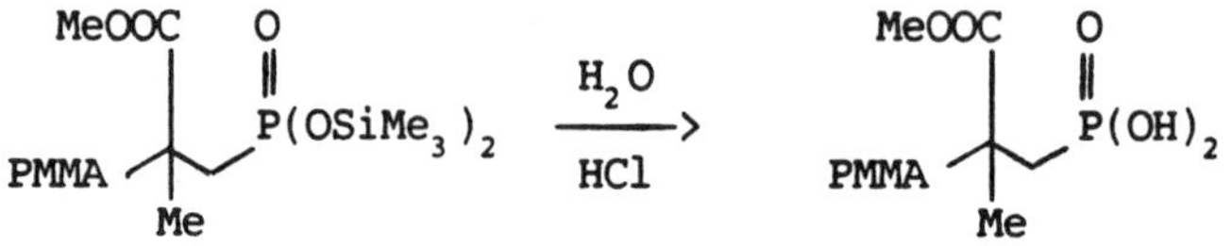

$$\text{PMMA}\!-\!C(\text{Me})(\text{COOMe})\!-\!CH_2\!-\!P(=O)(OSiMe_3)_2 \xrightarrow[\text{HCl}]{H_2O} \text{PMMA}\!-\!C(\text{Me})(\text{COOMe})\!-\!CH_2\!-\!P(=O)(OH)_2$$

Scheme 15

As mentioned earlier, polymers with sulfide end groups are obtained by initiation with Me_3SiSMe[10] and polymers with cyanide end groups, by initiation with Me_3SiCN[10].

Another way to synthesize chelic polymers is to react the living silyl ketene end group with electrophilic reagents. Examples are benzaldehyde to give a masked hydroxyl group 21, bromine to give an α-bromoester[2,31], 22, and 4-(bromomethyl)styrene to give a styryl ended macromonomer, 23 (Scheme 16). A disadvantage to this type of functionalization is that any polymer chains that have died during the polymerization will not react.

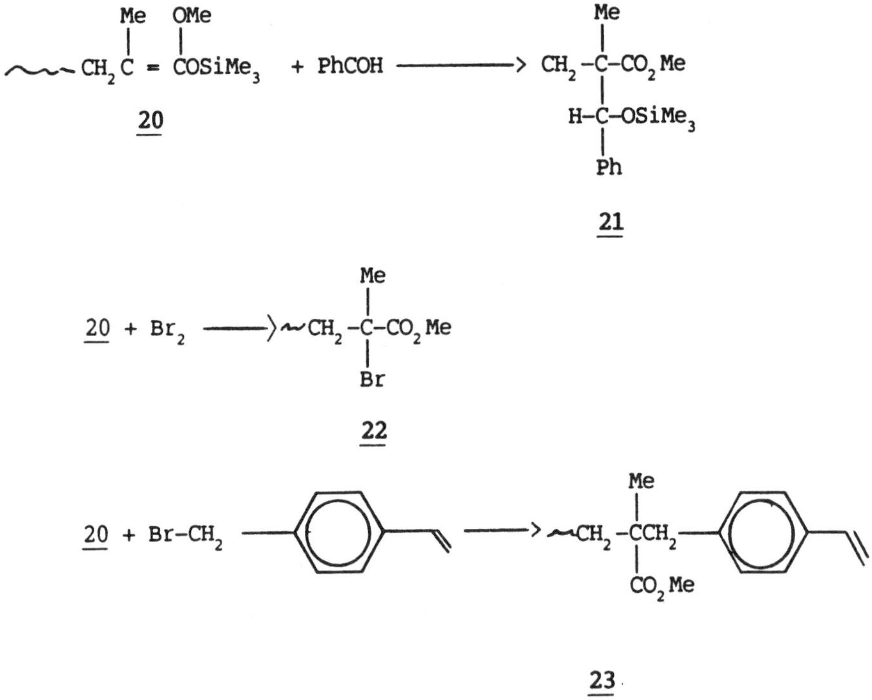

$$\sim\!CH_2C(Me)\!=\!C(OMe)OSiMe_3\ (\underline{20}) + PhCOH \longrightarrow \sim\!CH_2\!-\!C(Me)(CO_2Me)\!-\!CH(Ph)\!-\!OSiMe_3\ (\underline{21})$$

$$\underline{20} + Br_2 \longrightarrow \sim\!CH_2\!-\!C(Me)(Br)\!-\!CO_2Me\ (\underline{22})$$

$$\underline{20} + Br\!-\!CH_2\!-\!C_6H_4\!-\!CH\!=\!CH_2 \longrightarrow \sim\!CH_2\!-\!C(Me)(CO_2Me)\!-\!CH_2\!-\!C_6H_4\!-\!CH\!=\!CH_2\ (\underline{23})$$

Scheme 16

5.1.2. Telechelic Polymers. α,ω-Difunctional (telechelic) polymers are useful for construction of block, as well as, chain extended polymers. They can be readily synthesized by coupling living GTP chelic polymers (see previous section). Suitable coupling agents are p-xylylene dibromide or Br_2 (Scheme 17). With low molecular weight PMMA (in the 2000 range) one can obtain over 99% pure telechelic polymer[2]. However, it is difficult to obtain polymer in the 10,000 to 20,000 molecular weight range of greater than 95% telechelic purity.

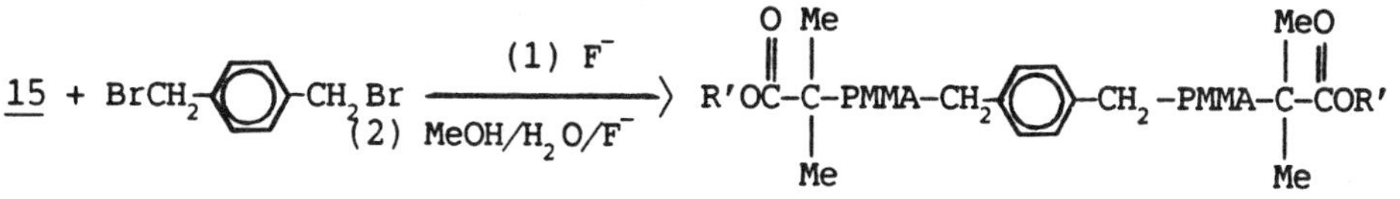

$$\underline{15} + BrCH_2\text{-}C_6H_4\text{-}CH_2Br \xrightarrow[(2)\ MeOH/H_2O/F^-]{(1)\ F^-} R'OC(=O)\text{-}C(Me)_2\text{-PMMA-}CH_2\text{-}C_6H_4\text{-}CH_2\text{-PMMA-}C(Me)_2\text{-}C(=O)OR'$$

a. $R' = CH_2CH_2OH$

b. $R' = H$

Scheme 17

5.2. Random Copolymers

As long as one uses monomers belonging to the same family (all methacrylate or all acrylate, for example), one can make random copolymers simply by adding a mixture of the monomers to the initiator and catalyst. The large difference in reactivity between the various acrylic monomer types (e.g., acrylonitrile, methacrylonitrile, acrylates and methacrylates) prevents random copolymer formation by GTP of their mixtures. At this point, reactivity ratios are not available.

Simultaneously feeding MMA and butyl methacrylate into a THF solution of initiator and solid TAS bifluoride gave a quantitative yield of copolymer of butyl methacrylate (65%) and MMA (35%) with M_n 22,100, M_w 24,500, D 1.11 (Theor. M_n 20,215)[10].

A random terpolymer of MMA (50 wt.%), butyl methacrylate (20 wt.%), and glycidyl methacrylate (30 wt.%) was prepared by GTP with the only special precautions being that the polymerization temperature was kept near 0°C, and one equiv of MMA was added prior to beginning the feed of the mixed monomers to minimize the possibility of termination by reaction of the ketene silyl acetal with the epoxide group of glycidyl methacrylate. The reulting terpolymer, isolated in quantitative yield, had M_n 3910, M_w 4290, D 1.10, T_g -13°C and +47°C. Thus, even as reactive a monomer as glycidyl methacrylate can be polymerized to a narrow molecular weight polymer by GTP.

5.3. Block Polymers

A useful attribute of "living" polymer systems is the ability to prepare block copolymers of predetermined block length and sequence. GTP provides a particularly facile methodology for the sequencing of both homopolymer blocks and random copolymer blocks. One simply adds a new monomer or mixture of monomers when the first batch is depleted.

An AB block copolymer of lauryl methacrylate and MMA was prepared in two ways. First, MMA was polymerized and then lauryl methacrylate was fed to give quantitative conversion to polymer with M_n 6650, M_w 7070, D 1.06 (Theor. M_n 7043). The same AB block copolymer was obtained by first polymerizing lauryl methacrylate, and then changing the monomer feed to MMA to give a quantitative yield of polymer with M_n 6540, M_w 7470, D 1.14[10] (Theor. M_n 7470). In each case the design composition of 10 mol % lauryl methacrylate was confirmed by NMR analysis of the copolymers.

An ABC block terpolymer of MMA, butyl methacrylate, and allyl methacrylate was prepared by sequential feeding of the three monomers in the order: MMA, butyl methacrylate, allyl methacrylate. The resulting polymer, isolated in 95% yield, had M_n 3800, M_w 4060, D 1.07 (Theor. M_n 4100)[10]. The expected composition of 50 mol % MMA, 30 mol % butyl methacrylate, 20 mol % allyl methacrylate was confirmed by ^1H NMR analysis. The polymerization of allyl methacrylate to a soluble copolymer by GTP demonstrates an advantage of this relatively mild polymerization process. Radical polymerization of allyl methacrylate generally leds to gel formation.

An ABA block copolymer with A blocks consisting of random copolymer of MMA and 2-hydroxyethyl methacrylate and a B block of lauryl methacrylate was prepared by GTP. By using the difunctional initiator 24, the B block (lauryl methacrylate) was constructed first as a homopolymer with two living ends followed by addition of the random copolymer A blocks. The trimethylsilyl-protected monomer 13,

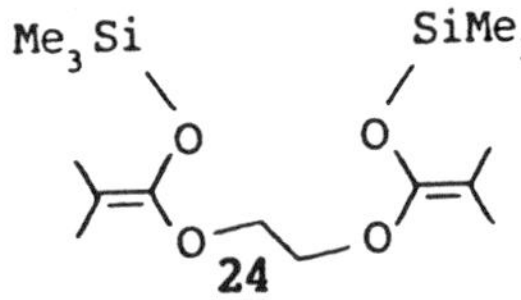

Me_3Si $SiMe_3$

24

was used in forming the A blocks. The resulting polymer had M_n 9340, M_w 11,200, D 1.20 (Theor. M_n 13,600). Following deprotection of the hydroxyl groups with methanolic tetrabutylammonium fluoride, the polymer was found to have the expected composition: 31 mol % lauryl methacrylate, 39 mol % MMA, and 30 mol % 2-hydroxyethyl methacrylate. Differential scanning calorimetry showed two T_g's at -43° and 95°C.

Superior pigment dispersants for acrylic finished have been made by forming AB block polymer with PMMA or PBMA blocks coupled with short glycidyl methacrylate blocks. The epoxy functions are then converted to polar absorbing groups by reaction with carboxylic acids or amines (Scheme 18)[30].

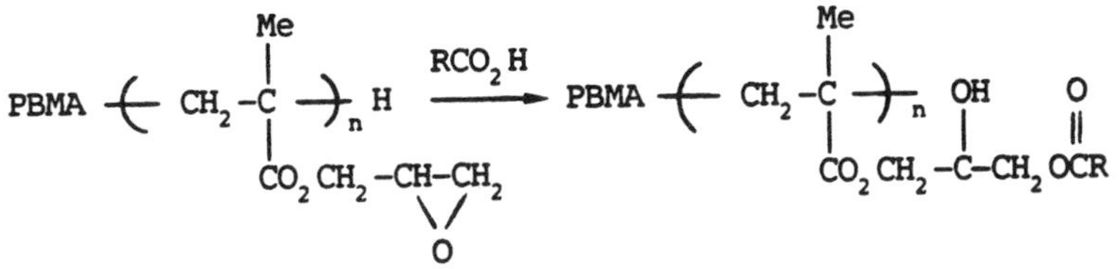

$$\text{PBMA}-\!\left(CH_2-\underset{CO_2CH_2-CH-CH_2\ (\text{epoxide, O})}{\overset{Me}{C}}\right)_n\!-H \xrightarrow{RCO_2H} \text{PBMA}-\!\left(CH_2-\underset{CO_2CH_2-\underset{}{\overset{OH}{C}}-CH_2O\overset{O}{\overset{\|}{C}}R}{\overset{Me}{C}}\right)_n$$

Scheme 18

AB type block polymers with methyl methacrylate and ethyl acrylate segments can be prepared by polymerizing the less reactive methacrylate monomer first and then adding the acrylate monomer[1].

A block copolymer containing a nonacrylic block was prepared by first converting polycaprolactone to a GTP initiator, followed by initiation of MMA polymerization[10]. The polymeric difunctional initiator, 26 was prepared by the rapid reaction of trimethylsilyl cyanide with the diacrylate of polycaprolactone diol, 25 (M_n 1250, M_w 2200, D 1.76) in the presence of tetraethylammonium cyanide catalyst. The resulting polymeric difunctional GTP initiator initiated polymerization of MMA to give a block copolymer, 27 with M_n 4120, M_w 7340, D 1.78 (Scheme 19).

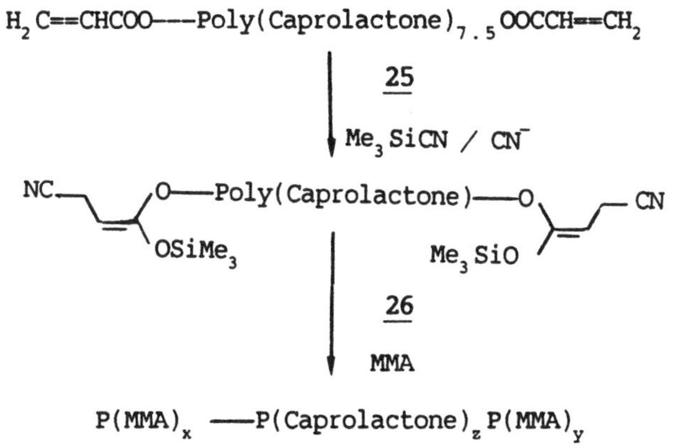

$H_2C{=}CHCOO$—Poly(Caprolactone)$_{7.5}$$OOCCH{=}CH_2$

25

Me_3SiCN / CN^-

NC–CH=C(OSiMe$_3$)–O—Poly(Caprolactone)—O–C(OSiMe$_3$)=CH–CN

26

MMA

P(MMA)$_x$ —P(Caprolactone)$_z$P(MMA)$_y$

27

Scheme 19

Another method for obtaining acrylic/nonacrylic block polymers was devised by Eatmond. α-Bromoester ended polymethyl methacrylate was made by GTP[2] and this was mixed with dimanganesedecacarbonyl and irradiated in the presence of styrene to produce a mixture of AB and ABA block polymers (A = PMMA)[31].

5.4. Comb Polymers from GTP Macromonomers

Comb or graft polymers have been made by two routes involving GTP. First, methacryl ended polystyrene[26] or polyphenylene oxide[24] were polymerized by GTP. Second, GTP was used to make the macromonomer which was then polymerized by free radical initiation. Styryl PMMA (see IV A) was homopolymerized in 95% conversion by AlBN in benzene at 60°C[31]. In other work[30] blocked hydroxy initators were used to prepare methacrylate polymers by GTP (homopolymer, random copolymers and block copolymers). The hydroxy group was then unblocked and reacted with acryloyl or methacryloyl chloride or isocyanoethyl methacrylate to obtain macromonomers. These were used to synthesize non-aqueous dispersion (NAD's) by copolymerization with other acrylic and styrene monomers. NAD's are used for rheology control and reinforcement of coating systems.

5.5. Star Polymers

Initiation of GTP with polyfunctional initiators to produce star polymers presents the problem of synthesis of large, reactive, polyfunctional silyl ketene acetals. An Alternative approach, which is described here, is to generate polyfunctional initiators in situ by the Michael addition reaction of polyfunctional monomers with silicon reagents. When one equivalent of an n-functional monomer is allowed to react with n equivalents of silyl ketene acetal initiator, 1, in the presence of a Lewis acid catalyst[3] (especially a dialkylaluminum chloride or dialkylaluminum oxide) followed by addition of excess monofunctional monomer, crosslinking does not occur, but instead, star polymers are formed. Thus, reaction of 1 with a 0.33 molar equiv of trimethylolpropane triacrylate, 28 at -78°C followed by 10 molar equiv of ethyl acrylate, gave a quantitative yield of a soluble star polymer containing no residual unsaturation with M_n 2190, M_w 3040, D 1.39 (Theor. M_n 3300)(Scheme 20). Similarly, treatment of 1 at -78°C with 0.25 molar equiv of pentaerythritol tetraacrylate, 29 in the presence of diisobutylaluminum oxide followed by 10 equiv of ethyl acrylate gave a quantitative yield of soluble polymer with M_n 2400, M_2 2970, D 1.24 (Theor. M_n 4752). As expected for star polymers, the molecular weights measured by size exclusion chromatography are lower than the theoretical values for linear polymers.

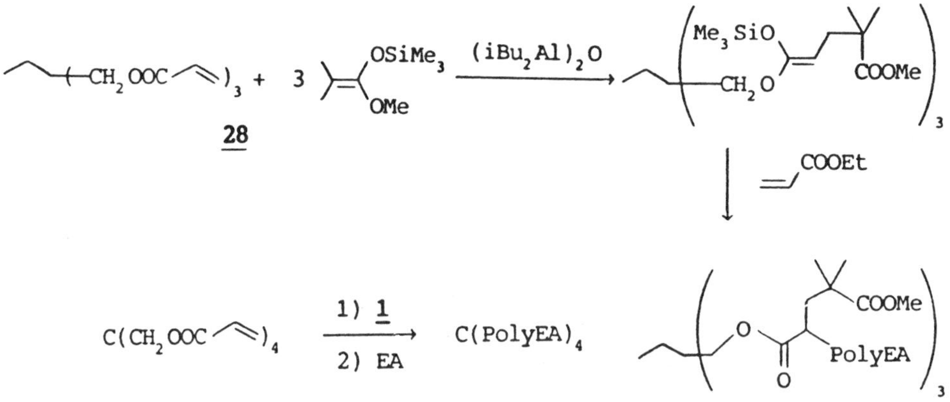

$(CH_2OOC-CH=CH_2)_3$ + 3 $Me_2C=C(OSiMe_3)(OMe)$ $\xrightarrow{(iBu_2Al)_2O}$ $(CH_2O-C(=\ldots)-C(Me_3SiO)\ldots COOMe)_3$

28

$\xrightarrow{CH_2=CH-COOEt}$ $(CH_2-O-CO-CH(PolyEA)-CH_2-C(Me)_2-COOMe)_3$

$C(CH_2OOC-CH=CH_2)_4$ $\xrightarrow[2)\ EA]{1)\ 1}$ $C(PolyEA)_4$

29

Scheme 20

Multiarmed star polymers can be made by addition of a difunctional methacrylate, for example ethylene glycol dimethacrylate, 30, to a living GTP polymer[30] (Scheme 21).

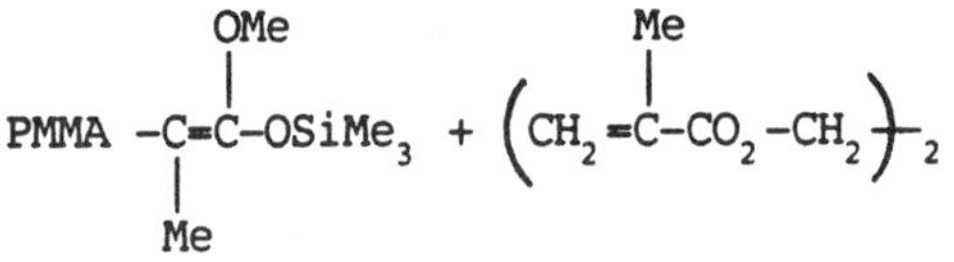

$$\text{PMMA}-C(Me)=C(OMe)-OSiMe_3 \; + \; (CH_2=C(Me)-CO_2-CH_2)_2$$

30

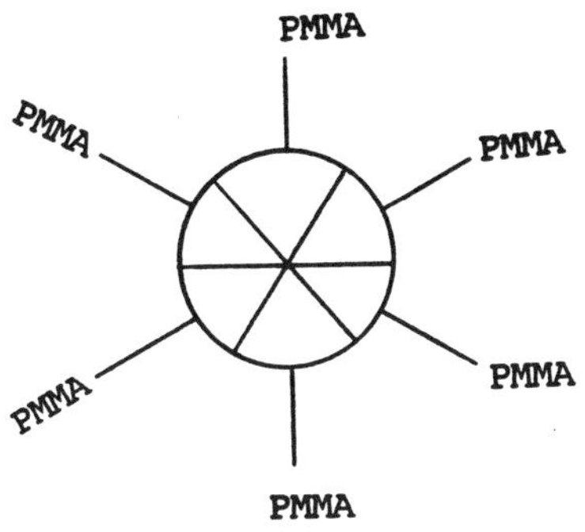

Scheme 21

One end use for these stars is in toughening plastics and coatings. The addition of hydroxy functional stars to other polyols and then co-crosslinking produces two phase films which are clear, flexible and hard[30]. Another end use for stars is in rheology control for application of coatings.

6. CONCLUSIONS

At present, most acrylic polymers are made by free-radical polymerization. Random copolymers are made by controlled addition of monomers; molecular weight is controlled with chain-transfer agents. The control of structural features, such as end groups and segmentation of chains into blocks, is difficult, if not impossible. For the preparation of high molecular-weight homopolymer, free radical initiation remains the method of choice.

Michael addition polymerization, initiated with alkoxide-alcohol mixtures, operates at room temperture or above and is useful for the synthesis of methacrylate polymers containing free-radical sensitive groups, such as allyl, but only low molecular-weight polymers and random copolymers are obtained[33].

Low temperature anionic polymerization of methacrylates, initiated by hindered alkyl lithium compounds, give the same structural control as GTP. However, the process is uneconomical because very low temperatures must be maintained (-20 to -70°C)[34]. At higher temperatures, the strongly basic living ends attack the ester groups on monomer and on polymer chains, and polymerization ceases[35]. In addition, monomers containing base-sensitive functional groups cannot be used.

The large number of functional methacrylates available, coupled with the numerous structural variations possible, makes GTP an extremely useful tool for polymer research. In addition, even though GTP is a complex procedure, requiring very pure reagents, dry conditions and much fine turning, it appears that many industrial applications will be forthcoming.

References

1. O. W. Webster, W. R. Hertler, D. Y. Sogah, W. B. Farnham, and T. V. RajanBabu, J. Amer. Chem. Soc., **105**, 5706 (1983).

2. D. Y. Sogah and O. W. Webster, J. Polym. Soc. Lett. Ed. **21** 927 (1983).

3. W. R. Hertler, D. Y. Sogah, O. W. Webster, and B. M. Trost, Macromolecules, **17**, 1417 (1984).

4. O. W. Webster, W. R. Hertler, D. Y. sogah, W. B. Farnham, and T. V. RajanBabu, J. Macromol. Sci-Chem., **A21(8&9)**, 943-960 (1984).

5. D. Y. Sogah, W. R. Hertler, and O. W. Webster, ACS Polymer Preprints; **25** (2) 3(1984).

6. D. Y. Sogah and W. B. Farnham, "Organosilicon and Bioorganosilicon Chemistry: Structures, Bonding, Reactivity and Synthetic Application", H. Sakurai, Ed., John Wiley & Sons, N.Y. 1985, Chapter 20.

7. O. W. Webster, ACS Polymer Preprints, **27**, 161 (1986).

8. W. R. Hertler, ACS Polymer Preprints, **27**, 165 (1986).

9. W. B. Farnham and D. Y. Sogah, ACS Polymer Preprints, **27**, 167 (1986).

10. D. Y. Sogah, W. R. Hertler, O. W. Webster and G. M. Cohen, Macromolecules, in Press.

11. O. W. Webster, U. S. Patents, 4,417,034 (11-22-83); 4,508,880 (4-2-85). W. B. Farnham, D. Y. Sogah, U. S. Patents 4,414,372 (11-8-83); 4,524,196 (6-18-85); 4,581,428 (4-8-86). O. W. Webster, W. B. Farnham, D. W. Sogah, EPA 68,887 (1983); ibid EPA 145,263 (1984).

12. I. B. Dicker, W. B. Farnham, W. R. Hertler, E. D. Laganis, D. Y. Sogah, T. W. Del Pesco and P. H. Fitzgerald and I. B. Dicker, U. S. Patent 4,588,795 (5-13-86).

13. F. Bandermann, H. P. Sitz and H. D. Speikamp, ACS Polymer Preprints, **27**, 169 (1986).

14. F. Bandermann and H. D. Speikamp Makromol. Chem., Rapid Commun., **6**, 336 (1985).

15. M. T. Reetz, Pure and Apl. Chem., **67**, (12) 1785 (1985).

16. M. T. Reetz, R. Ostarek, and K-E Piejko Ger. Patent Application 3,504,168 (2-7-85). U. S. Patent, 4,626,579 (12-2-86).

17. W. J. Brittain and D. Y. Sogah, Unpublished results.

18. T. G. Fox, and H. W. Schnecko, Polymer, 3, 575 (1962).

19. T. Otsu, B. Yamada and M. Imoto, J. Macromol. chem., 1, 61 (1966).

20. Y. Wei and G. Wnek, unpublished results.

21. M. A. Muller and M. Stickler, Makromol. Chem., Rapid Commun., 7, 575 (1986).

22. D. A. Dixon, D. Y. Sogah, W. B. Farnham, to be published. Ab initio molecular orbital calculations using double zeta basis sets augmented by polarization functions have been performed on model systems on the GTP potential energy surface. The results show a preference for a pentavalent anionic silicon intermediate and a hexavalent transition state for transfer of the silicon.

23. See papers by R. Corriu, on 'The Mechanism of Nucleophilic Activation at Silicon", by T. E. Hogen-Esch on "Sterochemistry of Group Transfer Polymerization' and by A. H. E. Muller on "Kinetics of Group Transfer Polymerization".

24. C. Pugh and V. Percec, ACS Polymer Preprints, 26, 303 (1985).

25. W. Kreuder, O. W. Webster, H. Ringsdorf, Makromol. Che., Rapid Commun., 7, 5 (1986).

26. R. Asami, M. Takaki and Y. Moriyama, Polymer Bulletin, 16, 125 (1986).

27. M. K. Akkapeddi, Macromolecules, 12, 546 (1979).

28. J. Suenaga, D. M. Sutherlin, and J. K. Stille, Macromolecules, 17, 2913 (1984).

29. T. V. RajanBabu and W. R. Hertler, unpublished results.

30. J. A. Simms and H. J. Spinelli, Results presented at Federation of Societies for Coatings Technology, November 5, 1986, Atlanta, Georgia. WO 86/00626.

31. G. C. Eastmond and J. Grigor Makromol. Chem., Rapid Commun, 7, 375 (1986).

32. R. Assami, Y. Kondo, and M. Takaki, ACS Polymer Preprints, 27, 186 (1986).

33. R. A. Haggard and S. N. Lewis, Prog. in Org. Coatings, 12, (1984).

34. M. Morton, Anionic Polymerization, Principles and practice, Academic Press, Inc., New York, 1983, Page 101.

35. K. Hatada, T. Kitayano, K. Fumikano, K. Onto and H. Yaki, ACS Symp. Ser., 166, 327 (1981).

KINETICS OF GROUP TRANSFER POLYMERIZATION

Axel H. E. Müller
Institut für Physikalische Chemie
Johannes Gutenberg-Universität Mainz
D-6500 Mainz
Federal Republic of Germany

ABSTRACT. The group transfer polymerization (GTP) of methyl methacrylate (MMA) and *tert.*-butyl methacrylate (TBMA), catalyzed by bifluoride, is of first order with respect to monomer and catalyst concentrations. A slightly negative order with respect to initiator concentration is observed, indicating an inhibitive function of the initiator. Termination reactions occur at lower catalyst concentrations. They are very pronounced for TBMA and result in incomplete conversions. For MMA, at lower catalyst and higher initiator concentrations, induction periods are observed which increase when lowering the temperature. This is related with the formation of active chains. The Arrhenius plots for the propagation rate constants of MMA and TBMA are very similar to those found in the anionic polymerization using bulky counterions. Based on the kinetic results and on the microstructure of the polymers, a mechanism for the addition of monomer is proposed.
The non-linear dependence of the degrees of polymerization on conversion shows that the rates connected with the "activation equilibrium"

catalyst + "dormant" polymer $\rightleftharpoons$ activated polymer

influence the molecular weight distribution.

1. INTRODUCTION

Group transfer polymerization (GTP) is a new polymerization process discovered and patented by E. I. du Pont de Nemours & Company in 1983[1-7]. It allows the polymerization of polar aprotic monomers, such as methacrylates and acrylates by initiation with silyl compounds, such as ketene silyl acetals, in the presence of a nucleophilic or electrophilic catalyst.

Especially for methacrylates, GTP combines the important advantages of "living" polymerization (e. g. narrow molecular weight distributions, control of molecular weight and the facile preparation of block and functionalized polymers), with the added advantage that the reaction can be carried out at ambient or above temperatures. In the contrast, anionic

M. Fontanille and A. Guyot (eds.), Recent Advances in Mechanistic and Synthetic Aspects of Polymerization, 23–40.

polymerization of methacrylates (with the exception of *tert.*-butyl methacrylate) leads to "living" polymers only at low temperatures[8].

For acrylate polymerization, catalysis by Lewis acids, e. g. zinc iodide, results in polymers of much lower polydispersity than nucleophilic catalysis[9]. However, as of yet, no kinetic studies have been published on the electrophilic catalysis of GTP. It is assumed that the Lewis acid catalyst activates the monomer by increasing the polarization of the vinyl double bond.

Only few studies have been published, so far, on the mechanism of nucleophilic catalysis. An associative mechanism was proposed by the du Pont group[2,10,11]. It is outlined in Scheme 1:

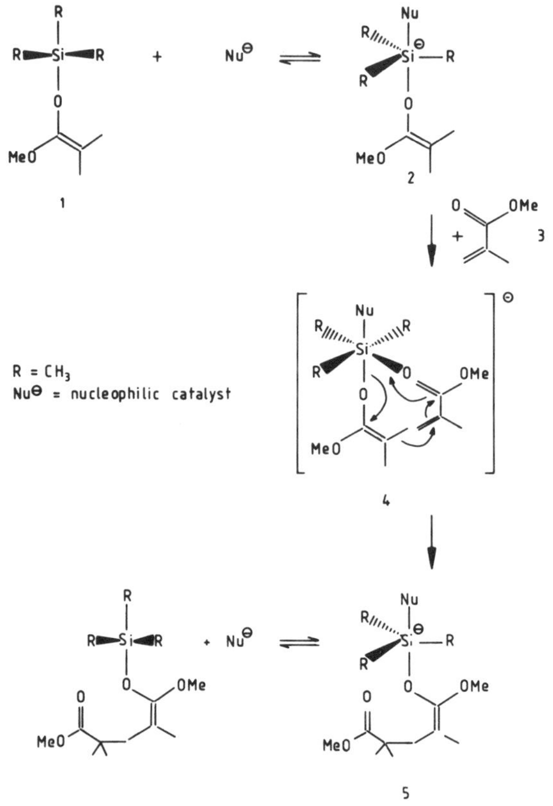

R = CH_3
Nu$^{\ominus}$ = nucleophilic catalyst

Scheme 1

The nucleophilic catalyst (e.g. HF_2^-) coordinates to the silicon atom of the initiator, 1-methoxy-1-(trimethylsiloxy)-2-methyl-1-propene **1** (MTS) to provide a pentacoordinate species **2**. The activated initiator and monomer **3** are proposed to form a hypervalent silicon intermediate **4**. A new

C–C bond is created between initiator and monomer and the trimethylsilyl group is transferred to the carbonyl oxygen of the monomer under formation of a new Si–O bond and cleavage of the old Si–O bond. This transfer regenerates a structure similar to that of the initiator. The propagation reaction 2 ——> 5 should proceed as long as monomer is present. However, there is no direct evidence for a "concerted" mechanism symbolized by structure **4**.

Bandermann et al. reported on the kinetics of the GTP of methyl methacrylate[12,13] and acrylonitrile[14] in acetonitrile as the solvent. Various initiators and catalysts were used. It was found that the kinetics of the propagation reaction is obscured by a variety of side reactions, typical ones being the reaction of the initiator or active chains with the solvent and the initiation of the polymerization by the catalyst alone. As a consequence, incomplete conversions and bimodal molecular weight distributions were reported. Thus, acetonitrile can be regarded as an unfavourable solvent for GTP.

The kinetic results reported below were all obtained in THF as the solvent. Although even in this solvent the reaction is not completely free of side reactions, the deviations from ideal behaviour are not severe enough to inhibit the determination of kinetic parameters.

2. GENERAL CONSIDERATIONS

A kinetic scheme was developed by Mai and Müller[15] which is consistent with the mechanism shown in Scheme 1:

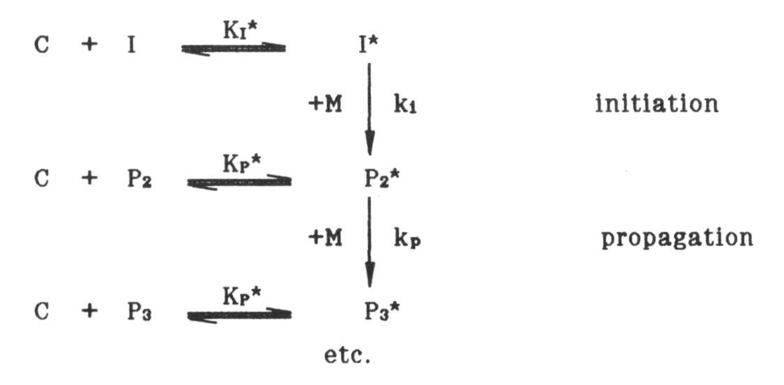

I = initiator, I* = activated initiator,
P_i = "dormant" polymer of DP = i, P_i^* = activated polymer,
C = catalyst, M = monomer.

Scheme 2

If initiator and polymer have the same chemical structure,

$$I = P_1 \quad \text{and} \quad I^* = P_1^*,$$

then to a first approximation it is assumed that

$$K_I^* = K_P^* = K^* \quad \text{and} \quad k_i = k_p.$$

Scheme 2 then simplifies to Scheme 3:

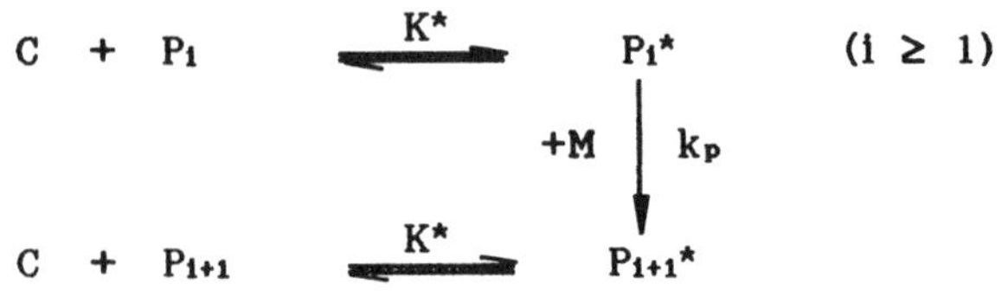

$$C + P_i \underset{}{\overset{K^*}{\rightleftharpoons}} P_i^* \qquad (i \geq 1)$$

$$\downarrow +M \quad k_p$$

$$C + P_{i+1} \underset{}{\overset{K^*}{\rightleftharpoons}} P_{i+1}^*$$

Scheme 3

The rate of polymerization is given by

$$R_p = -\frac{d[M]}{dt} = k_p \cdot [M] \cdot \sum_{i=1}^{\infty}[P_i^*] = k_p \cdot [M] \cdot [P^*] \qquad (1)$$

For a "living" system the concentration of activated polymers $[P^*]$ = const. and integration leads to

$$\ln \frac{[M]_0}{[M]} = k_p \cdot [P^*] \cdot t = k_{app} \cdot t \,. \qquad (2)$$

The apparent rate constant k_{app} is the slope of a first-order plot of conversion vs. time.

Application of the mass action law renders[15]

$$[P^*] = \frac{K^* \cdot [I]_0}{1 + K^* \cdot [I]_0} \cdot [C]_0 \qquad (3)$$

and

$$k_{app} = k_p \cdot \frac{K^* \cdot [I]_0}{1 + K^* \cdot [I]_0} \cdot [C]_0. \qquad (4)$$

Two limiting cases can be discussed:
(a) $K^* \cdot [I]_0 \gg 1$, i.e. the activation equilibrium is shifted to the right. This leads to

$$[P]^* = [C]_0 \quad \text{and} \quad k_{app} = k_p \cdot [C]_0. \qquad (5a)$$

(b) $K^* \cdot [I]_0 \ll 1$, i.e. the activation equilibrium is shifted to the left. This leads to

$$[P]^* = K^* \cdot [I]_0 \cdot [C]_0 \quad \text{and} \quad k_{app} = k_p \cdot K^* \cdot [I]_0 \cdot [C]_0. \qquad (5b)$$

Thus, it follows from the proposed mechanism that the kinetic orders of the reaction, defined by

$$R_p = \text{const} \cdot [M]^\alpha \cdot [I]_0^\beta \cdot [C]_0^\tau, \qquad (6)$$

are $\quad \alpha = 1, \quad 0 \leq \beta \leq 1 \quad$ and $\quad \tau = 1.$

Entering eqs. (1) and (2) leads to

$$k_{app} = const \cdot [I]_0^{\beta} \cdot [C]_0 = k_p' \cdot [C]_0. \qquad (7)$$

For $K^* \cdot [I]_0 \gg 1$ (i.e. $\beta = 0$) the "pseudo" rate constant of propagation k_p' is equal to the "true" rate constant k_p:

$$k_p' = k_{app}/[C]_0 = k_p. \qquad (8a)$$

For $K^* \cdot [I]_0 \ll 1$ (i.e. $\beta = 1$)

$$k_p' = k_{app}/[C]_0 = k_p \cdot K^* \cdot [I]_0 \ll k_p. \qquad (8b)$$

In the general case ($0 < \beta < 1$)

$$k_p' = k_p \cdot \frac{K^* \cdot [I]_0}{1 + K^* \cdot [I]_0} < k_p. \qquad (8c)$$

3. KINETIC INVESTIGATIONS IN THF

3.1. Methyl Methacrylate (MMA)

For the GTP of MMA, the reaction orders with respect to reagents were investigated by Mai and Müller[15,16]. 1-methoxy-1-(trimethylsiloxy)-2-methyl-1-propene (MTS) was used as the initiator and tris(dimethylamino)sulfonium bifluoride ($TASHF_2$) as the catalyst. The kinetic methods used are identical to those employed in the anionic polymerization of methacrylic esters[8]. However, the mode of addition of reagents was changed compared to the original procedure for GTP[1-7] (i.e. dropwise addition of monomer to a pre-mixed solution of initiator and catalyst). All experiments were carried out in a stirred tank reactor[17], allowing for the withdrawal of samples at intervals of 1 s or longer. At the start of each experiment, equal volumes of catalyst solution in THF (without addition of acetonitrile) and a pre-mixed solution of monomer and initiator in THF were mixed within 0.4 s. The standard conditions were T = 20°C, $[M]_0 = 0.18$ mol/l, $[I]_0 = 1 \cdot 10^{-3}$ mol/l and $[C]_0 = 2 \cdot 10^{-5}$ mol/l. In the various sets of experiments only one parameter was varied at a time, the others being kept constant.

3.1.1. Effect of catalyst concentration. Fig. 1 shows first-order plots for the conversion of monomer obtained at various catalyst concentrations. For $[C]_0 \geq 3 \cdot 10^{-5}$ the reaction proceeds in an ideal manner; initiation is fast as compared to propagation and the propagation reaction follows first-order kinetics (indicating the absence of termination reactions). For $[C]_0 < 3 \cdot 10^{-5}$ mol/l low reaction rates are observed initially ("induction period") followed by increased rates in a linear manner. Finally, the rates decrease again, indicating the presence of termination reactions. In those cases the apparent rate constants, k_{app}, were estimated from the maximum slope of the first-order time-conversion curves. Nevertheless, monomer conversion always was complete.

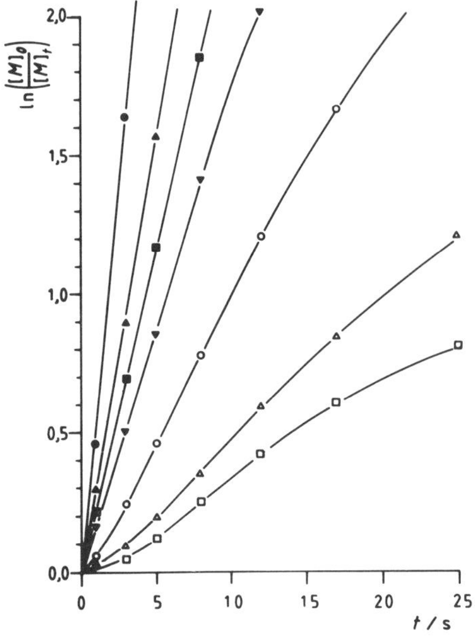

Fig. 1: First-order time-conversion plots for the group transfer polymerization of MMA with MTS and TASHF₂ in THF as a function of catalyst concentration.[15]
$10^5 \cdot [C]_0$/mol·l⁻¹: ● = 7.5; ▲ = 5.1; ■ = 4.0; ▼ = 3.0; ○ = 2.0; △ = 1.5; □ = 1.0

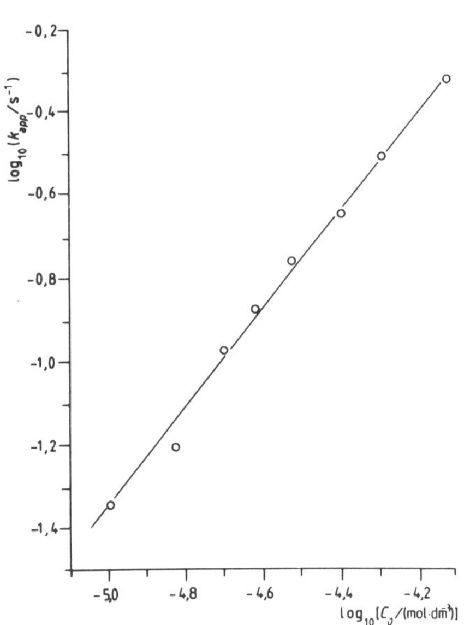

Fig. 2

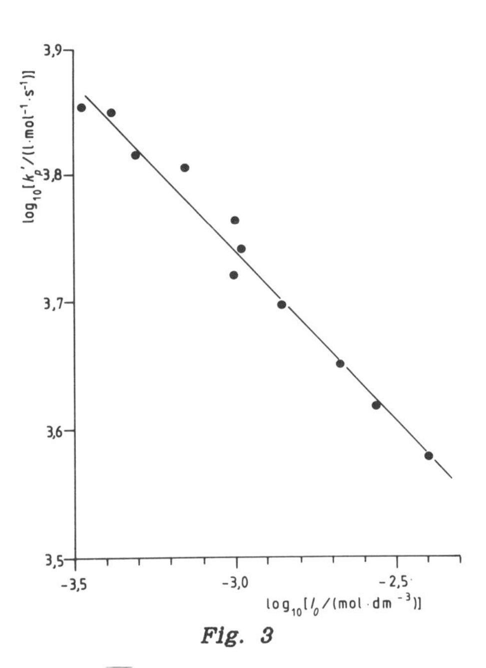

Fig. 3

Fig. 2: Bilogarithmic plot of the apparent rate constants k_{app} vs. catalyst concentrations.[15] *Slope = 1.17.*
Fig. 3: Bilogarithmic plot of the rate constants k_p' vs. initiator concentrations.[15] *Slope = −0.27.*

Fig. 2 shows a plot of log k_{app} vs. log $[C]_0$. The linearity is good with a slope of 1.17. This indicates that the order of the reaction with respect to catalyst concentration is equal to 1, as was expected. However, for lower catalyst concentrations (i.e. for runs having an "induction period") the rate constants k_p' (as calculated from eq. 8) are smaller than expected. It is conceivable that during the "induction period" some deactivation of active centres occurs, decreasing the rate constants k_p' calculated from the stoichiometric catalyst concentration. This is discussed below. In the initial stage of polymerization (up to the trimer) a kinetic order of 2 was found by Brittain and Sogah[26] and attributed to a pre-equilibrium of formation of fluoride ions from bifluoride. This may also be related to the "induction periods".

<u>3.1.2. Effect of initiator concentration.</u> In a second set of experiments the influence of the initiator concentration on the reaction rate was investigated. When increasing the initiator concentrations the time-conversion curves are characterized by a slight "induction period" at the beginning of the reaction followed by a linear phase. At higher conversions the slopes again decrease, indicating termination reactions. Fig. 3 shows a plot of log k_p' vs. log $[I]_0$. The slope of −0.27 is not in accordance with the kinetic scheme proposed. Rather, it indicates that the initiator is inhibiting the reaction. This may be explained by the extended kinetic scheme 4:

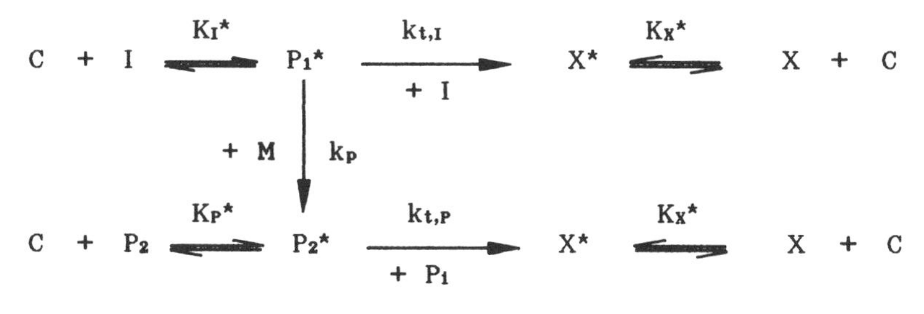

Scheme 4

Initiator or "dormant" polymer may react with activated initiator or polymer leading to an inactive product X capable of binding catalyst ($K_X^* \geq K_I^* \approx K_P^*$). The reduced actual catalyst concentration will be reflected in lower apparent rate constants. It is conceivable that the "induction periods" observed relate to the same phenomenon.

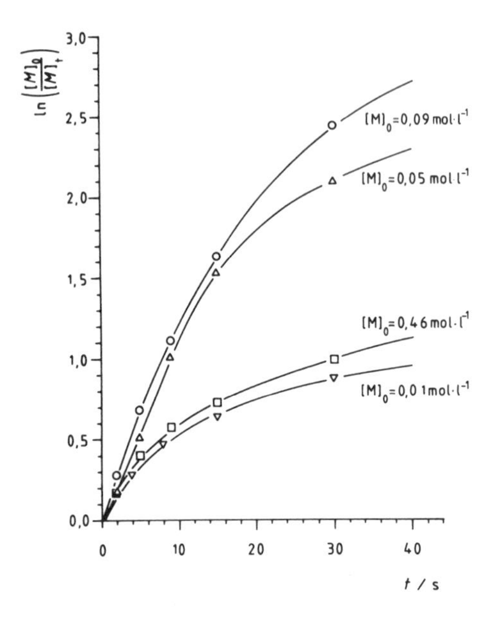

Fig. 4: First-order time-conversion plots for the group transfer polymerization of MMA with MTS and $TASHF_2$ in THF as a function of monomer concentration.[16]

3.1.3. Effect of monomer concentration. Fig. 4 shows first-order time-conversion plots for different initial monomer concentrations. The plots are linear to a certain extent. Deviations from linearity are observed in the later stages of polymerization, indicating the presence of termination reactions. The extent of termination depends on the initial monomer concentration. An increase of termination is observed for $[M]_0 < 0.09$ mol/l and for $[M]_0 > 0.18$ mol/l rendering the determination of the initial slope, k_{app}, difficult and reducing its accuracy. The complex dependence of termination on monomer concentration may be the result of the coexistence of different termination mechanisms, the nature of which remains open to future investigations.

By using eq. 8 the rate constants of polymerization, k_p', were calculated from the initial slopes of the first-order time-conversion plots. Within experimental error the rate constants are independent of monomer concentration. This is in accordance with the kinetic scheme proposed.

3.1.4. Effect of temperature. Valuable information on the mechanism of the reaction can be obtained from an investigation of the temperature dependence of the rate constants and from the comparison of the activation parameters with those obtained in the anionic polymerization of MMA in THF. Moreover, it was expected that the extent of side reactions is less pronounced at lower temperatures.

In two set of experiments the temperature was varied from ambient to –82°C. Fig. 5 shows selected first-order time-conversion plots. In order to avoid the occurrence of induction periods at ambient temperature all experiments of the temperature study were performed at a relatively high

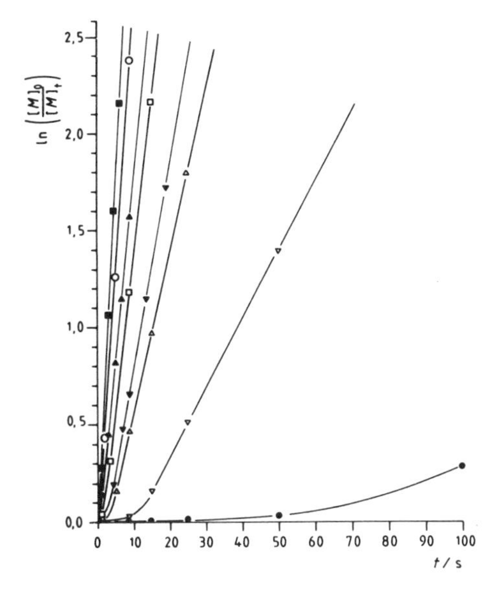

Fig. 5: First-order time-conversion plots for the group transfer polymerization of MMA with MTS and $TASHF_2$ in THF at various temperatures.[16]
■ = +23.7; ○ = +3.4;
▲ = +6.8; ▼ = −13.8;
□ = −17.6; △ = −36.5;
▽ = −58.5; ● = −81.5 °C.

catalyst concentration ($[C]_0 \approx 6 \cdot 10^{-5}$ mol/l). However, at lower temperatures the induction periods appear again, becoming more pronounced with decreasing temperature. With the exception of this deviation all runs show ideal first-order behaviour, including the absence of termination reactions, leading to complete monomer conversion.

A quasi-Arrhenius plot for the reciprocal induction times gives a straight line. This may indicate that the reciprocal induction time represents the rate of formation of active centres.

By using eq. 8a the rate constants k_p were calculated from the maximum slopes of the time-conversion curves. The assumption that $K_p \cdot [I]_0 \gg 1$ seems to be justified by the results on the dependence of reaction rates on initiator concentration. Even if this assumption should not hold true the data given represent the lower limit of the rate constants (cf eq. 8c). Fig. 6 shows the corresponding Arrhenius plot. The linearity of this plot renders evidence for the existence of only one active species. It is compared to the Arrhenius plots for the anionic polymerization of MMA in THF using various counterions[18-21]. The plot for GTP is very similar to those of the anionic polymerization with larger counterions like sodium or caesium. Table 1 shows that frequency exponents and activation energies are of comparable size, indicating similar mechanisms for the two processes.

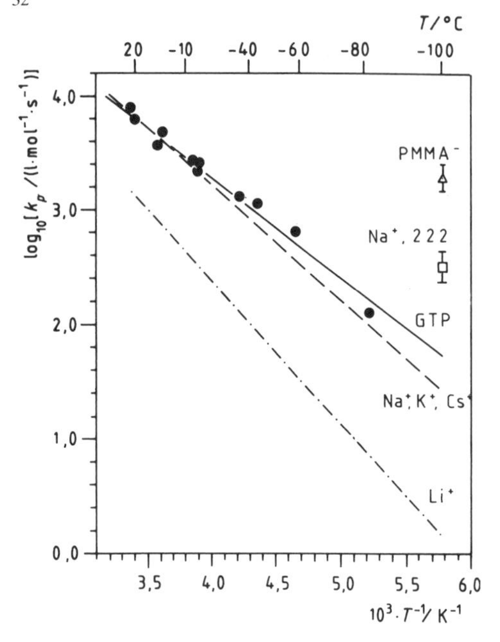

Fig. 6: Arrhenius plot for the polymerization rate constants in the group transfer polymerization of MMA with MTS and $TASHF_2$ in THF (•) and in the anionic polymerization of MMA in THF using different counterions (References, cf. Table 1).

Table 1: Rate constants and activation parameters in the group transfer and anionic polymerization of MMA using different counterions

	GTP	free anion	$(Na^+, 222)$ [a]	Cs^+	Na^+	Li^+
E_a/kJ/mol	16.9			19.5	18.3	24.0
log A	6.8			7.3	7.0	7.4
k_p^{175}	56	2100	270	30	34	2
Ref.	16	18	18	20	20,21	19

[a] Na^+ cryptated by the bicyclic ligand 222

As was shown earlier[8], in the anionic polymerization with large counterions the monomer addition proceeds via a direct attack of the carbanion to the monomer vinyl group, similar to radical polymerization. Coordination of the monomer carbonyl group to the hypervalent silicon atom in the transition state (cf. Scheme 1) should result in much higher activation entropies (corresponding to lower frequency exponents) as those observed in anionic polymerization. Consequently, it seems more likely that monomer addition is a two-step process. This invokes the addition of the vinyl group to the activated initiator or chain end as the rate-determining step. This is followed by the transfer of the silyl group to the newly formed enolate, as is shown in Scheme 5:

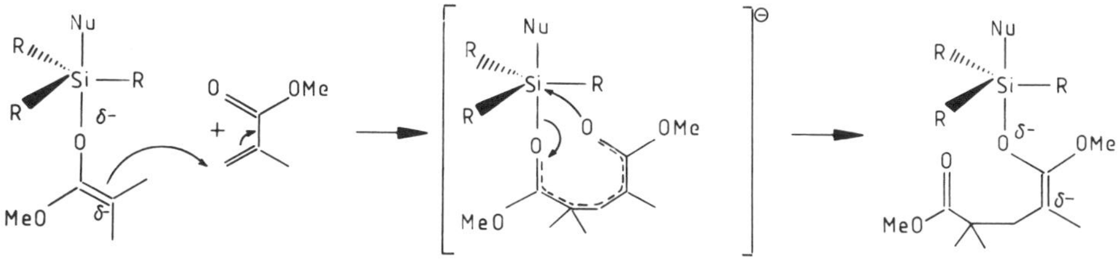

Scheme 5

Alternatively it may be discussed that the active species is an enolate or the corresponding ion pair formed in the (reversible or irreversible) reaction reported by Noyori et al.[22] (Scheme 6):

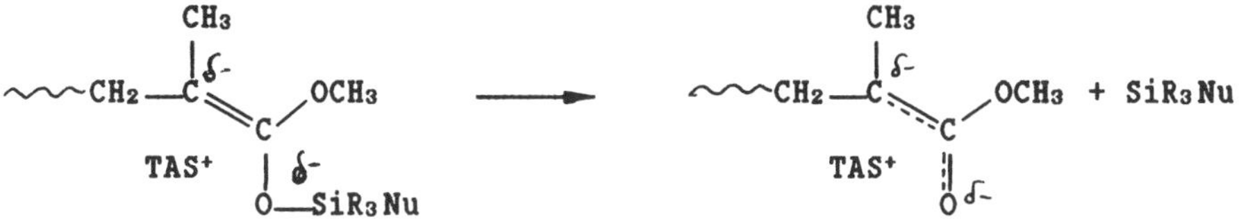

Scheme 6

According to this scheme the induction periods would represent the time of formation of the free enolate. However, the rate constants found for GTP are significantly smaller than those for the anionic polymerization via the free anion or the cryptated sodium ion pair (which from the interionic distance should have comparable rate constants as the TAS^+ ion pair). Through labelling experiments it was shown also by Farnham and Sogah[10,11] that such a "dissociative" mechanism is not very likely. Thus, at present a mechanism via the free enolate can be regarded as improbable.

3.1.5. Microstructure of polymers. The two-step "associative" mechanism of monomer addition for GTP is very similar to that for anionic or radical polymerization. Thus, the stereochemistry of monomer addition should be similar, too. This is supported by a comparison of the tacticities of polymers prepared by GTP[23], radical[23] and anionic polymerization[8] (using the free anion, or bulky or strongly solvated counterions) at comparable temperatures. Table 2 shows that all polymers are rather syndiotactic, having very similar triad distributions which obey Bernoullian statistics within experimental error. On the other hand, the "concerted" mechanism of monomer addition is similar to that assumed for anionic polymerization with Li^+ counterion in non-polar solvents which is known to lead to isotactic polymers[24]

TABLE 2: **Tacticities of PMMA polymerized in THF by GTP, radical and anionic polymerization using different counterions.**

Counter-ion	Temp [°C]	mm	mr	rr	r	ρ[a]	Ref.
	-70	.01	.23	.76	.88	.93	23
GTP	-50	.01	.29	.70	.85	.91	23
	-30	$.01_5$	.30	$.68_5$	$.83_5$	.92	23
radical	-29	$.00_5$	.24	$.75_5$	$.87_5$	.91	23
anion	-98	.01	.20	.79	.90	.97	18
Na^+,222	-98	.01	.19	.80	.89	1.05	18
Cs^+	-53	.05	.52	.42	.68	.82	25
Na^+	-51	.04	.38	.58	.77	.93	21
Li^+	-45	.01	.22	.77	.88	.96	19

[a] persistence ratio $\rho = 2 \cdot m \cdot r/mr$ (error ca. 10%)

Thus, from kinetic and tacticity data it may be concluded that the mechanism of monomer addition in GTP is a two-step rather than a "concerted" process. The primary role of the nucleophile may be seen in providing the carbon α to the acetal group with the negative charge needed to attack the polarized vinyl group of the monomer.

3.2 *Tert.*-Butyl Methacrylate

The kinetic orders observed in the polymerization of *tert.*-butyl methacrylate (TBMA) are very similar to those found for MMA.[27] However, as is seen in Figs. 7 and 8 the first-order time-conversion plots exhibit two important differences:

(i) "Induction periods" are not observed, except at temperatures below –20°C.

(ii) Termination is much more pronounced, rendering the determination of the rate constants of polymerization very difficult at ambient temperature. Only for comparatively high catalyst concentrations or low temperatures full monomer conversion is achieved. This is in contrast to the anionic polymerization of TBMA. Here, even at ambient temperature, the polymerization proceeds without termination[8,28]. It is possible that the higher extent of termination is due to the lower rate constants of polymerization (*vide infra*) in combination with similar rate constants of termination. Nevertheless, the molecular weight distributions appear to be somewhat narrower than those of PMMA.

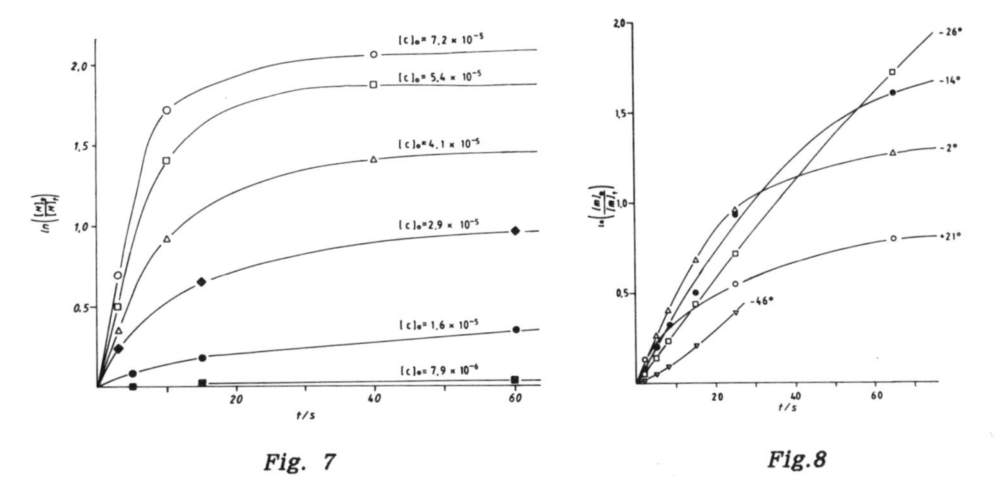

Fig. 7: First-order time-conversion plots for the group transfer polymerization of TBMA with MTS and $TASHF_2$ in THF as a function of catalyst concentration.

Fig. 8: First-order time-conversion plots for the group transfer polymerization of TBMA with MTS and $TASHF_2$ in THF at various temperatures.

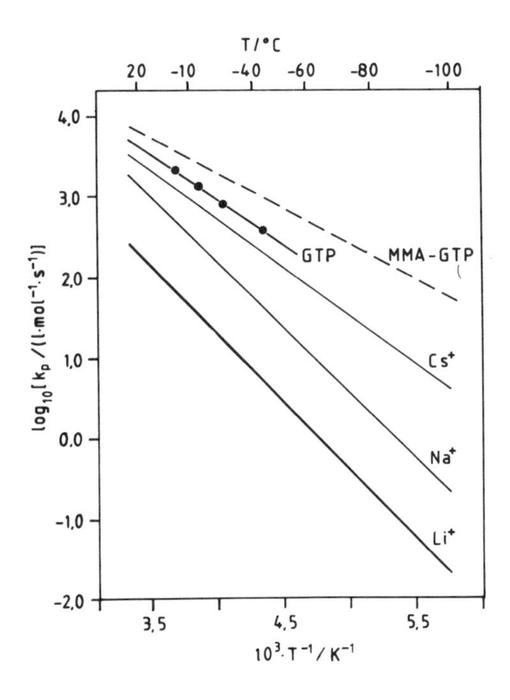

Fig. 9: Arrhenius plot for the polymerization rate constants in the group transfer polymerization of TBMA and MMA with MTS and $TASHF_2$ in THF and in the anionic polymerization of MMA in THF using different counterions (References, cf. Table 3).

Fig. 9 shows an Arrhenius plot of the rate constants of polymerization. It is compared to the corresponding plots for MMA and for the anionic polymerization of TBMA[8,28]. It is seen that the GTP of TBMA proceeds slower than that of MMA by a factor of 1.5-5 depending on temperature. The rate constants for GTP of TBMA correspond to those observed in anionic polymerization using bulky counterions (Unfortunately no data for cryptated sodium or the free anion are available). The similarity in activation parameters is seen in Table 3:

Table 3: Rate constants and activation parameters in the group transfer and anionic polymerization of TBMA using different counterions

	GTP	Cs^+	Na^+	Li^+	GTP,MMA
E_a/kJ/mol	21.7	19.5	18.3	24.0	16.9
log A	7.5	7.3	7.0	7.4	6.8
k_p^{175}	10	4	0.21	0.021	56
Ref.	27	28	28	8	16

These results again confirm the mechanism shown in Scheme 5.

4. MOLECULAR WEIGHT DISTRIBUTIONS

The molecular weight distributions of polymers prepared by GTP considerably depend on the mode of addition of reagents. The "standard" method introduced by the du Pont group[1-7] uses slow addition of monomer to a solution of initiator and catalyst and leads to polydispersities as low as M_w/M_n = 1.05. The *actual* monomer concentration always is rather low.

In order to perform kinetic experiments it is favourable to add the monomer in one batch. It was observed that addition of a monomer batch to a solution of initiator and catalyst leads to unsatisfactory kinetic results, due to a deactivation of the initiator/catalyst system.[29] Thus, a catalyst solution was added to a pre-mixed solution of monomer and initiator. In those experiments the monomer concentration is high at the beginning and decreases with conversion.

When adding monomer in a batch the molecular weight distributions of the resulting polymers are broader than is expected ($M_w/M_n \geq 1.3$). Especially at low monomer conversions (i.e. high concentrations) the polydispersities are very high ($M_w/M_n > 2$; *cf.* Fig. 10). Fig. 10 shows that a non-linear dependence of the number-average degrees of polymerization *vs.* conversion is observed. However, at complete monomer conversion the calculated and experimental degrees of polymerization usually agree within experimental error.[29]

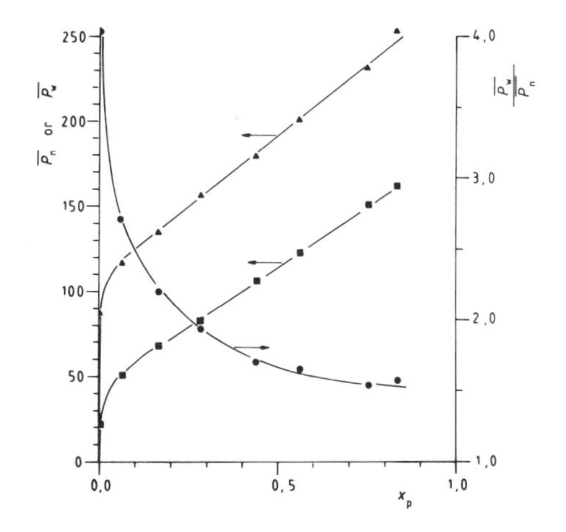

Fig. 10: Number and weight averages of the degree of polymerization and polydispersity as a function of monomer conversion in the GTP of MMA in THF using MTS as the initiator and $TASHF_2$ as the catalyst.

In order to explain these observations it is necessary to take into account the rates of the equilibrium between "dormant" and activated polymers (or initiator):

$$C + P_i \underset{k_{-1}}{\overset{k_1}{\rightleftharpoons}} P_i^*$$

$$P_i^* \xrightarrow[+M]{k_p} P_{i+1}^*$$

$$C + P_{i+1} \underset{k_{-1}}{\overset{k_1}{\rightleftharpoons}} P_{i+1}^*$$

$$K^* = k_1/k_{-1}$$

Scheme 7

The probability of a given activated chain to add monomer is

$$p = \frac{\text{rate of polymerization}}{\text{rate of deactivation}} = \frac{k_p\cdot[M]\cdot[P^*]}{k_{-1}\cdot[P^*]} = \frac{k_p\cdot[M]}{k_{-1}} \tag{9}$$

The number-average degree of polymerization is given by

$$P_n = \frac{\text{number of polymerized monomers}}{\text{total number of polymer chains}} = \frac{[M]_0\cdot x_p}{[\text{chains}]} \tag{10}$$

Thus, P_n/x_p is reciprocal to the *actual* total chain concentration at a given conversion x_p.

Two limiting cases can be discussed:

(a) fast exchange of catalyst (low monomer concentrations):

$$k_{-1} \gg k_p \cdot [M]$$

$$[\text{chains}] = [I]_0$$

$$P_n = \frac{[M]_0 \cdot x_p}{[I]_0} = P_{n,th}$$

(b) no exchange of catalyst (high monomer concentrations):

$$k_{-1} \ll k_p \cdot [M]$$

$$[\text{chains}] = [C]_0$$

$$P_n = \frac{[M]_0 \cdot x_p}{[C]_0} \gg P_{n,th} \qquad (11)$$

As usually $[C]_0 \ll [I]_0$, the plot of P_n *vs.* x_p will have a much higher slope for case (b) than that expected for case (a). For both cases the molecular weight distribution should be of the Poisson type.

In case that the exchange of catalyst *is comparable* to the rate of polymerization ($k_{-1} \approx k_p \cdot [M]$) and [M] decreases during polymerization (i.e. in batch experiments) the number of chains will increase with conversion:

$$[C]_0 \leq [\text{chains}] \leq [I]_0.$$

According to eq. 10 the slope of the plot of P_n *vs.* x_p will decrease until $[\text{chains}] = [I]_0$. This is seen in Fig. 10. Due to the increase in the number of chains the polydispersity can be considerably higher than 1.

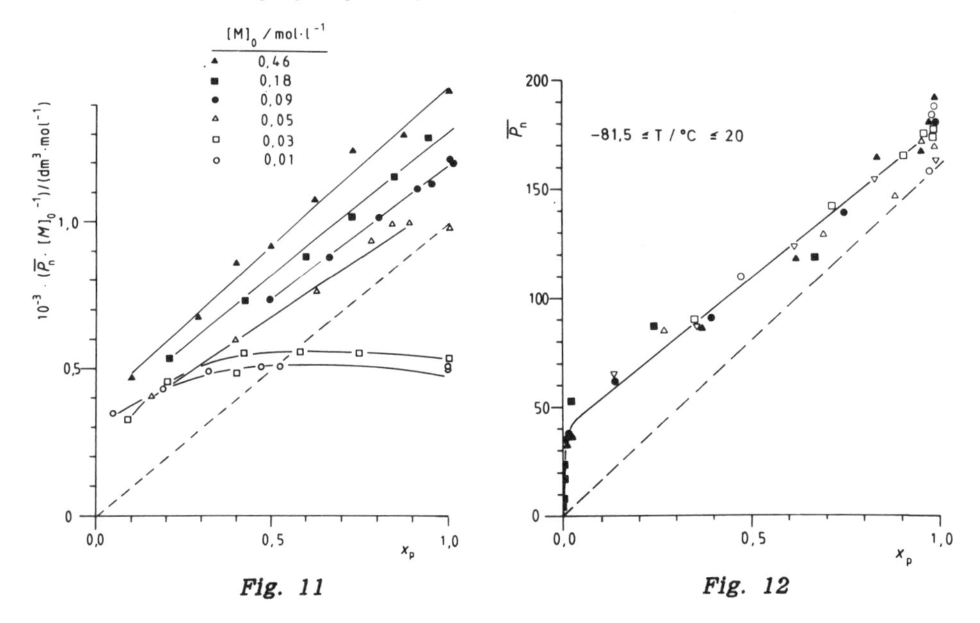

Fig. 11: *Normalized number-average degrees of polymerization vs. monomer conversion as a function of initial monomer concentration.*

Fig. 12: *Number-average degrees of polymerization vs. monomer conversion as a function of temperature.*

According to eq. 9 the deviations from linearity should increase with $[M]_0$ and k_p/k_{-1}. Fig. 11 shows that this is true for the monomer concentration. The deviations at very low initial monomer concentrations may be attributed to side reactions also seen in the corresponding time-conversion plots. As expected, the effect does not depend on initiator and catalyst concentrations. Fig. 12 shows that it also does not depend on temperature indicating that propagation and "deactivation" have similar activation energies.

5. CONCLUSIONS

It now seems well established that the mechanism of GTP is an associative one, the addition of monomer being a two-step process. Information is lacking on the processes in the initial stages of polymerization and on the mechanisms of termination. The influence of the equilibration rates on the molecular weight distribution also needs further clarification.

ACKNOWLEDGEMENT: This work is supported by the *Dr. Otto Röhm Gedächtnisstiftung, Darmstadt* and by the *Deutsche Forschungsgemeinschaft* within the *Sonderforschungsbereich 41 "Chemie und Physik der Makromoleküle", Mainz/Darmstadt.*

REFERENCES:

1) O. W. Webster, W. B. Farnham, D. Y. Sogah; E. P. O. Patent 68887 (1983), E. I. du Pont de Nemours & Company
2) O. W. Webster, W. R. Hertler, D. Y. Sogah, W. B. Farnham, T. V. RajanBabu; *J. Amer. Chem. Soc.* **105**, 5706 (1983)
3) D. Y. Sogah, O. W. Webster; *J. Polym. Sci., Polym. Lett. Ed.* **21**, 927 (1983)
4) O. W. Webster, W. R. Hertler, D. Y. Sogah, W. B. Farnham, T. V. RajanBabu; *Polym. Prepr., ACS Div. Polym. Chem.* **24(2)**, 52 (1983)
5) D. Y. Sogah, O. W. Webster; *Polym. Prepr., ACS Div. Polym. Chem.* **24(2)**, 54 (1983)
6) W. B. Farnham, D. Y. Sogah; U. S. Patent 4414372 (1983), E. I. du Pont de Nemours & Company
7) O. W. Webster, U. S. Patent 4417034 (1983), E. I. du Pont de Nemours & Company
8) A. H. E. Müller in: "Recent Advances in Anionic Polymerization", T. E. Hogen-Esch, J. Smid, Eds.,Elsevier, New York 1987
9) W. R. Hertler, D. Y. Sogah, O. W. Webster, B. M. Trost; *Macromolecules* **17**, 1415 (1984)
10) D. Y. Sogah, W. B. Farnham, in : "Organosilicon and Bioorganosilicon Chemistry", H. Sakurai, Ed., Wiley, New York 1986, p. 219
11) W. B. Farnham, D. Y. Sogah; *Polym. Prepr., ACS Div. Polym. Chem.* **27(1)**, 167 (1986)

12) F. Bandermann, H. D. Speikamp; *Makromol. Chem., Rapid* Commun. **6**, 335 (1985)
13) F. Bandermann, H. D. Sitz, H. D. Speikamp; *Polym. Prepr., ACS Div. Polym. Chem.* **27(1)**, 169 (1986)
14) F. Bandermann, R. Witkowski; *Makromol. Chem.* **187**, 2691 (1986)
15) P. M. Mai, A. H. E. Müller, *Makromol. Chem., Rapid Commun.* **8**, 99 (1987)
16) P. M. Mai, A. H. E. Müller, *Makromol. Chem., Rapid Commun.* **8** (1987) in press
17) V. Warzelhan, G. Löhr, H. Höcker, G. V. Schulz; *Makromol. Chem.* **179**, 2211 (1978)
18) C. Johann, A. H. E, Müller; *Makromol. Chem., Rapid Commun.* **2**, 687 (1981)
19) H. Jeuck, A. H. E. Müller; *Makromol. Chem., Rapid Commun.* **3**, 121 (1982)
20) R. Kraft, A. H. E. Müller, V. Warzelhan, H. Höcker, G. V. Schulz; *Macromolecules* **11**, 1093 (1978)
21) V. Warzelhan, H. Höcker, G. V. Schulz; *Makromol. Chem.* **179**, 2221 (1978)
22) R. Noyori, I. Nishida, J. Sakada, M. Nishizawa; *J. Amer. Chem. Soc.* **102**, 1223 (1980)
23) M. A. Müller, M. Stickler; *Makromol. Chem., Rapid Commun.* **7**, 575 (1986)
24) W. Fowells, C. Schuerch, F. A. Bovey, F. P. Hood; *J. Amer. Chem. Soc.* **89**, 1369 (1967)
25) A. H. E. Müller, H. Höcker, G. V. Schulz; *Macromolecules* **10**, 1086 (1977)
26) W. J. Brittain, D. Y. Sogah; private communication
27) M. A. Doherty, A. H. E. Müller, in preparation
28) A. H. E. Müller, *Makromol. Chem.* **182**, 2863 (1981)
29) P. M. Mai, A. H. E. Müller, in preparation

GROUP TRANSFER POLYMERIZATION OF METHYL METHACRYLATE WITH BASIC CATALYSTS

Hans-Dieter Sitz, Friedhelm Bandermann
Institut für Technische Chemie
Universität-Gesamthochschule Essen
Universitätsstraße 5, D-4300 Essen 1, FRG

ABSTRACT

A screening of catalysts for the group transfer polymerization (GTP) of methyl methacrylate (MMA) with (1-methoxy-2-methyl-1-propenyloxy)-trimethylsilane (MTS) as initiator (I) was performed. The use of tris(dimethylamino)sulfonium difluorotrimethylsilicate ($TASF_2SiMe_3$), one of the most preferred catalysts of Du Pont, needs the addition of acetonitrile as solvent, which unfortunately reacts with the active centres in the presence of the catalyst under termination. This difficulty can be overcome using tris(piperidino)sulfonium bifluoride ($TPSF_2H$) as catalyst. It is soluble in pure tetrahydrofuran and only traces are sufficient for the catalysis of a very rapid GTP of MMA. But it catalyses a rapid oligomerization of the initiator, too. So, tetrabutylammonium cyanide (Bu_4NCN) is recommended as catalyst. It is soluble in tetrahydrofuran, an as powerful catalyst as $TPSF_2H$, but less active in the oligomerization reaction of MTS. Its initiation of an anionic polymerization of MMA can be neglected under the reaction conditions of the GTP.

INTRODUCTION

In 1983 Du Pont claimed a new type of living polymerization, termed group transfer polymerization (GTP) (1). Especially derivatives of acrylic and methacrylic acid may be polymerized. Besides a silicon containing initiator (I), a catalyst (C) is necessary. As catalysing agents, Du Pont proposed a number of basic compounds, tris(dimethylamino)sulfonium difluorotrimethylsilicate ($TASF_2SiMe_3$) and tris(dimethylamino)sulfonium bifluoride ($TASF_2H$) being the most preferred ones.

In this paper we would like to compare some catalysts with concern to their catalytic activity in the GTP of methyl methacrylate (MMA) with (1-methoxy-2-methyl-1-propenyloxy)trimethylsilane (MTS) as initiator.

M. Fontanille and A. Guyot (eds.), Recent Advances in Mechanistic and Synthetic Aspects of Polymerization, 41–47.

MATERIALS AND METHODS

MTS was prepared according to Ainsworth et al. (2), $TASF_2SiMe_3$ and tris(piperidino)sulfonium bifluoride ($TPSF_2H$) according to Middleton (3). Tetrabutylammonium cyanide (Bu_4NCN) was purchased from Fluka AG, FRG. Tetrahydrofuran was treated with CuCl, then dried over KOH and sodium wire, distilled under Argon and condensed from n-Butyllithium before use. The purification of MMA and acetonitrile and the measuring of time conversion curves and of average molecular weights are described elsewhere (4).

RESULTS AND DISCUSSION

$TASF_2SiMe_3$ is insoluble in THF. Therefore, small amounts of acetonitrile about 3 - 5% by volume have got to be added to the system, to obtain a homogeneous solution. Recently, we reported (5), that in the presence of the catalyst, acetonitrile can react with MTS and, as we assume, also with the growing chains under formation of trimethylsilyl acetonitrile (Me_3SiCH_2CN). These reactions reduce the concentration of the active centres, so that complete conversions cannot be obtained at all experimental conditions.

Fig. 1 shows time conversion curves for different MTS concentrations in a THF/acetonitrile mixture in dependence on the catalyst concentration. The numbers at the curves give the ratio of the initiator to the catalyst concentration.

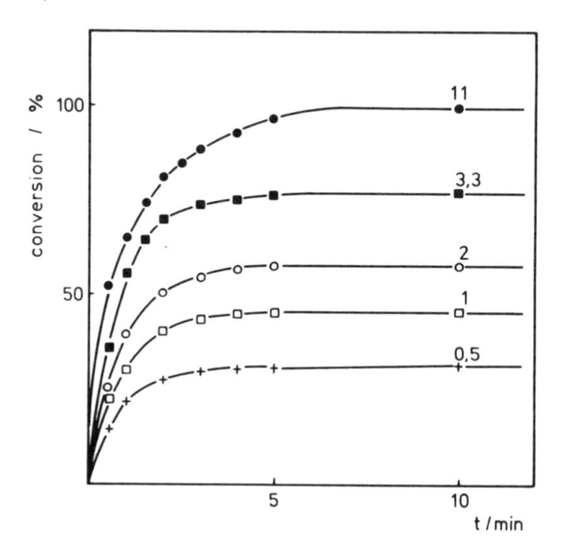

Figure 1. Time conversion curves for the polymerization of MMA with MTS (I) and $TASF_2SiMe_3$ (C) in a THF/acetonitrile mixture.
$(MMA)_o$ = 1.58 mol/l; $(I)_o$ = 0.008 mol/l; $(C)_o$ (mol/l) : (•) 0.00073; (■) 0.0024; (o) 0.004; (□) 0.008; (+) 0.016;

With increasing catalyst concentration the reaction rate and the end conversions decrease. The polymerization can only be started again by the addition of fresh initiator. An overview of the dependence of the end conversions on the catalyst concentration for different initiator concentrations gives Fig. 2. Only for low catalyst concentrations high conversions up to 100 % can be obtained.

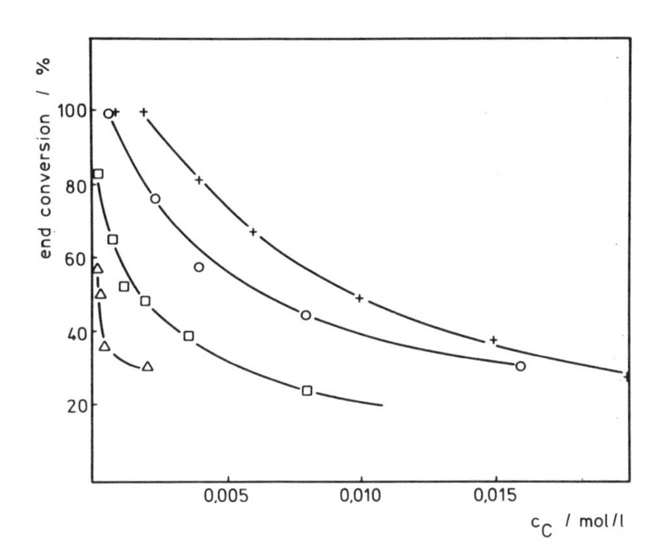

Figure 2. End conversions in the polymerization of MMA with MTS (I) and $TASF_2SiMe_3$ (C) in dependence on the catalyst concentration for different initiator concentrations in a THF/acetonitrile mixture. $(MMA)_o$ = 1.5 mol/l; $(I)_o$ (mol/l): (+) 0.01; (o) 0.008; (□) 0.004; (△) 0.002;

The number average molecular weights (Fig.3) decrease with the catalyst concentration for all initiator concentrations, being lower especially at high catalyst concentrations than the theoretical ones (points at the right ordinate), calculated under the assumption of an instantaneous initiation and a living system.

So, to avoid the termination reaction between the active centres and the acetonitrile we have got to look for a catalyst, which is soluble in pure THF. Such catalysts are tris-(piperidino)sulfonium difluorotrimethylsilicate ($TPSF_2SiMe_3$) and tris-(piperidino)sulfonium bifluoride ($TPSF_2H$),the letter being preferred, because it can be synthetized in higher purity. Both catalysts are not expressively mentioned in the original Du Pont patent but incorporated in its claims (6). At reaction conditions, comparable to those in Fig. 1, the GTP with $TPSF_2H$ is extremly rapid. So, we had got to lower its concentration in order to be able to measure time conversion curves (Fig. 4).

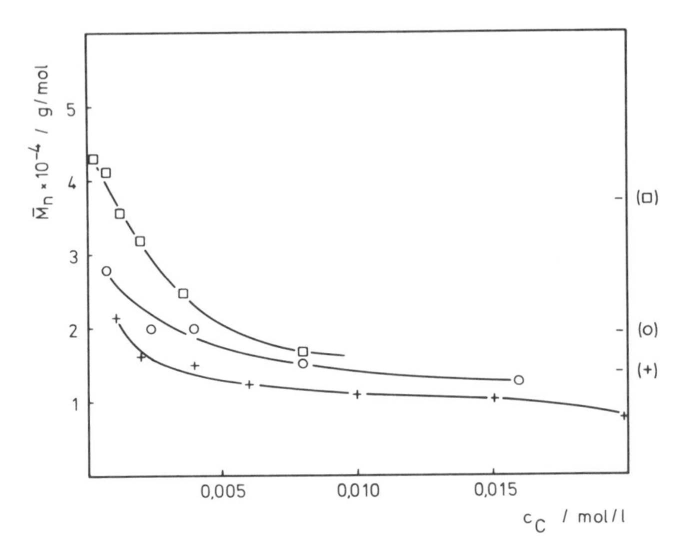

Figure 3. Number average molecular weight ($\bar{M}_n$) in dependence on the catalyst concentration for different initiator concentrations in the polymerization of MMA with MTS (I) and $TASF_2SiMe_3$ (C) in a THF/acetonitrile mixture. The points at the right ordinate give the theoretical number average molecular weights for a living polymerization with complete conversion.
$(MMA)_o$ = 1.5 mol/l; $(I)_o$ (mol/l): (□) 0.004; (o) 0.008; (+) 0.01;

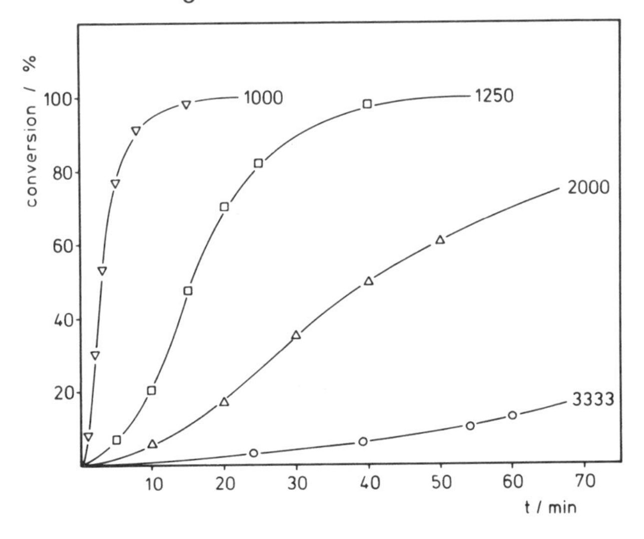

Figure 4. Time conversion curves for the polymerization of MMA with MTS (I) and $TPSF_2H$ (C) in tetrahydrofuran.
$(MMA)_o$ = 1.5 mol/l; $(I)_o$ = 0.01 mol/l; $(C)_o$ (mol/l): (o) 0.000003; (Δ) 0.000005; (□) 0.000008; (▽) 0.00001;

Here, in contrast to the TAS catalysts, we observe an increase in the overall reaction rate, when the catalyst concentration is raised. Complete conversions are obtained for all employed conditions.
The curves show induction periods, which point to a more complicated mechanism.

The number average molecular weights are in the range of those expected for a living system. At higher initiator concentrations, they are nearly independent on, at lower initiator concentrations, they slightly increase with the catalyst concentration. The dispersion indices are low (about 1.3) at low initiator concentrations for all catalyst concentrations. They increase with the catalyst concentration at higher initiator concentration, but, here, it must be taken into account, that at catalyst concentrations above 0.00005 mol/l the reaction system is heated up, so that the solvent begins to boil.

We found furthermore, that in the presence of $TPSF_2H$, MTS is not only rapidly isomerized to its C-silylized form to an extent of nearly 25 %, but that it also reacts to its corresponding ester, thereby releasing Me_3SiF, and to its dimer and higher oligomers. By these reactions, the concentration of active centres is decreased, this effect being especially pronounced at high initiator concentrations.

Therefore we looked for an other catalyst, soluble in THF and being less active in catalysing the above mentioned side reactions. We found, that tetrabutylammonium cyanide (Bu_4NCN), also not expressively cited in the original patent of Du Pont but incorporated in its claims, can advantageously be used as catalyst in the GTP of MMA. Fig. 5 shows some time conversion curves.

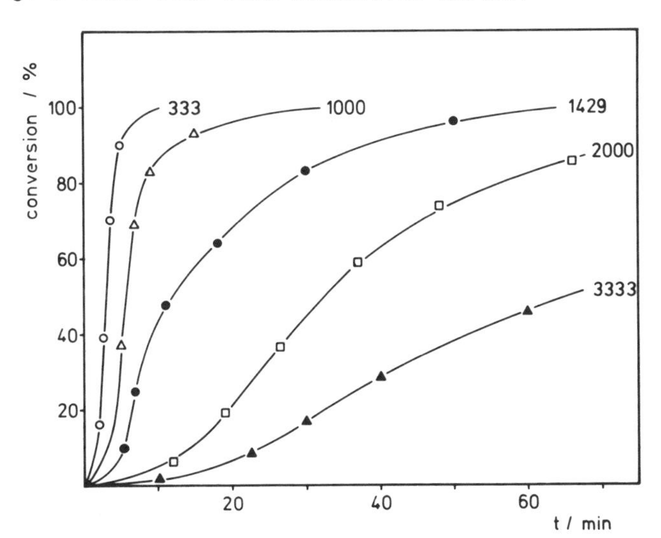

Figure 5. Time conversion curves for the polymerization of MMA with MTS (I) and Bu_4NCN (C) in tetrahydrofuran.
$(MMA)_o$ = 1.5 mol/l; $(I)_o$ = 0.01 mol/l; $(C)_o$ (mol/l: (▲) 0.000003; (□) 0.000005; (●) 0.000007; (△) 0.00001; (o) 0.00003;

As in the TPS system, the catalyst concentrations, necessary for a rapid GTP, are extremly low, the reaction rate increases with the catalyst concentration, the curves show induction periods and complete conversions are obtained for all our experimental conditions. Because it is known, that cyanide anions may initiate an anionic polymerization of MMA, we repeated these investigations without MTS and observed no relevant conversion of MMA within the time, in which the GTP, catalysed by Bu_4NCN, is finished. So, we can neglect the pure anionic pathway of the conversion of MMA. This may also be deduced from the dependence of the number average molecular weights on the catalyst concentration (Fig.6).

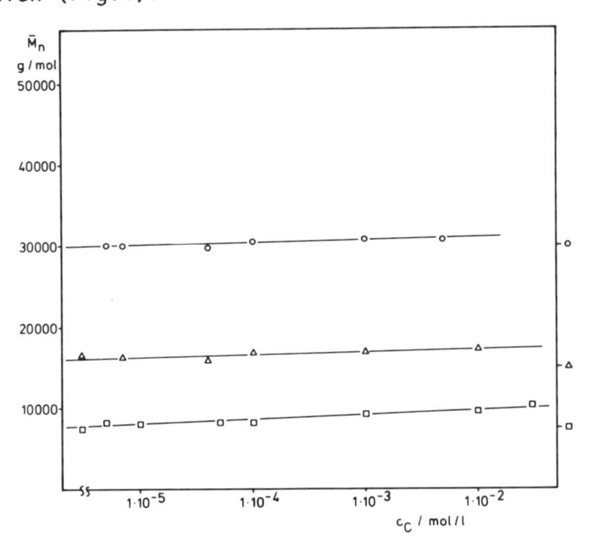

Figure 6. Number average molecular weight ($\bar{M}_n$) in dependence on the catalyst concentration for different initiator concentrations in the polymerization of MMA with MTS (I) and Bu_4NCN (C) in tetrahydrofuran. The points at the right ordinate give the theoretical number average molecular weights for a living polymerization with complete conversion. $(MMA)_o$ = 1.5 mol/l; $(I)_o$ (mol/l): (o) 0.005; (Δ) 0.01; (□) 0.02;

The $\bar{M}_n$ values increase a few with the catalyst concentration and don't appreciably differ from those expected for a living system. The dispersion indices are low (about 1.3) and increase with the catalyst concentration, presumably due to the high temperature rise in the reaction system as already discussed in the TPS catalyst.

MTS gives with Bu_4NCN the same side reactions as with $TPSF_2H$. But they are very much slower than with the latter one, so that they don't play any dominant role within the time, in which the GTP is going to complete conversion.

CONCLUSION

Bu_4NCN seems to us to be a very recommendable catalyst in the GTP of MMA. It is soluble in THF. Only traces of it are sufficient to catalyse a very rapid reaction. Its initiation of an anionic polymerization of MMA can be neglected under these conditions, its catalysis of oligomerization reactions of MTS, too. Furthermore, it is much cheaper than the bifluorides, preferred by Du Pont.

This work was supported by the Deutsche Forschungsgemeinschaft and by the Forschungspool of the Universität-Gesamthochschule Essen.

REFERENCES

1. O.W.Webster, W.R.Hertler, D.Y.Sogah, W.B.Farnham, T.V.RajanBabu, J.Am.Chem.Soc. 105, 5706 (1983)

2. C.Ainsworth, F.Chen, Y.-N.Kuo, J.Organomet.Chem. 46, 59 (1972)

3. U.S.3950.402 (1976), E.I.Du Pont de Nemours and Co., inv. W.J.Middleton; Chem.Abstr. 85, P 6388j (1976)

4. F.Bandermann, H.-D.Speikamp, Makromol.Chem., Rapid Commun. 6, 335 (1985)

5. F.Bandermann, H.-D.Sitz, H.-D.Speikamp, ACS Polymer Preprints 27 (1) 169 (1986)

6. EPO 0068887 (1983), E.I.Du Pont de Nemours and Co., inv.O.W.Webster, W.B.Farnham, O.Y.Sogah, Chem.Abstr.98, P 144031g (1983)

THE MECHANISM OF NUCLEOPHILIC ACTIVATION AT SILICON.

R.J.P. CORRIU
Heterochimie et amino-acides UA(CNRS) 1097
Institut de Chimie Fine
Université des Sciences et Techniques du Languedoc
34060 MONTPELLIER Cedex, FRANCE.

ABSTRACT. This lecture gives a presentation of hypervalent species of silicon. The occurrence of these complexes in the course of reaction at silicon is discussed.

The two main differences in the chemistries of silicon and carbon are :

1) - the great difficulty to obtain low valent species of silicon ($>Si=O$, $>Si=CH_2$, $>Si=Si<$, etc..) because of their very high reactivity (1-2),

2) - the ability of silicon to form hypervalent species (3). The penta and hexacoordinated species at silicon have been recognized since a long time. Some examples of neutral, anionic and cationic species are given in the Table 1.

TABLE 1.

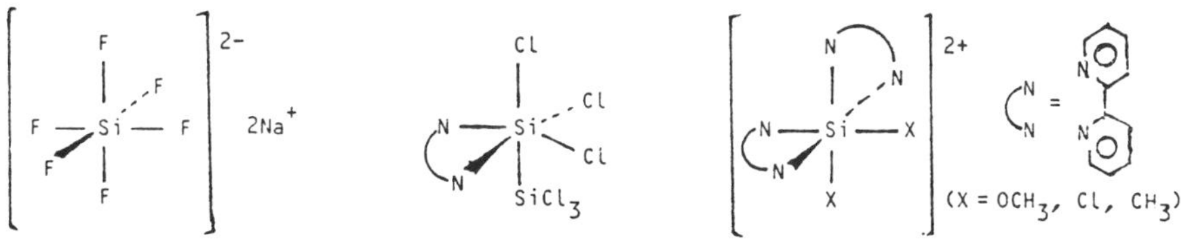

$[SiF_6]^{2-}\ 2Na^+$ (1)

$(N\frown N)SiCl_3(SiCl_3)$ (2)

$[(N\frown N)_2SiX_2]^{2+}$, $N\frown N$ = 2,2'-bipyridyl (X = OCH_3, Cl, CH_3) (3)

M. Fontanille and A. Guyot (eds.), Recent Advances in Mechanistic and Synthetic Aspects of Polymerization, 49–60.

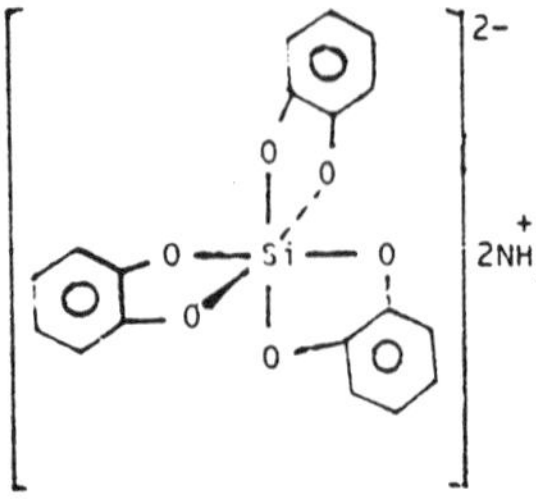

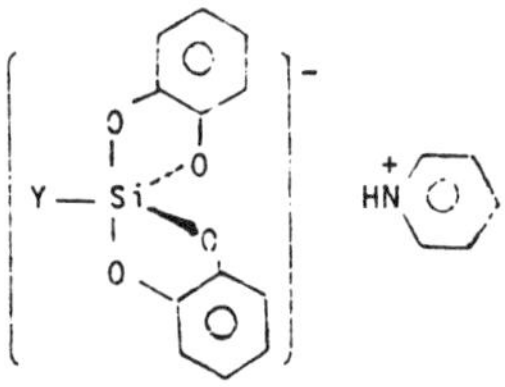

4

5

The formation of adducts between a Lewis base and an organosilicon compound was also extensively studied and illustrated very well by the silatranes 8 (Table 2) which are the first example in which it was possible to study the coordination of nitrogen atom (3). Following the nature of the trans apical group Y, the Si-N bond lenght changes from 2.02 Å (Y = Cl) to 2.21 (Y = CH_2CH_3)

TABLE 2.

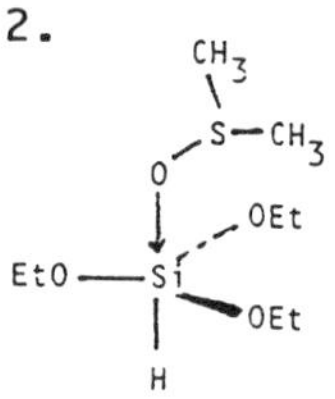

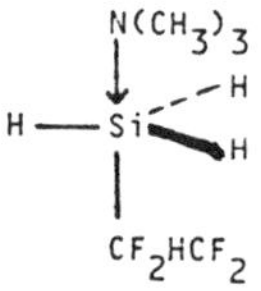

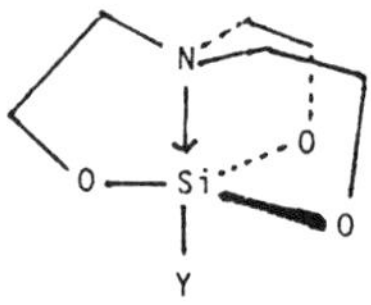

6 7 8

More recently, it was shown that the pentacoordination at silicon is possible even when the silicon is not surrounded by electronegative groups. In the Table 3, some structures are reported, for which the pentacoordination has been established unambiguously. The case of 10 is very interesting since the pentacoordinated silicon is bound only to carbon atoms.

TABLE 3.

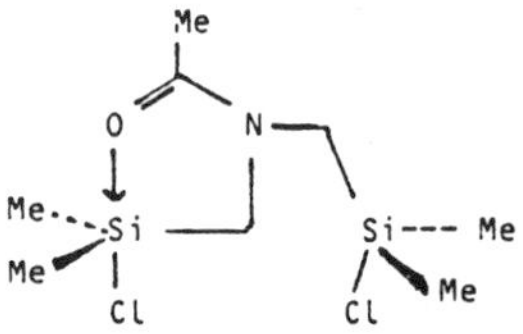

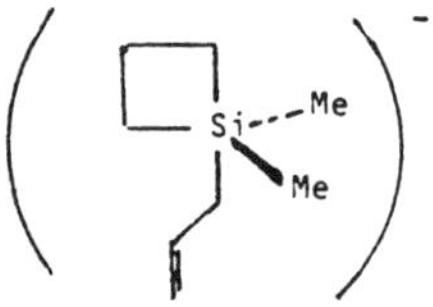

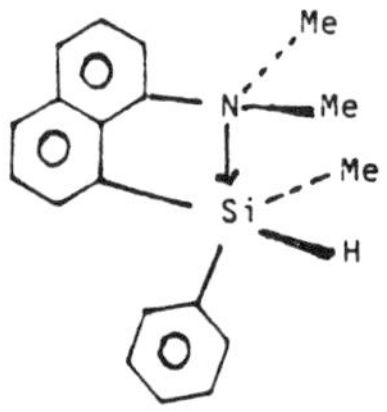

9 (4) 10 (5) 11 (6)

This case and the connected examples strongly support the mechanism of nucleophilic substitution at silicon, which has been described as "Addition-Elimination" process :

$$R_3Si\,X + Nu^{\ominus} \xrightarrow{R.D.S.} \left[R_3\text{-}\underset{\vdots}{\overset{}{Si}}\text{-}X\right] \longrightarrow R_3Si\,Nu + X^-$$
$$\qquad\qquad\qquad\qquad Nu$$

The rate-determining step (R.D.S.) is the addition of the nucleophile at silicon, with formation of the pentacoordinated silicon as an intermadiate. This intermediate reacts quickly with elimination of $X^{\ominus}$ (1, 7, 8).

As extensively studied in the case of Phosphorus compounds (9), the pentacoordinated structures are fluxional compounds. They undergo a permutational isomerisation ; the pseudo-rotation which is a pure intramolecular process. The studies performed with pentafluorosilicates show the F^{19} NMR equivalence of fluorines (10). However as pointed out by Janzen (11), the equivalence can be explained either by the pseudo-rotation process or by a possible hexacoordination since the coalescence is sensitive to the presence of acids and bases.

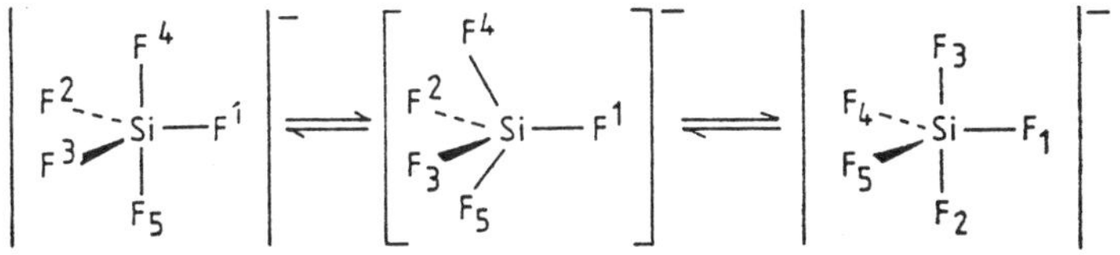

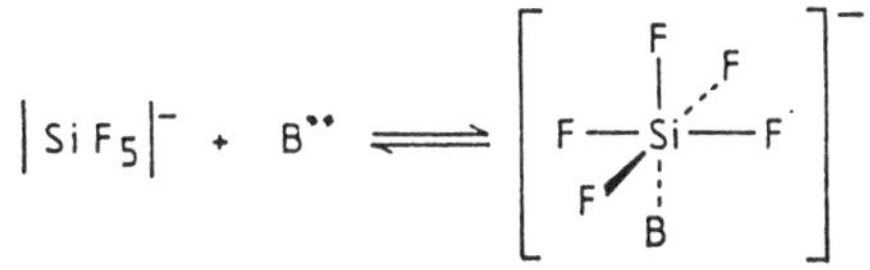

$$|SiF^1F^2F^3F^4F_5|^- \rightleftharpoons [SiF^1F^2F_3F^4F_5]^- \rightleftharpoons |SiF_1F_2F_3F_4F_5|^-$$

$$|SiF_5|^- + B^{\cdot\cdot} \rightleftharpoons [F_5Si{\cdots}B]^-$$

All F are equivalents

During the process, the coupling constant between Si^{29} an F^{19} is always present showing there is no cleavage of Si-F bond.

In fact, pseudorotation has been demonstrated with the compounds presented in Table 4.

Table 4.

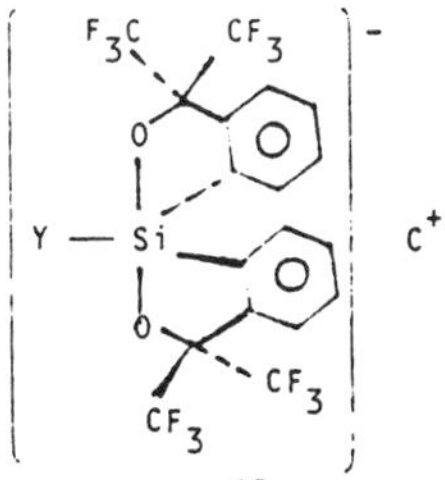

12 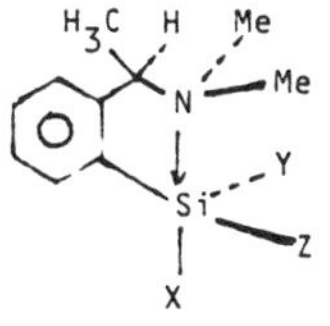 13 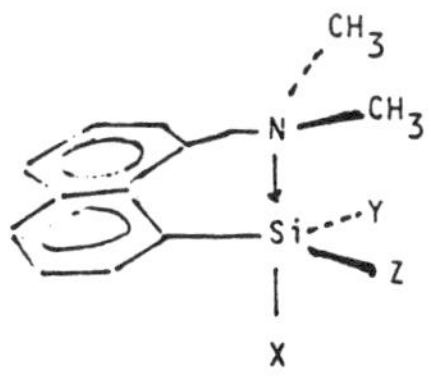 14 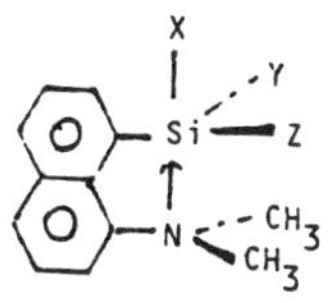 15

In the case of 12 (12), J.C. MARTIN observed $\Delta G^{\neq}$ values for pseudorotation at Si, similar or even lower, than those observed with the corresponding phosphorus compounds. With 13 (13) thanks to the both chiralities at carbon and silicon, it is possible to discriminate between pseudorotation and the process corresponding to the opening of the ring by Si-N bond breaking. The chirality at Si depends only of the pseudorotation process. The chirality at carbon does not influence the diastereotopism of the methyl groups at nitrogen when this nitrogen can be free. Interestingly when X = Y = Z = H, OR , $\Delta G^{\neq}$ for the pseudorotation of 13, 14 and 15 is very low, <7 kcal/mole, showing an almost free pseudorotation at silicon (13 -15).

It is now interesting to consider how the coordination of nucleophiles at silicon can change the reactivity. It is possible to show that in pentacoordinated structures, the Si-H bond is more reactive than in tetracoordinated ones (16). This bond behaves as silicon-hydride and the following reactions have been observed (Table 5).

Table 5.

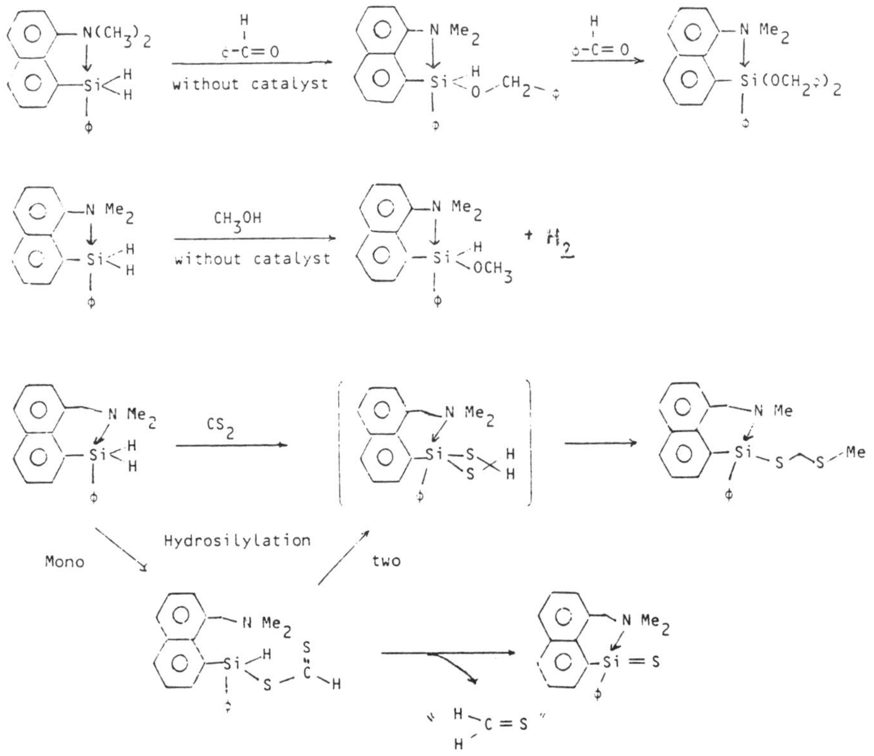

On the other end, the activation of organosilicon compounds by nucleophiles is extensively used especially in the field of the applications of organosilicon compounds in organic synthesis. The Fluoride Ions and HMPA ($O = P(NMe_2)_3$) are the most efficient activating agents (18). It is possible to review briefly the reactions which have been already reported.

1. - Activation of Si-H Bond (18, 19). There is an efficient way for the reduction of >C=O groups. The activation takes place in heterogeneous conditions using salts (fluorides and caboxylates).
CsF appears to be the most efficient, certainly because of the weak interaction between F^- and Cs^+

$$\equiv SiH + F^{\ominus} \rightleftharpoons \left[>\underset{F}{Si}-H \right]^- \xrightarrow[R.D.S.]{>C=O} >\overset{H}{C}-O-Si\equiv$$

In the pentacoordinated intermediate, the SiH bond behaves as silicon hydride and the addition on the carbonyl group takes place. The salt can be used as an heterogeneous catalyst and reused indefinitly.
This efficient method allows selective reductions of carbonyl groups.
Some examples are reported in Table 6.

Table 6.

$CH_2=CH(CH_2)_8-CO_2$-menthyl ⟶ $CH_2=CH(CH_2)_8CH_2-OH$ crowded ESTER

$O_2N-C_6H_4-CO-CH_3$ ⟶ $O_2N-C_6H_4-CH(OH)-CH_3$

citronellal (CHO) ⟶ citronellol (CH_2OH)

$\phi-CO-CH=CH-\phi \xrightarrow[r.t.]{\phi_2SiH_2,\ FCs} \phi-CH(O-Si\equiv)-CH=CH-\phi \xrightarrow{CsF} \phi-C(O-Si\equiv)=CH-CH_2-\phi$

$\downarrow H_2O$: $\phi-CH(OH)-CH=CH-\phi$; $\downarrow H_2O$: $\phi-CO-CH_2CH_2-\phi$

1-2 Reduction 1-4

It is possible to reduce aldehydes in presence of ketones without any reduction of the ketones and to observe the same chimio-selectivity between ketones and esters.

2. - Activation of Si-C bond. The nucleophilic activation of allyl and benzyl silanes is performed with fluoride ions. It permits the formation of carbon-carbon bonds in the reaction with carbonyl groups (Table 7) giving homoallylic alcohols (20).

Table 7.

Synthesis of homoallylic alcohols

Me_3Si-allyl + >C=O —($nEt_4^+F^-$)→ >C(OSiMe$_3$)(allyl) ——→ >C(OH)(allyl)

Mechanism.

a. Me_3Si-allyl + $nEt_4^+F^-$ ——→ Me_3SiF + allyl$^-$ NEt_4^+

b. Me_3Si-allyl + F^- ——→ [F—Si(Me)(Me)(Me)-allyl]$^-$

[F—Si(Me)(Me)-allyl ... O=C<]$^-$ ——→ >Si—O—C< (cyclic) + F^-

The mechanism cannot be the formation of a free anion (a) since, when $NEt_4^-F^-$ is used as a catalyst, an acid-base reaction should take place very quickly.

allyl$^-$ NEt_4^+ ——→ $CH_2{=}CH{-}CH_2$-H + $CH_2{=}CH_2$ + NEt_3

More probably, the fluoride ion is able to be coordinated to silicon (b). F and allyl are both in apical position in the pentacoordinated intermediate ; the Si-C bond is weak and reaction with the electrophilic center of >C=O group can occur. It is very interesting to underline the generality of this nucleophilic activation which can be observed even on a pentacoordinated, anionic, silicon atom (22) :

>C = O + [allyl—Si(O$_2$C$_6$H$_4$)$_2$]$^-$ N Et_4^+ —(F K)→ >C(OH)(allyl) ; —(AlCl$_3$ or TiCl$_4$)↛

The same reaction does not take place with electrophilic activation. The activation of an anionic pentacoordinated silicon species illustrate very well the great possibility of coordination at silicon.

3. - Activation of Si-O bond. This activation is the most popular since the deprotection of alcoxy and enoxysilanes by fluoride ions was reported many years ago :

$$\equiv Si{-}OR \xrightarrow[H_2O]{KF} ROH + \geq Si{-}O{-}Si \leq$$

The activation of silyl enol ethers (or enoxysilanes) by $F^{\ominus}$ permits interesting reactions. The most famous are the cross-aldolisation, the Michaël reaction and, of course, the group transfer polymerisation. Table 8 illustrate some examples.

Table 8.

. Cross aldolisation (22 - 23)

. Michaël reaction (18 - 23 - 24)

It was even possible to obtain a generalisation of Michaël reaction with an "in situ" formation of enoxysilanes (Table 9).

Table 9.

4. - Activation of Si-Cl bond. This activation correspond in fact to a mechanistic study (26 - 28). It was possible to show that alcoholysis or hydrolysis of Si-Cl bond can be activated by the presence of nucleophiles having a good affinity for silicon like dimethylformamide (DMF), dimethylsulfoxyde (DMSO) or even better hexamethylphosphotriamide (HMPA). These agents are catalysts ; they do not enter in the composition of the final products. The presence of these agents accelerate the reaction by a factor of 10^3. As an illustration of this reactivity enhancement, the system Ph_3SiCl/HMPA can be used for the titration of water in organic solvents (28). This titration is made by the polarographic measurement of the HCl wave. In presence of HMPA, the amount of water reacts completely with Ph_3SiCl. It is not the case when the reaction is performed without HMPA. We have studied the kinetics. The rate law is

$$v = k\,(HMPA)_0\,(R_3SiCl)(H_2O)$$

and very interestingly this reaction is controlled by the entropy of activation ($-50 < \Delta S^{\neq} < -70$ u-e). At the opposite, the energy of activation is very low ($-3 < \Delta^{\neq} < +3$ kcal/mole). This energy can be negative. That means, experimentally, that the rate increases when the temperature decreases. There is a very good illustration of the control of the reaction by the entropy.

Furthermore, we have studied the stereochemistry of the process. When the Si-Cl bond is hydrolysed or alcoholysed without any activation agent, the normal stereochemistry is Inversion of configuration. At the opposite, in presence of HMPA (or DMF, DMSO), the stereochemistry observed is Retention of configuration (Table 10).

Table 10.

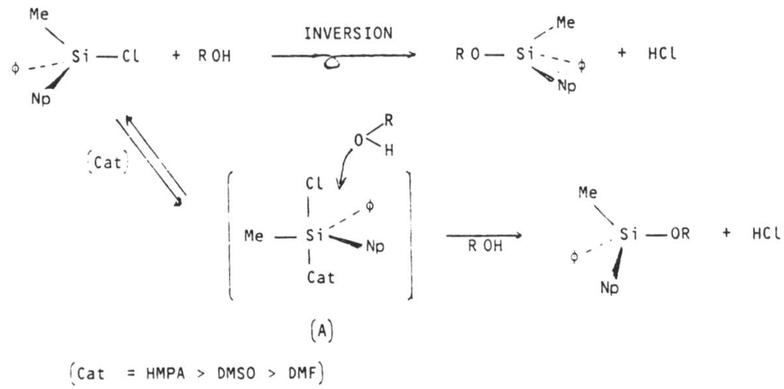

(A)

(Cat = HMPA > DMSO > DMF)

The direct substitution is a SN_2-Si process similar to walden Inversion. However, the nucleophilic activated process involves a pre-equilibrium in which the activating nucleophile which acts as a catalyst, coordinates at silicon in apical position opposite to Si-Cl bond, giving the most stable trigonal bipyramid. In the rate determining step, the water (or the alcohol) displace the chlorine by nucleophilic attack at silicon. This attack is a nucleophilic attack on a pentacoordinated chlorosilane in which the Si-Cl bond is longer and weaker than in the tetracoordinated one.

It can be surprising, if we refer to carbon chemistry to consider that a pentacoordinated complex like (A) is more reactive than a tetracoordinated compound, since " a priori" these complex could appear as having a silicon less electrophilic and more crowded than in a tetracoordinated state. In fact, we have recently reported that pentacoordinated molecules can react better than tetracoordinated one, for instance in the case of Si-H bond (16). Furthermore, we have also refer to the reactivity of hypervalent anionic silicon species. We have shown that penta and hexacoordinated anionic silicon species can undergo a nucleophilic attack at silicon (29). Table 11 illustrate some results.

Table 11.

(B) $[\phi\text{-Si(O}_2\text{C}_6\text{H}_4)_2]^- K^+$ — 1) EtMgBr, 2) $LiAlH_4$ → $\phi\text{-SiHEt}_2$

(B) — $LiAlH_4$ → $\phi\text{-SiH}_3$

(C) $[\text{Si(O}_2\text{C}_6\text{H}_4)_3]^{2-}\ 2K^+$ — $LiAlH_4$ → SiH_4

(C) — 1) 3 EtMgBr, 2) HCl → Et_3SiCl

(C) — 1) 3 EtMgBr, 2) ϕ_3GeLi → $Et_3Si\text{—}Ge\phi_3$

In these reactions, even if the silicon is crowded and even if the complex is negatively charged, we observe a very good nucleophilic reaction. The reaction on complexes (B) and (C) are performed in very mild conditions (0°C in diethylether during 30 mn).

Another interesting observation is the possibility of heptacoordination. Table 12 show the structure of germanium compound in which the Ge atom is heptacoordinated. The same compound with Si as central atom syncrystallized (solid solution with it) (30).

Table 12.

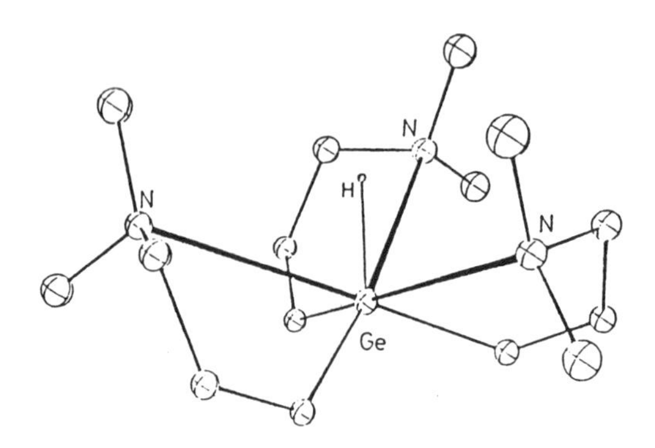

Then, the Ge...N bondings are much more longer (3.05 Å) than the Ge - N.. in a pentacoordinated Ge (2.5 Å).

We can assume that these structure show the possibility to write as an intermediate the attack of a nucleophile on an hexacoordinated complexe.

References.

1. L.H. SOMMER, Stereochemistry, mechanism and silicon, Mc Graw Hill New York (1965)

2. C. EABORN, Organosilicon compounds, London Butterworths Scientific Publications (1960)

3. S.N. TANDURA, M.G. VORONKOV, N.V. ALEKSEEV, 'Molecular and Electronic Structure of penta and hexacoordinate silicon compounds', Topics in current chemistry (1986) 131

4. K.D. ONAN, A.T. McPHAIL, C.H. YODER, R.W. HILLYARD, J. chem. Soc. Commun. (1978) 209 ; R.W. HILLYARD, C.M. RYAN, C.H. YODER, J. Organometal. Chem. (1978) 153 369 ; C.Y. YODER, C.M. RYAN, C.F. MARTIN, P.S. HO, ibid. (1980) 190 1

5. S.A. SULLIVAN, C.H. de PUY, R. DAMRAUER, J. Amer chem. Soc. (1981) 103 408

6. C. BRELIERE, F. CARRE, R.J.P. CORRIU, M. POIRIER, G. ROYO, Organometallics (1986) 5 388

7. L.H. SOMMER, Intra Sci. chem. Rep. (1973) 7 1

8. R.J.P. CORRIU, Journal of organometallic chemistry Library, Organometallic chemistry Reviews (1980) 9 357-407 ; R.J.P. CORRIU, C. GUERIN, J. Organometal.chem.(1980) 198 231-320 ; R.J.P. CORRIU, C. GUERIN, Advances Organometal.chem. (1982) 20 265-311

9. R. LUCKENBACH, Dynamic stereochemistry of pentacoordinated phosphorus and related elements, Stuttgart : Georg Thieme Verlag (1973)

10. F. KLANBERG, E.L. MUETTERTIES, Inorg. chem. (1968) 7 155

11. J.A. GIBSON, D.G. IBOTT, A.F. JANZEN, Canad. chem. (1973) 51 3203

12. W.H. STEVENSON III, S. WILSON, J.C. MARTIN, W.B. FARNHAM, J. Amer. chem. Soc. (1985) 107 6340

13. R.J.P. CORRIU, A. KPOTON, M. POIRIER, G. ROYO and J.Y. COREY J. organometal.chem. (1986) 277 C 25

14. J. BOYER, R.J.P. CORRIU, A. KPOTON, M. MAZHAR, M. POIRIER, G. ROYO, J. organometal.chem. (1986) 301 131

15. R.J.P. CORRIU, M. MAZHAR, M. POIRIER, G. ROYO, J. organometal.chem. (1986) 306 C5

16. J. BOYER, C. BRELIERE, R.J.P. CORRIU, A. KPOTON, M. POIRIER, G. ROYO, J. organometal.chem. (1986) 311 C39

17. R.J.P. CORRIU, G. LANNEAU, A. PRABHAT, unpublished work.

18. R.J.P. CORRIU, C. PERZ, C. REYE, Tetrahedron, (1983) 39 999

19. J. BOYER, R.J.P. CORRIU, R. PERZ, C. REYE, J. organometal. chem. (1978) 141 C31, 157 153 ; J. BOYER, R.J.P. CORRIU, R. PERZ, C. REYE Tetrahedron, (1981) 37 2165 ; J. BOYER, R.J.P. CORRIU, R. PERZ, M. POIRIER, C. REYE, Synthesis (1981) 558 ; C. CHUIT, R.J.P. CORRIU R. PERZ, C. REYE, Synthesis (1982) 981

20. A. HOSOMI, A. SHIRAHATA, H. SAKURAI, Tetrahedron Letters, (1978) 3043

21. G. CERVEAU, C. CHUIT, R.J.P. CORRIU, C. REYE, unpublished work.

22. R. NOYORI, K. YOKOYAMA, J. SAKATA, I. KUWAJIMA, E. NAKAMURA, M. SHIMIZU, J. Amer. chem. Soc. (1977), 99 1265

23. J. BOYER, R.J.P. CORRIU, R. PERZ, C. REYE, J. organometal. Chem. (1980) 184 157

24. J. BOYER, R.J.P. CORRIU, R. PERZ, C. REYE, J. chem Soc. Chem. Comm. (1981) 122 ; Tetrahedron (1983) 39 117

25. R.J.P. CORRIU, R. PERZ, C. REYE, Tetrahedron Letters (1985) 26 1311 ; Tetrahedron, (1986) 42 2293

26. R.J.P. CORRIU, M. HENNER, J. Organometal. chem., (1974) 24 1

27. R.J.P. CORRIU, G. DABOSI, M. MARTINEAU, J. Organometal. chem., (1980) 186 19 ; (1980) 186 25 ; (1980) 188 63

28. R.J.P. CORRIU, G. DABOSI, M. MARTINEAU, J. Organometal. chem. (1981) 222 195

29. A. BOUDIN, G. CERVEAU, C. CHUIT, R.J.P. CORRIU, C. REYE, Angew. chem. Int. Ed. Engl., (1986) 25 473 ; ibid. 474

30. C. BRELIERE, F. CARRE, R.J.P. CORRIU, G. ROYO, unpublished work.

ALDOL-GTP IN CONTROLLED SYNTHESIS OF VINYL ALCOHOL POLYMERS

Dotsevi Y. Sogah and Owen W. Webster
Central Research and Development Department
E. I. du Pont de Nemours and Company, Inc.
Experimental Station
Wilmington, DE 19898 USA

ABSTRACT. Sequential silyl aldol condensation involving aldehydes and silyl vinyl ethers gives monodisperse poly(silyl vinyl ether) whose molecular weight ($\bar{M}_n$ 1000-160,000) is controlled by the aldehyde initiator. The new process, termed aldol-group transfer polymerization (aldol-GTP) involves a silyl group transfer from monomer to the carbonyl oxygen of either the initiator or the living polymer, leading to generation of a new terminal aldehyde functional group. The reaction is catalyzed by Lewis acids and can be initiated by other electrophiles, e.g., alkyl halides and acetals. The living polymers are stable, neutral materials whose hydrolytic stability depends on the bulkiness of the sily group. In general, aromatic aldehydes tend to react more cleanly as initiators than do aliphatic aldehydes. Unlike the GTP of methyl methacrylate in which the silyl group is transferred from the initiator to the monomer, aldol-GTP involves a transfer of silyl group from monomer to initiator. Some of the advantages of aldol-GTP over existing methods such as cationic polymerization include operability over a broad temperature range, complete monomer conversion, living polymer formation, very good molecular weight control, and facile block copolymer synthesis. It permits control of hydrophilicity of block copolymers.

I. INTRODUCTION

We recently reported application of Michael addition to controlled synthesis of acrylic polymers.[1,2] The method, named group transfer polymerization (GTP), involves addition of silyl ketene acetals (as Michael donors) to α,β-unsaturated carbonyl and nitrile compounds (as Michael acceptors) in the presence of a suitable catalyst (Scheme 1).

Scheme 1. Group Transfer Polymerization

$$Me_2C{=}C(OR')(OSiMe_3) + n\,CH_2{=}C(Me)(CO_2R) \xrightarrow{HF_2^-} Me_2C(CO_2R')\text{—}(CH_2C(Me)(CO_2R))_{n-1}\text{—}CH_2C(Me){=}C(OMe)(OSiMe_3)$$

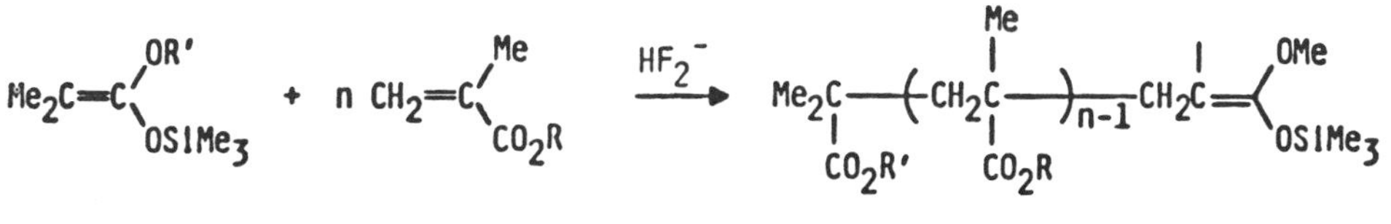

M. Fontanille and A. Guyot (eds.), Recent Advances in Mechanistic and Synthetic Aspects of Polymerization, 61–72.

GTP gives monodisperse living methacrylate polymers of well-controlled molecular weights. Another reaction that has been critical in making carbon-carbon bonds is aldol condensation. However, it has been exploited in preparation of polymers. In our earlier communication, we described the first application of aldol condensation to direct synthesis of living polymers whose structure and molecular weight are well-controlled.[3] To accomplish this, we took advantage of the fact that aldehydes react with silyl vinyl ethers to give silylated cross-aldol products containing a new terminal aldehyde group.[4] Thus, addition of excess silyl vinyl ether to benzaldehyde as initiator gives a silylated vinyl alcohol polymer (Scheme 2). This method was named "aldol-GTP".

Scheme 2. Aldol-GTP

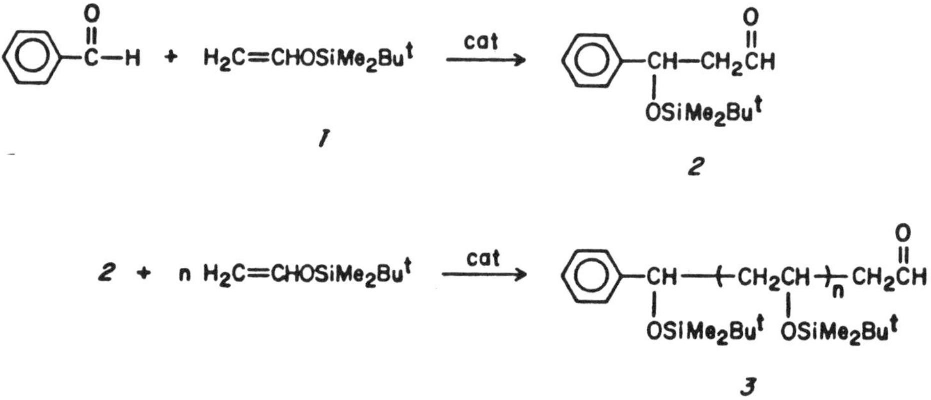

$$C_6H_5\text{-}CHO + H_2C{=}CHOSiMe_2Bu^t\ (\mathbf{1}) \xrightarrow{cat} C_6H_5\text{-}CH(OSiMe_2Bu^t)\text{-}CH_2CHO\ (\mathbf{2})$$

$$\mathbf{2} + n\ H_2C{=}CHOSiMe_2Bu^t \xrightarrow{cat} C_6H_5\text{-}CH(OSiMe_2Bu^t)\text{-}\left(CH_2CH(OSiMe_2Bu^t)\right)_n\text{-}CH_2CHO\ (\mathbf{3})$$

2. RESULTS AND DISCUSSION

In GTP of methacrylates (Scheme 1), the silyl group is transferred from the initiator to the monomer, thereby regenerating the silyl ketene acetal functionality. In contrast, aldol-GTP involves a transfer of silyl group backwards from monomer to initiator. The aldehyde functionality is regenerated. The adduct is stable, and upon further addition of monomer, reacts with the excess silyl vinyl ether to give a stable polymer.

2.1 Solvents and Catalysts

The catalysts that are operative include Lewis acids such as di(isobutyl)aluminum chloride, titanium tetrachloride, tin tetrachloride, boron trifluoride, zinc bromide, zinc iodide and zinc chloride. The preferred catalysts are the zinc halides. The results are summarized in Table I. Whereas $TiCl_4$ gives low yields, the zinc halides tend to give quantitative yields of polymers of narrow molecular weight distribution. Anion catalysts, such as tris(dimethylamino)sulfonium bifluoride, tetrabutylammonium bifluoride and tetrabutylammonium acetate give less satisfactory results. These anions strongly coordinate to the backbone siloxy groups. This makes them less efficient catalysts for the aldol-GTP. In contrast, the Lewis acids are presumed to preferably coordinate to the aldehyde group of the initiator or the living polymer.

Non-reactive chlorinated and non-polar aromatic solvents, such as dichloromethane, 1,2-dichloroethane, toluene, xylenes and benzene, are preferred since they are more compatible with the Lewis acids. However, tetrahydrofuran and acetonitrile are required if anion catalysts are used. Another difference between the GTP of methacrylates and aldol-GTP is that for Lewis acid catalysis, the former requires 10-20 mole percent of catalyst relative to monomer,[5] whereas in aldol-GTP, only about 10^{-4} to 10^{-2} mole percent relative to monomer is needed. In the case of anion catalysis, the reverse is true.

Table I. Polymerization of t-Butyldimethyl Silyl Vinyl Ether[a]

Run No.	Initiator *(mmol)*	Catalyst *(mmol)*	Solv. *(ml)*	Temp. (°c)	Conv. (%)	$\bar{M}_n$ (theory)[b]	$\bar{M}_n$[c]	$\bar{M}_w$[c]	$\bar{M}_w/\bar{M}_n$[c]
1	ØCHO *(1.88)*	$ZnBr_2$ *(0.26)*	DCM *(20)*	30	100	4100	3610	4390	1.21
2	ØCHO *(2.49)*	$ZnCl_2$ *(0.26)*	DCM *(20)*	30	100	3120	2180	2800	1.28
3	$pC_6H_4(CHO)_2$ *(2.02)*	$ZnBr_2$ *(2.50)*	DCM *(20)*	30	100	3840	4150	5130	1.24
4	Me_2CHCHO *(2.40)*	$ZnBr_2$ *(0.20)*	DCM *(20)*	30	70	3200	2900	2950	1.02
5	tRetinal *(0.62)*	$ZnBr_2$ *(0.20)*	DCM *(50)*	30	100	12130	10400	15500	1.49
6	$Ø_2CHCl$ *(0.97)*	$ZnBr_2$ *(0.10)*	DCM *(20)*	30	100	7900	6500	8000	1.23
7[d]	ØCHO *(0.20)*	$ZnBr_2$ *(0.20)*	DCM *(50)*	30	100	160000	78900	121000	1.51
8	ØCHO *(2.45)*	$TASHF_2$ *(0.10)*	THF *(20)*	-65	85	2650[e]	2400	3200	1.33
9	ØCHO *(2.42)*	iBu_2AlCl *(0.25)*	DCM *(20)*	30	95	3200	6640	11300	1.70
10	ØCHO *(2.45)*	$TiCl_4$ *(0.25)*	DCM *(20)*	30	10	300[e]	476	801	1.68
11	$ØCH(OMe)_2$ *(1.81)*	$ZnBr_2$ *(0.26)*	DCM *(20)*	30	100	4300	1490	3690	2.48
12	tØCH=CHCHO *(0.40)*	$ZnBr_2$ *(0.40)*	DCM *(20)*	0	98	10130	7350	11000	1.50

a. Each run performed on 7.50 g of monomer. *b.* Theoretical $\bar{M}_n$ = (# moles of monomer/# moles initiator) x Mol. Wt. of Monomer. *c.* Determined by gel permeation chromatography. *d.* Performed on 32.0 g of monomer. *e.* Based upon isolated yield of polymer.

2.2 Initiators and Monomers

Most aldehydes will initiate aldol-GTP. In general, aromatic aldehydes are preferred to aliphatic ones. *In living polymerizations, it is necessary for the rate of initiation to be greater than or, at least, equal to that of propagation in order to achieve molecular weight control.* This condition is best satisfied by the aromatic aldehyde initiators. Other initiators include alkyl halides, acetals, and anhydrides.

In line with the enhanced hydrolytic stability of t-butyldimethylsilyl (BDS) ethers compared to the corresponding trimethylsilyl (TMS) ethers[6], we find that the best control of molecular weight is obtained with t-butyldimethylsilyl vinyl ethers (BDSVE) over a wide temperature range (-80° to 70°C). Other monomers that are polymerized include phenyldimethylsilyl, diphenylmethylsilyl, triethylsilyl and isopropyldimethylsilyl vinyl ethers. All of these give polymers in good yield with good control of molecular weight and are readily converted into polyvinyl alcohol. The latter process is achieved by either[6] refluxing in THF/methanol with Bu_4NF or hydrochloric acid.[6]

2.3 Novel Copolymers

Since aldol-GTP is living, synthesis of block copolymers by sequential addition of monomers is facile. Thus the block copolymer of BDSVE and trimethylsilyl vinyl ether (TMSVE) are prepared by first adding BDSVE followed by TMSVE (Scheme 3). Treatment with methanolic HCl selectively removes the TMS group since the BDS group is removed preferentially by fluoride treatment and therefore not cleaved under the present conditions.[6] This gives the desired partially hydrophilic and partially hydrophobic block copolymer, poly(t-butyldimethylsilyl vinyl ether-b-vinyl alcohol) (Scheme 3). Since aldehydes are very reactive, it is important to treat the living polymer with a suitable silyl ketene acetal to convert the end group into an ester via a Reformatsky process as shown in Scheme 3.

Scheme 3

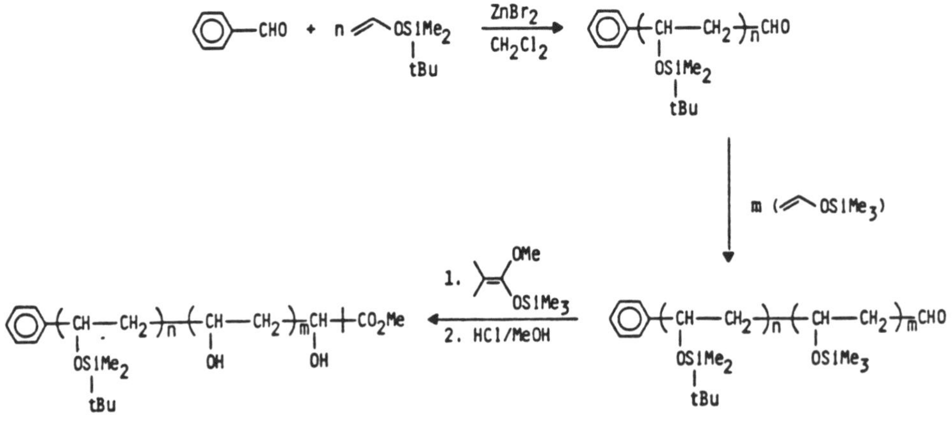

$$PhCHO + n\ CH_2{=}CH{-}OSiMe_2(tBu) \xrightarrow[CH_2Cl_2]{ZnBr_2} Ph{-}[CH(OSiMe_2tBu){-}CH_2]_n{-}CHO$$

$$\xrightarrow{m\ (CH_2{=}CH{-}OSiMe_3)} Ph{-}[CH(OSiMe_2tBu){-}CH_2]_n{-}[CH(OSiMe_3){-}CH_2]_m{-}CHO$$

$$\xrightarrow[2.\ HCl/MeOH]{1.\ Me_2C{=}C(OMe)(OSiMe_3)} Ph{-}[CH(OSiMe_2tBu){-}CH_2]_n{-}[CH(OH){-}CH_2]_m{-}CH(OH){-}CO_2Me$$

Silyl ketene acetals are known to react with aldehydes to give Reformatsky products.[6a,b] This reaction can be used to prepare unique block copolymers. For example, treatment of the aldehyde-terminated polymer ($\bar{M}_n$ 1080, $\bar{M}_w$ 1300) with living poly(methyl methacrylate) ($\bar{M}_n$ 1080, $\bar{M}_w$ 1390) made by GTP in the presence of $ZnBr_2$ in CH_2Cl_2 gives a block copolymer ($\bar{M}_n$ 1610, $\bar{M}_w$ 1910) of BDSVE and MMA. Instead of $ZnBr_2$ in CH_2Cl_2, TAS bifluoride in THF can be used as the catalyst for the coupling reaction (Scheme 4). Upon cleavage of the silyl groups with fluoride ion[6c] in THF/methanol, a copolymer containing hydroxyl groups, is obtained. This is a unique combination of GTP and aldol-GTP to prepare polymers of controlled structure, molecular weight and degree of hydrophilicity. These types of polymers are difficult to prepare using previously reported methods.[7-9]

Scheme 4. Block Copolymers

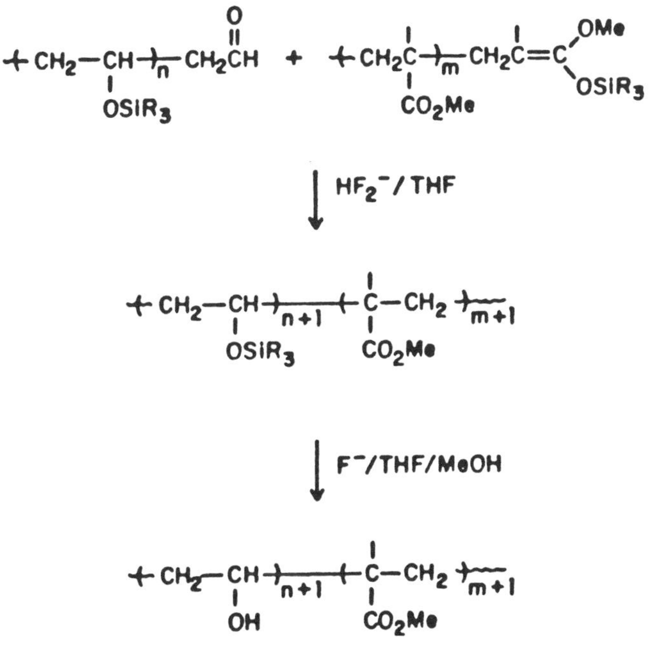

$$-\!\!\left(CH_2-\underset{OSiR_3}{CH}\right)_n\!-CH_2\overset{O}{\overset{\|}{C}}H \;+\; -\!\!\left(CH_2\underset{CO_2Me}{C}\right)_m\!-CH_2C=C\begin{matrix}OMe\\OSiR_3\end{matrix}$$

$$\Big\downarrow\; HF_2^-/THF$$

$$-\!\!\left(CH_2-\underset{OSiR_3}{CH}\right)_{n+1}\!\!-\!\!\left(\underset{CO_2Me}{C}-CH_2\right)_{m+1}$$

$$\Big\downarrow\; F^-/THF/MeOH$$

$$-\!\!\left(CH_2-\underset{OH}{CH}\right)_{n+1}\!\!-\!\!\left(\underset{CO_2Me}{C}-CH_2\right)_{m+1}$$

Just as GTP living polymers are used, one can also use living polymers prepared by other processes, such as anionic and cationic polymerization, in conjuction with aldol-GTP to prepare block copolymers of well-defined segments. Poly(isoprene-b-vinyl alcohol) and poly(styrene-b-vinyl alcohol) are, thus, prepared readily. Poly(isoprene) is prepared by butyllithium-initiated anionic polymerization. The living polymer is treated with N-formylpiperidine *in situ* to convert the end group into an aldehyde (Scheme V). Using

Scheme 5. Polyisoprene-PVA Block Copolymer

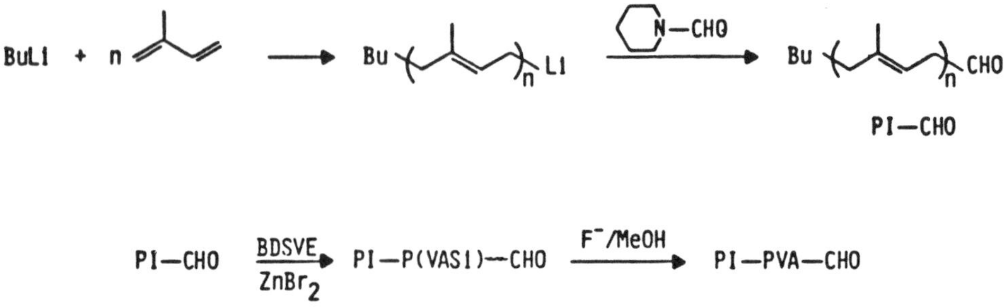

BuLi + n (isoprene) ⟶ Bu–(polyisoprene)$_n$–Li —(N-formylpiperidine)⟶ Bu–(polyisoprene)$_n$–CHO

PI—CHO

PI—CHO $\xrightarrow[ZnBr_2]{BDSVE}$ PI—P(VASi)—CHO $\xrightarrow{F^-/MeOH}$ PI—PVA—CHO

T_{g_1} -64.7, T_{g_2} ±69.5°C, T_m 220

this as a polymeric initiator for aldol-GTP of BDSVE gives a block copolymer which shows two glass transition (Tg) temperatures, Tg_1 -65 and Tg_2 +70, corresponding to poly(isoprene) and polyBDSVE segments, respectively. Upon removal of the silyl groups, followed by exhaustive extraction with water, a block copolymer of isoprene and vinyl alcohol is obtained with Tg_1 -65 and Tm 220. Applying the same techniques, styrene/vinyl alcohol block copolymer is prepared (Scheme 6).

Scheme 6. PS-PVA Block Copolymer

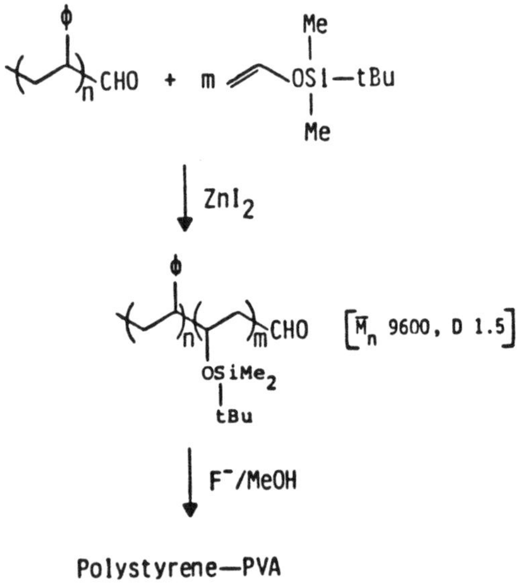

(polystyrene, ϕ)$_n$–CHO + m CH$_2$=CH–OSi(Me)$_2$–tBu

$\downarrow ZnI_2$

(polystyrene, ϕ)$_n$–(CH$_2$–CH(OSiMe$_2$tBu))$_m$–CHO [$\bar{M}_n$ 9600, D 1.5]

$\downarrow F^-/MeOH$

Polystyrene—PVA

2.4 Comb Polymers

Using multi-functional polymeric initiators, it is possible to graft polymer chains to an appropriate backbone. GTP and aldol-GTP are nicely suited for this since both are capable of giving living polymers of well-defined structures and molecular weights. This is illustrated for a GTP random copolymer of methyl methacrylate (50 mol%) and 5-methacryloxyvaleraldehyde protected as the acetal (Scheme 7).

Scheme 7. Functionalized Copolymers

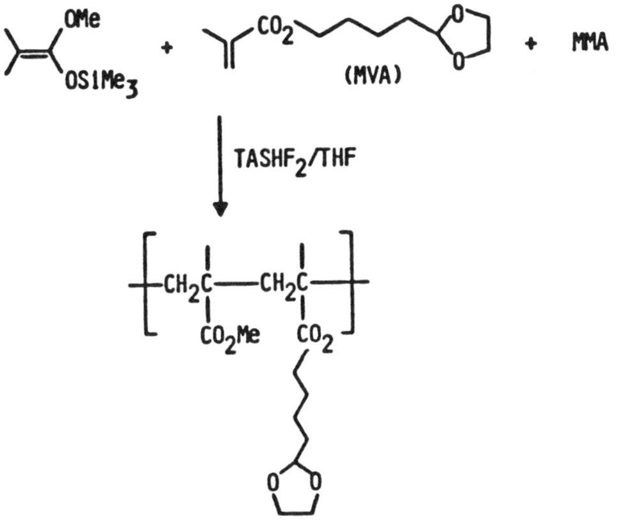

OMe, $OSiMe_3$ + (MVA) + MMA

$TASHF_2/THF$

$-CH_2C-CH_2C-$; CO_2Me ; CO_2

Random Copolymer

($\bar{M}_n/\bar{M}_w/D$ = 5770/8160/1.41)

Upon hydrolysis of the acetal, a polyaldehyde is obtained (Scheme 8).

Scheme 8. Deprotection

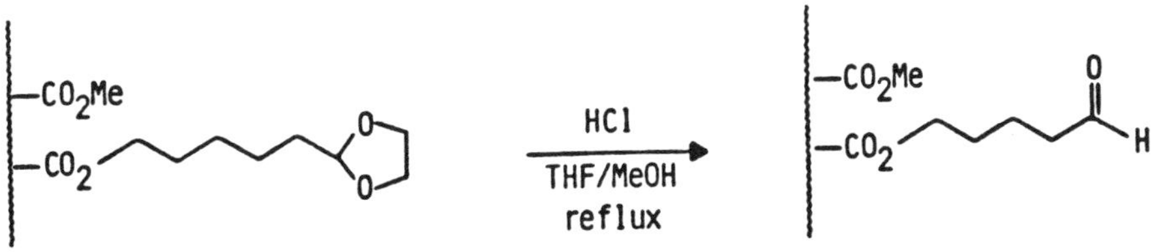

$-CO_2Me$, $-CO_2$ → HCl, THF/MeOH, reflux → $-CO_2Me$, $-CO_2$ … H

($\bar{M}_n$ 5770, $\bar{M}_w$ 8160, D 1.41)

Using this as the initiator, a comb polymer containing a methacrylate backbone and pendent silylated polyvinyl alcohol is obtained. This then is desilylated to give the final comb polymer (Scheme 9).

Scheme 9. Comb Polymers by MCP

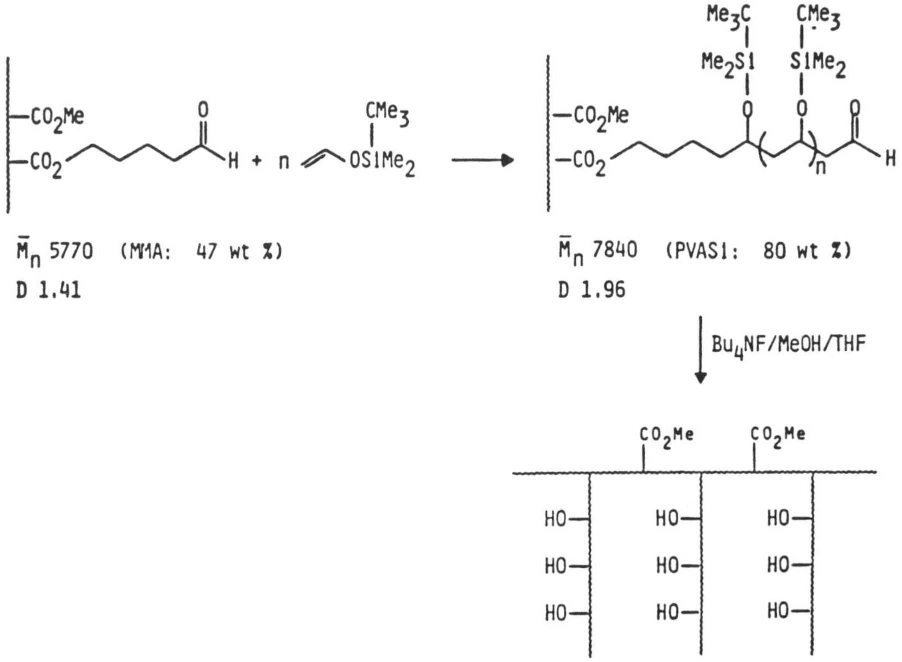

$\bar{M}_n$ 5770 (MMA: 47 wt %)
D 1.41

$\bar{M}_n$ 7840 (PVASi: 80 wt %)
D 1.96

$Bu_4NF/MeOH/THF$

2.5 Star Polymers

Use of either 1,4-bis(bromomethyl)benzene or terephthaldehyde (Scheme 10) enables polymer chain growth in two directions. When this

Scheme 10. Polymer Growth in Two Directions

$$OHC-C_6H_4-CHO + 2n\ H_2C{=}CHOSiMe_2CMe_3 \xrightarrow{ZnBr_2} OHC\text{-}(CH_2CH(OSiMe_2CMe_3))_n\text{-}C_6H_4\text{-}(CH(OSiMe_2CMe_3)CH_2)_n\text{-}CHO$$

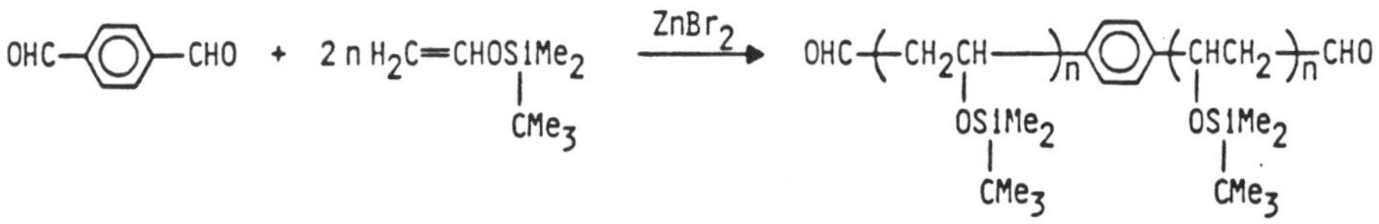

[$\bar{M}_n$ 4150, $\bar{M}_w$ 5130, $\bar{M}_w/\bar{M}_n$ 1.24

$\bar{M}_n$ (VPO) = 4000 (THF, 37°C)

$\bar{M}_n$ (Theory) = 3840]

is extended to tris(bromomethyl)benzene and tetrakis(bromomethyl)-benzene, star polymers containing three and four arms, respectively, are obtained (Table II). The molecular weight distributions are

Table II. Star Polymers

$$C_6H_{6-y}\left[CH_2\left(CH_2\underset{\substack{|\\ OSiMe_2\\ |\\ CMe_3}}{CH}\right)_{n-1}CH_2\overset{\substack{O\\ \|}}{C}H\right]_y$$

y	n	$\bar{M}_n$	$\bar{M}_w$	D	$\bar{M}_n$ (Calc.)
2	50	16,500	37,900	2.29	16,000
3	42	13,600	30,900	2.26	20,000
4	42	15,500	35,300	2.27	26,700

broader than linear polymers obtained via initiation by aldehydes, presumably due to the fact that benzylation is slower than the aldol initiation and propagation. Upon desilylation, a poly(vinyl alcohol) star is obtained. To our knowledge, this is the first report of PVA stars.

2.6 Aldol-GTP vs. Cationic Polymerization

The aldol-GTP should be distinguished from cationic polymerization of silyl vinyl ethers reported by Murahashi and co-workers.[7,8] Their process is best carried out at low temperatures, because above 0°C their method does not give any polymer. Results of our comparative studies are reported in Table III. Entries 1 and 3 clearly show that in the absence of an aldehyde, even at low temperatures, the cationic process does not lead to molecular weight control by the mole ratio of monomer to initiator. Contrastingly, entries 2 and 4 show that as the monomer/aldehyde ratio increases at constant concentration of $ZnBr_2$, molecular weight increases. Furthermore, the polydispersities are smaller in aldol-GTP than in the cationic process. Using the Murahashi method, we could not prepare the type of block, star and comb polymers described earlier.

Table III. Comparison of Aldol-GTP and Cationic Polymerization[a]

Run No.	Initiator (mmol)	Catalyst (mmol)	Temp. (°C)	Conv. (%)	$\bar{M}_n$ (Theory)	$\bar{M}_n$	$\bar{M}_w$	$\bar{M}_w/\bar{M}_n$
1	-	$ZnBr_2$ (0.25)	-60	59	9440[b]	7610	12000	1.58
2	C_6H_5-CHO (2.46)	$ZnBr_2$ (0.25)	-60	100	1730[c]	2450	3080	1.25
3	-	$ZnBr_2$ (0.79)	-60	33	1675[b]	21600	38800	1.80
4	C_6H_5-CHO (0.79)	$ZnBr_2$ (0.25)	-60	100	5100[c]	3930	5330	1.35
5	-	$ZnBr_2$ (2.50)	25	3	1600[c]	7850	17400	2.22

a. All runs performed in 1,2-dichloroethane (25 ml) on 4.0 g of *t*-butyldimethylsilyl vinyl ethers. *b.* Based upon monomer consumed. *c.* Based upon initial monomer and initiator concentrations.

Higashimura and co-workers[9] recently reported a living HI/I_2-initiated cationic polymerization. This method is reported to give living polymers of alkyl vinyl ethers at low temperatures. However, it has not been shown to work for polymerization of silyl vinyl ethers. The aldol-GTP and Higashimura/Savamoto process complement each other.

3. MECHANISM

Mechanistically, Lewis acids are presumed to catalyze the reaction by activating the initiator through coordination to the carbonyl functional group. Reaction of the coordinated species with the silyl vinyl ether, followed by the silyl transfer, gives the desired product (Scheme 11). The silyl transfer could occur either in a concerted fashion via the 6-membered cyclic β-silyl-stabilized carbocation intermediate, or as a two-step process. In the case of the latter, a Zn-alkoxide and a silyl halide intermediate are produced which then react with each to form the observed products. Work is in progress to determine which mechanism is operating.

Scheme 11. Mechanism of Lewis Acid Catalysis

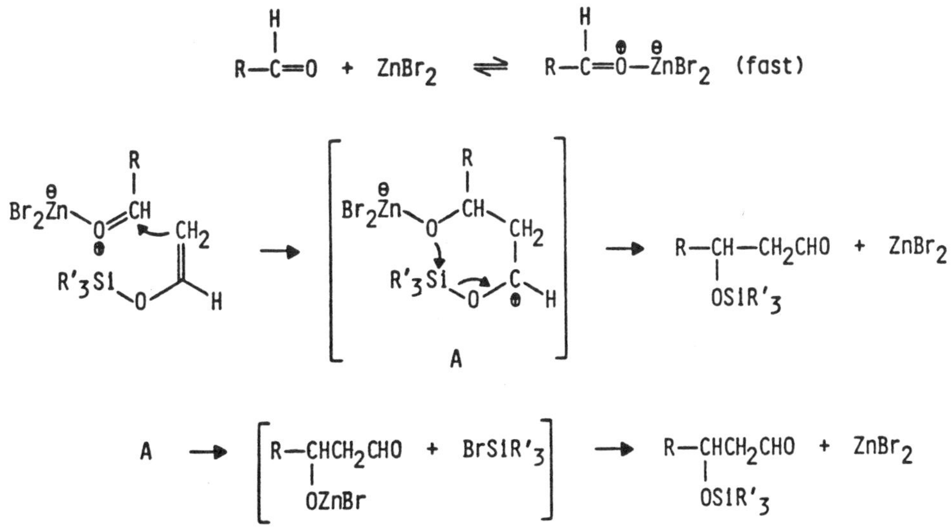

$$R\text{—}C(H)\text{=}O + ZnBr_2 \rightleftharpoons R\text{—}C(H)\text{=}\overset{\oplus}{O}\text{—}\overset{\ominus}{Zn}Br_2 \quad \text{(fast)}$$

$$Br_2\overset{\ominus}{Zn}\text{—}\overset{\oplus}{O}\text{=}CH(R) + CH_2\text{=}C(H)\text{—}O\text{—}SiR'_3 \longrightarrow \left[Br_2\overset{\ominus}{Zn}\text{—}O\text{—}CH(R)\text{—}CH_2\text{—}\overset{\oplus}{C}(H)\text{—}O\text{—}SiR'_3 \right]_{A} \longrightarrow R\text{—}CH(OSiR'_3)\text{—}CH_2CHO + ZnBr_2$$

$$A \longrightarrow \left[R\text{—}CH(OZnBr)CH_2CHO + BrSiR'_3 \right] \longrightarrow R\text{—}CH(OSiR'_3)CH_2CHO + ZnBr_2$$

4. CONCLUSIONS

The results of aldol-GTP demonstrate another instance where organic synthetic methods can be applied to the controlled synthesis of living polymers. The important distinction is that most carbon-carbon bond-forming reactions are not quantitative enough to be used in the formation of high polymers. The aldol-GTP clearly has advantages over the previously known methods. These advantages include complete monomer conversion over a broad temperature range, living polymer formation, facile block copolymer preparation, good molecular weight control and synthesis of polymers of unusual structures.

5. REFERENCES

5.1. Webster, O. W.; Hertler, W. R.; Sogah, D. Y.; Farnham, W. B.; RajanBabu, T. V. J. Am. Chem. Soc., 1983, **105**, 5706.

5.2. Sogah, D. Y.; Farnham, W. B. Organosilicon and Bioorganosilicon Chemistry. Structures, Bonding, Reactivity and Synthetic Application, H. Sakurai, Ed., John Wiley & Sons, New York, 1985, Ch. 20.

5.3. a. Sogah, D. Y.; Webster, O. W. Macromolecules, 1986, **19**, 1775. b. Sogah, D. Y.; Webster, O. W. United States Patent No. 4,544,724, October 1, 1985.

5.4. a. Colvin, E. Silicon in Organic Synthesis, Butterworths: London 1981. b. Brownbridge, P. Synthesis, 1983, 1; ibid, 1983, 85. c. Jung, M. E.; Blum, R. B. Tet. Letters, 1977, 3791.

5.5. Hertler, W. R.; Sogah, D. Y.; Webster, O. W.; Trost, B. M. Macromolecules, 1984, **17**, 1415.

5.6. a. Ackerman, E. Acta Chem. Scand. 1957, **11**, 373. b. Reference 4a, pp 184-185 and references cited therein. c. Corey, E. J.; Snider, B. B. J. Amer. Chem. Soc. 1972, **94**, 2549.

5.7. a. Murahashi, S.; Nozakura, S.; Sumi, M. J. Polym. Sci., 1965, **B3**, 245. b. Solaro, R.; Chiellini, E., Gazz. Chim. Ital., 1976, **106**, 1037.

5.8. Nozakura, S.; Ishihara, S.; Inaba, Y.; Matsumura, K. J. Polym. Sci., 1973, **11**, 1053.

5.9. Miyamoto, M.; Sawamoto, M.; Higashimura, T. Macromolecules, 1984, **17**, 265.

REPORT OF DISCUSSIONS AND RECOMMENDATIONS

Mechanistic Considerations

The group at Du Pont is studying the kinetics of the addition of the first, as well as second, monomer unit to initiator and the overall rate of propogation. Miller's group and Bandermann's group are studying the overall rate of polymerization. Miller and Bandermann have encountered induction periods; Du Pont does not. To get better agreement, we need to purify monomer, solvents, initiators, and catalyst the same way and check each other's experiments. Absolute rates should be compared.

Synthesis of an analog of the proposed cyclic intermediates will be attempted by reaction of the dimethyl ketene acetal 1 with methyl methacrylate.

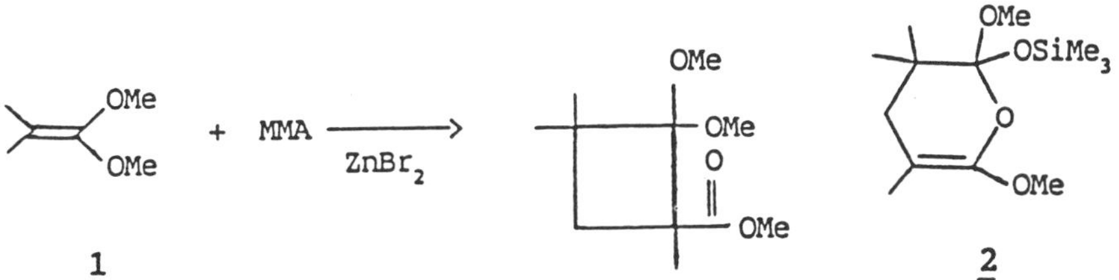

OMe, OMe (1) + MMA —ZnBr$_2$→ OMe, OMe, O, OMe; OMe, $OSiMe_3$, O, OMe (2)

The cyclic intermediate 2 is unlikely since it would open to the Z isomer. In practice, a mixture of E and Z isomer is obtained.

The question was raised whether the existence of both E and Z end groups can be explained by the addition of a monomer which is coordinated to the hexavalent silicon. Sogah pointed out that in the transition state the monomers can add in an S-cis and in an s-trans configuration. Of the four possible transition states only two are probable for steric reasons:

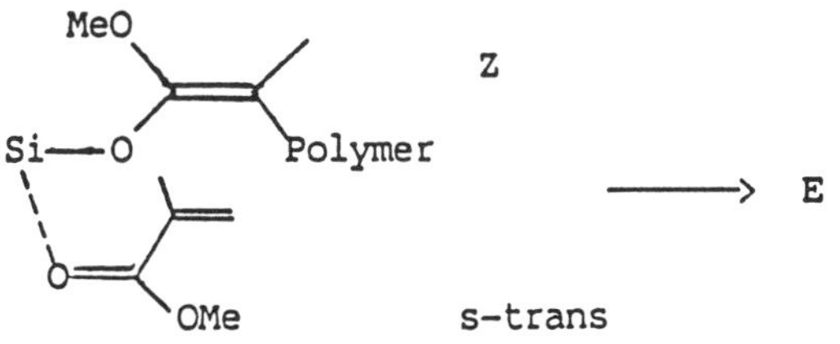

MeO, Si—O, Polymer, O, OMe — Z —→ E

s-trans

M. Fontanille and A. Guyot (eds.), Recent Advances in Mechanistic and Synthetic Aspects of Polymerization, 73–76.

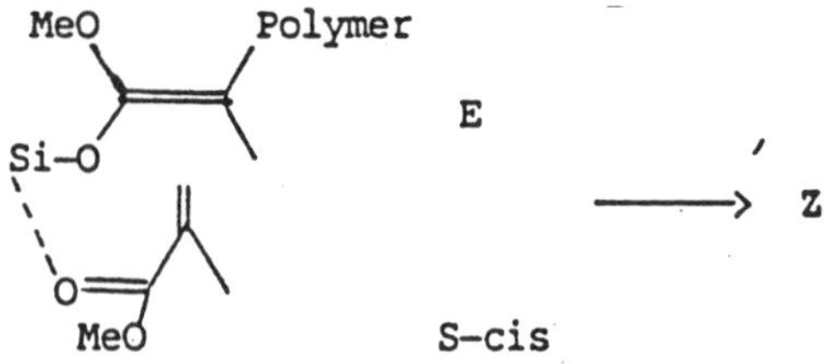

MeO Polymer

E

Si–O

→ Z

MeO

S–cis

Thus, the "concerted" mechanation of monomer addition cannot be excluded on the basis of stereochemical considerations.

A question was asked by Nolte about the effect of pressure on GTP. Sogah answered that GTP can be conducted under high pressure without any catalyst. This result is an important consideration for the mechanism and supports a cyclic intermediate of some kind.

Corriu noted that hexacoordinated silicon intermediates are slightly lower in energy than pentacoordinated intermediates and this should be taken into account in mechanistic proposals. The group suggests more consultation with the organosilicon community.

The group agreed that catalyst complexation and transfer rates should be studied to determine how they affect molecular weight dispersity.

The affect of larger or different functional groups on the silicon should be studied. Can one slow down the polymerization of acrylates, for example, and keep the system living longer?

Process Studies

Bandermann plans to study dual initiators and process scale up. The group agreed that trimethylsilyl will not likely be the only transfer group which will operate and that others should be examined both for methacrylates and other monomers. (It was felt that a listing of candidates would inhibit the search rather than foster it.)

General

The group feels that GTP scientists need to consult more with anionic, cationic, Zigler-Natta and other "living polymer" research workers.

Aldol-GTP (Sogah, Wulff)

In responce to a question by Chiellini, Sogah noted that the stereochemistry of Aldol-GTP was random.

Wulff stated that he had prepared polymers with a hydroxy group on each carbon of a chain by polymerization of 1,3,2-dioxaboroles.

We were able for the first time to prepare 1,3,2—dioxaboroles having substituents only in the 2-position[1]. Reactions with aldehydes did not yield the desired monoaddition products, but gave instead diaddition and oligomeric products as shown in the scheme:

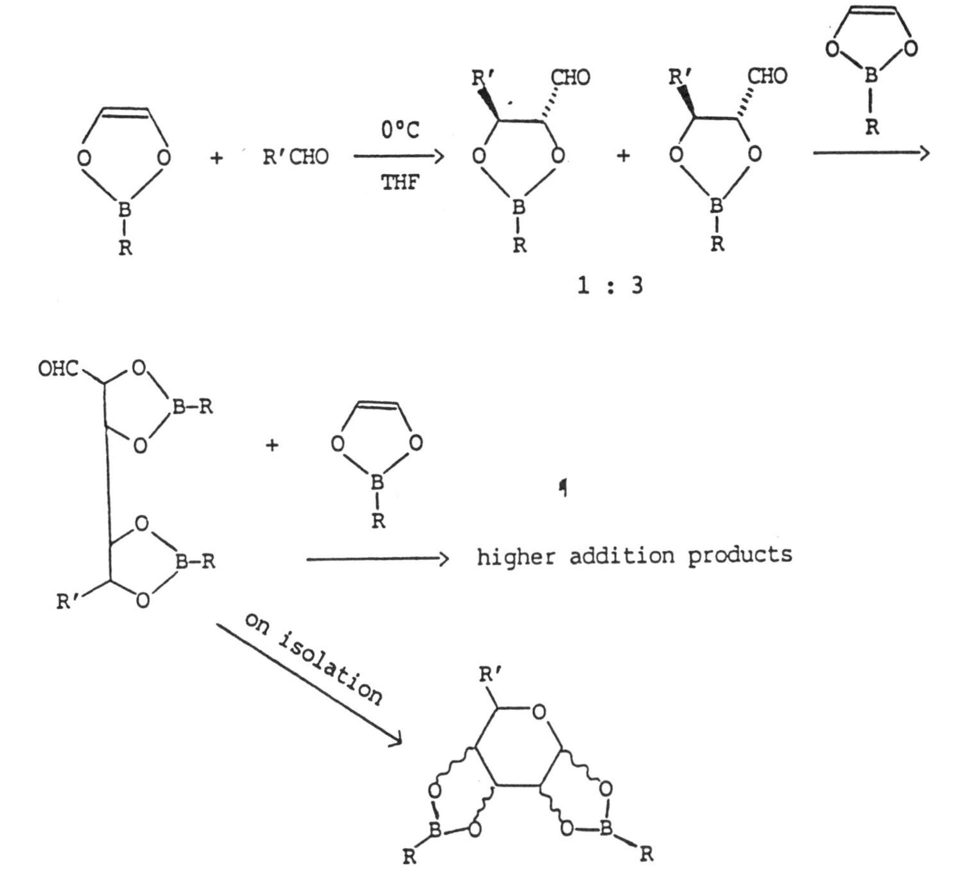

0°C
THF
1 : 3

higher addition products

on isolation

This reaction is an aldol-type rection and mechanism should be as follows:

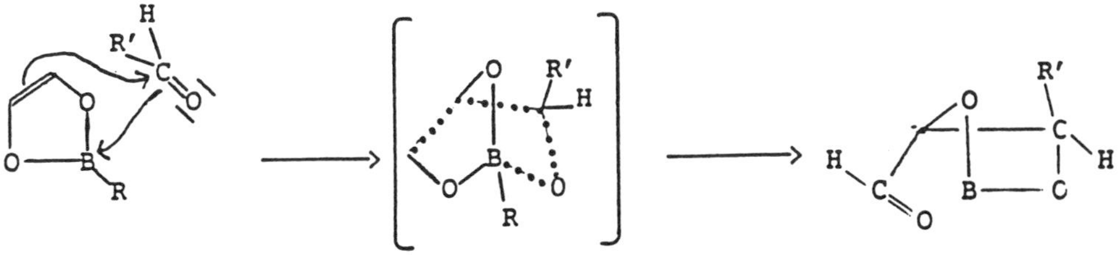

The reaction yields monoaddition products only with a large excess of aldehyde. Monoaddition products can be obtained by using immobilized 1,3,2-dioxaboroles where site separation allows clean monoaddition[1].

We have now used an excess of the 1,3,2-dioxaborole and were able to get group transfer polymerization (work together with T. Birnbrich and A. Hausen). In this case a living system resulted and the molecular weight could be controlled by the ratio of initiator aldehyde to 1,3,2-dioxaborole, as shown in the next scheme:

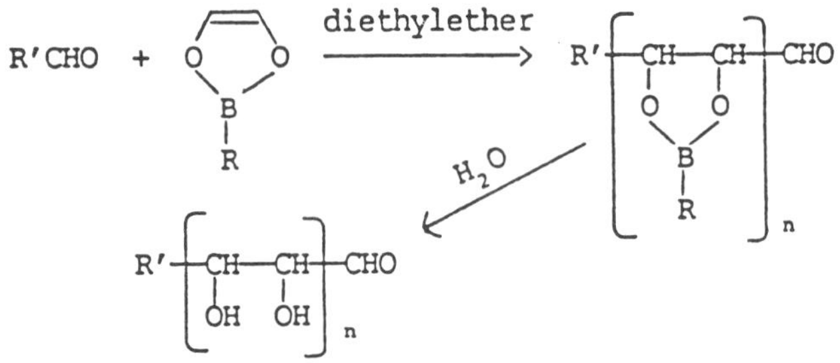

R'CHO + (1,3,2-dioxaborole, B–R) —diethylether→ R'–[CH–CH (O–B(R)–O)]$_n$–CHO

H_2O ↓

R'–[CH(OH)–CH(OH)]$_n$–CHO

In contrast to the aldol-type group transfer polymerization reported by D. Sogah, et al., there are three different features:

1. In this case, not an acetaldehyde but a glycolaldehyde derivative is used, thus resulting in a polymer having an OH-group on every carbon atom. In effect, it is the synthesis of oligomeric or polymeric monosaccharides.

2. Instead of a silane blocking group, a boron containing group is used This provides a built-in catalyst so that the reaction can be performed at room temperature during a few hours without addition of catalyst. It can, however, be speeded up by the addition of $ZnCl_2$, $ZnBr_2$, or ZnI_2. The choice of the solvent (e.g. diethylether is suitable) and the concentration of the reactants is critical.

3. Due to the ring structure, the reaction proceeds stereochemically to yield predominantly an isotactic structure.

---

1) G. Wulff and A. Hausen, Angew. Chem. Int. Engl. Ed 25, 560 (1986).

II - STABILITY OF ACTIVE CENTERS IN POLYMERIZATION

Chairman : J.E. Mc GRATH

CONTROLLED SYNTHESIS OF VARIOUS POLY(ALKYL METHACRYLATES) BY ANIONIC TECHNIQUES

T. E. Long, R. D. Allen[1], and J. E. McGrath
Department of Chemistry and Polymer Materials and Interfaces Laboratory
Virginia Polytechnic Institute and State University
Blacksburg, VA 24061
USA

ABSTRACT. Homopolymer synthesis investigations have further defined the requirements for the controlled anionic synthesis of various poly(alkyl methacrylates). Small variations in polymerization conditions have a profound influence on the nature of the propagating lithium enolate. However, the effects depend largely upon the size of the ester alkyl group involved. The delicate relationship among such variables as ester alkyl group size, polymerization temperature and polymerization solvent have been explored and will be discussed. Drastically different enolate stabilities have been observed for the bulkier branched alkyl methacrylates compared to those based upon linear ester alkyl groups. This has been primarily attributed to the significantly modified steric and electronic environment of the ester carbonyl. Thus, the constraints of classical alkyl methacrylate polymerizations, i.e., low temperatures and polar solvents, may be overcome by the judicious choice of the ester alkyl group. Extensive thermal, microstructural, and mechanical characterization reveal that a wide range of polymer physical properties can be achieved by the careful alteration of several variables.

1. INTRODUCTION

The anionic polymerization of various alkyl methacrylates has received sparse attention in comparison to nonpolar hydrocarbon monomers such as styrenes or the dienes (1-3). The two major reasons for the sluggish synthetic development of this class of polar monomers are the protic impurities present in most commercially available grades of monomer and the inherent side reactions of the propagating enolate with the ester carbonyl during anionic polymerization (4-6). However, by very carefully controlling various synthetic parameters and utilizing carefully purified monomers, one can take advantage of the

[1]Current address: IBM Almaden Research Center, 650 Harry Road, San Jose, CA 95120-6099 USA

M. Fontanille and A. Guyot (eds.), Recent Advances in Mechanistic and Synthetic Aspects of Polymerization, 79–100.

living nature of this polymerization to synthesize a variety of poly(alkyl methacrylates). In addition, these polymers will possess such desirable properties as predictable molecular weights, narrow molecular weight distributions and controlled stereochemistry. Once the conditions for controlled synthesis have been established, the role of important synthetic variables can be analyzed. The primary focus of our recent efforts has been on the investigation of enolate stabilities during anionic polymerization. In particular, the relationship among such variables as initiator sterics and basicity, polymerization temperature, polymerization solvent, and ester alkyl group size has been investigated. All of these factors play a critical role in determining the efficiency of initiation and the stabilities of the initiator anion and the propagating metal enolate. However, by subtle modifications in these synthetic variables, a wide spectrum of chemical, physical, thermal, microstructural and mechanical properties can be achieved.

Our earlier investigations have dealt with the introduction of a novel purification methodology involving trialkyl aluminum and dialkyl aluminum hydride reagents (3,6,7). These techniques take advantage of the enhanced reactivity of protic species with various aluminum reagents. Thus, one can quantitatively remove deleterious protic impurities such as alcohols from the monomer prior to anionic polymerization. The capability of readily obtaining ultrapure, anionic-grade alkyl methacrylate monomers allows for the careful investigation of anionic initiation and mechanisms. For example, small variation in initiator or enolate stability can safely be attributed to factors inherent in the chemistry and not to the presence of side reactions associated with monomer impurities. In addition, in order to maximize ultimate polymer properties, the molecular weight should be greater than 10^5 g/mole with a molecular weight distribution less than 1.25. These optimum properties are only possible by anionic polymerization when the monomers are free of terminating impurities. This present discussion will demonstrate the applicability of this purification technique to a wide range of alkyl methacrylate monomers. In particular, special attention will be given to the branched alkyl methacrylates. These monomers are more difficult to purify than the normal ester alkyl methacrylates, but contribute interesting insight into the nature of the propagating anionic species.

The choice of anionic initiators for alkyl methacrylate polymerizations is important from several viewpoints. Firstly, the anion should be sterically hindered in order to avoid carbonyl attack (1,2 addition) of the alkyl methacrylate monomer (8,9). This undesirable situation would lead to deactivation of a portion of the initiator and subsequent loss of molecular weight control. Hatada, et. al., have devoted significant effort to the elucidation of the termination mechanisms that arise when employing less hindered initiators (10,11). On the other hand, by selecting a monomer with a sterically and electronically "protected" carbonyl, the steric nature of the initiator can be compromised. Secondly, the anion should be a strong enough base to generate the propagating, delocalized enolate. Despite these two restrictions, the highly delocalized, sterically hindered, weak diphenylhexyl lithium anion has been routinely utilized as an efficient initiator for a variety of alkyl methacrylate polymerizations (10,12). Finally, the selection of initiator defines at least one end group of the polymer chain. Consequently, one can utilize this initiator fragment as a means of characterization. For example, one can utilize Nuclear Magnetic Resonance to determine the relative integration of end group to an integration of a chemical shift associated with the repeat unit, e.g., ester alkyl (13). This allows for determination of the number average molecular weight ($\bar{M}n$). One can imagine another similar approach based on Ultraviolet-Visible spectroscopy if the fragment absorbs in this wavelength region. In fact, both of these techniques have been extended to a wide range of poly(alkyl methacrylates). Each technique has inherent limitations such as maximum molecular weight, but together are quite complimentary in applicability and utility.

Variation in polymerization temperature affects both product yield and polymer microstructure. It is well known that the stability of the propagating enolate is strongly influenced by the polymerization temperature. Generally, an increase in temperature decreases the enolate stability and subsequent side reactions associated with ester carbonyl functionalities begin to occur. Some workers have stated earlier (13) that if the temperature exceeds approximately -65°C for a methyl methacrylate polymerization, the side reactions begin to occur. However, the maximum temperature that defines enolate stability is a function of the ester alkyl group. Significantly different behavior is observed for the bulkier ester groups such as isopropyl or t-butyl. In fact, it has been realized that t-butyl methacrylate lithium (3), cesium and sodium (14) enolates are stable at temperatures higher than 35°C. An increase in polymerization temperature also dictates a decrease in the level of syndiotactic addition. This has been attributed to an increase in polymer chain mobility during polymerization due to additional thermal energy (15). The magnitude of the microstructural change is influenced by the size of the ester alkyl group. Alkyl methacrylates with larger, branched ester groups behave differently than normal esters when polymerization temperature is increased. Similar trends have been seen in the Group Transfer Polymerization (GTP) of alkyl methacrylates, i.e., a decrease in polymerization temperature results in an increase of the syndiotactic triad composition (16,17).

The nature of the polymerization solvent has been shown to have a larger effect in ionic polymerizations by altering both the reactivity and identity of the propagating species (18,19). For alkyl methacrylate polymerizations, solvent polarity dictates the mode of stereochemical addition (20). Polar solvents such as THF favor syndiotactic addition, while polymerizations in hydrocarbon solvents, e.g., toluene or hexane, result in isotactic triad formation. Combinations of polar and nonpolar solvents provide for a wide range of dielectric media. Thus, one can finely control the tactic composition of poly(alkyl methacrylates) by systematic variations of solvent polarity. But the range of control is governed again by the size of the ester alkyl group. In general, larger groups restrict the highest level of syndiotacticity possible, but promote higher levels of isotacticity. Solvent polarity also controls the rate of polymerization due to the interaction of the solvent with the living anionic chain end. Polar solvents separate the ion pair leading to higher reactivities and shorter polymerization times (minutes). On the other hand, the absence of peripheral solvation by the hydrocarbon solvent results in decreased reactivity and longer polymerization times (hours).

It is obvious that the ester alkyl group size has a profound effect on all of the synthetic variables described. The influence is generally attributed to the steric and electronic deactivation of the carbonyl. Consequently, the stability of the propagating enolate or, in other words, the inability to undergo deleterious side reactions, is enhanced when the group is large and donates electronically to the partial positive carbonyl carbon. An attempt has been made to elucidate the interdependence of the ester alkyl group and a variety of synthetic parameters. Thus, one can further define the initiator, solvent, and temperature requirements necessary for anion stability and ultimate syntheses.

A variety of polymer properties are influenced by the alteration in polymer microstructure and monomer structure. By systematic variations in synthetic parameters, one can generate a wide spectrum of physical and chemical properties. For instance, the glass transition temperature (Tg) for poly(methyl methacrylate) can range from 54°C for the isotactic (~82%) material to 131°C for the syndiotactic (~80%) polymer. Similar observations can be made for a variety of alkyl methacrylates. However, the range, Tg syndiotactic minus Tg isotactic, is a function of the ester alkyl group and the stereochemical composition. In most instances, it is difficult to compare thermal properties because equivalent microstructural compositions are not easily obtained. Despite this limitation, elastomeric, semi-crystalline and thermoplastic materials can readily be prepared. In many cases, the ester group also determines the thermal degradation mechanism of the polymer chain. This observation is easily made by utilizing thermogravimetric analysis (TGA) coupled with mass spectroscopy and infrared spectroscopy. In addition, the anionically prepared poly(alkyl methacrylates) exhibit enhanced thermal stability over the corresponding polymer synthesized by free radical techniques. This has been associated with the absence of thermally weak groups such as head-head linkages or unsaturated end groups which arise due to free radical mechanisms (21,22). Preliminary investigations indicate that the microstructure affects the thermal stability of poly(alkyl methacrylates). However, this observation may be the result of thermally weak linkages generated during pseudo-living polymerizations in hydrocarbon solvents.

Mechanical relaxations are also affected by microstructure and ester alkyl. The presence of the β-relaxation, which is associated with side group motion, is determined by the microstructural composition. The position of the β-relaxation is affected by the ester alkyl group. Recent efforts in the area utilizing dielectric analysis have shed fundamental insight into these effects (23).

The variation in ester alkyl group also results in a change of the chemical properties of the polymer. Quantitative transformation reactions have also been demonstrated (24,25). Significant effort has been dedicated to the hydrolysis of certain esters in order to generate carboxylic acid and metal carboxylate functionalities. These quantitative conversions permit the synthesis of a whole new class of polymers and further structure-property correlation (26).

The primary focus of this work is on the investigations of factors that influence the stability of the propagating enolate. Interesting correlations have been made regarding the nature of the active center with various synthetic departures. Polymer characterization has revealed a wide spectrum of physical properties and has helped to substantiate the identity of the active center.

2. MATERIALS AND METHODS

2.1. Materials

The polymerization solvent tetrahydrofuran (Fisher, Certified Grade) was purified by distillation under nitrogen from the purple sodium/benzophenone ketyl. Polymerizations were also performed in toluene (Fisher, Certified Grade) which was purified first by distillation under nitrogen from finely dispersed sodium, followed by vacuum distillation from either a low molecular weight polystyryl anion (<2000 g/mole) or a diphenylhexyl lithium anion. The anion was in contact with the toluene for 2-3 days prior to distillation.

The initiator, diphenylhexyl lithium (DPHL), was formed by the reaction of s-butyl lithium and 1,1-diphenylethylene (DPE). These reagents are available from the Lithco Division of FMC and Eastman Kodak Co., respectively. Sec-butyl lithium (1.35M in cyclohexane) was used as received, but the exact molarity of the solution can be ascertained by the Gilman "double titration" method (27). DPE can be distilled quite easily under good vacuum (<0.1mm) from sec-butyl or n-butyl lithium. The yellow color associated with the crude DPE disappears after distillation. In addition, the distilled DPE can be stored under nitrogen in the cold for several weeks.

The alkyl methacrylate monomers (available from Rohm and Haas Co. and Rohm Tech) were first distilled from finely ground CaH_2 to remove water and dissolved oxygen. Immediately prior to use, the monomers were distilled from trialkyl aluminum (TAA) reagents which were provided by the Ethyl Corporation as 25 weight percent hexane solutions. Extreme caution must be taken when using these reagents due to their highly reactive and pyrophoric nature. Best results were obtained if the monomer was degassed immediately prior to TAA addition. This step ensured removal of all oxygen dissolved in the monomer. The TAA solution was added drop-wise to the neat, cold monomer until a yellow-green colored complex formed. This complex is the result of the trialkyl aluminum interacting with the carbonyl oxygen. The complex formation indicates that all impurities have been scavenged and a sufficient amount of the TAA solution has been added. Typically, a 20-

30% excess of TAA was added after the endpoint was reached. For the branched esters, small amounts of the more reactive dialkyl aluminum hydride were added to the TAA reagent to facilitate purification. This could be accomplished without reduction of the monomer ester carbonyl. The complex was allowed to stand due to slower reacting impurities which require longer contact times between the monomer and purification reagent. The monomer was subsequently distilled under vacuum and used immediately. The ultrapure monomer cannot be stored for long periods of time due to its propensity to polymerize.

2.2. Polymerization

All polymerizations were performed in scrupulously cleaned and dried glassware under a prepurified nitrogen atmosphere. The polymerization solvent was transferred via a double-ended needle to a round bottom flask pressured with nitrogen and equipped with a rubber septum. The amount of solvent charged was based on the weight percent solids and typically 10-12 wt. % is desired. After the reactor containing solvent has reached thermal equilibrium at the desired polymerization temperature, a 3-4 molar excess of DPE was added. Depending on the alkyl methacrylate monomer of interest, the polymerization temperature ranged from -78°C to above room temperature. The calculated amount of sec-butyl lithium was syringed into the DPE solution which quickly generated the red anion of DPHL in ether solvents. The length of time required for this reaction is a function of both polymerization solvent and temperature, i.e. toluene required a much longer time, at higher temperatures. The purified monomer is then slowly added to the initiator solution to avoid any undesirable exotherms, which could lead to premature termination in some cases. Initiation is rapid and is evident by the rapid disappearance of the red color and the formation of the colorless enolate. The polymerization time is a function of both polymerization solvent and reaction temperature. For example, at -78°C, the polymerization of MMA (though probably quickly completed) usually is allowed to proceed for 20-30 minutes in polar solvents and several hours in hydrocarbons. Scheme 1 illustrates the chemistry involved and the sequential addition of the necessary reagents.

The living polymers can be terminated at the reaction temperature with degassed, HPLC-grade methanol after complete conversion. Isolation is possible by precipitation in a large excess (10X) of an appropriate nonsolvent. Petroleum ether, methanol or methanol/water can be used as coagulation solvents, depending on the solubility characteristics of the polymers. However, the reaction solvent must be soluble in the precipitation media.

Scheme 1. Anionic Polymerization of Alkyl Methacrylates

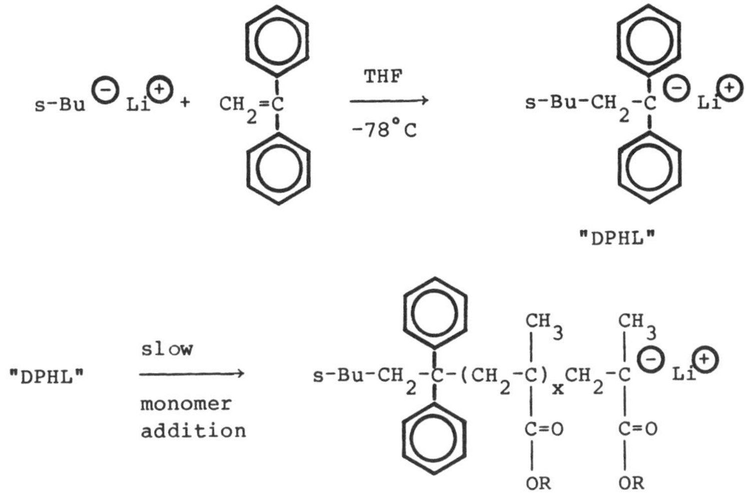

$$\text{s-Bu}^{\ominus}\,\text{Li}^{\oplus} + CH_2=C(C_6H_5)_2 \xrightarrow[-78^{\circ}C]{THF} \text{s-Bu-}CH_2\text{-}C(C_6H_5)_2^{\ominus}\,\text{Li}^{\oplus}$$

"DPHL"

$$\text{"DPHL"} \xrightarrow[\text{monomer addition}]{\text{slow}} \text{s-Bu-}CH_2\text{-}C(C_6H_5)_2\text{-}(CH_2\text{-}C(CH_3)(C{=}O)(OR))_x CH_2\text{-}C(CH_3)(C{=}O)(OR)^{\ominus}\,\text{Li}^{\oplus}$$

2.3. Characterization

2.3.1. Thermal Analysis. The glass transition temperatures of the polymers were determined using a Perkin-Elmer Model 4 Differential Scanning Calorimeter. The heating rate was 10°C/min. Thermogravimetric analysis (TGA) was accomplished using either a Perkin-Elmer Model 2 or Model 4. Experiments were conducted in a nitrogen atmosphere and a heating rate of 5°C/min was utilized.

2.3.2. NMR Analysis. High resolution 50 MHz ^{13}C NMR spectra of the polymers were obtained in $CDCl_3$ solutions (~15 % wt./vol) using a Bruker WP 200 SY spectrometer. Inverse-gated proton decoupling in combination with the addition of a small amount (0.1 M) of $Cr(AcAc)_3$ to the samples assured quantitative accuracy of the data. Typically, 1000 free induction decays with a recycle delay of 5.0 seconds and a flip angle of ~40° were accumulated and Fourier transformed. Electronic integrals were used in the calculation after ensuring adequate digitization of the spectra.

2.3.3. Molecular Weight and MWD. All values were determined primarily by size exclusion chromatography (SEC). A variable temperature Waters 590 GPC equipped with Ultrastyragel columns with porosities of 500Å, 10^3Å, 10^4Å and 10^5Å was employed. PMMA and PS standards (Polymer Laboratories) were utilized for the construction of calibration curves. A Waters 490 programmable wavelength detector and a Waters R401 differential refractive index detector were utilized.

3. RESULTS AND DISCUSSION

A very critical, but not always appreciated, aspect of homopolymer synthesis by living polymerization techniques is monomer purity. This consideration is particularly important when high molecular weight polymers (>5.0x10^4 g/mole) are desired, and the corresponding anion concentration is quite low (<0.20 mmoles). Any impurities present in the alkyl methacrylate monomer will quickly terminate the diphenylhexyl lithium (DPHL) anion. Our earlier investigations have outlined a novel purification methodology for alkyl methacrylate monomers (28). This technique employs trialkyl aluminum reagents which quickly react with any acid, alcohol or water impurities present in many methacrylate monomers.

Recent developments in this purification methodology involve the purification of branched alkyl methacrylate monomers, e.g. t-butyl methacrylate (TBMA). Special attention must be given to their purification, since the branched alcoholic impurities are more difficult to remove, due to the lower reactivity of branched alcohols with R_3Al (28). In particular, we have utilized two approaches to obtain anionic polymerization grade branched alkyl methacrylates. First, longer contact times between the alkyl methacrylate monomer and triethylaluminum (TEA) are necessary for complete conversion of the alcohol to the aluminum alkoxide. Although this approach seems quite feasible, it is complicated by the fact that the possibility of premature polymerization increases prior to distillation. This undesirable complication arises due to the inherent instability of the ester carbonyl-trialkyl aluminum complex. Despite attempts using longer reaction times, molecular weight distributions range from 1.3 to 1.5. The second approach involves the addition of small amounts (50 wt% or less) of diisobutyl aluminum hydride (DIBAH) to the TEA solution. It is known that the aluminum hydrogen bond is significantly more labile to nucleophilic substitution than the aluminum alkyl bond (29). This mixed reagent approach is necessary for two reasons. First, the aluminum hydride, unlike the trialkyl aluminum reagent, does not form a colored complex with the α,β-unsaturated ester. Second, the mixed system is believed to tone down the reactivity of the aluminum hydride. A TEA/DIBAH mixture can be added to cold (-78°C) monomer until the stable colored complex forms. The purification reaction is then allowed to proceed for ~60 minutes at room temperature. This procedure allows for removal of impurities without reduction of the ester. Significantly narrower gel permeation chromatograms ($\bar{M}w/\bar{M}n$ <1.20) are observed for poly(t-butyl methacrylate) samples prepared from TEA/DIBAH purified monomer.

Various alkyl methacrylates have been polymerized in a controlled manner by the anionic techniques described herein. All samples demonstrated good molecular weight control and narrow molecular weight distributions (<1.25). Table I depicts representative poly(methyl methacrylate) (PMMA) and poly(t-butyl methacrylate) (PTBMA) samples. The number average molecular weights and molecular weight distributions were calculated based on polystyrene standards for those polymers described in Table I. Structural characterization has been performed

TABLE I. ANIONIC POLYMERIZATION OF ALKYL METHACRYLATES: MOLECULAR WEIGHT CONTROL

SAMPLE	$\bar{M}n$(th)*	$\bar{M}n$(gpc)**	$\bar{M}w/\bar{M}n$
PMMA-1	7,400	10,000	1.18
PMMA-2	30,000	31,100	1.15
PTBMA-1	50,000	60,500	1.12
PTBMA-2	100,000	136,000	1.18

* $\bar{M}n(th) = \dfrac{\text{g monomer}}{\text{moles initiator}}$

** PS standards

by a variety of techniques including Fourier Transform Infrared (FTIR) spectroscopy and Nuclear Magnetic Resonance (NMR) (^{13}C and ^{1}H). Figure 1 describes the chemical shift regions of interest for poly(t-butyl methacrylate) samples. Similar regions apply to all poly(alkyl methacrylates) studied and Figure 2 shows ^{13}C analysis for poly(methyl methacrylate). Stereochemical information is obtainable from both the quaternary and carbonyl carbon regions. In particular, triad and pentad sequences can be ascertained by studying the quaternary and carbonyl carbon chemical shifts, respectively. While ^{1}H NMR is a convenient and well-established method for the determination of tacticity for PMMA, it is not so for the higher alkyl esters due to overlapping signals in the α-methyl region. Although some peak suppression pulse sequences have been developed and shown to work well for poly(ethyl methacrylate) PEMA (30,31), we have used ^{13}C as a straightforward and less time consuming method. Secondly, ^{13}C NMR discriminates up to pentad sequences quite easily. High field ^{13}C NMR published recently (32) reveals that for PMMA, well resolved pentad and triad sequences are easily identified for the carbonyl, α-methyl and quaternary carbon regions. Thus, triad fractions can be determined either from proton or carbon spectra. The values from ^{1}H and ^{13}C spectra for PMMA were compared and found to agree within 2%. FTIR can also be utilized for characterization. The carbonyl absorbance ranges from 1723 cm^{-1} for PTBMA to 1731 cm^{-1} for PMMA. This wide range demonstrates the alkyl group's effect on the carbonyl environment. In fact, poly(isopropyl methacrylate) (PIPMA) has a carbonyl absorbance at a similar wavenumber to PTBMA. This observation prompted further analysis into the similar polymerizability of PIPMA and PTBMA. This aspect will be discussed in greater detail later.

The polymerization temperature plays a very critical role in defining the enolate stability. Recent investigations have shown that the variation in polymerization temperature for the synthesis of PMMA affects both product yield and polymer tacticity. Table II depicts the stereochemical effects of polymerization temperature on polymer tacticity. All samples were prepared in THF using DPHL as an initiator.

FIGURE 1. 50 MHz ^{13}C NMR - PTBMA

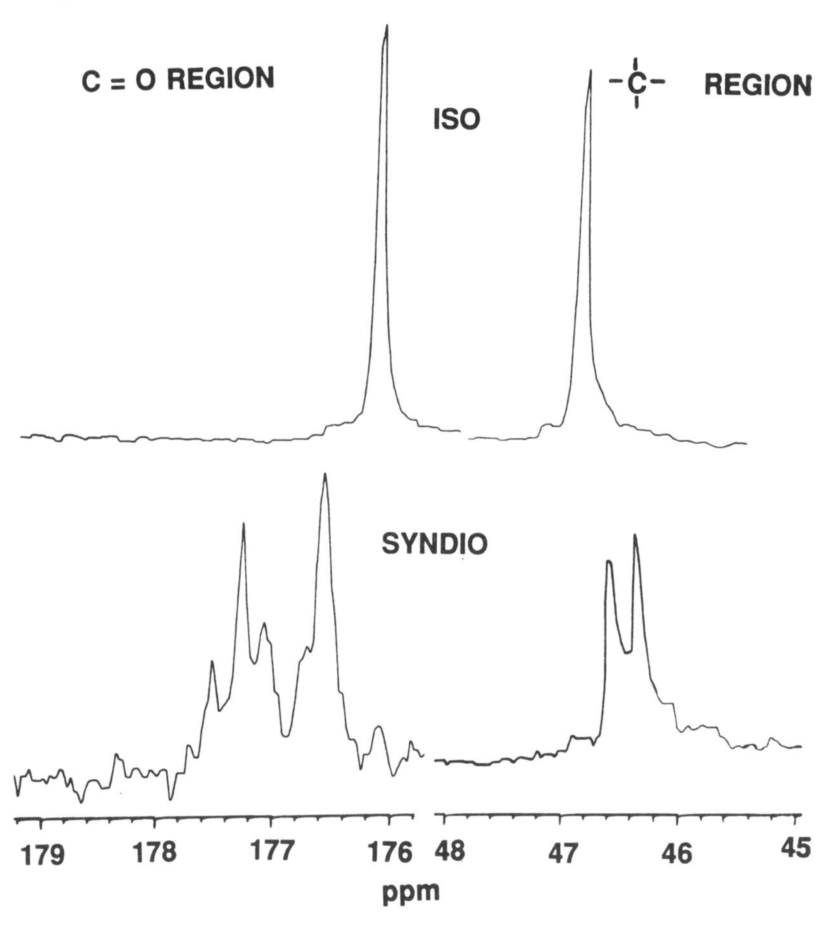

TABLE II. EFFECT OF POLYMERIZATION TEMPERATURE ON PMMA TACTICITY

T(°C)	%S	%H	%I
-90	82	17	1
-78	81	18	1
-48	72	26	2
-20	69	28	3
0	63	33	4

As one can see, the syndiotactic triad composition decreases at the expense primarily of the heterotactic sequences. Insignificant changes can be seen in the isotactic levels. All values in Table II were determined by ^{1}H NMR utilizing the α-methyl resonance. Interestingly,

FIGURE 2. 50 MHz ^{13}C NMR - PMMA

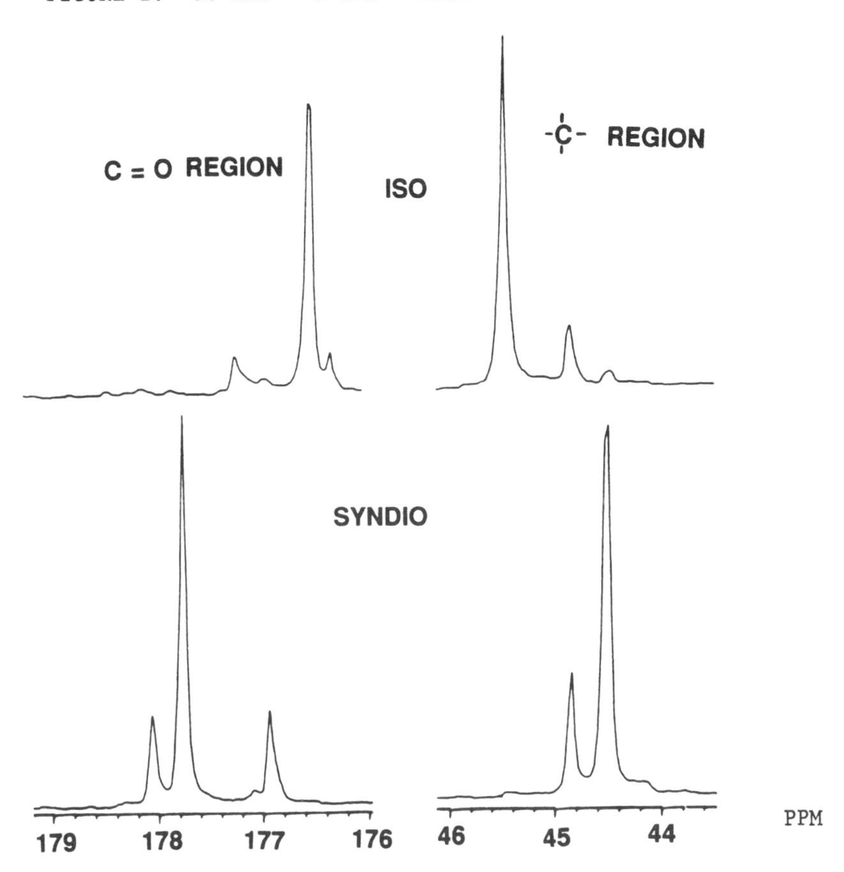

if one plots polymerization temperature versus % syndiotacticity, a linear relationship exists (correlation coefficient 0.99). It should be noted that the polymers obtained from polymerizations at temperatures greater than -48°C show significant low molecular weight tailing in the gel permeation chromatogram. This can be attributed to enolate-ester carbonyl side reactions. Despite the broad distributions (>1.5), stereochemical analysis was still performed. Table III illustrates the effect of increased PMMA polymerization temperature on molecular weight control. In each case, a theoretical $\bar{M}n$ = 30,000 g/mole was desired based on initiator and monomer charges. The data clearly depict a loss in molecular weight control with increasing temperature while the yield decreases to <10% at 0°C. The $\bar{M}n$ values were determined by vapor phase osmometry utilizing toluene as a solvent. Molecular weight distributions also broaden, but may appear artificially narrow due to dissolution of lower molecular weight chains in the coagulation solvent (petroleum ether). This is especially true for the PMMA sample obtained at 0°C.

TABLE III. EFFECT OF TEMPERATURE ON PMMA POLYMERIZATION

T(°C)	% YIELD	$\bar{M}n^a$
-90	100	32,500
-78	100	29,900
-48	86	40,700
-20	80	23,500
0	7.5	5,300

[a]Osmometry; Mw(th) = 30,000 g/mole

The stereochemical and yield dependence on polymerization temperature are also a function of the ester alkyl group. Preliminary stereochemical investigations indicate that PTBMA behaves quite differently than PMMA. Table IV describes the change in stereochemistry with an increase in polymerization temperature from -78°C to 44°C.

TABLE IV. EFFECT OF POLYMERIZATION TEMPERATURE ON PTBMA MICROSTRUCTURE

T(°C)	%S	%H	%I
-78	52	48	0
0	46	50	4
25	41	49	10
44	38	48	14

Tactic compositions for PTBMA samples are based on ^{13}C NMR analysis due to overlapping signals in ^{1}H NMR spectra. In a similar fashion as the PMMA study, DPHL was utilized in THF. Similarly, one can note the decrease in syndiotacticity as the temperature is increased, but this occurs at the expense of the isotactic levels rather than the heterotactic levels. More importantly, all polymers obtained over this range demonstrated quantitative yield, good molecular weight control and narrow molecular weight distributions. Although at 44°C (initiation temperature), the MWD begins to broaden significantly (1.76) which indicates that termination reactions are beginning to occur. Table V describes the enhanced enolate stability over a wide thermal range. Narrow molecular weight distributions are obtained at temperatures as high as 37°C. Also, molecular weight control is good, indicating the absence of termination mechanisms. Figure 3 depicts a gel permeation chromatogram (GPC) of a PTBMA sample polymerized with DPHL as an initiator in THF at room temperature. Also, shown for reference are polystyrene and polymethyl methacrylate standards. As one can see, the distribution appears narrow with no evidence of high or low molecular weight tailing.

The polymerizability of PTBMA is in sharp contrast to PMMA. Other alkyl methacrylates with linear, normal ester alkyl groups behave in a similar fashion to PMMA. For example, PEMA polymerizations at 0°C also

TABLE V. POLY(t-BUTYL METHACRYLATE) SYNTHESIS AT HIGHER POLYMERIZATION TEMPERATURES*

TEMP(°C)	$\bar{M}n$(th)	$\bar{M}n$(PS)	MWD(PS)
-78	100,000	139,000	1.18
0	10,000	7,800	1.32
22	30,000	31,000	1.14
26	50,000	60,500	1.12
44	30,000	23,600	1.76

*Reaction conditions: THF, DPHL, N_2 (6 PSI)

FIGURE 3. GEL PERMEATION CHROMATOGRAM OF PTBMA SYNTHESIZED AT ROOM TEMPERATURE IN THF

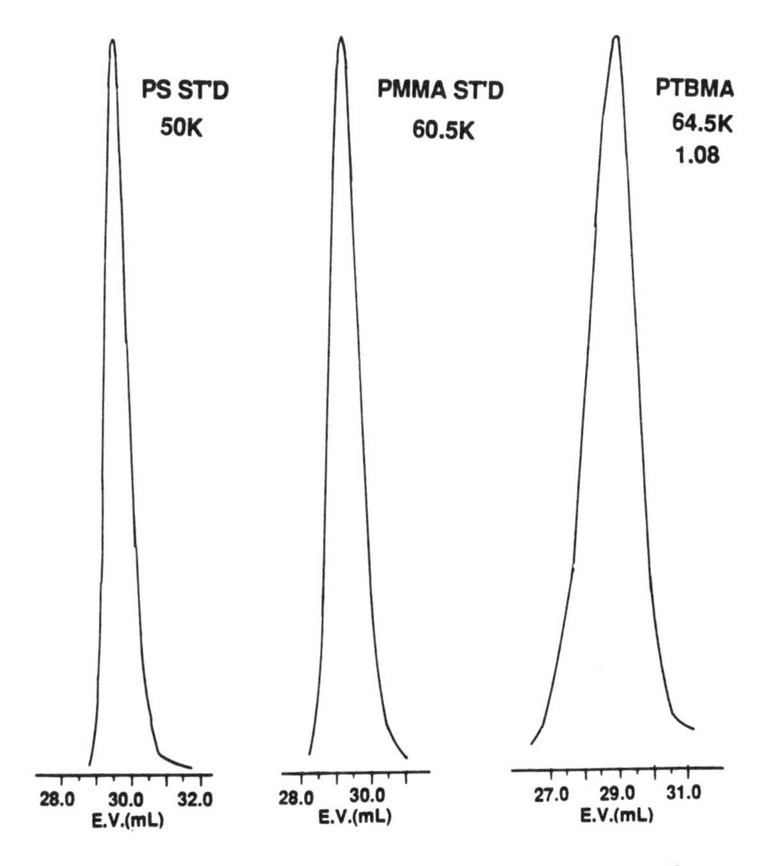

result in low yields. However, if one polymerizes PIPMA at 0°C, quantitative yields of polymer are obtained with narrow molecular weight distributions. Thus, PMMA, PEMA, PIPMA and PTBMA represent an excellent series for elucidating many effects on the enolate stability. Essentially, this series involves replacing each proton on the methyl ester group of PMMA i.e. replacement of the first proton results in PEMA, the second results in PIPMA and the third yields PTBMA. Table VI summarizes the polymer yields obtained at higher polymerization temperatures for the series of poly(alkyl methacrylates).

TABLE VI. POLYMERIZATION RESULTS FOR VARIOUS ALKYL METHACRYLATES AT HIGH TEMPERATURE*

SAMPLE	TEMP (°C)	YIELD (%)
PMMA	0	8
PEMA	0	15
PIPMA	0	quantitative
PTBMA	0	quantitative
PTBMA	22	quantitative
PTBMA	26-37	quantitative
PTBMA	44	90

*All reactions in THF using DPHL as the initiator

It is clear that as the ester group gets bulkier, the stability of the propagating enolate is greater. This realization has many important, experimental ramifications. Unlike in a PMMA polymerization at -78°C, one does not need to worry about large exotherms (>10°C) causing side reactions in the branched alkyl methacrylate polymerizations. This eliminates both the necessity to monitor polymerization temperature and the tedious slow addition of monomer.

In addition to the effect of polymerization temperature, the polymerization solvent influences the polymer tacticity. The mode of stereochemical addition of incoming monomer to the propagating enolate is altered by the presence of peripheral solvation. Polar solvents will separate the ion pair leading to higher levels of syndiotacticity and enhanced reaction rates. On the other hand, hydrocarbons do not effectively solvent separate the pair leading to isotactic addition and decreased reaction rates. Present research efforts involve the combination of polar and nonpolar solvents in an attempt to control the tactic composition over a wide range. Table VII describes the effects of solvent polarity and ester alkyl group. All polymers in Table VII were synthesized in THF employing DPHL as an initiator. In each case, THF promotes high levels of syndiotacticity, but as the ester group becomes branched, the steric control is diminished. PTBMA exhibits only ~52% syndiotacticity when polymerize in polar solvents. This is lower than the syndiotactic levels obtainable by most free radical mechanisms. Typically, a free radical polymerization of alkyl methacrylates yields ~60% syndiotacticity. When polymerizations are conducted in hydrocarbon media, the various poly(alkyl methacrylates) exhibit high levels of isotacticity. In fact, in each case, it is

TABLE VII. THERMAL AND MICROSTRUCTURAL CHARACTERIZATION OF POLY(ALKYL METHACRYLATES)

SAMPLE	PZN. SOLVENT	Tg(°C)	%TACTICITY[a] ISO	HETERO	SYNDIO
PMMA	THF	130	1	21	78
	TOL	54	82	15	3
PEMA	THF	86	2	21	77
	TOL	6	90	10	0
PIPMA	THF	86	6	29	65
	TOL	34	83	17	0
PTBMA	THF	118	2	46	52
	TOL	84	99	1	0

[a]Mole fractions (in %) of isotactic, heterotactic and syndiotactic triads from ^{13}C NMR.

easier to synthesize pure isotactic polymers and poly(t-butyl methacrylate) can be prepared with exclusively (100%) isotactic triads. This stereochemical control is beginning to approach the control obtainable by Ziegler-Natta catalysts. The 100% isotactic PTBMA samples also have very narrow molecular weight distributions as shown in Figure 4. Alkyl methacrylates generally do not undergo living polymerizations in toluene and distributions generally range from 2.0-4.0. Thus, PTBMA is a very unique monomer compared to other alkyl methacrylates.

The glass transition temperatures (determined at the transition midpoint) are also included for the various systems. From a quick glance, it is apparent that variations of the ester group and solvent enable one to modify the room temperature characteristics of the polymers from a glass (s-PMMA) to an elastomer (i-PEMA). Stress-strain analysis of these systems has demonstrated diverse mechanical behavior. We have shown earlier (33) that one could realize similar goals with merely PMMA by adding polar modifiers in catalytic amounts to the polymerization solvent. However, the result in that case was a non-Bernoullian polymer with heterotactic-like stereosequence distributions. In the present case, we have been able to maintain the stereochemical integrity of the polymers, as is clear from the tacticity distributions given in Table VII.

A closer inspection of this data reveals that with the branched alkyl methacrylates, there is a smaller difference between the syndiotactic and isotactic glass transition temperatures.

e.g. Tg(syndio)-Tg(iso) ≅ 80°C for PMMA
≅ 34°C for PTBMA

This aspect must be taken with caution since unequivalent tactic compositions are being compared. Another interesting observation is that the glass transition temperatures for methyl and t-butyl

FIGURE 4. GEL PERMEATION CHROMATOGRAMS OF POLY(t-BUTYL METHACRYLATE) SYNTHESIZED IN HYDROCARBON SOLVENT AT -78°C

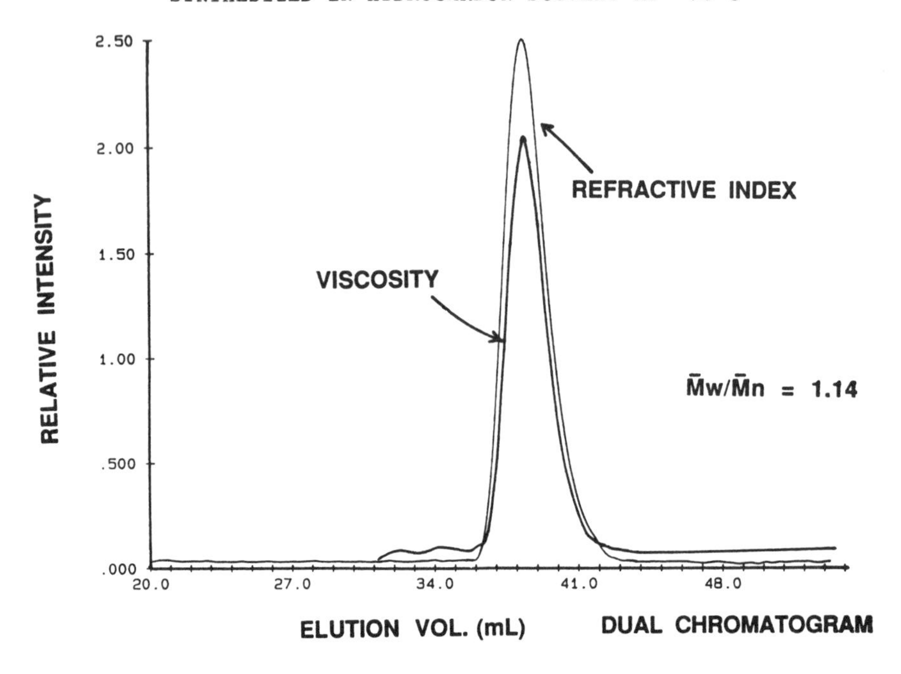

methacrylate which are prepared by free radical initiators are essentially the same, i.e., 123°C for PTBMA and 124°C for PMMA. Despite the bulkier t-butyl group, similar thermal energy is required for segmental motion of the polymer backbone. It is presumed that there are differences in free volume between the two samples due to packing of the bulkier ester group. Free radically prepared PMMA and PTBMA may be interesting to study from the perspective of a membrane application because of their similar Tg, but drastically different free volume.

The selection of initiator is important in alkyl methacrylate polymerizations for a variety of reasons. Most importantly, the initiator should be sterically hindered to avoid carbonyl attack of the monomer. Diphenylhexyl lithium has been shown to be an efficient initiator for alkyl methacrylates (8,9). In contrast, if sec-butyl lithium is used as an initiator for MMA polymerizations, significant termination occurs resulting in loss of molecular weight control and broadening of the distribution. However, the bulkiness of the initiator can be compromised if the ester alkyl group is branched. For example, TBMA can be initiated in THF using sec-butyl lithium at -78°C and the molecular weight distribution only broadens to approximately 1.3. This is in sharp contrast to most other alkyl methacrylates. Despite the highly delocalized nature of DPHL, the anion is still basic enough to initiate quickly. Styrene, on the other hand, cannot be initiated efficiently employing DPHL, and very high molecular weights generally result.

The fact that DPHL is a deep red color facilitates polymerization techniques. The reactor can be "titrated" in order to eliminate any impurities present in the reactor prior to monomer addition. Also, the formation of DPHL by reacting DPE and s-butyl lithium can be monitored very easily using Ultraviolet-Visible spectroscopy. For example, the DPHL anion absorbs very strongly at approximately 435 nm in toluene. The increase in absorbance as the reaction proceeds can be observed and one can be confident that the reaction is complete when the absorbance reaches a maximum. Typically, the DPHL formation is allowed to proceed for several hours in hydrocarbons at room temperature to ensure sample conversion. The initator reaction in polar solvents is complete after only several minutes at -78°C. It is important to realize that the monomer is added directly to the living initiator. This technique places stringent demands on monomer purity; consequently, a major focus of our work has involved the development of purification reagents for alkyl methacrylate monomers.

Lithium diisopropyl amide (LDA) has also been found to be an efficient initiator for methyl methacrylate monomers. LDA is a very common non-nucleophilic, strong base used in many organic reactions. Its bulkiness eliminates carbonyl attack, and narrow molecular weight distribution samples with predictable molecular weights (Table VIII) are obtained.

TABLE VIII. LITHIUM DIISOPROPYL AMIDE AS AN INITIATOR FOR ALKYL METHACRYLATE MONOMERS

SAMPLE	$\bar{M}n$(th)	$\bar{M}n$(gpc)	MWD
PMMA	30,000	23,100	1.16
PMMA	7,500	10,000	1.18
PIPMA	10,000	17,500	1.31

One disadvantage of this initiator is that it is not colored and does not allow for visual detection of initiation. In addition, the reactor systems cannot be titrated. Unfortunately, LDA does not initiate DPE. However, elemental analysis does confirm the presence of nitrogen and can actually be used to determine approximate number average molecular weights. The chain end can also be titrated with acid to generate telechelic ammonium salts. Current research involves the extension of LDA as an initiator for branched alkyl methacrylates. Preliminary evidence indicates that as the ester group becomes branched, the initiation reaction is less efficient. This may be due to the decreased reactivity of the double bond by ester alkyl electron donation.

Many reviews have been published describing the determination of number average molecular weights by end group analysis (39). In the anionic polymerization of alkyl methacrylates, at least one chain end is defined by the initiation fragment. Thus, one can develop various techniques to determine number average molecular weight based on the determination of the end group concentration. ^{1}H NMR has been utilized to determined $\bar{M}n$ for a variety of poly(alkyl methacrylates). This is easily accomplished by ratioing the integration associated with the 10 aromatic protons to an integration associated with the ester alkyl, e.g. the methyl ester at ~3.5 ppm for poly(methyl methacrylate). As the number of protons in the ester group decreases, the maximum $\bar{M}n$ that can be determined is increased. For example, PIPMA number average molecular weights can be determined routinely from 20,000 g/mole-30,000 g/mole by using the single proton in the ester alkyl group. On the other hand, the maximum molecular weight for PMMA is approximately 10,000 g/mole. A very sensitive technique based on UV-Visible spectroscopy is currently being developed. This approach takes advantage of the absorbance of the phenyl rings at ~254 nm. By constructing a Beers-Lambert plot, one can routinely determine $\bar{M}n$ for DPHL initiated poly(alkyl methacrylates).

Although not a major focus of this presentation, a wide range of physical properties can be obtained by variations of the major synthetic variables described herein. Present efforts include the dynamic mechanical thermal analysis (DMTA) and dielectric analysis of the various stereoregular materials. The microstructure plays an important role in determining the presence and position of α and β transitions. Figure 5 depicts the DMTA results for both isotactic and syndiotactic PMMA. The β-transition is evident for the syndiotactic material, but cannot be found in the corresponding isotactic PMMA. The

FIGURE 5. DYNAMIC MECHANICAL THERMAL ANALYSIS (DMTA) OF STEREOREGULAR PMMA (1 Hz)

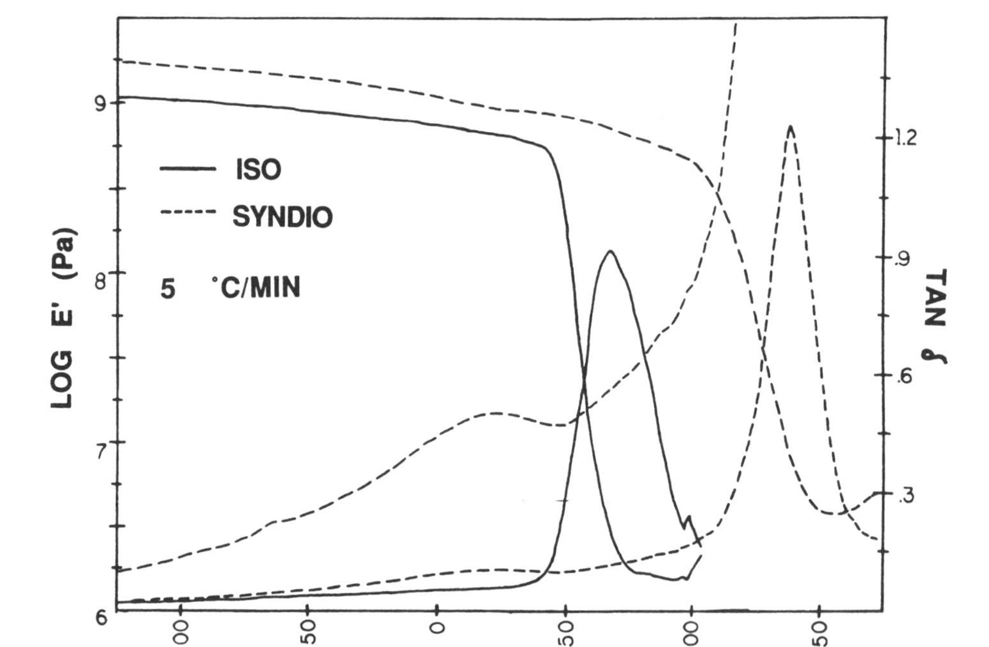

activation energies for these transitions have been calculated by dielectric techniques and reported earlier (23). Stress-strain analysis has also been performed and a variety of behaviors have been noted. In addition to high modulus glasses such as PMMA, elastomers such as isotactic PEMA have ultimate elongations over 400%.

Thermogravimetric analysis (TGA) of the various systems describes the thermal stability of the various materials. Figure 6 illustrates weight loss as a function of temperature for a variety of samples which were prepared in polar solvents. Interestingly, PTBMA reproducibly

FIGURE 6. THERMOGRAVIMETRIC ANALYSIS OF POLY(ALKYL METHACRYLATES)

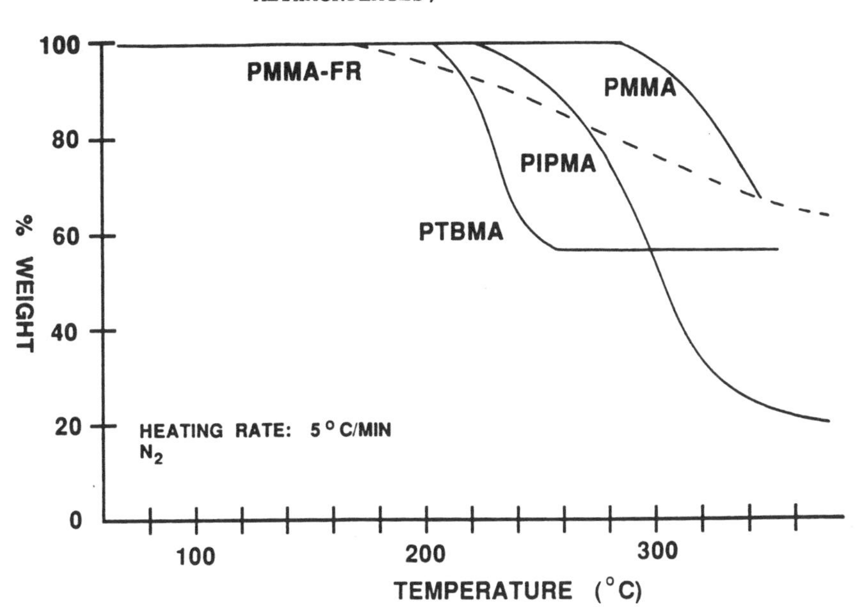

loses approximately 46% of its weight. FTIR and mass spectroscopy (MS) analysis attribute this loss to the liberation of isobutylene to form poly(methacrylic acid) followed by subsequent anhydride formation. Also shown, free radically prepared PMMA begins to lose weight at ~160°C. This early weight loss is the consequence of thermally labile linkages resulting from free radical polymerization. Anionic polymerization does not result in weak groups such as head to head linkages or unsaturated chain ends. Thus, the anionically prepared materials generally demonstrate enhanced thermal stability. Current research involves the application of TGA-MS to analyze the degradation products of these materials. In addition, the isotactic materials also demonstrate different thermal stability than the syndiotactic polymers. However, isotactic PTBMA degrades in a similar fashion as PTBMA prepared by anionic initiators in THF. This can be explained by the fact that PTBMA undergoes living polymerizations in hydrocarbon solvents and thermally weak linkages do not form. This is in contrast to other alkyl methacrylates which undergo pseudo-living polymerizations in hydrocarbons leading to weak linkages.

4. CONCLUSION

Trialkyl aluminum/dialkyl aluminum hydride purification reagents provide for ultrapure, anionic-grade alkyl methacrylate monomers. The controlled synthetic conditions have been established for a wide range of poly(alkyl methacrylates) using the rigorously purified monomers. Once controlled conditions have been established, investigations of various synthetic parameters such as initiator, solvent, temperature and monomer structure were investigated. Insight was obtained regarding the stability of the propagating enolate by systematic variations in these variables. It can be concluded that all alkyl methacrylates do not behave in a similar manner. In particular, the branched alkyl methacrylates are inherently easier to polymerize due to the enhanced stability of the propagating species. Thus, many compromises can be tolerated such as initiator size, higher temperatures and nonpolar solvents. Various characterization techniques can also be developed by taking advantage of the initiator residue which resides at the chain end. The resulting materials demonstrate very diverse properties by simple alterations in synthesis. Current research is directed towards the synthesis of various elastomeric poly(alkyl methacrylates) and their incorporation into multiphase systems. One can utilize the insight gained in these fundamental studies to synthesize a host of novel alkyl methacrylate containing polymeric systems.

5. ACKNOWLEDGMENTS

The authors would like to thank Mr. R. H. Bott and Ms. B. E. McGrath for various thermal measurements, Dr. R. Subramanian for Nuclear Magnetic Resonance characterization and Dr. W. Wunderlich (Rohm Tech, W. Germany) for the generous donation of various acrylic monomers. They also appreciate the support of the Exxon Foundation, the Army Research Office (ARO) and the Defense Advanced Research Projects Agency (DARPA).

6. REFERENCES

1. P. Lutz, P. Masson, G. Brenert and P. Rempp, Polymer Bull., **12**, 79 (1984).
2. T. Q. Nguyen and H. H. Kausch, Makromol. Chem., Rapid Commun., **6**, 391 (1985).
3. T. E. Long, R. Subramanian, T. C. Ward and J. E. McGrath, Polymer Preprints, **27(2)**, 258 (1986).
4. K. Hatada, T. Kitayama, K. Fumikana, K. Onta and H. Yuki, Anionic Polymerization: Kinetics, Mechanisms and Synthesis, ed. J. E. McGrath, ACS Symposium Series, **No. 166**, 327 (1981).

5. R. D. Allen, S. D. Smith, T. E. Long and J. E. McGrath, Polymer Preprints, **26(1)**, 247 (1985).
6. R. D. Allen, T. E. Long and J. E. McGrath, Polymer Bull., **15(2)**, 127 (1986).
7. R. D. Allen, T. E. Long and J. E. McGrath, Advances in Polymer Synthesis, Eds. B. M. Culbertson and J. E. McGrath, **31**, p. 347-362, Plenum Pub. Corp., 1985.
8. P. M. Wiles and S. Bywater, Trans. Fara. Soc., **61**, 150 (1965).
9. D. Freyss, P. Rempp and H. Benoit, J. Polym. Sci. Letters, **2**, 217 (1964).
10. H. Yuki, K. Hatada, K. Onta and Y. Okamoto, J. Macromol. Chem., A9(6), 983 (1975).
11. K. Ute, T. Kitayama and K. Hatada, Polymer Journal, **18(3)**, 249 (1986).
12. J. E. McGrath, R. D. Allen, J. M. Hoover, T. E. Long, A. D. Broske, S. D. Smith and D. K. Mohanty, Polymer Preprints, **27(1)**, 183 (1986).
13. B. C. Anderson, G. D. Andrews, P. Arthur, Jr., H. W. Jacobson, L. R. Melby, A. J. Playtis and W. H. Sharkey, Macromolecules, **14**, 1599 (1981).
14. A. H. E. Müller, Makromol. Chem., **182**, 2863 (1981).
15. P. E. M. Allen and D. R. G. Williams, Ind. Eng. Prod. Res. Dev., **24**, 334 (1985).
16. A. M. Hellstern and J. E. McGrath, unpublished results.
17. W. R. Hertler, D. Y. Sogah and O. W. Webster, Macromolecules, **17**, 1415 (1984)
18. G. Odian, Principles of Polymerization, Wiley and Sons, New York, 2nd edition, 1981, p. 624.
19. H. Yuki and K. Hatada, Advances in Polymer Sci., **31**, 1 (1979).
20. R. Kraft, A. H. E. Müller, H. Höcker and G. V. Schulz, Makromol. Chem., Rapid Commun., **1**, 363 (1980).
21. J. C. Bevington, H. W. Melville and R. P. Taylor, J. Polym. Sci., **12**, 449 (1954).
22. G. Moad, D. H. Solomon, S. R. Johns and R. I. Willing, Macromolecules, **17**, 1094 (1984).
23. A. M. Walstrom, R. Subramanian, T. E. Long, J. E. McGrath and T. C. Ward, Polymer Preprints, **27(2)**, 135 (1986).
24. T. E. Long, R. D. Allen and J. E. McGrath, Polymer Preprints, **27(2)**, 54 (1986).
25. T. E. Long, R. D. Allen and J. E. McGrath, Reactions on Polymers, ACS Symposium Series, 1987, to be published.
26. T. E. Long, G. L. Wilkes and J. E. McGrath, unpublished results.
27. H. Gilman and F. K. Cartledge, J. Organometal. Chem., **2**, 447 (1964).
28. T. Mole and E. A. Jeffrey, Organoaluminum Compounds, Elsevier, 1972, Ch. 12.
29. T. Mole and E. A. Jeffrey, Organoaluminum Compounds, Elsevier, 1972, Ch. 2.
30. K. Hatada, Y. Okamoto, K. Ohta and H. Yuki, J. Poly. Sci., Polym. Lett. Ed., **14**, 51 (1976).
31. K. Hatada, T. Kitayama, Y. Okamoto, K. Onta, Y. Umemura and H. Yuki, Makromol. Chem., **178**, 617 (1977).

32. R. C. Ferguson and D. W. Overnall, Polymer Preprints, **26(1)**, 182 (1985).
33. R. Subramanian, R. D. Allen, T. C. Ward and J. E. McGrath, Polymer Preprints, **26(2)**, 238 (1985).
34. T. C. Ward, J. Chem. Ed., **58**, 867 (1981).

THE ANIONIC POLYMERIZATION OF ALKYL ACRYLATES : A CHALLENGE

R.JEROME, R.FORTE, S.K.VARSHNEY, R.FAYT, Ph.TEYSSIE*
Laboratory of Macromolecular Chemistry and Organic Catalysis
University of Liège - Sart-Tilman B6 - 4000 LIEGE - BELGIUM

ABSTRACT. Kinetics and mechanism of anionic polymerization processes can be modified on purpose by ligands specific of the active anionic species. The reactivity enhancement promoted by cation-binding agents is well-documented, whereas complexing the well-known alkyllithium initiators to non reactive alkyl metals or metal alkoxides can affect the kinetics and improve the stereoregulation of the polymerization. Quite interestingly, some inorganic salts, such as lithium chloride, exhibit an important regulating effect in the anionic polymerization of alkyl acrylates. Indeed, side-reactions are now controlled to such an extent that the propagation is a living process and leads to polymers with an unimodal and narrow molecular weight distribution.

1.INTRODUCTION

Since the discovery by Szwarc of the living anionic polymerization (1), spectacular advances have been reported in the fine tailoring of synthetic polymers and therefore of their final properties. Although this technique works well for nonpolar monomers, such as styrene and dienes, its application to polar monomers is most often a challenge. Methacrylate and acrylate esters are examples of polar monomers of great industrial interest, the polymerization of which is generally restricted to poorly controlled radical processes. Indeed, their anionic polymerization gives rise to side reactions mainly due to the nucleophilic attack of the carbonyl group and the acidity of the α-hydrogen atom of the acrylates (2). Some time ago, experimental conditions have been defined that prevent successfully the

*to whom all inquiries should be addressed.

M. Fontanille and A. Guyot (eds.), Recent Advances in Mechanistic and Synthetic Aspects of Polymerization, 101–117.

occurrence of transfer and termination reactions in the anionic polymerization of alkyl methacrylates (3-5), but not in that one of alkyl acrylates. A substantial progress has been achieved recently, when a Du Pont Company research team has demonstrated that nucleophile-assisted "group transfer" reaction of silyl ketene acetals could be turned into a "living" polymerization process for alkyl acrylates (6).Nevertheless, it is still quite a problem to block polymerize alkyl acrylates with monomers such as styrene and dienes by a multistep living polymerization. In order to cope with that problem we have considered the use of ligands as a tool to minimize the relative importance of the secondary reactions in the anionic polymerization of alkyl acrylates.In this regard, the current literature provides examples of ligands which deeply modify the reactivity of anionic species and perturb the mechanism of the anionic polymerization. In the next section, pertinent applications of this approach are briefly reported.

2. EXAMPLES OF ANIONIC SPECIES MODIFIED BY VARIOUS TYPES OF LIGANDS

2.1. Cation-binding agents

Since about two decades, it is well stated that the activity of alkyllithium is greatly enhanced by the addition of equimolar amounts of a chelating amine, such as tetramethylethylenediamine. In that way, Langer, in 1965, was able to oligomerize ethylene through an anionic pathway (7). In addition to the reactivity enhancement, the stereoregulation of the diene polymerization can also be modified to such an extent that polybutadiene containing over 80% 1.2 structures is obtained in conditions where largely 1.4 polymers are formed in the absence of the Li-solvating diamine (8). Today it is of usual practice to increase the reactivity of carbanion pairs by cation-binding ligands, such as ethers, crown ethers, polyamines and cryptands (9,10).

2.2. Alkyl metals as ligands

Alkyllithium is also known to form complexes with other metal alkyls such as those of Mg, Zn and Cd (11).

It has been reported that dialkylmagnesium, which is not an active initiator for diene and styrene polymerization, participates in polymerization when complexed either with the alkyllithium initiator or with the propagating polymer-lithium molecules (12). Apparently, the lithium and magnesium alkyls form the complex instantly upon mixing, and the rapid exchange of all alkyls coordinated in the complex provides one unique organometallic species. When one refers to the butadiene polymerization initiated by butyllithium (BuLi) in a cyclohexane/THF solvent mixture, table I shows that the molecular weight of polybutadiene decreases as a function of increasing amounts of dibutylmagnesium relative to BuLi. This observation has been tentatively rationalized by assuming the participation of C-Mg bonds in the anionic polymerization. The participating factor (PF) of these bonds has been calculated from the comparison of the experimental molecular weight and the theoretical value based on the monomer to BuLi molar ratio. As schematized in figure 1, an equilibrium between a 1:1 and a 2:1

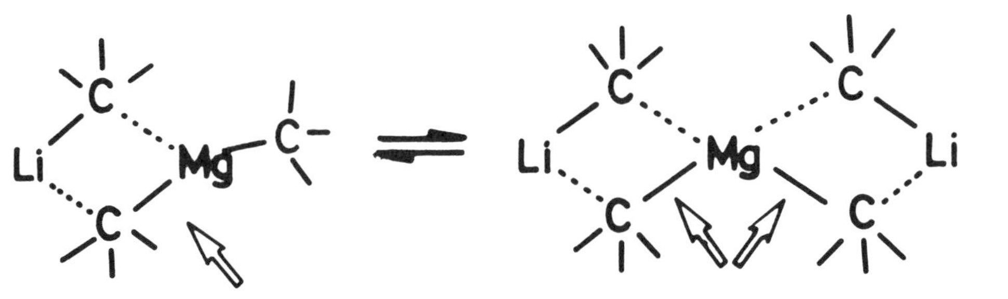

⇨ : **NEW PROPAGATING SPECIES**

Figure 1. Schematic representation of alkyllithium / dialkylmagnesium complexation.

BuLi / dibutylMg complex has been proposed. As a rule, the dialkylmagnesium does not change the stereochemistry of the diene polymerization, but it decreases the polymerization rate in strong contrast to the kinetic enhancement due to the cation-binding agents.

As a further example, NMR experiments have shown the existence of a 1:1 stoechiometric alkyllithium/diethylzinc complex (13). This complex has been used as an initiator in the low temperature polymerization of methylmethacrylate (MMA) in toluene and toluene/THF

TABLE 1. Effect of dibutylmagnesium on the anionic polymerization of butadiene initiated by butyllithium.

$[BuLi/MgBu_2]$	$\bar{M}_n \times 10^{-3}$	P.F.
1/0	74	0
1/0.15	61	1.5
1/0.30	49	1.7
1/0.44	43	1.6
1/0.59	36	1.8
1/0.74	32	1.6

Experimental conditions. Butadiene : 100 g; BuLi : 1.35 mmol; Solvent : cyclohexane/THF; Temperature : 50°C

mixtures (14). The strong complexing properties of diethylzinc for the growing chains results in a noticeable stereoregulating effect. Indeed, the content in isotactic triads has remarkably increased in toluene containing less than 3% THF (Table II), i.e. when the growing chains are essentially associated as supported by viscometric measurements. Although the molecular weight of PMMA increases proportionally with the degree of conversion, the polydispersity of the seemingly living polymer remains quite high in toluene ($\bar{M}_v/\bar{M}_n$=5.4).

It is worth noting that during the initiation step, a considerable fraction of the catalyst is lost by side reactions with the ester function of the monomer (Figure 2). The experimental number-average molecular weight ($\bar{M}_n$) is indeed much greater than expected, and the initiation efficiency (f), calculated as the polymer to initiator molar ratio, amounts to 0.34 independently of the monomer and initiator concentrations and the THF content. In contrast to the initiator, the living chain end of the polymer does not react with the ester groups as supported by the living character of the propagation step. This could mean that diethylzinc is able to balance properly the reactivity of the propagating species (methacrylate Li carbanion) but not that one of the initiator (BuLi). In toluene-THF mixtures, a kinetic scheme based on ion pairs and THF-solvated ion pairs in equilibrium with each other is consistent with the experimental data.

TABLE II. Anionic polymerization of MMA initiated by a 1:1 $BuLi/ZnEt_2$ complex. Tacticity of PMMA (14)

THF	Tolu	Cata.	I,%	H,%	S,%
0	100	BuLi	76	13	11
		$BuLi/ZnEt_2$	95		
1	99	BuLi	38	27	35
		$BuLi/ZnEt_2$	71	21	8
3	97	BuLi	15	39	46
		$BuLi/ZnEt_2$	61	23	16
5	95	BuLi	18	36	46
		$BuLi/ZnEt_2$	18	39	43

Experimental conditions : [MMA]=0.46 mol l^{-1}; [Cat.]=2.3 x 10^{-2} mol l^{-1}; - 60°C

$$MMA + BuLiZnEt_2 \xrightarrow{k_i} R\text{-}CH_2\text{-}C^{(-)}(CH_3)(COOCH_3)\ Li^{(+)}ZnEt_2$$

f = 0.34

$$MMA + BuLiZnEt_2 \xrightarrow{k_e} CH_2{=}C(CH_3)\text{-}C(=O)\text{-}R + MeOLi + ZnEt_2$$

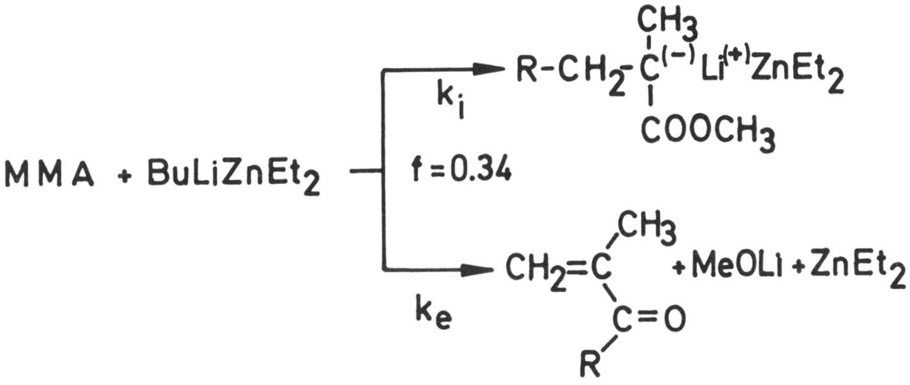

Figure 2. Competitive reactions of the 1:1 $BuLi/ZnEt_2$ complex with MMA. k_i and k_e are the rate constants of initiation and transfer, respectively. f is the initiation efficiency.

2.3. Metal alkoxides as ligands

As shown in figure 2, alkali metal alkoxides are commonly formed as byproducts in the anionic polymerization of MMA. It is the reason why attention has been paid to the effect that alkali alkoxides can have on the polymerization of MMA when initiated by either BuLi (15) or α-metalloesters (16,17). In the former case (toluene, at -30°C), the system to which alkoxide has been added polymerizes faster to form higher molecular weights. Let us recall that α-metalloesters result from the substitution of esters of isobutyric acid by Li, Na and K in the α-position and accordingly foreshadow the growth centre in the anionic polymerization of methacrylate esters. Starting with α-metalloesters and tert-butoxide of alkali metals, under conditions that leave the α-metalloester as the main active species, the actual growth centre does yield polymers with a microstructure close to that usually found for the alkoxide alone. These results indicate interaction of alkali tert.butoxide with the growing chain. Furthermore the IR analysis of the catalyst supports the formation of an adduct which has been tentatively schematized as follows.

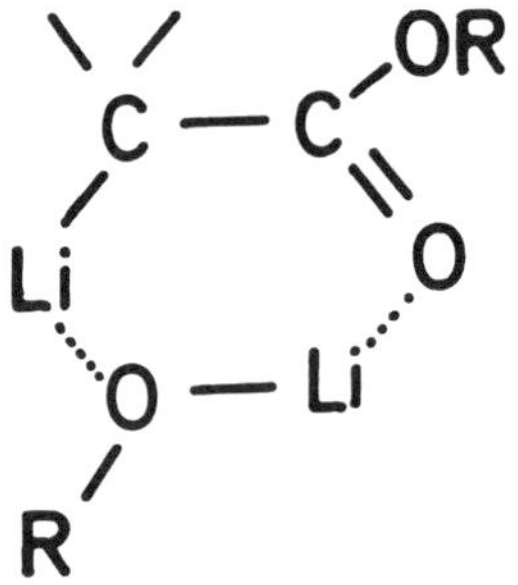

Similarly some adducts of carboxylic acid esters with Na and K alkoxides have been isolated and characterized by IR spectroscopy. The coordination of the monomer to the alkoxide may be assumed to be of importance in the propagation reaction. Finally the alkaline alkoxides affect the polymer yield favourably and confirms that the adduct formed by the alkoxide with the growth centre stabilizes the latter under the experimental conditions used, i.e. toluene at 20°C.

It is worth stressing that alkali metal alkoxides have also been used to activate the very insoluble alkali metal amides in solvents such as THF and toluene (18). It is proposed that the alkoxide complexes $NaNH_2$ on the crystal surface and brings it into the solvent under a complexed form. The ability of the so-called "Complex Bases" to initiate the polymerization of vinyl monomers has been evaluated, they influence favourably the initiation and propagation rates (19,20), but they have no significant stereoregulation effect, e.g. in the MMA polymerization (21). It must be emphasized that inorganic salts, such as nitrites, and thiocyanides, can also activate sodamide and lead to initiators active when neither alkali amides nor alkali salts alone initiate the polymerization of vinyl monomers (22). Finally it has been demonstrated that anions and cations are rapidly exchanging in the "salt complex bases" with formation of new species.

2.4. Inorganic salts as ligands

Due to the easy handling of inorganic salts, their possible ability in improving the control of some polymerization processes deserves interest. The strong interactions between lithium chloride and polyamides which cause a relevant melting point depression of the polymers have led some authors to investigate the influence of the salt on the progress of the anionic polymerization of ε-caprolactam (ε-CL) (23). First of all, a crystalline complex containing four molecules of lactam and one molecule of LiCl has been isolated. Li cation is actually coordinated to the carbonyl oxygen of each of the four lactam molecules, the NH hydrogens of which are hydrogen bonded to the salt anions. Above the melting point of the adduct, a red shift of the carbonyl stretching bands of lactam indicates that a direct interaction persists at 155°C, i.e. in the mid-range of temperatures required by the anionic polymerization of ε-CL. This process is significantly retarded by the addition of LiCl, but the most relevant role of the salt has to be found in the strong reduction of the irregular structures of nylon-6 due to Claisen-type condensations. This example supports that an inorganic salt can regulate the activity of anionic species in such a way that the main reaction is not prevented while the extent of the secondary reactions is strongly minimized. Further examples can also be mentioned as the complexation equilibrium between MMA and $MgCl_2$ (characterized by ^1H NMR analysis) which influences the thermodynamics of the MMA polymerization initiated by tert.butylmagnesium chloride in THF (24).

Complexation of MMA with $ZnCl_2$ (25), organochloroaluminum compounds (26) and other Lewis acids (27,28) is also known to increase the rate of the radical polymerization and to modify the reactivity ratios in copolymerization with donor monomers leading to nearly alternating copolymers.

This brief survey of the scientific literature convincingly supports that kinetics and mechanism of the anionic polymerization of vinyl monomers, dienes and polar monomers, such as MMA and lactams, can be extensively controlled by metal compounds able to interact with the active species and sometimes with both carbanion and monomer. Let us discuss now how this approach is helpful in mastering the polymerization and block polymerization of both the alkyl methacrylates and acrylates.

3. EFFECT OF LiCl ON THE BLOCK POLYMERIZATION OF STYRENE AND MMA

The sequential addition of MMA to living polystyryl anions might be considered as a direct pathway to the controlled synthesis of poly(styrene-b-MMA). When performed in a 50/50 (v/v) benzene/THF mixture at -78°C in the presence of Li as a counterion, the block polymerization gives disappointing results since about half the initial polystyrene is recovered as homopolymer and the polydispersity of the block polymer is bimodal (Table III) . The situation is dramatically improved

Table III. Anionic block polymerization of styrene and MMA in a 50/50 (v/v) benzene/THF mixture at 78°C. $[PS^-Li^+]/[Li\ salt]=1/10$

Li salt	Homo PS (%)	$\bar{M}_w/\bar{M}_n$
-	45-50	bimodal
LiOAc	13	>2
LiCl	2	1.2-1.3

when a molar excess of Li acetate is added to polystyryl Li (PS^-Li^+) before starting the MMA block polymerization. The percentage of homo PS falls to 13% of the initial amount whereas the block polymer exhibits an unimodal but still broad distribution. A final improvement is reported when Li acetate is replaced by Li chloride, since more than 98% polystyrene is block polymerized and the polydispersity of the block polymer is in between 1.2 and 1.3. It is worth noting that the same result is observed in a 50/50 (v/v) toluene/THF mixture and that these experimental conditions can be successfully extended to methacrylates containing n.alkyl radicals of various length (e.g. n.butyl and n.hexyl) in the ester group.

These preliminary results clearly show that LiCl has a dual favourable effect. First of all LiCl decreases the reactivity of PS^-Li^+ towards MMA, resulting in a well-controlled initiation step free from side reactions on the monomer. Furthermore, the propagation of the anionic polymerization of MMA occurs now without deleterious transfer and termination reactions. Actually, in the presence of LiCl, the initiation by PS^-Li^+ and the propagation of the MMA polymerization are the two main reactions observed, the rate constants of which are in a ratio k_i/k_p great enough to lead to a narrow molecular weight distribution. This remarkable effect of LiCl is strongly supported by the dramatic change in the polydispersity of PMMA anionically synthesized in a 90/10 (v/v) toluene/THF mixture. As already stated (29), PMMA is obtained in limited yields with a high polydispersity

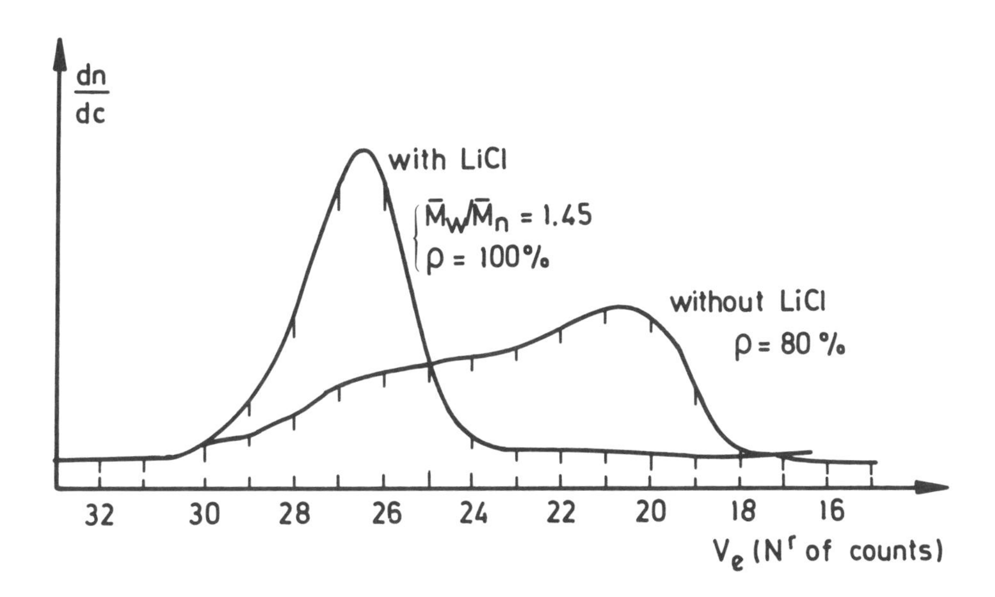

Figure 3. Exclusion chromatography of PMMA synthesized in a 90/10 (v/v) toluene/THF mixture, at -78°C, with α-methylstyryl Li as an initiator.

when synthesized in toluene/THF mixtures. Figure 3 demonstrates that LiCl is an efficient ligand of the carbanionic species active in the anionic polymerization of MMA; this conclusion is consistent with the observation previously reported by Russo and coworkers for the anionic polymerization of ε-caprolactam (23).

4. EFFECT OF LiCl ON THE ANIONIC POLYMERIZATION OF tert.BUTYLACRYLATE

The control provided by LiCl on the progress of the anionic polymerization of alkyl methacrylates has prompted us to investigate the beneficial effect that an inorganic salt could have when the polymerization of alkyl acrylates is considered. Tert.butylacrylate (t.BuA) has been selected as a monomer for two main reasons. Both the bulkiness and the positive inductive effect of the tert.butyl group are expected to decrease the extent of the usual secondary reactions and make less uncertain the research for a living anionic process. The second interest has to be found in the easy hydrolysis of the related polymer to give polyacrylic acid as well as in the possible transalcoholysis leading to a range of alkyl polyacrylates of the same molecular features.

It is worth recalling that, when performed in THF, at - 78°C, the anionic polymerization of MMA initiated by styryl or α-methylstyryl Li is essentially a living process leading to samples of relatively narrow polydispersity. When the same conditions are applied to t.BuA, the monomer conversion is still complete but the molecular weight distribution is bimodal, and especially broad when styryl Li is the initiator (Figure 4).
These results are observed although the monomer is added to the polymerization medium as a dilute solution (10% in benzene or THF) rather than as a pure monomer. Addition of LiCl has again a spectacular effect on the kinetic scheme, since unimodal polyacrylates with a polydispersity of ca. 1.2 are obtained (Figure 4). It means that LiCl is able to balance properly the reactivity of both the styryl and α-methylstyryl Li initiating species and that one of the propagating t.butylacrylate carbanion.

Figure 4 provides a piece of information suggesting that the coordination of the ion pairs to a μ-type hindering ligand (LiCl) is an efficient means of avoiding the transfer and termination side-reactions. In order to ascertain that opportunity, resumption experiments have been performed. A first monomer feed has been

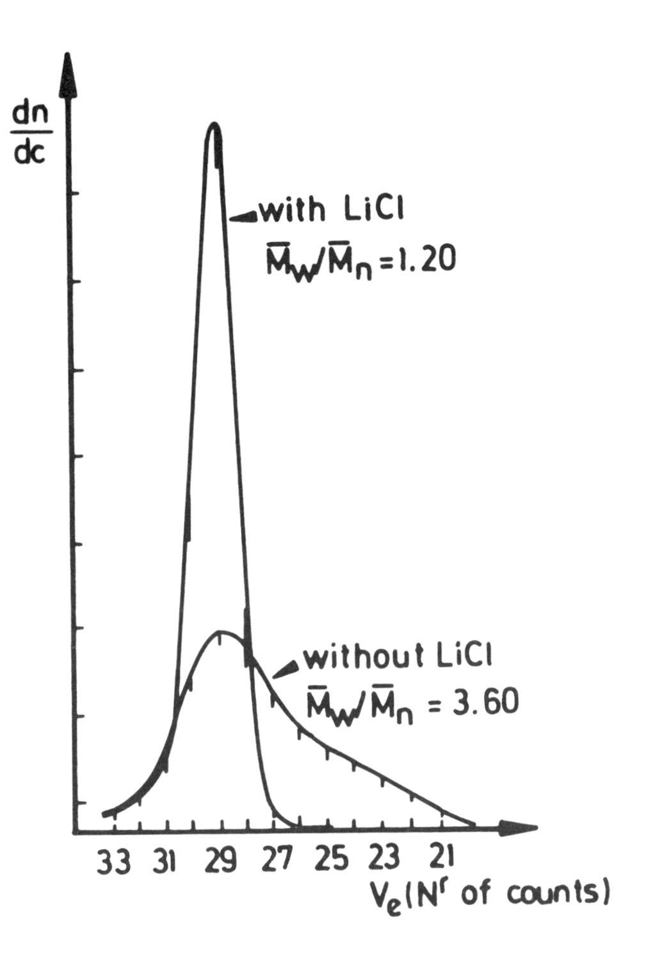

Figure 4. Effect of LiCl on the exclusion chromatogram of poly(tert.butylacrylate) prepared in THF, at - 78°C, with styryl Li as the initiator ([Initiator]/[LiCl]-=1/10).

quantitatively polymerized in THF at - 78°C, and a second monomer feed corresponding to the 2/3 of the initial one has been then completely converted without any significant change in the molecular weight distribution. As shown in table IV the experimental molecular weight determined after each polymerization step increases in the same ratio as the amount of the added monomer.
Another convincing argument can be found in Table IV where, at constant monomer concentration, the change in the initiator concentration dictates the change in the experimental molecular weight, as determined by exclu-

Table IV. Two step polymerization of tert.butylacrylate initiated by α-methylstyryl Li in THF, at -78°C.

t.BuA(mol)	$\bar{M}_n$	$\bar{M}_w/\bar{M}_n$
0.06	11,000	1.25
x 2/3	x 2/3	
+ 0.04	7,500	
0.10	18,500	1.25

Table V. Dependence of the molecular weight of poly(tert.butylacrylate) on the monomer to initiator molar ratio.

α-Mst^-Li^+(mol)	$\bar{M}_n$	$\Delta\ \bar{M}_n$	$\bar{M}_w/\bar{M}_n$
x 0.75	22,000	0.73	1.30
x 1	30,000	-	1.30
x 1.5	40,000	1.33	1.25
x 2.0	58,000	1.93	1.25

Experimental conditions : Initiator :α-methylstyryl Li; THF; -78°C

sion chromatography calibrated with polystyrene standards; the molecular weight distribution remains nearly constant.

It must be stressed that, in contrast to MMA, t.BuA has to be added to the polymerization medium, not as a pure monomer, but as a dilute solution (10 ωt %) in either benzene or THF. The molecular weight distribution becomes broader as the concentration of the t.BuA solution increases and a fraction of very high molecular weight is ultimately formed which is reponsible for the gelation of the reaction medium.

A last interesting point to be mentioned is that LiCl is a ligand able to counterbalance the unfavourable effect due to an increase in the polymerization temperature. In the absence of LiCl, a temperature increase from -78°C to 0°C results in a three to fourfold increase of the experimental molecular weight (all the other conditions being kept constant) compared with the value obtained with LiCl as a ligand (Table VI). It is to be concluded that a fraction of the initiator is lost at 0°C, most likely by side-reactions with the monomer. The addition of LiCl as a ligand is thus an obvious means of avoiding that drawback and to maintain an essentially living polymerization; this point should be ascertained in the near future.

Table VI. Effect of temperature on the anionic polymerization of tert.butylacrylate initiated by α-methylstyryl Li in a 25/75 (v/v) THF/toluene mixture ($\alpha\text{-Mst}^-\text{Li}^+$ = 1/10).

Temp. (°C)	LiCl	$\bar{M}_n$	$\bar{M}_w/\bar{M}_n$
- 78	Yes	14,500	1.25
0	Yes	17,500	1.65
0	No	40,000	2.00

5. PRELIMINARY INVESTIGATION OF THE MODE OF ACTION OF LiCl

A systematic change of the LiCl to initiator (α-methylstyryl Li) molar ratio results in a change in the molecular weight distribution of poly(t.BuA) synthesized in THF at - 78°C. Table VII shows that $\bar{M}_w/\bar{M}_n$ decreases sharply from 3.6 to 1.3 when LiCl is progressively added to the initiator until a 1/1 molar ratio is reached. Beyond that ratio, the polydispersity does not change to a large extent anymore. Thus a 1/1 or slightly greater LiCl/ion pair adduct ratio should have the most beneficial effect in mastering the anionic polymerization of t.BuA. This is of course a rough information on the stoechiometry of the possible salt/ion pair complexes.

In order to shed light on the possible interaction between the organolithium species and LiCl, the ^{7}Li NMR spectra of polystyryl Li and PMMA^-Li^+ have been recorded before and after the addition of various amounts of LiCl in THF at - 60°C. The chemical shifts reported in Table VIII are relative to 0.02 mol l^{-1} LiCl in THF.
When LiCl is added to PS^-Li^+, a single line is observed the chemical shift of which is not the weighted average of that one for LiCl and PS^-Li^+; it means that the

Table VII. Effect of the LiCl to α-methylstyryl Li molar ratio on the molecular weight distribution of poly(tert.butylacrylate) synthesized in THF at -78°C.

$[LiCl]/[\alpha\text{-}Mst^{-}Li^{+}]$	$\bar{M}_w/\bar{M}_n$
0.0	3.60
0.2	1.85
0.6	1.50
1.0	1.30
5.0	1.20
10.0	1.20

Table VIII. ^{7}Li chemical shift in $PS^{-}Li^{+}$ and $PMMA^{-}Li^{+}$ in THF at - 60°C, added with LiCl or not.

Li^{+} species	^{7}Li shift(ppm)	$[Li^{+}](ml^{-1})$
LiCl	0	0.02
$PS^{-}Li^{+}$	0.59	0.02
$LiCl/PS^{-}Li^{+}$(1/2)	0.19	0.06
$LiCl/PS^{-}Li^{+}$(1/1)	0.21	0.04
$LiCl/PS^{-}Li^{+}$(2/1)	0.06	0.06
$PMMA^{-}Li^{+}$	0.42	0.02
$LiCl/PMMA^{-}Li^{+}$(1/2)	0.42-0.11	0.06
$LiCl/PMMA^{-}Li^{+}$(1/1)	0.15	0.04
$LiCl/PMMA^{-}Li^{+}$(2/1)	0.06	0.06

observed resonance cannot be attributed to a fast exchange of Li cations between the salt and the organo compound. It is most likely due to a complexed species in equilibrium with an excess of one or both of the starting Li compounds. Since the chemical shift is not very different for the 1/1 and 1/2 $LiCl/PS^{-}Li^{+}$ ratio the adduct is most likely to accomodate exceeding amounts of $PS^{-}Li^{+}$ relative to LiCl. The reverse is obviously not true : the chemical shift decreases sharply while going from a 1/1 to a 2/1 molar ratio.

In contrast to PS^-Li^+, two resonance lines are observed when $PMMA^-Li^+$ is used in a twofold excess relative to LiCl. One of these two resonances corresponds to free $PMMA^-Li^+$ and the other one most likely to a salt/initiator adduct. The resonance of the suspected adduct is not very different from that one of the 1/1 combination.

In conclusion LiCl does interact as well with the initiating species (styryl-type anion) as with the propagating carbanion derived from MMA. Compared with the chemical shift reported for PS^-Li^+ and $PMMA^-Li^+$, the value for the suspected adduct is systematically smaller in agreement with a loss of reactivity of the anionic species.

Acknowledgements. The authors are indebted to IRSIA for a fellowship to one of them (R. Forte), and to NORSOLOR (France) and CIPARI (Luxembourg) for their interest and support.

6. REFERENCES

1. M. SZWARC, Ions and Ion Pairs in Organic Reactions Wiley-Interscience, New-York, NY, vol. 1 (1972); vol. 2 (1974).

2. D.M. WILES, Structure and Mechanism in Vinyl Polymerization, T. Tsuruta and K.F. O'Driscoll, Eds, Marcel Dekker, New-York, NY, 1969 p. 233.

3. F. WENGER, Chem. Ind. (London) 1959, 1094.

4. R.K. GRAHAM, D.L. DUNKELBERGER and E.S. COHN, J. Polym. Sci., 42, 501 (1960).

5. A. ROIG, J.E. FIGUERUELO and E. LLANO, J. Polym. Sci. B, 3, 171 (1965).

6. D.W. WEBSTER, W.R. HERTLER, D.Y. SOGAH, W.B. FARNHAM, and T.V. RAJAN BABU, J. Am. Chem. Soc., 105, 5706 (1983).

7. A.W. LANGER, Trans. N.Y. Acad. Sci., 27, 741 (1965).

8. P. BRES, M. VIGUIER, J. SLEDZ, F. SCHUE, P.E. BLACK, D.J. WORSFOLD, and S. BYWATER, Macromolecules, 19, 1325 (1986).

9. J. SMID, Anionic Polymerization, J.E. Mc Grath, Ed., ACS Symp. Series, 166, 79 (1981).

10. S. BOILEAU, Anionic Polymerization, J.E. Mc Grath Ed., ACS Symp. series, 166, 283 (1981).

11. L.M. SELTZ, and G.F. LITTLE, J. Organomet. Chem., 18, 227 (1969).

12. H.L. HSIEH, and I.W. WANG, Macromolecules, 19, 299 (1986).

13. S. TOPPET, G. SLINCKX and G. SMETS, J. Organomet. Chem., 9, 205 (1967).

14. G. L'ABBE, and G. SMETS, J. Polym. Sci., Part A1, 5, 1359 (1967).

15. D.M. WILES, and S. BYWATER, J. Phys. Chem., 68, 1983 (1964).

16. L. LOCHMANN, D. DOSKOCILOVA, and J. TREKOVAL, Coll. Czech. Chem. Commun., 42, 1355 (1977).

17. L. LOCHMANN, J. KOLARIK, D. DOSKOCILOVA, S. VOZKA, and J. TREKOVAL, J. Polym. Sci., Polym. Chem. Ed., 17, 1727 (1979).

18. P. CAUBERE, Topics in Current Chemistry, 73, 50 (1978).

19. G. NDEBEKA, P. CAUBERE, S. RAYNAL, and S. LECOLIER, Polymer, 22, 347 (1981).

20. S. RAYNAL, S. LECOLIER, G. NDEBEKA, and P. CAUBERE, Polymer, 22, 356 (1981).

21. S. RAYNAL, G. NDEBEKA, P. CAUBERE, J. SLEDZ, and F. SCHUE, J. Macromol. Sci., Chem., A18, 313, 1982.

22. S. RAYNAL, S. LECOLIER, G. NDEBEKA, and P. CAUBERE, Polymer, 23, 283 (1982).

23. G. BONTA, A. CIFERRI, and S. RUSSO, ACS Symp. Ser., 59, 216 (1977).

24. R. PETIAUD, and QUANG-THO PHAM, Europ. Polym. J., 12, 449 and 455 (1976).

25. M. IMOTO, T. OTSU and Y. HARADA, Makromol. Chem., 65, 180 (1963).

26. J. AFCHAR-MOMTAZ, A. POLTON, M. TARDI and P. SIGWALT, Eur. Polym. J., 21, 583 (1985).

27. H. HIRAI, and M. KOMIYAMA, J. Polym. Sci., Part B, 10, 925 (1972).

28. H. HIRAI, K. TAKAUCHI, and M. KOMIYAMA, J. Polym. Sci., Polym. Chem., 20, 159 (1982).

29. M. SZWARC, Ions and Ion Pairs in Organic Reactions, Wiley - Interscience, New York, NY, vol. 2, p 401 (1974).

STABILITY OF ACTIVE CENTRES IN POLYMERIZATION - METHODS OF STUDY OF SHORT-LIVED ACTIVE CENTRES

P. SIGWALT
Université P. et M. Curie - Laboratoire de Chimie Macromoléculaire - Tour 44
4, place Jussieu - 75252 Paris Cédex 05
France

ABSTRACT. The problem of the identification and of the measurement of concentrations of active species with short lifetimes is examined, and is particularly important in radical and cationic polymerizations. The use of indirect methods (such as trapping of intermediates) and the direct detection of radicals (by e.s.r) and of carbocations (by UV spectrophotometry) are particularly discussed.

In other lectures during this meeting the determination of active centres concentration in Ziegler-type polymerizations and in anionic polymerizations has been examined. In many of these systems, a relatively constant concentration of active centres has been observed, at least during the time needed to study the polymerization kinetics, and this generally results from the stability of the active species during this period. Some cationic polymerizations occurring by ring-opening show a similar behaviour, even if some unanswered questions may remain about the relative concentrations of the various species. In carbocationic polymerizations and in radical polymerizations,their lifetimes τ are of the ordre of the second and their concentration very small, and most kinetic data (τ, k_p, k_t...) have been obtained by indirect methods. These are often not enough to come to definite conclusions and the development of direct techniques seems necessary to identify the mechanisms involved. We shall examine successively radical and cationic polymerizations.

1. RADICAL POLYMERIZATION

For the determination of concentrations of radicals and of their reactivities (particularly k_p and k_t) in radical polymerizations, methods such as the rotating-sector one are still used, the data of which are compared with those in the quasi-stationary state, giving both the lifetime and the concentration of the radicals. We shall see however that direct measurements of the radicals concentration in quasi-stationary state did not give exactly the same results. The direct method

M. Fontanille and A. Guyot (eds.), Recent Advances in Mechanistic and Synthetic Aspects of Polymerization, 119–132.

would be particularly useful for the study of copolymerizations for which controversies have occurred about the various radicals concentrations. Another important problem is that of the nature of the initiating radicals and of their reaction with monomers, that determines the types of end-groups in the polymers.

However, the e.s.r technique, which is generally used for radical identification, is generally not sensitive enough with commercial apparatus to give precise values of their concentration, which is of the order of 10^{-9} to 10^{-7} in most polymerization conditions. An improved apparatus permitting to measure the concentration of radicals in various polymerizations of vinyl monomers in bulk has been described by Bresler et al (1) and recently more highly resolved e.s.r. spectra were obtained by Nozakura et al (2), the results of whom shall be described here.

1.1. E.S.R. observation in the stationary state of methacrylate and vinyl acetate radicals.

Polymerizations were carried out in the e.s.r. spectrophotometer and initiated by UV irradiation, the monomer being in a quartz cell of 1mm width (2). The cavity of the spectrophotometer was of a special type and about 10 times more sensitive for detection than a conventional TE_{011} mode cavity.

For the methacrylates, the reaction was carried out in concentrated benzene solutions (1ml benzene, 0.01g Bz_2O_2 ; for methylmethacrylate (MMA) : 1ml monomer ; for tritylmethacrylate (TPMA) : 0,316g). The yields after 15 nm were 4.1 % for MMA and 1.8 % for TPMA. The measured radical concentrations and the corresponding k_p are given in table I, together with data obtained by the rotating sector technique.

The directly measured k_p for MMA was about 1/3 of that obtained by the same authors using the rotating-sector method. The reasons are not known but the authors suggested that some error might have been made during the polymer recovery. No mol.wts were given and it is difficult to have an opinion. An interesting result was the much lower k_p and k_t observed for TPMA. For this monomer, the lifetime of the radical and k_t could be measured by following the decay of the ESR spectrum. For MMA, the lifetime of the radical was too short to be measured.

TABLE I

Monomer	Radical 10^7	by E S R		by rotating-sector	
	(M^{-1})	k_p $(M^{-1}s^{-1})$	$k_t.10^{-5}$ $(M^{-1}s^{-1})$	k_p $(M^{-1}s^{-1})$	$k_t.10^{-5}$ $(M^{-1}s^{-1})$
MMA	2.7	187	-	450	420
TPMA	10.5	26	3.0	-	-

Kinetic data for methylmethacrylate, triphenylmethylmethacrylate and vinyl acetate (2,3) at 30°C.

More recent results for vinyl acetate (VA) have been published by the same authors (3), using similar experimental conditions. The radical concentration was about the same as for MMA. It was however found that the k_p was about three times higher than that measured by the rotating sector technique, but no explanation could be offered.

1.2. Identification by spin-trapping of radicals involved in the initiation reaction

Free radicals formed by fragmentation of the initiator are usually called "primary" radicals and are assumed to add to the monomer giving a new radical, in the following way :

$$R{-}R \longrightarrow 2R\cdot$$

$$R\cdot + CH_2 = \underset{\displaystyle X}{\underset{|}{C}}H \longrightarrow RCH_2{-}\underset{\displaystyle X}{\underset{|}{C}}H$$

The fraction $\underline{f}$ of radicals derived from the initiator that initiate directly the formation of kinetic chains is called the efficiency of the initiator, the fragment R- being assumed to be incorporated into the polymer in the initiation step. However, various anomalies in the values of the efficiencies have been observed for some monomers, and it had been considered for some time that the real initiation could sometimes occur after a transfer involving the solvent SH or the monomer, such as:

$$R\cdot + SH \longrightarrow RH + S\cdot$$

$$R\cdot + CH_2{=}\underset{\displaystyle X}{\underset{|}{C}}{-}CH_3 \longrightarrow RH + CH_2{=}\underset{\displaystyle X}{\underset{|}{C}}{-}CH_2^{\cdot} \qquad (1)$$

the new radicals eventually initiating the formation of a macromolecule. Spin-trapping is a technique in which a reagent is added to the polymerizing solution and traps either the radical formed by decomposition of the initiator, or its addition product to monomer, or the product of its reaction with another molecule, giving new species that may be identified and the concentration of which may be measured. Two main types of traps have been used in recent years. Nitroso compounds give birth to stable nitroxy radicals :

$$R_1^{\cdot} + R'{-}NO \longrightarrow \begin{matrix} R_1 \\ R' \end{matrix}\!\!>N{-}O\cdot$$

that may be identified by e.s.r, and their concentration measured, but the interpretation of the results has been sometimes difficult (4,5). Apparently simpler results have been obtained using nitroxy radicals as traps :

$$R_1^{\cdot} + \cdot O{-}NR_2' \longrightarrow R_1{-}O{-}NR_2'$$

The products that accumulate may be recovered, identified by NMR and their concentrations measured. We shall examine some recent data about the initiation of methacrylates and that of styrene.

1.2.1. Initiation of the polymerization of methacrylates and other isopropenyl monomers with tert-butoxy radicals (5,6,7). Polymerization of various methacrylates initiated by tert-butoxy radicals have given values of the efficiency $\underline{f}$ that varied widely with the size of the ester group. This had been tentatively attributed to a steric hindrance for the addition to the CH_2, that led to increased hydrogen abstraction on the CH_3, as shown on equation (1).

This assumption has been partly verified since the occurrence of this abstraction has been shown with methylmethacrylate (MMA) (6) and also with other isopropenyl monomers such as α-Mestyrene and isopropenyl acetate (5). The source of the primary radicals was di-tert-butylperoxalate and the spin trap was an isoindoline nitroxide (1,1,3,3 tetramethylisoindolin-2 oxyl):

[Structure: 1,1,3,3-tetramethylisoindolin-2-oxyl, N–O·, with four CH_3 groups] or [ring N–O· abbreviation]

An advantage of this type of radical is that it adds only on hydrocarbon radicals and not to tert-butoxy or benzoyloxy radicals.

With αMe-styrene for example :

$$(CH_3)_3CO\cdot + CH_2{=}C(Ph){-}CH_3 \nearrow (CH_3)_3C{-}O{-}CH_2{-}\dot{C}(Ph){-}CH_3 \rightarrow (CH_3)_3COCH_2C(CH_3)(Ph){-}O{-}N\langle \quad \text{I}$$

$$\searrow (CH_3)_3COH + CH_2{=}C(Ph){-}\dot{C}H_2 \rightarrow CH_2{=}C(Ph){-}CH_2{-}O{-}N\langle \quad \text{II}$$

When solvents were used, there could be competition between solvent and the monomer :

$$(CH_3)_3C{-}O\cdot + RH \rightarrow (CH_3)_3COH + R\cdot \rightarrow R{-}O{-}N\langle \quad \text{III}$$

In the bulk and in solution, the relative ratio of product II to product I has been found similar : 0,17 to 0,20. For MMA in bulk, it was about 0,3 (6).

In solution, the formation of III with α Mestyrene did not occur for example in benzene and CCl_4 but was significant in acetone and acetonitrile. Similar results were obtained in the initiation of the polymerization of isopropenylacetate $CH_2 = C(CH_3)OCOCH_3$ which could suffer hydrogen abstraction from the two CH_3 groups.

With methylmethacrylate a wider range of solvents was used (see table II).

TABLE II
Polymerization of MMA initiated with di-tert-butyl peroxalate (7). Relative proportion of solvent derived products formed in the initiation step and recovered after reaction with a radical trapping agent.

Solvent	Mole fraction of solvent derived products
CCl_4	0
Benzene	0
Ph Cl	0
Acetone	0.33
Ph CH_3	0.79
THF	0.79
Cyclohexane	0.83

The solvent derived products were consisting mainly of the type III adduct or of a type IV resulting from the intermediate addition of MMA on R·, e.g. $(CH_3)_3O\text{-}CH_2\text{-}C(CH_3)(CO_2CH_3)O\text{-}N\langle$

In some solvents, the majority of the $(CH_3)_3C\text{-}O\cdot$ radicals had undergone "transfer" to the solvent before reacting with the monomer. This shows the possibility of controlling end groups in a polymer by the choice of the solvent used in the polymerization.

1.2.2. Initiation of the polymerization of styrene. The same technique using the isoindoline nitroxide has shown the differences in addition of various radicals to styrene (8).
With $(CH_3)_3CO\cdot$, about 99 % of the radical resulting from addition to monomer was recovered as $(CH_3)_3COCH_2CH(Ph)\text{-}ON\langle$

The remaining product resulted from the "normal" addition of a $CH_3^{\cdot}$ radical formed by fragmentation of $(CH_3)_3CO\cdot$. So both its reaction and that of the $CH_3^{\cdot}$ on the CH_2 are extremely selective.

Phenyl radicals (from triphenylmethylazobenzene as initiator) give 99 % of "normal" addition, with 4 % of Ph· trapped directly and 95 % as Ph-M-O-N$\langle$. The remaining 1 % were aromatic substitution products.

With benzoyloxy radicals (from benzoylperoxide as initiator) the reaction in bulk was complicated by some direct reactions involving the nitroxide. The results were simpler in acetone solution and are given in scheme 1. The products were analyzed by HPLC (by comparison with samples prepared independently) and identified by NMR. A majority of the products consisted of the adduct of the nitroxide on the second radical as shown with the formula of scheme 1, but a small quantity of dimer derivative (e.g. $PhCO_2(CH_2\underset{|}{C}H)_2O-N$) was also recovered, (with Ph on the CH), the quantity of which decreased for higher nitroxide concentration, which explains the values 80-85 %, etc. These results show that the "normal" addition occurs only for about 85 % for $PhCO_2^{\bullet}$ and 7 % for Ph· . The main "abnormal" head addition on the substituted carbon of the double bond accounts for about 6 % and that on the benzene nucleus for 3 %.

SCHEME 1
Selectivity of the reaction of radicals with styrene (8).
Initiation with benzoyl peroxide in acetone solution at 60°C.

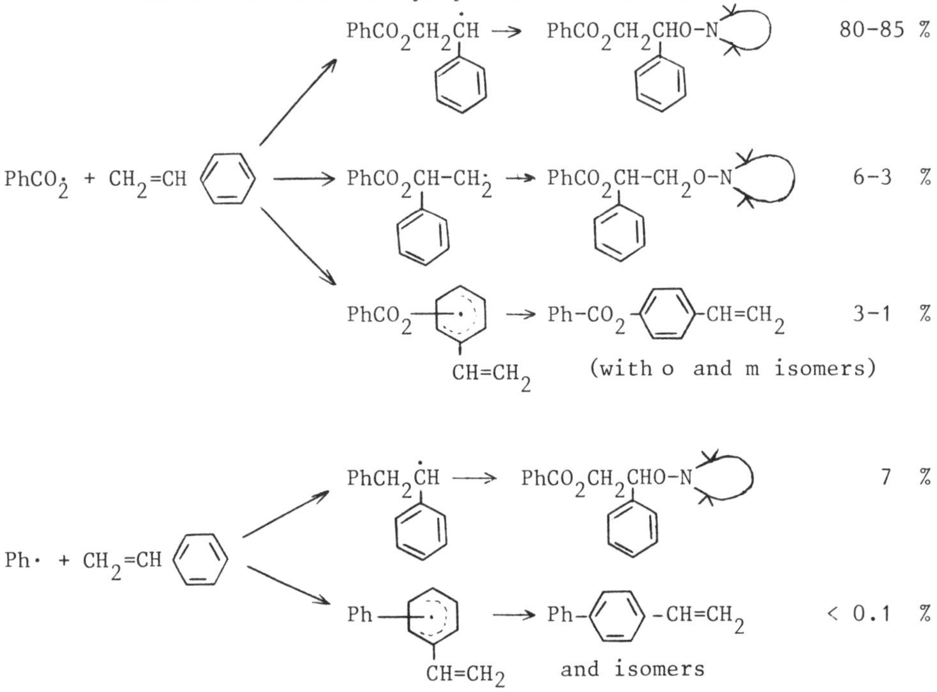

$PhCO_2^{\bullet} + CH_2=CH-C_6H_5$ →

- $PhCO_2CH_2\dot{C}H(C_6H_5)$ → $PhCO_2CH_2CH(C_6H_5)O-N$ 80-85 %
- $PhCO_2CH(C_6H_5)-CH_2^{\bullet}$ → $PhCO_2CH(C_6H_5)-CH_2O-N$ 6-3 %
- $PhCO_2-(C_6H_5^{\bullet})-CH=CH_2$ → $Ph-CO_2-C_6H_4-CH=CH_2$ (with o and m isomers) 3-1 %

$Ph^{\bullet} + CH_2=CH-C_6H_5$ →

- $PhCH_2\dot{C}H(C_6H_5)$ → $PhCO_2CH_2CH(C_6H_5)O-N$ 7 %
- $Ph-(C_6H_5^{\bullet})-CH=CH_2$ → $Ph-C_6H_4-CH=CH_2$ and isomers < 0.1 %

It has been speculated (8) that the relative instability of benzoyl peroxyde initiated polystyrene versus thermally initiated polystyrene might result from the pyrolytic elimination of the secondary benzoate end groups leading to the formation of less stable unsaturated end groups.

2. CATIONIC POLYMERIZATION

The situation has been found generally different for monomers polymerizing by double bond opening or by ring-opening. In the first case, if carbocationic species are involved, they are extremely unstable and with short lifetimes. In ring-opening polymerizations the active species are generally of the onium type, which stability may be much higher. Living polymers could be obtained in specific conditions, which made possible the determination of the various active centres concentrations, and of their reactivities (e.g. for tetrahydrofuran). We shall mainly discuss in the present article the problem of the identification of the carbocationic species, but shall first examine briefly some results about ion-trapping of active species in ring-opening polymerizations.

2.1. Ion-trapping of active species in ring-opening polymerizations

This technique has not been used until now to trap unstable active species but to differentiate various types of oxonium ions. This is sometimes possible by the direct NMR technique but quite often it is not sensitive enough to determine the respective concentrations of the various onium ions present in the reaction mixture. The chemical shifts of phosphorous in phosphonium ions are very sensitive to the chemical environment and a better separation is observed than for oxonium ions. This led Penczek et al. to use phosphines as ion-trapping agents since the basicity of phosphines is higher than that of ethers.

This method has been used to determine the relative proportion of secondary (2 and 3) and tertiary (1) oxonium-ions in the polymerization of cyclic acetals initiated by triflic acid, and particularly of 1,3 dioxolane (9):

$$\sim\!\!\sim OCH_2-\overset{+}{O}\!\!< \; + PR_3 \longrightarrow \; \sim\!\!\sim OCH_2\overset{+}{P}R_3 \quad \text{chemical shift : -31.4}$$

(1)

$$H-\overset{+}{O}\bigcirc \; + PR_3 \longrightarrow O\bigcirc \; + H\overset{+}{P}R_3$$

(2)

$$-CH_2\underset{\substack{|\\H}}{\overset{+}{O}}-CH_2- \; + RP_3 \longrightarrow \; \sim\!\!\sim CH_2OCH_2- \; + H\overset{+}{P}R_3$$

(mainly)

(3)

(2) and (3): Chem.shift : -11.9

The chemical shifts of the phosphine and of derived phosphonium ions (relatively to PO_4H_3) are quite different : + 32.5 ppm for $P(nC_4H_9)_3$, -31.4 for the quaternary phosphonium and -11.9 for the tertiary phosphonium ion. The relative proportion of these two ions was found to change considerably according to the initial $|dioxolane|_0/|CF_3SO_3H|_0$ ratio. The proportion of quaternary ions (derived from the tertiary oxonium one) decreases considerably with a decrease of this ratio, that is for a polymer of low degree of polymerization. This is the result of the formation of more secondary oxonium ions by end-to-end cyclisation involving terminal OH groups and giving species (2).

This technique may also give informations on the relative proportions of active species involved in copolymerization reactions. Triphenyl phosphine was used in the copolymerization of 1,3,5-trioxane with 1,3-dioxolane (initiated by $PhCO^+,SbF_6^-$) and it was possible to differentiate the two phosphoniums $\sim OCH_2OCH_2\overset{\oplus}{P}Ph_3$ (from trioxane) and $\sim OCH_2CH_2OCH_2\overset{\oplus}{P}Ph_3$ (from dioxolane), the latter in much higher concentration(10). In the terpolymerization of oxetane, THF and oxepane with the same initiator, the following ions were detected in decreasing quantities : $O(CH_2)_6\overset{\oplus}{P}Ph_3$, $O(CH_2)_4\overset{\oplus}{P}Ph_3$ and $O(CH_2)_3\overset{\oplus}{P}Ph_3$ (10).

2.2. Carbocationic polymerization

2.2.1. Various types of systems. Cationic polymerization of ethylenic monomers is probably still that with the largest number of unsolved problems. The nature of the active species - not to speak of their concentrations - is unknown in many systems, and particularly for the most interesting ones from the preparative point of view, which are those initiated by Friedel Crafts catalysts and their derivatives. Reactions are generally written as involving carbocations, and this is probably often true, but there has been most often no direct observation of their presence (11).

Carbocations derived from olefins (e.g. from isobutylene) have never been identified and even detected during polymerization. The situation is better for those derived from vinyl aromatics (e.g. from styrene), since these carbocations absorb in the near UV with extinction coefficients higher than 10^4, and UV spectrophotometry is the most sensitive method for their detection. However, it cannot be used when initiation is made by Friedel Crafts type initiators that generally give complexes with the monomers absorbing in the same wavelength range. This type of initiation could be studied quantitatively only with model monomers such as 1,1-diphenylethylene, the carbocation of which is absorbing around 430 nm. This is why the experiments with styrene and related monomers have been made using strong protonic acids such as ClO_4H and CF_3SO_3H which are transparent in the UV. But as we shall see, carbocations were observed only in specific conditions, and in usual polymerization conditions giving high polymer (high $[M]_0$ / [acid] ratio) no cationic species had been seen.

The methods used to determine the concentration of active species in cationic polymerizations have been different, according to the kinetics of the reaction, that may be of two broad different types, according as active centres concentration P* is stable or variable with conversion.

1) Systems with a stationary concentration of active centres. This is the case when a linear first order plot for monomer consumption is observed, showing that k_p [P*] is constant during the reaction. But such a situation may result either from the presence of a relatively large concentration of stable active centres (as in anionic polymeriza-

tions) with a low rate constant of propagation,or from a very small concentration of P* with a large rate constant. The two opposite views are still held for many systems and the proof in this matter is not easy to get directly, attempts at the detection of the active species having failed.

The only type of initiation for which the situation is apparently simple is that of the radiation-induced polymerizations, from which $P^{\oplus}$ concentrations of free ions in a quasi-stationary state could be evaluated indirectly from conductivity or inhibition studies, and were found of the order of 10^{-10} M.

For polymerizations initiated by $SnCl_4$ (with a cocatalyst) and with $TiCl_4$, the two above possibilities could not be distinguished. The situation is the same for polymerizations initiated by CF_3SO_3H and ClO_4H near room temperature (and with high $[M]_o / [acid]_o$ ratio) for which no UV absorption was observed during the polymerization. This had led to the theory of pseudocationic polymerization (12) according to which propagation occurs on a covalent ester species, but the view that carbocations in very small and undetected concentration are responsible for the propagation has also been held. It was concluded from recent investigations that the second possibility is the most likely for CF_3SO_3H initiated polymerizations (13), but the situation may be more complex with ClO_4H (13,14).

2) Systems with a variable concentration of active centres

The study of the non-stationary kinetics of fast polymerization reactions has given until now most of the information on initiation, propagation and termination rate constants in carbocationic polymerizations, and some examples will be given now.

2.2.2. Investigation of "medium fast" reactions. A simultaneous measurement of the polymerization rate and of the active species concentration permit to calculate the rate constants of initiation k_i (when it is not too high) of propagation k_p and of termination k_t. The high vacuum apparatus used has been described (15). This method was applied to the polymerizations of vinyl ethers (16) and of p-methoxystyrene (17) in real polymerization conditions. The polymerization rate was measured by calorimetry and the initiation rate was measured by following the initiator consumption ($Ph_3C^+, SbCl_6^-$) by UV spectrophotometry. Most of the initiator was generally consumed during the reaction, and the active species had a half life of the order of 10 sec at 0° for the ethers A simple kinetic treatment gave k_i, k_p and k_t between 20°C and -40°C (16,17).

However, this method did not permit to see the various types of species active for the polymerization of the vinyl ethers, while the large variations in rate constants, $\Delta H^{\neq}$ and $\Delta S^{\neq}$ according to the size of the substituent of the monomer and to the solvent showed a large variation in their respective concentrations (18). But they will not be easy to measure directly since the ionic species involved don't absorb in UV.

The situation is better with p-methoxystyrene, for which an absorption corresponding to carbocationic species may be observed at 380 nm when the initiation is made using triflic acid (30,32). This wavelength however doesn't seem to correspond to that expected from the "normal" $\sim CH_2-\overset{+}{C}H$ carbocation, and reactions made using p-methoxy α-methylstyrene as a model monomer (21) have given at -63° a similar absorption that NMR permitted to assign to the following stable species :

$$CH_3-\underset{C_6H_4OCH_3}{\overset{CH_3}{C}}-CH_2-\underset{C_6H_4OCH_3}{\overset{\oplus}{C}}-CH_2-\underset{C_6H_4OCH_3}{\overset{CH_3}{C}}-CH_3$$

This might mean that the species observed by UV with p-MeOstyrene correspond to an "isomerized" form in equilibrium with the real active species :

$$\sim CH_2-\underset{C_6H_4OCH_3}{CH}-CH_2-\underset{C_6H_4OCH_3}{\overset{+}{CH}} \rightleftharpoons \sim CH_2-\underset{C_6H_4OCH_3}{\overset{+}{C}}-CH_2-CH_2(C_6H_4OCH_3)$$

Other unpublished data obtained with p substituted αMe-styrenes (22) (with CH_3, $C(CH_3)_3$ and OCH_3 substituents) have shown that the "normal" species resulting from the addition of a carbocation to monomer had their main UV maxima in the 345-368 range. The stable monomeric cation derived from $CH_2 = C(CH_3)-C_6H_4-C(CH_3)_3$ could be prepared and studied at -73°C and was shown to have a single maximum at 350 nm while the carbocation observed with polymeric species has 2 maxima at 363 and 490 nm. The monomeric cation and the mixture of monomeric and dimeric cations of αMe p-methoxystyrene have also only one maximum at 368 nm whereas the higher oligomeric cations have 2 maxima at 382 and 478nm (21). The situation is similar for the p-methyl and the p-isopropyl α-Me styrenes. All this seems to show that the maxima observed during polymerization are relative either to "isomerized" species or to normal species having strong interactions with electron donors.

It is obvious that if isomerization reactions may occur rapidly in such systems, informations about these processes can only be obtained by using methods of investigation permitting to follow the concentration and if possible the change in nature of the active species on a much shorter time-scale. The technique available is of the stopped-flow type and has been used widely in kinetic studies (23), but we shall see in the next section that its use in cationic polymerization is difficult. Even if interesting results have been obtained, there are still many unsolved problems.

2.2.3. Investigations of very fast reactions involving carbocationic species. Informations about the UV spectra and the reactivities of carbocationic species have been obtained by L.M. Dorfman et al (19) who used (at room temperature) submicrosecond pulses of high energy electrons to generate the formation of carbocations in chlorocarbon solvents from selected solute molecules such as dibenzylmercury (to form benzyl-cations) benzhydrilbromide (to form benzylhydril cations), trityl chloride (to form trityl cations) etc. These carbocations are formed by reaction with the solute of the cation radicals derived from the solvent :

$$S^{\dot{+}} + (Ph\ CH_2)_2Hg \xrightarrow{k_i} S + PhCH_2^{\oplus} + PhCH_2Hg\cdot$$

These cations eventually recombine with Cl^- ions formed simultaneously but the rate of recombination even if it is rapid, is of the same order of magnitude as that of the above reaction (k_i from 4.10^8 to $1.6.10^{10} M^{-1} sec^{-1}$ in dichloroethane) and a "pulse" of carbocations may be observed, as may be also its decay in the presence of variable quantities of nucleophilic reagents. A wavelength of 364 nm was thus assigned to the free benzyl cation, of 445 nm to the benzhydril cation, etc. and rate constants were obtained for their reactions with amines,ethers, water, etc..that were of the order of 10^6 to 10^{10} M^{-1} sec^{-1}. There has been also some data for the rate constants of the reaction of the benzyl cations with alkenes (20) (e.g. $1.9.10^7$ for isobutylene, 9.10^5 for butadiene, in CH_2ClCH_2Cl at 24°C). The benzhydryl cation was only a little less reactive.

The pioneering work in the investigation of carbocationic polymerization by a stopped-flow technique has been that of Pepper et al. In 1973 (24,25) they used an apparatus with detection at one wavelength, stored with an oscilloscope with memory, successive experiments being made at the various wavelengths, in order to obtain the spectrum of the cationic species. They studied the polymerization of styrene initiated by perchloric acid at -80°C (with $[M]_o \leqslant 0.2$ M). A more elaborate apparatus with a "dead-time" of 20ms was used later (26) in which the reagents were cooled before reaching the mixing chamber. In these conditions, carbocationic species absorbing at 340 nm and assigned to the styryl carbocation appeared in 0.5 second and disappeared in 5 to 10 seconds. But the maximum yield in carbocations based on the acid concentration was only of about 1 %. After the very fast and limited polymerization corresponding to the apparition of the carbocations, a much slower polymerization took place. By assuming an extinction coefficient of 10^4 for the polystyryl cation, the rate constant for the fast reaction was found to be $k_p \sim 5.10^3$ ($l.mole^{-1}sec^{-1}$) at -80°C.

Another series of experiments was realized with styrene by Kunitake et al (27) who used trifluoromethanesulfonic (triflic) and sulfuric acids as initiators. A more sophisticated stopped-flow apparatus was used (28), of the rapid-scan type with image dissector, with a rapid scanning in wavelength on the photomultiplier cathode (150 nm in 1 millisec.). The dead times varied from 3 to 24 ms according to the rate of injection by syringes. The data were obtained only between 30°C and -1°C. Here again, carbocationic species could be observed at 340 nm, with styrene in methylene chloride solutions initiated by triflic acid. At 30°, their concentration passed through a maximum in 50 ms (about 1% yield) and decreased near zero in 200 ms. By lowering the temperature (to -1°C) the yield increased up to about 5-10 %. Initiation rate constants were derived from the initial rate of formation of the carbocationic species P^+ and were of the order of 10 $M^{-1}s^{-1}$. The k_p values varied from 5.10^4 at -1°C to 2.10^5 at 30°C. The lifetimes $\tau = [P^+]_{max}/R_i$ of the carbocations varied from 0.1 sec (at -1°C) to 5.10^{-3} sec (at 30°C).

Spectroscopic studies of p-methoxystyrene (PMOS) polymerization have been made by Sawamoto and Higashimura (29,30) using a stopped-flow apparatus with a photomultiplier detector, an image dissector and accumulation, the latest data (30) being obtained with a multichannel detector (with 256 photodiodes) permitting the successive acquisition of 10 spectra with only 10 ms between each spectrum. Monomer and initiator solutions were prepared under dry nitrogen and were driven by pressurized nitrogen into a thermoregulated mixer and an observation cell. The water concentration in the mixture was estimated lower than 10^{-3} M at the outlet of the flow system.

Comparative data in $(CH_2Cl)_2$ at 30°C were obtained with various initiators (29), but in all cases a maximum in the near UV was observed at 380 nm and assigned to the carbocation P^+, the intensity of which passed through a maximum and then decreased. With BF_3,Et_2O as initiator, and assuming an ε_{380} of $2.8.10^4$ $M^{-1}cm^{-1}$, yields of carbocationic species up to 5 % were observed. They were much lower for iodine and CF_3SO_3H, but higher wich $SnCl_4$ (about 25 %). The lifetimes of these species were of the order of 0.2 sec (CH_3SO_3H) to 5 sec (I_2). The most detailed investigations (30) have been made in $(CH_2Cl)_2$ solutions using again triflic acid (and also acetyl perchlorate) that led to the highest initiation and propagation rate constants. A complete discussion of these results is outside the scope of this review and has been partly made elsewhere (11) but the main results were the following :

The efficiency of initiation (assuming the same ε_{380}) was much higher than with other initiators and was near 40 % for CF_3SO_3H and near 100 % for CH_3COClO_4. The absorption peak at 380 nm reached a maximum within 20-40 ms and then decayed gradually, the monomer being consumed in about 1 second. A considerable decrease in the maximum concentration of P^+ was observed with a lowering of the dielectric constant of CH_2Cl_2/CCl_4 solvent mixtures, while the rate did not decrease significantly. This was explained by the presence in solvents of low polarity of active but "invisible species", but would mean that their reactivity is higher than that of the visible ones.

About the visible species, one may note discrepancies in wavelengths between this 380 nm maximum and those observed earlier by the same authors (31) for the reaction products of the corresponding carbinol in sulfuric acid (348 nm, ε_{348} : 28 300) and with the monomer and BF_3,OEt_2 (in $(CH_2Cl)_2$ λ = 365 nm). The 348 nm value might correspond to that of the monomeric cation (compare with the recent data with the corresponding αMe derivative (21)), but the position at 380 nm is not easily explained. We also observed (after 1 sec, see above) the same species (at 380 nm) at the same temperature and below but found that in high vacuum conditions their concentrations passed through a maximum and then remained quite stable for long times (11,32). At lower temperatures, there was a simultaneous appearance of a strong shoulder around 340 nm. It is still not clear whether the 380 nm absorption is that of an active carbocation the wavelength of which has been modified by solvation (eventually involving penultimate units) or that of an isomerized species in equilibrium with the really active species. The differences in stability according to experimental conditions are probably linked with the difficulty to operate in high purity conditions with a stopped-flow system.

We have tried to overcome these difficulties and to build an apparatus that might work at the lowest possible temperature (to increase the cations stability), in the absence of impurities in the reagent and without "pushing" gas, and that would permit seeing carbocationic species of very short lifetimes.

A first model has been used by J.P. Vairon et al (33) to study p-isopropyl α-Mestyrene polymerization in the -60 to -20°C range. The carbocationic species appeared at 362 nm (with a second smaller maximum at 490 nm) and were unstable at -20°C but stable at -40°C and below. The yields (uncertain by a factor of two) were very high (50 to 100 %), and the initiation was extremely rapid and complete in less than 20 milliseconds, before an important propagation took place. However, quantitative kinetic results could not be obtained owing to various technical problems, and a new and more efficient type of stopped-flow apparatus has been developed (by J.P. Vairon in collaboration with A. Persoons and M. Van Beylen, University of Leuven) that shall be described now.

The one block pyrex-quartz system is of the piston-driven type with a rear-stopping and triggering syringe. The pistons are moved either by hand pushing or with a pneumatic device. This system is equiped with Teflon O-rings and allows the filling under high vacuum and handling of solutions from sealed phials. The reference cells, the four-jets mixer and the observation cell are in quartz and allow short dead times (3-5 ms). They are immersed in a thermoregulated bath together with helical coils preceding the solution, allowing to work at low temperatures.

Plurichromatic light is supplied to the reaction cell and then to the monochromator by optical fibers, the detection system being an array of 1K silicon photodiodes mounted on a 3-gratings monochromator. Multichannel data acquisition and control of the experiment are obtained from a P.A.R. optical analyzer, and this permits instantaneous sampling of a 50-600 nm spectrum which may be repeated more than a hundred times with intervals as low as 2 milliseconds. The overall monitoring of the experiment and data processing is obtained from a microcomputer and its peripherics.

It is hoped that the use of this apparatus will permit to give answers to some of the many unsolved problems of carbocationic polymerizations and that it might also be useful for the investigation of fast reactions in anionic polymerizations.

1) S.E. Bresler, E.N. Kozbekov, V.N. Fornichev, V.N. Shadrin, Makromol. Chem. 157, 167 (1972), ibid. 175, 2875 (1974)
2) M. Kamachi, M. Kohno, Y. Kuwae, S. Nozakura. Polymer J., 14, 749 (1982)
3) M. Kamachi, Y. Kuwae, M. Kohno, S. Nozakura. Polymer J., 15, 541 (1985)
4) T. Sato, T. Otsu, Makromol. Chem, 178, 1941 (1977)
5) R.D. Grant, E. Rizzardo, D.H. Solomon. Makromol.Chem. 185, 1809 (1984)
6) R.G. Griffiths, E. Rizzardo, D.H. Solomon. J. Macromol.Sci.-Chem-A17, 45 (1982)
7) R.D. Grant, P.G. Griffiths, G. Moad, E. Rizzardo, D.H. Solomon Aust. J. Chem. 36, 2447 (1983)
8) G. Moad, E. Rizzardo, D.H. Solomon. Macromolecules, 15, 909 (1982)

9) P. Kubisa, S. Penczek, Makromol. Chem., 180, 1821 (1979)
10) K. Brzezinska, W. Chwialkowska, P. Kubisa, K. Matyjaszewski, S. Penczek, Makromol. Chem. 178, 2491 (1977)
11) P. Sigwalt, Polymer J. 17, 57 (1985)
12) A. Gandini, P.H. Plesch, Proceed. Roy. Soc. 246 (1964)
13) K. Matyjaszewski, P. Sigwalt, Makromol. Chem, 187, 2299 (1986)
14) K. Matyjaszewski, Polymer Preprints 27 (2),112 (1986)
15) J.C. Favier, P. Sigwalt, M. Fontanille, Europ. Pol.J. 10, 717 (1974)
16) F. Subira, G. Sauvet, J.P. Vairon, P. Sigwalt, J. Polym. Sci. Symp. n° 56, 221 (1976)
17) R. Cotrel, G. Sauvet, J.P. Vairon, P. Sigwalt, Macromolecules, 9, 931 (1976)
18) F. Subira, J.P. Vairon, P. Sigwalt, IUPAC Macromol, Symp, (Amherst 1982) ; Proceedings p. 151
19) L.M. Dorfman, V.M. de Palma, Pure and Appl. Chem., 51, 123 (1979)
20) Y. Wang, L.M. Dorfman, Macromolecules, 13, 63 (1980)
21) M. Moreau, K. Matyjaszewski, P. Sigwalt, submitted to Macromolecules
22) K. Matyjaszewski, P. Sigwalt, to be published.
23) K. Hiromi, Kinetics of fast enzyme reactions, chapter 2, Kodansha (Tokyo 1979)
24) J.P. Lorimer, D.C. Pepper, International Symposium on cationic polymerization, Rouen 1973, comm. C 23
25) D.C. Pepper, Makromol. Chem., 175, 1077 (1974)
26) J.P. Lorimer, D.C. Pepper, Proc. Roy. Soc. A 351, 551 (1976)
27) T. Kunitake, K. Takarabe, J. Polym. Sci. Symp n° 56, 33 (1976)
28) T. Kunitake, K. Takarabe, Macromolecules, 12, 1061 (1979) ; ibid., 12, 1067 (1979)
29) M. Sawamoto, T. Higashimura, Macromolecules, 11, 328 (1978)
30) M. Sawamoto, T. Higashimura, Macromolecules, 12, 581 (1979)
31) T. Higashimura, N. Kanoh, S. Okamura, J. Macromol. Sci., Chem. 1, 109 (1966)
32) M. Moreau, thèse 3è cycle, Paris 1981
33) D. Teyssié, M. Villesange, J.P. Vairon, Polymer. Bull. 11, 459 (1984)

LIVING CATIONIC OLEFIN POLYMERIZATION

J. P. Kennedy, R. Faust, M. K. Mishra and A. Fehervari
Institute of Polymer Science
The University of Akron
Akron, Ohio 44325

ABSTRACT. Recently we have discovered that various tertiary ester and ether boron trichloride complexes induce truly living polymerization of isobutylene, 1,3,5-trimethylstyrene and copolymerization of isobutylene/isoprene in CH_3Cl and CH_2Cl_2 diluents over a wide range of temperatures (i.e., from 0° to -70°C). The $\bar{M}_n$s are determined by $[M_o]/[I_o]$, conversions are ∿100%, and the initiation efficienty is ≃100%. The living nature of the processes is demonstrated by linear $\bar{M}_n$ versus W_p (g polymer formed) plots starting at the origin and horizontal N (number of polymer moles) versus W_p plots. These living polymerizations give rise to tert.-chloro terminated products. The mechanism of termination has been studied. The use of difunctional esters and ethers produces α,ω-difunctional (telechelic) materials. Very recently three-arm star telechelic prepolymers have also been synthesized by a trifunctional ether· BCl_3 complex. The polymerization rates are rapid, and the experiments can be carried out under conventional laboratory conditions, i.e., by the use of stirred vessels under a blanket of nitrogen, in the presence of traces of moisture. Ease of operations has been demonstrated by the preparation of tert.-chloro telechelic liquid polyisobutylenes on the semicommercial scale (12ℓ reactor). It appears that the stability of living cationic centers is due to undissociated species. The nature of these centers will be discussed.

1. INTRODUCTION

Systematic research on "living" polymerizations has started in 1956 with Szwarc's epochal discovery of living styrene polymerization induced by the sodium/naphthalene system (1,2). Since that time the living polymerization of styrene, conjugated dienes and their derivatives has become the cornerstone of several industrial processes for the production of various commodities, e.g., glassy/rubbery/glassy triblock copolymers of styrene/butadiene/styrene.

The rigorous definition of a truly living polymerization stipulates a chain-transferless terminationless polymerization i.e., a polymerization that involves only initiation and propagation, and in which chain

M. Fontanille and A. Guyot (eds.), Recent Advances in Mechanistic and Synthetic Aspects of Polymerization, 133–143.

transfer and termination are absent, i.e., the rate of chain transfer and termination are zero, $R_{tr} = 0$ and $R_t = 0$. In various living polymerization systems the rate of initiation is much faster than propagation, $R_i >> R_p$, which leads to narrow molecular weight dispersities (i.e., $\bar{M}_w/\bar{M}_n$s are close to unity). However, $R_i >> R_p$ is not a mandatory requirement for truly living polymerizations. Indeed, for a variety of reasons (e.g., ease of processing) very narrow molecular weight distributions may be undesirable.

Technological and scientific interest in living polymerizations is justified because these processes lead to predictable molecular weights, tailored end groups, and well-defined block copolymers.

Despite sustained fundamental research in the field of cationic polymerizations truly living cationic polymerizations could not be obtained until the '80s except with a few ring-opening polymerization systems, e.g., tetrahydrofuran, t-butyl aziridine (3,4).

Quasiliving carbocationic polymerizations, i.e., systems in which termination was reversible and chain transfer to monomer was strongly suppressed by very slow monomer addition to active charges, approached but did not completely attain synthetically exploitable living conditions (5).

Recently the first truly living vinyl cationic polymerization of alkyl vinyl ethers has been described by Japanese investigators (6). Thus Higashimura, Sawamoto and coworkers have shown that many alkyl vinyl ethers can be polymerized by the HI/I_2 initiating system and the living nature of these polymerizations was demonstrated by linear $\bar{M}_n$ versus conversion plots starting at the origin. By skillful choice of the alkyl moiety block copolymers including amphiphilic products have been prepared (7).

In the course of fundamental investigations in the field of isobutylene polymerizations we have recently discovered that tert. esters and ethers in combination with Lewis acids, in particular with BCl_3, are efficient initiating systems for the living polymerization of this monomer (8-10). Following this breakthrough of the first living carbocationic polymerization of an aliphatic olefin, research has been continued for the exploration of the scope and significance of this discovery. This paper concerns highlights of these investigations and an examination of conditions and parameters whose simultaneous presence seem to be necessary for truly living carbocationic polymerizations.

2. EXPERIMENTAL

Details of materials, manipulations and polymerization techniques have been published (10-12). Two key techniques have been used: 1) Conventional (all monomer in AMI) technique in which a quantity of initiating system (t-ester·BCl_3) or coinitiator (BCl_3) is added to monomer-solvent or monomer-t·ester-solvent mixtures respectively, or 2) the Incremental Monomer Addition (IMA) technique in which the initiating system or coinitiator is added to monomer-solvent or monomer-initiator-solvent mixtures, respectively, and after an arbitrary waiting period, usually 30 minutes, fresh monomer is added to the system;

incremental monomer addition can then be repeated for as many times as desired (12). The IMA technique was developed to follow the $\overline{M}_n$ versus g polymer formed profile of rapid polymerizations, that is for experiments where the withdrawing of samples at representative low or intermediate conversions would have been difficult if not impossible.

3. RESULTS AND DISCUSSION

3.1. System Selection

Most research described in this presentation focuses on isobutylene IB. This monomer was selected because 1) IB is an ideal monomer for cationic polymerization research as it rapidly polymerizes with all kinds of acids and the structure of polyisobutylene PIB is free from complications due to isomerization and branching; 2) the polymerization of IB is diagnostic for cationic processes and the polymerization of this monomer by other mechanisms has not been described; 3) the cost of this monomer is low, indeed IB is one of the cheapest olefins available; 4) its polymer, PIB, leads to a desirable combination of physical-mechanical-solution and chemical properties, and PIBs are valuable commercially available commodities from high mol. wt. rubbers to low mol. wt. liquids.

A limited amount of IB copolymerization research has also been carried out with isoprene. Finally mention will be made of the living polymerization of 2,4,6-trimethylstyrene, a monomer that yields a rather high Tg (162°C) polymer.

3.2 Basic Observations, The Concept and Monofunctional Products

The key finding on the road to truly living carbocation polymerizations was the observation that certain organic esters in the presence of Lewis acids give rise to complexes that rapidly and efficiently induce the polymerization of simple olefins at relatively high temperatures (8). Initial experiments with isobutylene and cumyl acetate/ BCl_3 complexes in CH_2Cl_2 and similar polar solvents at -10° and -30°C appeared promising, which prompted a more detailed examination of the kinetics (12). These investigations rapidly yielded solid diagnostic proof for the truly living nature of this and a series of similar systems. Thus by the IMA and AMI techniques we have shown that plots of $\overline{M}_n$ versus W_{PIB} (g PIB formed) were linear starting from the origin and that the $\overline{M}_w/\overline{M}_n$ of PIBs became narrower (from ∿1.9 to 1.1) with increasing $\overline{M}_n$. Also, corresponding plots of N(the number of moles of polymer formed i.e., g PIB/$\overline{M}_n$) versus W_{PIB} gave horizontal lines starting at I_o (the cumyl acetate initiator concentration). Figure 1 shows a representative plot. Under suitable reaction conditions rapid living polymerizations can be carried out at relatively high temperatures to yield predictable molecular weights ($\overline{DP}_n$ = [monomer]/[initiator]) at complete conversions and ∿100% initiator efficiencies. This combination of self-reinforcing facts can be explained only with a truly living polymerization system.

Figure 1

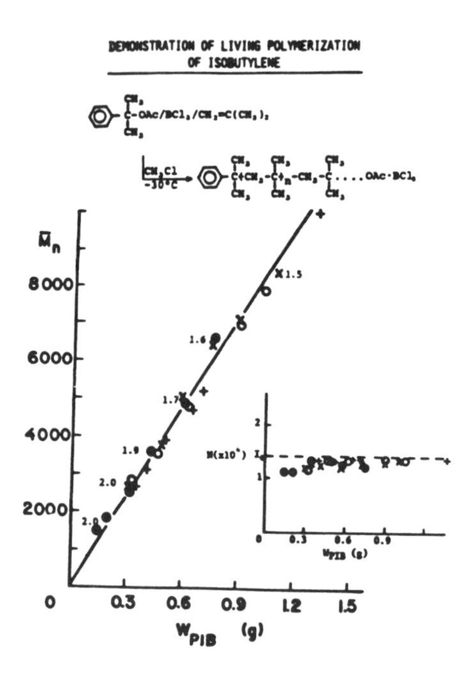

The initiation and propagation steps may be visualized by the insertion of isobutylene into the O-R bond of the ester·BCl_3 complex. The direction of electron flow weakens the O-R bond (indicated by the broken bond O--R in the equation below) and thus facilitates monomer insertion. It is postulated that free ions do not arise during monomer insertion, however, the degree or extent of the "ionicity" of the transition states is obscure. The insertion of the first monomer unit gives rise to a new tert.-ester·BCl_3 complex whose structure is very similar to that of the initiating complex. Indeed if instead of cumyl acetate 2-acetate-2,4,4-trimethylpentane is used, the structure of the newly formed tert.-ester after initiation and that of the latter initiator are for all practical purposes identical (12). Propagation is the repetition of the initiation step which may in principle proceed as long as monomer is available. In this sense this process may be regarded a cationic two component group transfer polymerization. The following equation helps to visualize the proposition:

$$R'\text{-}\overset{\overset{\displaystyle O}{\|}}{C}\text{-}O\text{-}R \xrightarrow{BCl_3} R'\overset{\overset{\overset{\displaystyle BCl_3}{\uparrow}}{\overset{\displaystyle O}{\|}}}{C}\text{-}O\text{--}R \xrightarrow{+nM} R'\overset{\overset{\overset{\displaystyle BCl_3}{\uparrow}}{\overset{\displaystyle O}{\|}}}{C}\text{-}O\text{--}MMMM\text{-}R$$

where R = $C_6H_5C(CH_3)_2$-, $C(CH_3)_3CH_2C(CH_3)_2$-, etc., and R' = CH_3, etc. The process is essentially similar with tert.-ether·BCl_3 complexes (13).

These living polymerizations can be carried out at temperatures remarkably high for isobutylene polymerizations i.e., in the range from 0° to -30°. Polymerizations at -10°C in CH_2Cl_2 are now routinely carried out in open systems in our laboratories while for those at 0°C we are using sealed reactors and thus operate under mild pressures (the boiling point of IB is -6.9°C) (14). Evidently the maximum temperature that can be used so as to attain living polymerizations is determined by the thermal decomposition point of the particular initiating complex employed.

The polarity of the medium strongly affects the rate and outcome of living polymerizations. Best overall results have been obtained by the use of the polar solvents CH_3Cl and CH_2Cl_2, while C_2H_5Cl gave less satisfactory systems in terms of rates, predictability of $\bar{M}_n$s, and product uniformity. However, PIB in excess of $\bar{M}_n \sim 6{,}000$ becomes insoluble in these solvents. CH_3Cl is a somewhat better solvent than CH_2Cl_2 for PIB. Since PIB is soluble in hydrocarbons, attempts have been made by the use of CH_3Cl/n-hexane mixtures to prepare higher molecular weight products. Alas, the rate of living polymerizations decreases in such mixtures; indeed in pure hydrocarbons the rates are unpracticably low. The effect of medium polarity on the rate is characteristic of ionic polymerizations, however the (ionic) nature of the intermediates remains unknown.

3.3 The End Group

An important facet of this research was the definition of end groups obtained after quenching a living polymerization. Model studies and quantitative spectroscopic analysis in combination with other processes showed that the end groups of PIBs prepared by tert.-ester or ether·BCl_3 complexes and quenched with CH_3OH were invariably and quantitatively $-CH_2-C(CH_3)_2Cl$ (12,13) e.g.,

$$\sim PIB \sim CH_2-\underset{CH_3}{\overset{CH_3}{\underset{|}{\overset{|}{C}}}}-CH_2-\underset{CH_3}{\overset{CH_3}{\underset{|}{\overset{|}{C}}}}-OAc\cdot BCl_3 \xrightarrow{CH_3OH} \sim PIB \sim CH_2-\underset{CH_3}{\overset{CH_3}{\underset{|}{\overset{|}{C}}}}-CH_2-\underset{CH_3}{\overset{CH_3}{\underset{|}{\overset{|}{C}}}}-Cl$$

Certainty in regard to the structure of the end groups was of paramount importance for the synthesis of telechelic PIB derivatives.

3.4 Telechelic Products

After having developed reliable methods for living polymerizations by monofunctional initiators, e.g., cumyl acetate, cumylmethyl ether (12,13), that led to α-phenyl-ω-tert.-chloride PIBs (asymmetric telechelics), research was turned toward the preparation of symmetric telechelics by the use of bi- and trifunctional initiators. These investigations culminated in the synthesis of linear and three-arm star tert.-chloride capped PIBs by BCl_3 complexes of, for example (11,14),

$$CH_3-\overset{O}{\overset{\|}{C}}-O-\underset{CH_3}{\overset{CH_3}{C}}-CH_2-\underset{CH_3}{\overset{CH_3}{C}}-CH_2-\underset{CH_3}{\overset{CH_3}{C}}-O-\overset{O}{\overset{\|}{C}}-CH_3$$

and

$$CH_3O-\underset{CH_3}{\overset{CH_3}{C}}-C_6H_3(-\underset{CH_3}{\overset{CH_3}{C}}-OCH_3)-\underset{OCH_3}{C}(CH_3)_2$$

According to spectroscopic analysis the structures of the tert.-chloride capped linear and three-armed star polymers are identical to those obtainable by binifers and trinifers, respectively (15).

Kinetic investigations indicated that the polymerizations giving rise to these telechelics were living in nature. Figure 2 shows a representative $\bar{M}_n$ versus W_{PIB} plot obtained with the system that yielded the linear tert.-chloride capped product:

Cl∿∿∿∿∿PIB∿∿∿∿∿Cl

Figure 2

PREPARATION OF LINEAR TERT.-CHLORO TELECHELIC PIB BY LIVING POLYMERIZATION

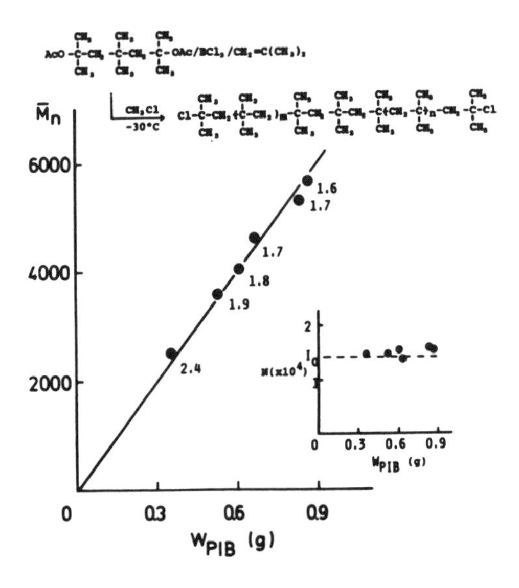

Independent research demonstrated that liquid linear telechelic PIBs can be conveniently prepared by a continuous process in a loop reactor (11).

3.5 Ring Expansion Polymerization of Isobutylene

Having ascertained the feasibility of living polymerizations with linear initiators, research was extended to explore the possibility of macrocyclic polymer synthesis induced by cyclic initiators. The first phase of this work concerned cyclic ester (lactone)$\cdot BCl_3$ complexes. According to kinetic-diagnostic experiments (i.e., $\bar{M}_n$ and $\bar{N}$ versus g PIB formed, trend of $\bar{M}_w/\bar{M}_n$ with conversions) isobutylene polymerization induced by certain lactone$\cdot BCl_3$ systems are living in nature (16). In these living polymerizations isobutylene inserts stepwise into the R--O bond of the activated lactore and this leads to a stepwise expanding macrocyclic species. Schematically:

$$C_6H_5\text{-}\underset{|}{\overset{CH_2 - CH_2}{CH}}\;\; \overset{|}{C}{=}O \;(\text{ring via } O) + BCl_3 + IB \longrightarrow C_6H_5\text{-}CH(CH_2)\text{-}CH_2\text{-}C(=O)\text{-}O\sim PIB\sim$$

We wish to term such polymerizations Ring Expansion Polymerizations REPs.

The presence of macrocyclics obtained by quenching with pyridine which for some reason does not yield tert.-chloride end groups, has been demonstrated by spectroscopy and GPC measurements. Thus investigations of a macrocyclic PIB and a linear PIB obtained from the former by hydrolysis have shown that the hydrodynamic volume of the cyclic species is smaller by a factor of about 0.6 than that of the linear product.

3.6 Copolymerization of Isobutylene with Isoprene

For a variety of reasons, it was of interest to examine the feasibility of copolymerization of IB with isoprene IP by the use of our recently discovered living initiating systems. Random IB-IP copolymers containing 1-4 mole % IP are commercial commodities available as high molecular weight butyl rubbers or low molecular weight liquids for sealant, potting, adhesive applications. The high molecular weight rubbers are made at -100°C and the liquid products are prepared from these by sheer degradation.

Experiments with the cumyl acetate$\cdot BCl_3$ complex and IB-IP charges containing from 2-12 mole % IP showed that living copolymerization readily occurs at -30°C and that statistical copolymers with up to $\bar{M}_n$ $\sim 10^4$ can be prepared. The copolymerizations were much slower than homopolymerizations so that sampling as a function of conversion was feasible. Figure 3 shows the data (17).

Figure 3

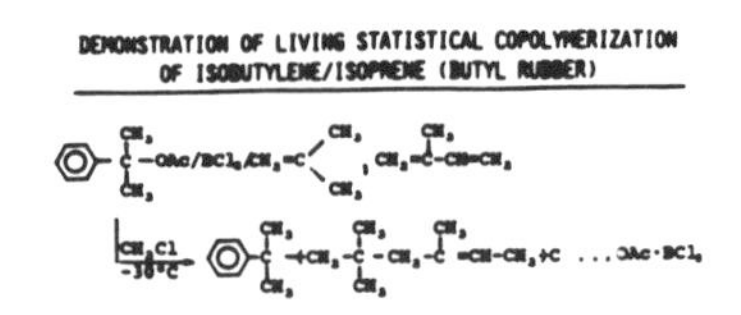

DEMONSTRATION OF LIVING STATISTICAL COPOLYMERIZATION
OF ISOBUTYLENE/ISOPRENE (BUTYL RUBBER)

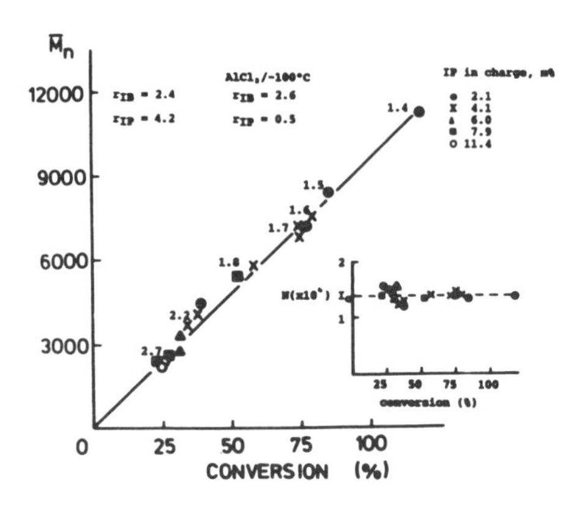

Separate experiments have been carried out to determine reactivity ratios. Interestingly, r_{IB} = 2.4 was quite similar to r_{IB} values obtained in conventional systems, while r_{IP} = 4.2 was much larger than customarily observed. Evidently the mechanism of living copolymerization is different from that operating in conventional systems. Although $r_{IB} \cdot r_{IP}$ was larger than unity, the copolymers were statistical and not blocky because of the small amount of IP in the charge (17).

3.7 Poly(2,4,6-Trimethylstyrene)

Living carbocationic polymerization has also been demonstrated to proceed with 2,4,6-trimethylstyrene by the use of the cumyl acetate/BCl_3 complex in CH_3Cl at -30°C. This monomer has been selected because the ortho CH_3-ring substituents prevent intramolecular self-alkylation (i.e., indane-skeleton formation), a disturbing side reaction that occurs with great ease with styrene derivatives in general, and particularly with α-methylstyrene. The high Tg (162°C) of the polymer was also an incentive.

Figure 4 shows the $\bar{M}_n$ and N versus g-polymer formed plots. The facts that the former is linear and starts at the origin, and that the latter is horizontal and starts at I_o, together with the narrow dispersities shown in the figure ($\bar{M}_w/\bar{M}_n$ = 1.3-1.17) indicate a living system. Poly(2,4,6-trimethylstyrene) is soluble in CH_3Cl at -30° so that this polymerization remained homogeneous up till the high viscosity of the system rendered further polymerization cumbersome (at $\bar{M}_n$ ∿30,000).

Figure 4

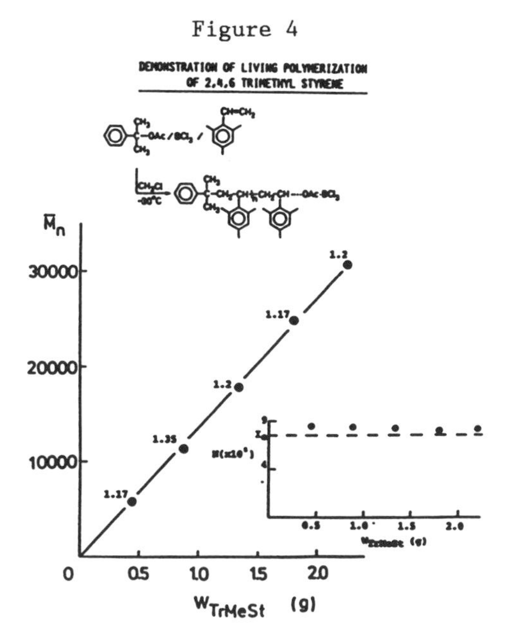

4. MECHANISTIC SPECULATIONS. REQUIREMENTS FOR LIVING CATIONIC POLYMERIZATION

A considerable body of evidence indicates for the first time the truly living polymerization of IB and 2,4,6-trimethylstyrene by certain tert.-ester and -ether·BCl_3 complexes in polar diluents in the 0° to -70°C range. Molecular weights obey $\overline{DP}_n = [M_o]/[I_o]$ and $\overline{M}_w/\overline{M}_n$ values are <2.0. Protic impurities (moisture) have very little effect on these living polymerizations.

These observations may be rationalized by postulating tert.-ester and -ether·BCl_3 complexes in which the R-O bond is weakened by the strong electron withdrawing effect of the Lewis acid, e.g.,

```
                 Cl                           ┌  Cl  ┐
                 |                            │  |   │ δ⊖
              Cl-B-Cl                         │Cl-B-Cl│
                 ↑                            │  ↑   │
        C        O                 C          │  O   │
        |        ‖                 |          │  ‖   │
  ~CH2-C--O-C        ⇌      ~CH2-Cδ⊕--     │-O-C   │
        |        |                 |          │  |   │
        C        C                 C          └  C   ┘

                 Cl                           ┌  Cl  ┐
                 |                            │  |   │ δ⊖
        C     Cl-B-Cl              C          │Cl-B-Cl│
        |        ↑                 |          │  ↑   │
  ~CH2-C ---- O       ⇌      ~CH2-Cδ⊕--      │ -O   │
        |        |                 |          │  |   │
        C        C                 C          └  C   ┘
```

The large rate enhancing effect of polar diluents may be due to a dynamic equilibrium connecting largely covalent (dormant) and polar or ionic (active) species. Monomer addition is most likely by insertion into the "activated" R-O bond, however, the extent of ionicity of the transition state is obscure.

Inspection of molecular models provided some insight into the mechanism of forced termination upon quenching with methanol or other strong nucleophiles. One of the Cl's of BCl_3 is in the proximity of the propagating site (see formula above) so that nucleophilic attack on the boron center may cause chlorination of the tertiary carbon (12).

The crucial difference between conventional carbocationic polymerizations and the living systems (i.e., those discussed here and those of Higashimara, Sawamoto, et al.) resides in the absence of chain transfer and termination in the latters. It appears that living carbocationic systems are characterized by an absence of free ionic or even strongly ionized species. This proposition and that of the Japanese authors are very similar and may be self-reinforcing. It now emerges that past efforts toward macromolecular engineering at least by carbocationic means were misdirected by a quest toward stable highly ionic initiating-propagating systems. Efforts in the future should rather be focused on the exploration of nondissociated and certainly non-free ionic systems, systems in which the relatively positive active sites remain concentrated exclusively on the carbon center and hyperconjugation is absent. Hyperconjugation tends to stabilize (spread) the positive charge by partial proton mobilization which in turn is viewed as a prelude to proton expulsion and therefore a bar to living carbocationic polymerizations.

ACKNOWLEDGEMENT

The material is based upon work supported by the National Science Foundation (Grant 84-18617).

REFERENCES

1. M. Szwarc, Nature, 178, 1168 (1956)
2. M. Szwarc, M. Levy and R. Milkovich, J. Am. Chem. Soc., 78, 2565 (1956)
3. P. Dreyfuss and M. P. Dreyfuss, Adv. Polym. Sci., 4, 528 (1967)
4. E. J. Goethals and M. Vlegels, Polym. Bull., 4, 521 (1981)
5. Quasiliving Carbocationic Polymerization, J. Macromol. Sci., Special Issue, J. P. Kennedy ed., A18 (1982-83)
6. T. Higashimura, M. Miyamoto, and M. Sawamoto, Macromolecules, 18, 611 (1985)
7. M. Sawamoto, and T. Higashimura, Makromol. Chem., Macromol. Symp., 3, 83 (1986)
8. J. P. Kennedy and R. Faust, patent applied
9. J. P. Kennedy and M. K. Mishra, patent applied
10. R. Faust and J. P. Kennedy, Polym. Bull., 15, 317 (1986)
11. A. Nagy, R. Faust and J. P. Kennedy, Polym. Bull., 15, 411 (1986)

12. R. Faust and J. P. Kennedy, J. Polym. Sci., Polym. Chem. Ed., in press
13. M. K. Mishra and J. P. Kennedy, J. Macromol. Sci., Chem., in press
14. M. K. Mishra, B. Wang and J. P. Kennedy, J. Polym. Sci., Polym. Chem. Ed., in press
15. J. P. Kennedy, L. R. Ross, J. E. Lackey and O. Nuyken, Polym. Bull., 4, 67 (1981)
16. A. Fehervari, R. Faust, and J. P. Kennedy, Polym. Prepr., in press
17. R. Faust, A. Fehervari, and J. P. Kennedy, Polymer J., in press

LIVING AND 'IMMORTAL' POLYMERIZATIONS OF EPOXIDE WITH METALLOPORPHYRIN CATALYST

Shohei Inoue and Takuzo Aida
Department of Synthetic Chemistry,
Faculty of Engineering, University of Tokyo
Hongo, Bunkyo-ku, Tokyo 113
Japan

ABSTRACT. By using aluminum porphyrin as catalyst for the polymerization of epoxide, the synthesis of polyether with well-defined molecular weight can be accomplished. Successive polymerization of two or more epoxides results in the formation of block copolymer with polyether chains of controlled lengths. By taking advantage of the spectroscopic characteristics of porphyrin, the structure of the growing species of the polymerization can be investigated in detail. The polymer with narrow molecular weight distribution can be obtained even in the presence of protic compound such as hydrogen chloride, carboxylic acid, alcohol, and water. The number of the molecules of polymer can be increased with the amount of protic compound added, retaining the narrow molecular weight distribution of polymer. These behaviors are much different from 'living' polymerization, and a new concept of 'immortal' polymerization is presented. Unusual reactivity of the aluminum-axial group bond of aluminum porphyrin is the origin of "immortality" of the polymerization.

1. INTRODUCTION

Control of molecular weight in polymer synthesis is of primary importance for the molecular design of polymeric materials. Although this has been accomplished by living polymerization of some vinyl and cyclic monomers, a particular type of initiator has been succefully applied only for a limited type of monomers. We have recently found that aluminum porphyrin is an excellent initiator for the living polymerization of a variety of monomers such as epoxide (1), β-lactone (2), δ-lactone (3), ε-lactone (4), and lactide (5), and also for the alternating copolymerization of epoxide and cyclic acid anhydride (6) or carbon dioxide (7), to give the polymers and copolymers with narrow molecular weight distribution.

In the living polymerization to give the product with narrow molecular weight distribution, uniform initiation and propagation with respect to all growing molecules are required, together with the absence of termination and chain transfer reactions. As the result,

M. Fontanille and A. Guyot (eds.), Recent Advances in Mechanistic and Synthetic Aspects of Polymerization, 145–153.

the number of the polymer molecules is the same, at the most, as that of the initiator molecule.

In contrast, 'immortal' polymerization described in the present article can afford polymers with narrow molecular weight distribution, with the number of the polymer molecules more than that of initiator. This could be accomplished in the polymerization of epoxide (8-10) and some cyclic esters (3,4) initiated by aluminum porphyrin in the presence of an appropriate protic compound. The present article deals with the living and 'immortal' polymerizations of epoxide initiated with aluminum porphyrin.

2. PREPARATION OF ALUMINUM PORPHYRIN

A representative of aluminum porphyrin used as the initiator is (tetraphenylporphinato)aluminum chloride ((TPP)AlCl: 1, X=Cl).

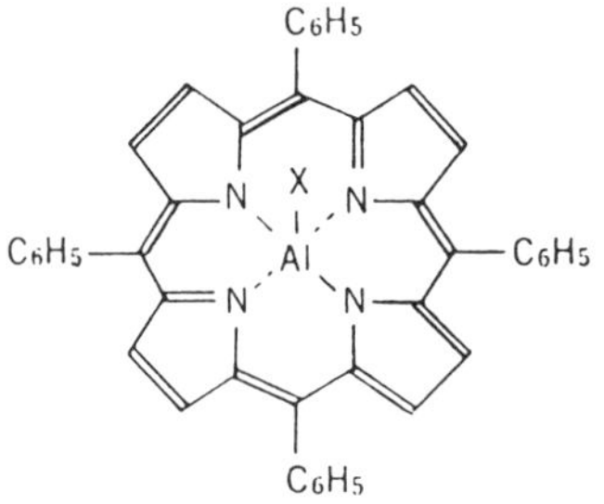

1

(TPP)AlCl is prepared by the equimolar reaction between tetraphenylporphine ((TPP)H_2) and diethylaluminum chloride (Et_2AlCl). The reaction of (TPP)H_2 with triethylaluminum gives (TPP)AlEt, which reacts with protic compounds such as alcohol, phenol, or carboxylic acid to afford various (TPP)AlX (X=OR, OAr, O_2CR, etc.).

$$(TPP)H_2 + Et_2AlCl \xrightarrow{-EtH} (TPP)AlCl \quad (1)$$

$$(TPP)H_2 + Et_3Al \xrightarrow{-EtH} (TPP)AlEt \quad (2)$$

$$(TPP)AlEt + XH \xrightarrow{-EtH} (TPP)AlX \quad (3)$$

3. LIVING POLYMERIZATION OF EPOXIDE

Living polymerization of epoxide with aluminum porphyrin can be conveniently carried out in usual flask, differently from the living polymerization of styrene with organoalkali metal compound, which usually requires all-sealed glass apparatus. In the polymerization of ethylene oxide with (TPP)AlCl in methylene chloride at room temperature, the reaction proceeds with heat evolution and completes in a few hours (Fig. 1).

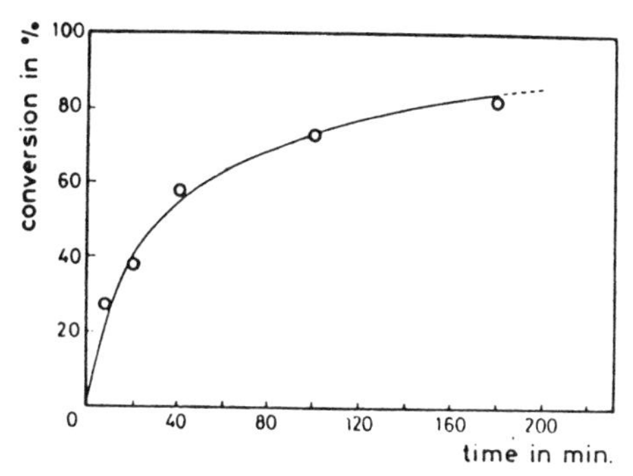

Fig. 1. Polymerization of ethylene oxide (EO) with (TPP)AlCl; room temp.; solvent, CH_2Cl_2; $[EO]_o/[(TPP)AlCl]_o=400$.

The reaction without solvent proceeds much more rapidly, and the reaction mixture becomes very viscous to make the stirring of the mixture impossible. Even under such conditions, the molecular weight distribution of the polymer is very narrow ($\overline{M}w/\overline{M}n=1.05-1.1$), and the molecular weight increases linearly with the consumption of the monomer (Fig. 2).

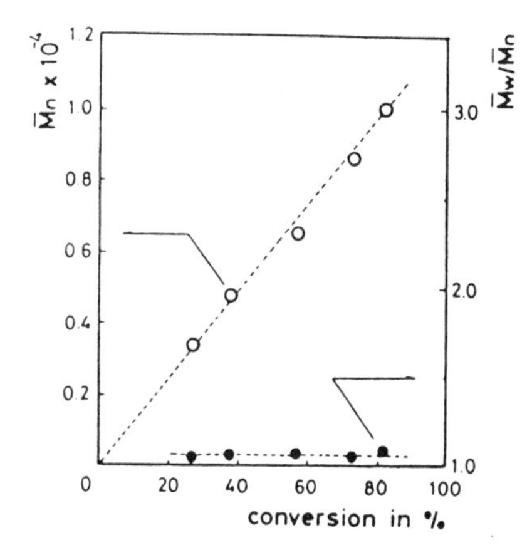

Fig. 2. Polymerization of ethylene oxide (EO) with (TPP)AlCl. Relationship between $\overline{M}n$ (o) or $\overline{M}w/\overline{M}n$ (•) and conversion; $[EO]_o/[(TPP)AlCl]_o=400$.

Similar results are obtained in the polymerizations of propylene oxide and 1-butene oxide. By changing the ratio of epoxide to the initiator, polyether with a prescribed molecular weight can be synthesized.

$$x\ \underset{\diagdown O \diagup}{CH_2-CHR} \xrightarrow{(TPP)AlCl} \left(CH_2-CHR-O\right)_x \qquad (4)$$

Based on the molecular weight ($\bar{M}n$) of the polymer and the conversion, the number of the polymer molecules (N) can be calculated. As shown in Fig. 3, the ratio of N to the number of the initiator molecules (Al) is close to one throughout the reaction, as exemplified in the polymerization of 1-butene oxide with (TPP)AlCl.

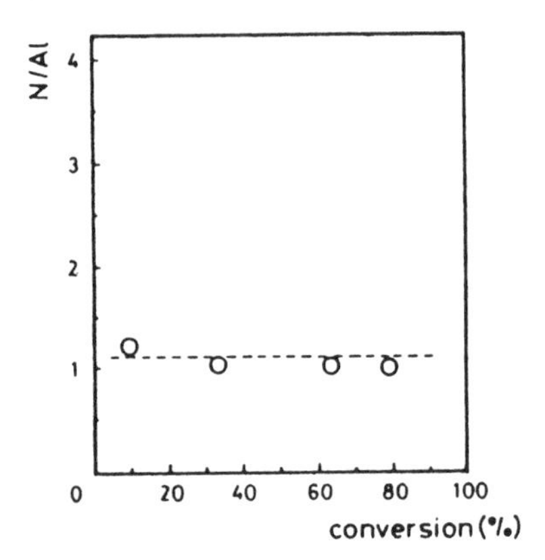

Fig. 3. Polymerization of 1-butene oxide (1-BO) with (TPP)AlCl. Relationship between (number of polymer molecules)/(number of aluminum atoms) (N/Al) and conversion; $[1\text{-BO}]_0/[(TPP)AlCl]_0=300$.

Thus, every molecule of (TPP)AlCl can initiate the polymerization of epoxide, and all molecules of the polymer formed remain alive, without any side reactions.

4. SYNTHESIS OF BLOCK COPOLYMER

Taking advantage of the living nature of the polymerization, block copolymer can be synthesized by successive polymerization of epoxides (11). Fig. 4 shows the gel permeation chromatogram (GPC) of the reaction mixture of the polymerization of propylene oxide with (TPP)AlCl followed by the addition of ethylene oxide. As seen in Fig. 4, the final product shows a sharp unimodal peak (b) in a higher molecular weight region than the prepolymer, and no trace of the prepolymer (a) is observed. Thus, the formation of block copolymer with narrow molecular weight distribution proceeds with a quantitative efficiency. Other binary and ternary block copolymers can be synthesized by a similar procedure. The length of each block can be regulated by the amounts of the epoxide reacted.

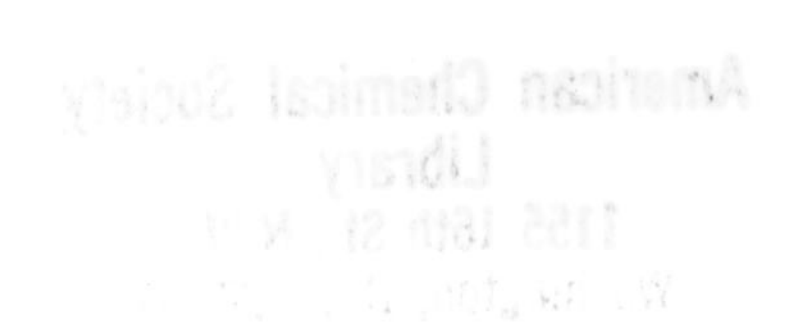

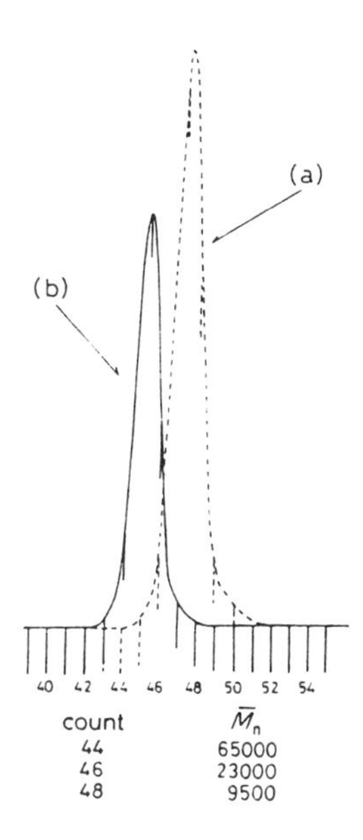

Fig. 4. GPC curves of a propylene oxide-ethylene oxide block copolymer and of the prepolymer of propylene oxide. (a) Prepolymer : $\overline{Mn}$=9300, $\overline{Mw}/\overline{Mn}$=1.12. (b) Block copolymer : $\overline{Mn}$=34000, $\overline{Mw}/\overline{Mn}$=1.18.

5. MECHANISM OF POLYMERIZATION

Detailed studies revealed that the polymerization of epoxide is initiated by the reaction of (TPP)AlCl with epoxide to form a (porphinato)aluminum alkoxide, followed by the repeated reaction of the alkoxide with epoxide (12)

$$\text{(TPP)Al-Cl} \quad + \quad \underset{\diagdown_{O}\diagup}{CH_2\text{-}\overset{\overset{CH_3}{|}}{CH}} \quad \longrightarrow \quad \text{(TPP)Al-O-}\overset{\overset{CH_3}{|}}{CH}\text{-}CH_2\text{-Cl} \qquad (5)$$

$$\underline{2}$$

$$\underline{2} \xrightarrow{(x-1)\ \underset{\diagdown_{O}\diagup}{CH_2\text{-}\overset{\overset{CH_3}{|}}{CH}}} \quad \text{(TPP)Al}\left(\text{O-}\overset{\overset{CH_3}{|}}{CH}\text{-}CH_2\right)_x\text{Cl} \qquad (6)$$

$$\underline{3}$$

NMR spectral investigation of the reaction mixture has been particularly useful for the elucidation of the structure of the growing species, since the nuclei in proximity above the porphyrin ring are strongly shielded by the ring current, and exhibit the corresponding signals at unusually high magnetic field.

For example, Fig. 5A shows the ^{1}H-NMR spectrum of the reaction mixture of (TPP)AlCl with two molar equivalents of propylene oxide. Signals a, b, and c are due to tetraphenylporphinato group, while broad signals d and e to oxy(methylethylene) unit. The most remarkable signal is that at δ -2.0 ppm (f). Since this signal disappears by the addition of acidified methanol to the reaction mixture, the signal is assigned to the methyl group of an aluminum alkoxide bound to porphyrin (3). Relative intensity of the signal to that of porphyrin is in conformity with the structure 3.

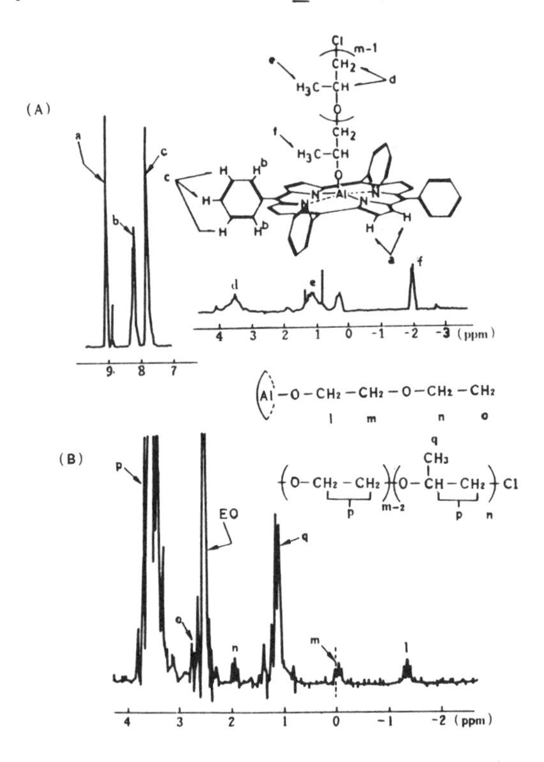

Fig. 5. ^{1}H-NMR spectra in CH_2Cl_2; (A) reaction mixture of propylene oxide with (TPP)AlCl, (B) reaction mixture after the addition of ethylene oxide to the system A.

When ethylene oxide is added to the above system (A), the reaction mixture shows ^{1}H-NMR spectrum as illustrated in Fig. 5B. Signal f disappears, while a new signal appears at δ -1.4 ppm (l), which is assigned to the growing species of the polymerization of ethylene oxide. This observation corresponds to the quantitative formation of block copolymer.

$$(TPP)Al\text{-}(O\text{-}CH(CH_3)\text{-}CH_2)_x\text{-}Cl \quad + \quad y\ CH_2\text{-}CH_2\ (\text{epoxide, }O\text{-bridged})$$

$$\longrightarrow \quad (TPP)Al\text{-}(O\text{-}CH_2\text{-}CH_2)_y\text{-}(O\text{-}CH(CH_3)\text{-}CH_2)_x\text{-}Cl \qquad (7)$$

6. 'IMMORTAL' POLYMERIZATION

Although the polymerization of epoxide with (TPP)AlCl is of nucleophilic or anionic nature as described above, the polymerization can not be 'killed' by the addition of a protic compound such as alcohol, water, phenol, carboxylic acid, and even hydrogen chloride.

For example, "living" polymer of ethylene oxide is prepared with (TPP)AlCl as initiator, and added by hydrogen chloride, followed by propylene oxide. If the "living" polymer is killed by hydrogen chloride as a strong acid, no further reaction will take place upon addition of propylene oxide. On the other hand, if hydrogen chloride does not participate in the reaction, the formation of ethylene oxide-propylene oxide block copolymer will result. In either case the gel permeation chromatogram (GPC) of the reaction mixture will show a single unimodal peak. Of particular interest, two unimodal peaks are observed in the GPC of the reaction mixture which moves to the higher molecular weight region with the progress of reaction. On the other hand, the peak due to the original poly(ethylene oxide) disappears. The fractions corresponding to the two peaks can be separated by the difference of solubility in $CHCl_3$/hexane. The insoluble fraction is identified by ^{1}H-NMR to be the block copolymer consisting of poly(ethylene oxide) and poly(propylene oxide), while the soluble fraction is the homopolymer of propylene oxide. The molecular weight estimated for the poly(propylene oxide) segment in the block copolymer is substantially the same as that observed for the homopolymer. The total number of the polymer molecules (N_p) as calculated on the basis of the molecular weight of the isolated homopolymer is in excellent agreement with the sum of the numbers of the molecules of the starting "living" poly(ethylene oxide) (N_{Al}) and of hydrogen chloride. The homopolymer of propylene oxide isolated is found by ^{13}C-NMR to carry a chlorine terminal ($Cl\text{-}CH_2CH(CH_3)O\text{-}$). These observations indicate that the following sequence of reactions takes place, besides the repeated reaction of (TPP)AlOR with epoxide to form polyether (eq. 6).

$$(TPP)Al\text{-}OR \quad + \quad HCl \quad \longrightarrow \quad (TPP)Al\text{-}Cl \quad + \quad HOR \qquad (8)$$

$$(TPP)Al\text{-}Cl \quad + \quad C\text{-}C\ (\text{O-bridged}) \quad \longrightarrow \quad (TPP)Al\text{-}O\text{-}C\text{-}C\text{-}Cl\ \ (OR') \qquad (9)$$

$$(TPP)Al\text{-}OR' \quad + \quad HOR \quad \rightleftharpoons \quad (TPP)Al\text{-}OR \quad + \quad HOR' \qquad (10)$$

The reaction between the growing species (TPP)AlOR and hydrogen chloride gives (TPP)AlCl, which can initiate the polymerization of epoxide. Although reaction (8) is irreversible, (TPP)AlOR' formed by reaction (9) exchanges with ROH, the "dead" polymer, which eventually revives to the growing species (eq. 10). Thus, in the above experiments, the polymerization of propylene oxide takes place uniformly from all the molecules of the "living" poly(ethylene oxide) and hydrogen chloride.

Thus, the polymerization of epoxide by aluminum porphyrin may not be killed even by hydrogen chloride, and may be regarded to have "immortal" nature. When compared, for example, with the corresponding polymerization of epoxide initiated with the alkali-metal alkoxide, an unusually high reactivity as nucleophile of the aluminum-chlorine bond of aluminum porphyrin (1, X=Cl) is most important to account for the origin of the "immortality" of polymerization. In contrast, the polymerization with alkali-metal alkoxide is killed by hydrogen chloride, since a comparable reactivity can hardly be expected for alkali-metal chloride.

Taking advantage of the principle of 'immortal' polymerization of epoxide with (TPP)AlCl in the presence of a protic compound, for example, methanol, the polymer of uniform molecular weight can be prepared with the number of the molecules (N_p) more than that of the initiator ((TPP)AlCl) (N_{Al}). The reaction may be applied to the polymerization of various epoxides such as ethylene oxide, propylene oxide, 1-butene oxide, and epichlorohydrin (Table I).

TABLE I. Polymerization of epoxide with (TPP)AlCl (1, X=Cl) in the presence of methanol[a)]

Epoxide (E)	$[E]_0/[1]_0$	$[MeOH]_0/[1]_0$	Time in hr	Conv. in %	$\bar{M}n$[b)]	$\bar{M}w/\bar{M}n$[b)]	N_p/N_{Al}[c)]
ethylene oxide	200	9	48	100	700	1.05	12.6
propylene oxide	200	9	48	100	1,300	1.08	8.9
1-butene oxide	200	9	96	100	1,200	1.10	12.0
epichlorohydrin	200	9	48	100	1,500	1.04	12.3

a) At room temperature without solvent. b) GPC c) Number of polymer molecules (N_p)/number of aluminum atoms (N_{Al}).

In the present system, the polymerization proceeds by the participation of reactions (6), (9) and (10). The narrow molecular weight distribution of the product corresponds to the fact that the reversible exchange reaction (10) proceeds much faster than the propagation reaction (eq. 6). The same principle can be applied to the synthesis of block copolymer with narrow molecular distribution starting from a polymeric diol, and to the synthesis of end-reactive polymer and oligomer.

REFERENCES

1. T. Aida, R. Mizuta, Y. Yoshida, and S. Inoue, Makromol. Chem., 182, 1073 (1981).
2. T. Yasuda, T. Aida, and S. Inoue, Macromolecules, 16, 1792 (1983).
3. K. Shimasaki, T. Aida, and S. Inoue, Polym. Prepr. Jpn., 35, 1636 (1986).
4. M. Endo, T. Aida, and S. Inoue, Polym. Prepr. Jpn., 35, 1628 (1986).
5. L. R. Trofimoff, T. Aida, and S. Inoue, Polym. Prepr. Jpn., 35, 251 (1986).
6. T. Aida and S. Inoue, J. Am. Chem. Soc., 107, 1358 (1985); T. Aida, K. Sanuki, and S. Inoue, Macromolecules, 18, 1049 (1985).
7. T. Aida, M. Ishikawa, and S. Inoue, Macromolecules, 19, 8 (1986).
8. S. Asano, T. Aida, and S. Inoue, J. Chem. Soc., Chem. Commun., 1148 (1985).
9. S. Inoue, Makromol. Chem., Macromol. Symp., 3, 295 (1986).
10. S. Inoue and T. Aida, Makromol. Chem., Macromol. Symp., in press.
11. T. Aida and S. Inoue, Macromolecules, 14, 1162 (1981).
12. T. Aida and S. Inoue, Macromolecules, 14, 1166 (1981); S. Asano, T. Aida, and S. Inoue, Macromolecules, 18, 2057 (1985); T. Yasuda, T. Aida, and S. Inoue, Bull. Chem. Soc. Jpn., 59, 3931 (1986); T. Aida, K. Wada, and S. Inoue, Macromolecules, 20, XXX (1987).

FROM ANIONIC TO ZIEGLER-NATTA POLYMERIZATION : APPLICATION TO THE DETERMINATION OF THE REACTIVITY OF ZIEGLER-NATTA ACTIVE CENTERS.

F. CANSELL*, A. SIOVE*, M. FONTANILLE**
* Laboratoire de Recherches sur les Macromolécules (associé au CNRS)
Université Paris-Nord - 93430 Villetaneuse - France.

** Laboratoire de Chimie des Polymères Organiques (associé au CNRS)
Université de Bordeaux I, 351, cours de la Libération
33405 Talence Cedex - France.

ABSTRACT. Living anionic lithium-ended polymer is transformed into a soluble Ziegler-Natta Ti (III)/Li-based system. The latter exhibits a high efficiency as at least 80% of C-Ti bonds participate in the homo- and co-polymerization of ethylene and propene, leading to vinylic-olefinic type block copolymers.
The "living character" of the Ziegler-Natta complexes is deduced from both the macromolecular features of the poly-olefinic blocks and the kinetic studies of C_2H_4 and C_3H_6 polymerizations.
From all the kinetic data obtained and on the basis of a two-steps propagation mechanism, the rate determining step is found to be complexation for C_3H_6 polymerization and insertion in the case of C_2H_4.

INTRODUCTION

During the last twenty years, drastic improvements in olefin polymerization have been made towards the optimization of Ziegler-Natta catalysts. The goals of the investigation have been to increase :

(a) the catalytic activity, via the efficiency of the systems, in order to reach a higher productivity. Important progress has been made in this field by using either supported heterogeneous catalysts (1) or homogeneous ones based on metallocene dichloride/methyl alumoxane systems (2).

(b) the isospecificity of the catalysts, thus yielding highly isotactic polymer in the polymerization of propene. This was achieved by using heterogeneous multi-components systems (including $TiCl_4/MgCl_2/AlR_3$ and electron donors as aromatic esters, amines...) and, more recently, homogeneous metallocene derivatives (2,3).

A numerous literature has dealt with these high-activity and high-stereospecificity catalysts. Nevertheless, from a fundamental point of view, although the general features of the mechanism of the polymeri-

M. Fontanille and A. Guyot (eds.), Recent Advances in Mechanistic and Synthetic Aspects of Polymerization, 155–170.

zation have been determined, important aspects of the reaction path of the Ziegler-Natta polymerization of olefins still remain obscure. The elucidation of the kinetic parameter is of a particular interest for the research of optimized activity systems as well as a better understanding on the origin of the structural and macromolecular features of the yielded polymers.

The problems posed for the determination of the reactivity of the active species have largely been reviewed and discussed (4,5). The major difficulties are connected with the heterogeneous nature of the polymerization system :

- which is partly responsible for the low efficiency of the catalyst, thus making its accurate determination and the identification of the nature of active species delicate.

- which involves adsorption and diffusion phenomena of the reactants in the solid phase significantly complicating the mathematical treatment of the kinetic expressions describing the polymerization. In order to overcome these problems, finding "ideal systems" such as those met in "living" anionic polymerization, is one of the target of Ziegler-Natta catalysis (6). Indeed, as has been shown in "living" polymerizations (7), the relevant systems provide a very interesting and useful tool for :

. determining reaction parameters governing the polymerization (active species and monomers reactivities).

. synthesizing well-defined polymers with controlled molar masses.

The first examples of "living polyolefinic systems" was found by Doï et al (8) in the syndio-specific polymerization of propene with a soluble V(III)/Al-based catalyst and by Siove et al.(9,10) in the polymerization of ethylene initiated with solubilized Ti(III)/Li-based macromolecular complexes. Whether these two living systems allow the determination of the reactivity of the relevant active sites (8,11), the latter offers in addition the advantage of being active in both the homo (10,12) -and co-polymerizations (13) of C_2H_4 and C_3H_6.

Indeed, it has been shown that such solubilized Ziegler-Natta catalysts can be obtained by reacting a living anionic lithium-ended polymer chain with $TiCl_4$ in apolar medium (9). The Ti III/Li complexes thus yielded exhibit a high efficiency (80-90%) in the homo-and co-polymerizations of C_2H_4 and C_3H_6 (11-13). Moreover the "living character" of these systems may be deduced from both the structural and macromolecular analyses of the polymers yielded and the kinetic studies of the polymerizations.

The present work is devoted to the utilization of these living systems to the determination of the reactivities of both the active species and the monomers in homo- and co-polymerization of C_2H_4 and C_3H_6 . On the basis of the obtained values for the related kinetic parameters, the relative influence of steric and electronic factors implied in the C_2H_4 and C_3H_6 polymerization is dicussed. The mechanism of propagation is debated from a kinetical point of view and the meaning of the rate constants is questioned.

1. FROM ANIONIC TO ZIEGLER-NATTA POLYMERISATION

1.1 Transformation of active anionic chain ends into Ziegler-Natta macromolecular complexes :

All the mechanisms which have been proposed for the activity of Ziegler-Natta Ti-halides/organometallic compound-based catalysts, imply substitution of a halogen atom by an alkyl group of the organometallic derivative. This point was checked out for polymer-Li/$TiCl_4$/apolar solvent system (where polymer = polybutadiene or poly (styrene-b-butadiene)). $\overline{DP}_n = x$ $\overline{DP}_n = y$ $\overline{DP}_n = 3$

Indeed addition of $TiCl_4$ to these living anionic chain ends (the reaction features of which have been previously described in details (7,14)) leads to the formation of brown-colored complexes, soluble or not in the reaction medium depending on the polymer-solvent interactions (10) (solvating power of the solvent, nature and DP of the macromolecular precursor etc...). For example, the lower limit of solubility of the polybutadienyl-lithium/$TiCl_4$/hexane system, at room temperature, lies between $30 < DP_n < 100$, regardless of the ratio $r = [Li]/[Ti]$.

The two physical states in which the catalytic system can occur, present advantages complementary to each other for the determination of its composition : when soluble, it can be studied by spectroscopic methods (E.S.R, U.V-visible) if not, it can be studied by gravimetry, elemental analysis, titration of Ti(III).

From all of the above physico-chemical analyses and in so far as the molar mass of the polymer-Li is precisely known, the overall molecular formula can be deduced for the Ziegler-Natta complexes :

$$\text{Polymer}-[TiCl_2, TiCl_3, 3\ LiCl]$$

The substitution of Ti (IV) and its subsequent reduction into Ti (III) would occur as follows :

$$\text{polymer-Li} + 2/3\ TiCl_4 \rightleftharpoons 1/3\ \text{polymer}-[TiCl_2, TiCl_3, 3\ LiCl] + 2/3\ \text{polymer}^{\bullet} \longrightarrow \text{termination}$$

U.V and GPC analyses of terminated polymer show that termination occurs by disproportionation and recombination reactions (9).

The solubilized poly styrene-b-butadiene —$(TiCl_2, TiCl_3, 3\ LiCl)$
$\overline{DP}_n=100$ $\overline{DP}_n=3$
complex (named $pS_{100}-B_3-Ti\langle$) was chosen to initiate the polymerizations of C_2H_4 and C_3H_6. Such a polymeric structure permits to avoid side reactions occurring in the transformation of the anionic sites into the Ziegler-Natta complex. Indeed, in so doing, Ti(III) chain ends exhibit a satisfactory stability at room temperature and the polybuta - diene block is sufficiently short to make improbable transfer reaction on residual double bonds (9).

1.2. " Living character" of the $pS_{100}-B_3-Ti\langle /C_2H_4, C_3H_6$ systems : activity and efficiency in homo- and co-polymerizations.

Addition of C_2H_4,C_3H_6 or a mixture, to the $pS_{100}-B_3-Ti$ complexes leads to the polymerization of olefins and yields block copolymers pS_{100}-B-olefin(s) (12,13,15).

Polymerizations occur in homogeneous phase as long as the length of the polyolefinic block is short enough to permit the solubility of the whole system. Kinetic measurements are only performed during this homogeneous stage, at the ratio r = [Li]/[Ti] = 2 corresponding to the maximum activity of the system (10).

Determination of the activity during this period is derived from the amount of alkene (s) consumed by measuring the variation of the total pressure. The activity "a" is expressed by :

$$a = \frac{\text{mass of olefin (s) consumed}}{[Ti].\ \text{Time}.\ \text{Pressure}} \qquad (\text{in g.l.mol}^{-1}.\text{hr}^{-1}.\text{atm}^{-1}.)$$

In the particular case of the E-P copolymerization, the variation of the E-P composition of the feed was deduced from the copolymer one. The latter was evaluated from ^{13}C NMR analyses according to the method of Knox et al(16) (see Fig 4, next section). It was verified that at low yields (which is the case for the homogeneous stage of the polymerization where conversion <20%), the copolymer composition as well as the feed composition remain constant (to within 5% error).

The results mentioned in Table I indicate roughly linear relationships between activity values and feed composition.

Table I : Activity and efficiency of $pS_{100}-B_3-Li/TiCl_4$ system in the E-P homo- and co-polymerizations.

E-P composition of the feed in % E	P	Activity during homogeneous phase ($g.mol^{-1}.l.h^{-1}.atm^{-1}$)	Efficiency (with respect to C-Ti bond)
100	0	1 700	0,88
40	60	900	0,85
30	70	730	0,82
20	80	550	0,81
10	90	410	0,80
0	100	280	0,80

At higher alkene conversion, the increase in the polyolefinic block length provokes the precipitation of the copolymer in the reaction medium thereby leading to a decrease in the activity (which has not been studied any further, in these conditions).

It is worth noting that in the case of the polymerization of C_3H_6, the precipitation of the p(S-Pro) copolymer in toluene discloses a substantial stereoregularity (and cristallinity) of the polypropene block (corroborated by DSC analyses (12)).

Precipitation of the p(S_{100} -B_3 -olefin$_x$) copolymers allows the separation of the pS_{100}-B_3 chains (remaining in solution) not bonded to a polyolefinic block. By weighing this inefficient pS-B, a direct calculation of the efficiency "f" of the system can be made :

$$f_{Li} = \frac{[pS\text{-}B\text{-}Li]_{initial} - [pS\text{-}B]_{inefficient}}{[pS\text{-}B\text{-}Li]_{initial}}$$

However, as a large part of the lithium compound is used for the reduction of Ti (IV) to Ti (III), it is more significant to evaluate efficiency with respect to titanium concentration or better with respect to carbon Ti(III) bonds concentration f_{C-Ti} :

$f_{C-Ti} = 2.[Li]/[Ti].f_{Li}$ (factor 2 representing the existence of 1 C-Ti bond for every 2 titanium atoms in the binuclear structure of active centers).

Table I indicates high f_{C-Ti} values for the pS_{100}-B_3-Ti$\langle$ /C_2H_4, C_3H_6 systems in homopolymerization as well as in copolymerization ($0.8 < f_{C-Ti} < 0.9$).

From both the measured values of efficiency and the amount of alkene polymerized, the theoretical length of the polyolefinic block can be calculated by assuming the absence of any transfer reaction as follows :

$$\bar{M}n_{theor.} = \frac{\text{mass of polyolefin(s) yielded}}{[pS_{100}\text{-}B_3\text{-Ti active}]_{mol.}}$$

Comparison of the calculated molar masses with those measured by G.P.C. after ozonization of the butadiene units of the p(S-B-Olefin) copolymers indicates a satisfactory agreement to within 30% (12,13,15) (See. a few examples in Table II).

TABLE II Molar masses of the polyolefinic blocks

E-P composition of the feed in % E	P	[C-Ti*] $mmol.l^{-1}$	Theoretical M_n of the polyolefinic block	Experimental $\bar{M}_n$	$\frac{\bar{M}_w}{\bar{M}_n}$
100	-	0.29	63000	41000	1.6
100	-	0.61	31000	20000	1.5
-	100	0.25	70000	52000	1.7
-	100	0.65	30000	21000	1.6
10	90	0.30	29000	20000	1.8
20	80	0.35	33000	22000	1.8

Moreover, the molar masses distribution are unimodal (Fig.1) and relatively narrow (P <1.8; see Table II), thus disclosing a fast initiation step and a satisfactory stability of the active centers during the course of the propagation step.

Figure 1 : example of GPC chromatograms of :
(a) $pS_{100}-B_3$-pEt copolymer ($\bar{M}n$ theoret = 73500); (b) : pEt block resulting from ozonolysis of (a) ($\bar{M}n$ theoret = 63000; $\bar{M}n$ exptl. = 41000; P = 1.6); (c) initial $pS_{100}-B_3$ ($\bar{M}n$ theor = 10500; $\bar{M}n$ exptl. = 11000; P = 1.06).

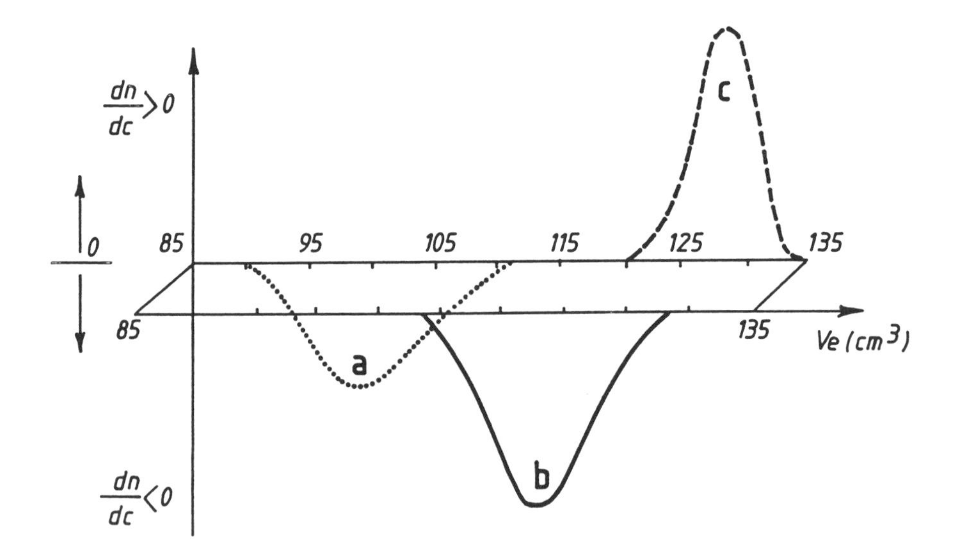

All of results mentioned above are consistent with E-P homo- and co-polymerization occurring under "living" conditions. Such conditions are favorable for determining the reactivity of the active sites involved in these systems.

2. APPLICATION TO THE DETERMINATION OF THE REACTIVITY OF ACTIVE SITES

2.1. Determination of the rate constants of the homo-propagations of C_2H_4 and C_3H_6 :

Experiments were performed in toluene, at 22°C with 3 $mmol.l^{-1} >$ [C-Li] > 0.2 $mmol.l^{-1}$ and [propene] = 0.45 $mol.l^{-1}$, [ethylene] = 0.15 $mol.l^{-1}$, at the ratio [Li]/[Ti] = 2.

It was found, in such conditions, that the kinetic order with respect to monomer concentration is equal to unity. Indeed, during the homogeneous stage $Log_e\ [M_o]/[M] = f(t)$ is linear (see Fig.2), thus showing up the stability and the constant reactivity of $p(S_{100}-B_3-Propylene_x)$ Ti and $p(S_{100}-B_3-Ethylene_y)$ Ti active sites.

Figure 2: Determination of kinetic order in monomer for $pS_{100}-B_3-Li/TiCl_4$ in toluene. P = Propene; E = Ethylene; [C-Li] = 1-3 $mmol.l^{-1}$; r = [Li]/[Ti] = 2; T = 22°C. (A) homogeneous stage; (B) : beginning of precipitation ; (C) : heterogeneous stage.

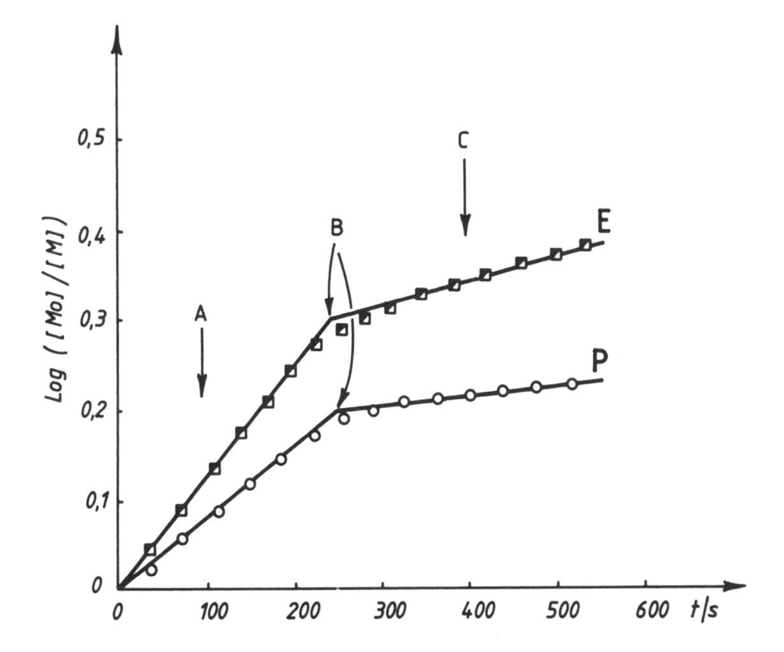

This kinetical behavior confirms the "living" character of the above systems.

When the $\overline{DP}_n$ of the polyolefinic block is high enough, the copolymer precipitates in the reaction medium, thus leading to a break of the slope of the kinetic curve (see Fig 2. part B.C).

The effect of the variation of the concentration in effective (really active) living ends on the rate of polymerization of C_2H_4 and C_3H_6 is represented Fig. 3.

Figure 3. Determination of kinetic order in active centers (C-Ti*) for pS_{100}-B_3-Li/$TiCl_4$ in toluene at 22°C. P = Propene , E = Ethylene.

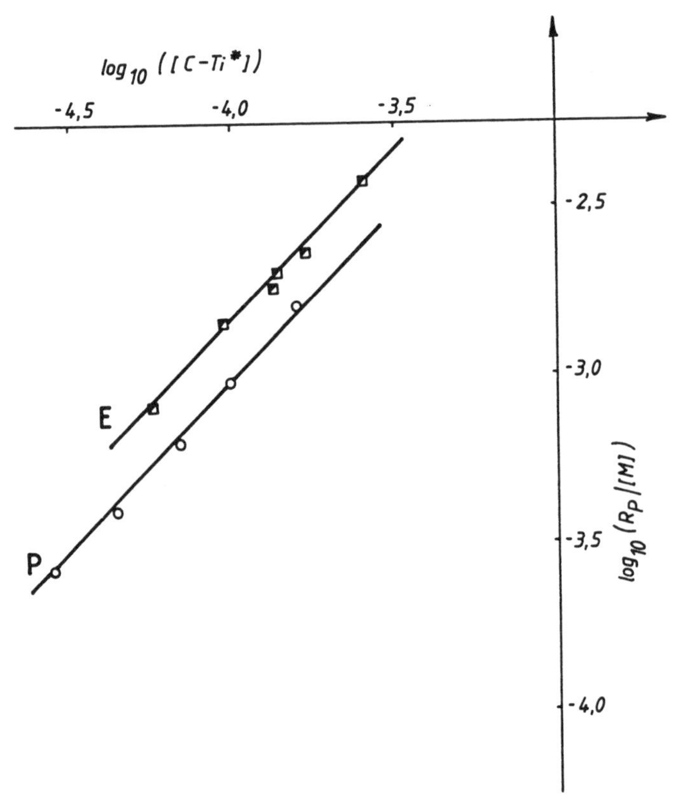

$\log_{10}$ (Rp/M) versus $\log_{10}$ ([C-Ti*] active) indicate a first order relationship with respect to the concentration of active centers, whatever their nature.

Hence the kinetic law of the relevant polymerizations can be expressed by :

$$R_{p(C_2H_4)} = k_{EE}\ [E]\ [\sim EE - Ti^*] \quad \text{and}$$

$$R_{P(C_3H_6)} = k_{PP}\ [P]\ [\sim PP - Ti^*]$$

where k_{EE} and k_{PP} measured the reactivity of the ethylenyl-titanium and propylenyl-titanium chain ends respectively.

From the comparison between the two k values measured for E homopolymerization (k_{EE} = 15.1.mol^{-1} .s^{-1}) and P homopolymerization (k_{PP} = 0.9 1.mol^{-1}.s^{-1}) initiated with the same solubilized complex, it may be concluded that the corresponding reactivities are not so different in contrast to the results reported in the literature for heteregeneous systems (6) (the meaning of the k values will be discussed in the following section).

However, kinetic data do not provide any information on the super molecular structure of active species and do not allow speculation on their aggregation state.

Measurements of the viscosity on living and deactivated species permitted to establish the absence of macromolecular aggregation (11).

Consequently, active species must have the following composition :

$$\sim(\text{olefin})_n\text{-}TiCl_2,\ TiCl_3,\ 3\ LiCl \xrightarrow{\text{olefin}}$$

$$\sim(\text{olefin})_{n+1}\text{-}TiCl_2,\ TiCl_3,\ 3\ LiCl$$

2.2. Determination of monomers reactivity ratios and copropagation rate constants(13):

Accounting k_{EE} and k_{PP} values and considering the four propagation rate constants involved in the copropagation steps k_{EE}, k_{EP}, k_{PP}, k_{PE}, the cross-propagation rate constants can be evaluated from the reactivity ratios r_1 and r_2.

$r_1 = k_{EE}/k_{EP}$; $r_2 = k_{PP}/k_{PE}$ were determined from ^{13}C N.M.R. analysis, evaluating the comonomer units distribution in the polyolefinic block of the copolymer (see ref. 16-18 for method of evaluation and detailed NMR assignments of the different carbon atoms). A typical spectrum of the olefinic block of p (S-B)-b-(E-co-P) copolymer is given Fig. 3, and chemical shifts of the different carbon atoms are reported in Table III

Figure 4 : 62,5 MHz ^{13}C N.M.R of a (S_{100}-B_3)-b- (E-co-P) copolymer at 120°C in O-dichlorobenzene.

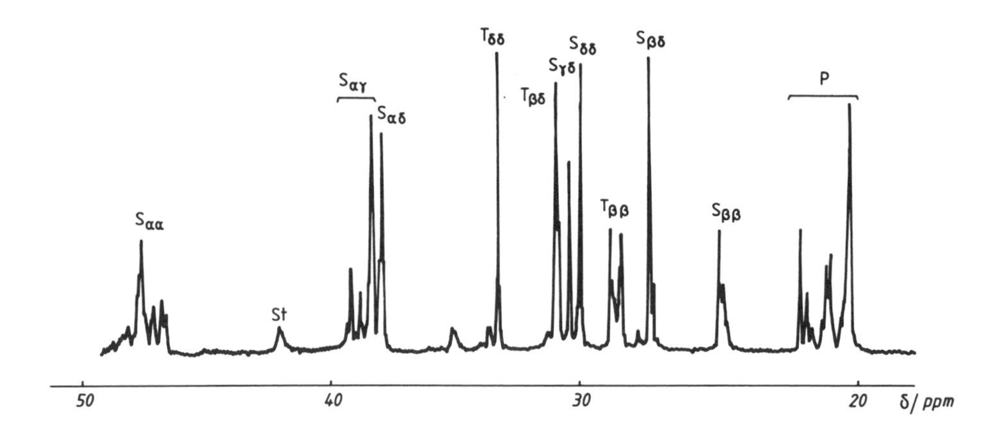

Table III : ^{13}C N.M.R assignments of styrene-b-(ethylene-co propylene) copolymer.

Sequence type [a)]	Carbon type [b)]	Chemical shift in ppm
PP	S	45.6-47.2
Styrene	CH and CH_2	41.2-41.9
2 PEP	S	38.0-39.0
PPE_n (n>1)	S	37.6
PEEP	S	30.8
$PEEE_n$ (n>1)	S	30.4
EEE_n (n>1)	S	30.0
PEE_n (n>1)	S	27.5
PEP	S	24.6-24.9
EPE	T	33.3
PPE	T	30.9
PPP	T	28.5-28.9
-	P	20.1-21.9

a) P=propene ; E=ethylene
b) assignments and terminology according to ref.17.

From the E-P compositions of both the copolymer and the feed (X), monomers reactivity ratios can be evaluated according to

$$r_1 = 2\ (EE)/(PE)X \ ;\ r_2 = 2\ (PP)X/(PE) \quad (\text{with } X = (E)/(P))$$

The results obtained (see Table IV) indicate average values of 18 and 0.24 respectively for r_1 and r_2 for different E-P compositions; this is not very different from the r values reported in the literature for E-P copolymerization initiated with Ti(III) based catalysts (6,19,20).

Table IV. Monomers reactivity ratios.

E-P composition of the feed in % E	P	E-P composition of the polyolefinic block in % E	P	r_1	r_2	r_1r_2
10	90	30	70	21	0.23	4.8
20	80	50	50	17	0.13	2.2
30	70	55	45	19	0.30	5.7
40	60	60	40	16	0.30	4.8

The $r_1 . r_2$ average value (4.4) together with the comparison between ^{13}C NMR signals intensities of the E-P sequences, disclose a random copolymerization with an ethylene tapered character (characterization of the copolymers will be presented in a forthcoming paper (21)).

From r_1, r_2, k_{PP}, k_{EE} the cross-propagation rate constants were calculated :

$k_{EP} = 0.8\ l.mole^{-1}.s^{-1}$

$k_{PE} = 4\ l.mol^{-1}.s^{-1}$

It must be kept in mind that: $k_{PP} = 0.9\ l.mol^{-1}.s^{-1}$
$k_{EE} = 15\ l.\ mol^{-1}.s^{-1}$

By comparing all of these values, several observations may be made :

- the propagation rate constants depend on the nature of the monomer entering a polymer chain end

- concerning the propylene addition onto ~E*-Ti and ~P*-Ti sites, the relevant propagation rate constants are independent of the nature of the active ends, whereas it is not the case with ethylene.

- ethylene is preferentially incorporated into the copolymer whatever the nature of the active site. This is in agreement with the relative reactivity of the two olefins observed in Ziegler-Natta copolymerization : C_2H_4 more reactive than C_3H_6 for which it is admitted that steric factors predominate.

Interpretation of these facts requires a deeper insight into both the reactivity of active sites and the reactivity of monomers.

Concerning active centers reactivity two cases can be distinghished :

- assuming a primary insertion of C_3H_6 into the C-Ti$\langle$ bond thus leading to a $\sim$CH(γCH$_3$)-CH$_2$-Ti$\langle$ site (found for TiIII -based catalysts(6)), one may reasonably admit that the electron donating effect of the methyl group on the C-Ti$\langle$ bond polarization, is weak. Thus, $\sim$E-Ti$\langle$ and $\sim$P-Ti$\langle$ would have close basicities and the difference between the reactivity of the two "carbanions" should arise from steric factors unfavorable to $\sim$P-Ti$\langle$ chain ends.

- secondly, taking into account a secondary insertion of C_3H_6 (an improbable situation herein which, however, must not be overlooked) one can expect a higher basicity of $\sim$CH$_2$-CH(βCH$_3$)-Ti$\langle$ site than $\sim$E-Ti$\langle$ one (due to the electronic effect of the β CH_3 but a steric effect unfavorable to the former.

Now, considering the monomers reactivity, it may be supposed that steric factors weakly influence the C_2H_4 reactivity. Concerning electronic interactions, the situation is more complicated; indeed, in the hypothesis of a two-steps propagation mechanism, the prior coordination of the monomer onto the electronic vacancy of the titanium atom is favorable to C_3H_6 (the more nucleophilic olefin) whereas the subsequent "anionic insertion" is favorable to C_2H_4 (the more electrophilic olefin).

Consequently, depending on the nature of the rate determining step, complexation or insertion, the rate constants of propagation will have a different meaning.

3. MEANING OF THE RATE CONSTANTS FOR A TWO-STEPS PROPAGATION MECHANISM.

It is now fairly accepted that the mechanistic scheme is that proposed by Natta et al.(22) and then by Cossee (23), which involved a two-step propagation proceeding by a reversible complexation of the ole-fin on the transition metal followed by an insertion into the C-Ti bond of the growing chain. A wide variety of evidence has been reported in the literature (see i.e ref. 1, 5) supporting this mechanism.

Concerning the pS_{100}-B_3-Ti$\langle$ /C_2H_4, C_3H_6 systems described herein, a two-step propagation mechanism would involve in the first elementary step a π complex formation between the Ti atom and the olefin. On the basis of the binuclear bioctahedral structure suggested for the active sites (11) the complexation of the olefin(s) might be formaly represented as follows :

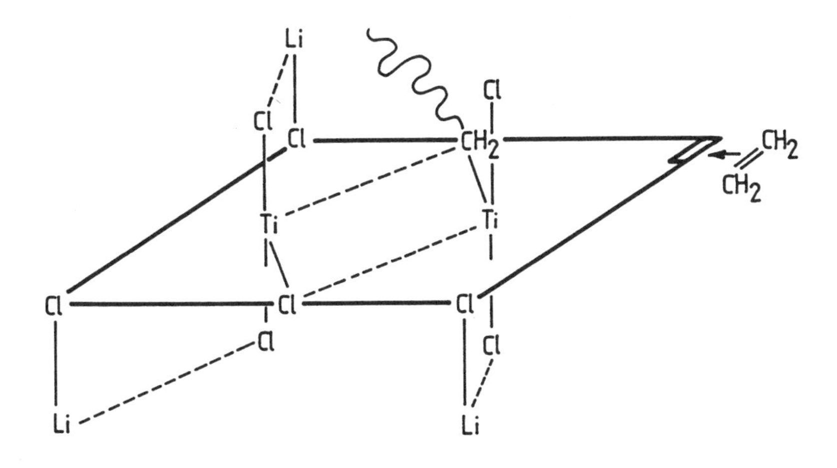

In the proposed structure the two titanium atoms are bonded to the "carbanion" and can play the same potential role in the monomer insertion step, nevertheless only one possesses a vacancy allowing the monomer complexation (comments on mechanistic considerations will be made elsewhere (24)).

From a kinetical point of view a two-step mechanism is described by the reactions :

$$\sim\!\sim\text{C}-\text{C}-\text{Ti} \quad + \underset{(\text{M})}{\text{C}=\text{C}} \quad \underset{k_{-1}}{\overset{k_1}{\rightleftarrows}}$$

$$\begin{array}{c} \sim\!\sim\text{C}-\text{C}-\text{Ti} \\ \uparrow \\ \text{C}=\text{C} \end{array}$$

$$\downarrow k_2$$

$$\sim\!\sim\text{C}-\text{C}-\text{C}-\text{C}-\text{Ti}$$

The experimental value of the rate constant of propagation k is given by (24):

$$k = \frac{Rp}{[C\text{-}Ti][M]_{total}} = \frac{k_1\, k_2}{k_1 + k_{-1} + k_1[M]}$$

where $[C\text{-}Ti]_{total}$ represents the concentration of active sites both complexed and free.

Since k is a composite term it is an apparent rate constant.

Depending on the ratio between the different constitutive k values of the elementary steps and on the monomer concentration several situations can be distinguished :

(I) $k_1\,[M] \gg k_2 + k$ $\quad$ $k = k_2/[M]$

(II) $k_2 \gg k_{-1} + k_1\,[M]$ $\quad$ $k = k_1$

(III) $k_{-1} \gg k_2 + k_1\,[M]$ $\quad$ $k_1 = k_2\,k\,/k_{-1}$

(IV) $k_2 + k_{-1} \gg k_1\,[M]$ $\quad$ $k = k_1\,k_2/k_{-1} + k_2$

First order dependence of the rate on the monomer concentration corresponds to the cases II, III, IV, where the propagation rate is respectively determined by the rate of olefin complexation ($k=k_1$), by the rate of complexed olefin insertion and its concentration ($k = k_1\,k_2\,/k_{-1}$) or by the rate of complexed olefin insertion in competition with desorption ($k = k_1\,k_2/k_{-1} + k_2$).

Taking into account all of the kinetics of homo- and co-polymerization of C_2H_4 and C_3H_6 initiated with the solubilized Li/TiIII based complexes, different situations can occur depending on the nature of olefin polymerized.

Indeed, for propene polymerization, since the nature of the last unit of the growing chain does not influence the rate of propagation ($k_{EP} \simeq k_{PP}$), the rate-determining step is the complexation and k represents the rate constant of C_3H_6 complexation (situation n°(II); $k = k_1$).

On the contrary, for ethylene polymerization, the nature of the living chain end notably influences the rate of propagation ($k_{EE} \simeq 4\,k_{PE}$), the rate-determining step being the complexed ethylene insertion more or less affected by desorption :

$k = k_1 \ k_2/k_{-1}$: insertion rate determining

$k = k_1 \ k_2/k_{-1} + k_2$: insertion-desorption competition.

In the present state of our work we have no plausible argument allowing the choice between these two situations. Nevertheless, from the above kinetics results it can be concluded that complexation is more favorable to C_3H_6 (the more nucleophilic olefin) whereas insertion is more favorable to C_2H_4 (the more electrophilic olefin), the meaning of the corresponding propagation rate constants being therefore different.

On the other hand the effect of strong electron-donors (such as tetra-methylethylenediamine) on the reaction parameters of the macromolecular complexes was recently tested in the polymerization of C_2H_4 (25). The depressor effect of the Lewis base on the propagation rate was interpreted in terms of blockage of titanium vacancy, thus reinforcing the concept of a two-step propagation mechanism which will be discussed elsewhere (24).

ACKNOWLEDGEMENTS

The authors are indebted to (CdF Chimie Company for financial support (to F.C.) and thank Pr. Ph. Teyssie for helpful discussions.

REFERENCES

1. See for example different contributions presented in "Catalytic Polymerization of Olefins", ed. by T. Keii and K. Soga, Elsevier. Sci. Publ. New-York (1986).

2. H. Sinn, W. Kaminsky, Adv. Organomet. Chem. **18**, 99 (1980)

3. J.A. Ewen, J. Am. Chem. Soc. 106, 6355 (1984)

4. See the papers of T. Keii and P.J.T. Tait in ref. 1

5. D.R. Burfield, Polymer **25**, 1645 (1984)

6. J. Boor. Jr., "Ziegler-Natta catalysts and Polymerizations", Academic Press, New-York (1979).

7. M. Szwarc, "Carbanions, Living Polymers and Electron Transfer Processes", Interscience, New-York (1968).

8. Y. Doi, S. Ueki, T. Keii, Macromolecules, **12**, 814 (1979)

9. A. Siove, M. Fontanille, Eur. Polym. J.,**17**, 1175 (1981)

10. A. Siove, M. Fontanille in "Transition Metal Catalyzed Polymerizations" ed. by R.P. Quirk, Harwood. Acad. Publ., New-York, P. 313 (1981).

11. A. Siove, M. Fontanille, J. Polym. Sci. Polym. Chem. Ed., **22** 3877 (1984)

12. F. Cansell, A. Siove, M. Fontanille, Makromol. Chem., **186**, 379 (1985)

13. F. Cansell, A. Siove, M. Fontanille, J. Polym. Sci. Polym. Chem. Ed. , **25**, 675 (1987)

14. J.E. Roovers, S. Bywater, Macromolecules, **8**, 3 (1975)

15. A. Soum, A. Siove, M. Fontanille, J. Appl. Polym. Sci., **28**, 961, (1983)

16. G.J. Ray, P.E. Johnson, J.R. Knox, Macromolecules, **10**, 773 (1977)

17. C.J. Carman, R.A. Harrington, C.E. Wilkes, Macromolecules, **10**, 536 (1977).

18. M. Kakugo, Y. Naito, K. Mizunuma, T. Miyatoke, Macromolecules, **15**, 1150 (1982).

19. G. Natta, G. Manzzati; A. Valvassori, G. Sartori, Chim. Ind (Milan) **40**, 896 (1958)

20. Y.V. Kissin, D.L. Beach, J. Polym. Sci. Polym. Chem. Ed., **21** 1065 (1983)

21. F. Cansell, A. Siove, G. Belorgey (to be published)

22. G. Natta, F. Danusso, D. Sianesi, Makromol. Chem, **30**, 238, 1959

23. P. Cossee, J. Catal, **3**, 30 (1964)

24. A. Siove (to be published)

25. C. Meverden, B. Brun, A. Siove, M. Fontanille, Makromol. Chem. **188** 103 (1987).

REPORT OF DISCUSSIONS AND RECOMMENDATIONS

I - Mechanisms of Polymerization

Considered in this discussion were mechanisms of chain growth polymerization including the following :

a) radical
b) cationic
c) anionic
d) group transfer
e) coordination

All the participants emphasized the important requirement of indentifiying and controlling the active center structure. This is certainly necessary to prepare well defined "living" polymers. Transfer and termination reactions must be also controlled, so as to allow for the preparation of predictable molecular weights and narrow molecular weight distributions. The various mechanisms listed above may be considered to represent a continuum where various polarizing conditions can be achieved either by electrophilic or nucleophilic activation of covalent bonds.

In recent years, it has become clear in a number of polymerization systems that the control of the active center can be best achieved by "moderating" its nature. For example, reliable data have been produced that demonstrate living polymerizations in the field of carbocationic polymerization as well as earlier anionic systems. Thus Higashamura and coworkers have recently described living polymerization of vinyl alkyl ethers by HI/I_2 combinations and Kennedy et al of isobutylene and other olefins by certain ester or ether/Lewis acid complexes. By skillful selection of specific alkyl vinyl ethers, for example, amphiphilic block and graft copolymers have been assembled. Also linear and three-arm star telechelic polyisobutylene capped by tert.-chlorine end groups have been conveniently obtained in the -30 to

M. Fontanille and A. Guyot (eds.), Recent Advances in Mechanistic and Synthetic Aspects of Polymerization, 171–175.

0^oC range. These advances in synthetic techniques together with the availability of these and similar well-defined new products bring carbocationic polymerizations again center-stage among the best-controlled polymer preparative methods.

The fundamental difference between conventional carbocationic polymerization systems and the recently described carbocationic living systems is that chain transfer and termination are absent in the latters. As conventional carbocationic polymerizations (i.e. polymerizations initiated by $H_2O.BF_3$, $H_2O.TiCl_4$, t-BuCl.Et_2AlCl, etc... combinations) are believed to proceed by solvated free ions or strongly ionized species, the recent living systems must involve different active species and most likely do not contain free ions or highly dissociated ion pairs.

While the fundamental differences in the kinetic profiles of conventional nonliving and new living polymerizations indicate some profound phenomenological differences, the nature of these differences remains obscure. Among our first objectives should be to gain increased insight into these differences, particularly in regard to a better definition of the active species involved. Our task however is rendered enormously difficult on account of the elusive and transient nature of the carbenium ion, carbanion or coordination intermediates, and because the activity of the true propagating species is determined not only by its primary structure but also by secondary external parameters, such as counter ions medium polarity, impurities, concentrations, aggregation states, temperature, addition sequence of ingredients and by the mutual interdependence of these factors.

In the past, researchers in the field of carbocationic and anionic polymerizations tended to concentrate on the exploration of highly ionic systems, which overemphasized the effect of traces of moisture. They were compelled to operate at cryogenic temperatures in order to "freeze out" undesirable side-reactions. These measures rendered ionic polymerizations rather cumbersome for the academic experimentalist and uneconomical for the industrialist. It now appears that the new living systems emerging in Japan and the U.S. not only provide heretofore unattainable synthetic-preparative advantages (i.e. new functional polymers, blocks, grafts) but also ease of operations in terms of relatively low moisture sensitivity and experimentation at close to or above ambient temperatures.

II - Kinetic aspects

Efforts should continue on establishing the reaction order with respect to monomer (S), initiator, catalyst, etc... as well as the influence of the medium and suspected impurities. There is a great need for improved methods of detection and quantification of the reactive intermediates and active chain ends. In radical polymerizations more sensitive electron spin resonance (ESR) instrumentation would be useful, especially for the investigation of various types of copolymerizations. In ionic and group transfer polymerizations, the nature of the active species is still under discussion and answer to the pending problems necessitate parallel studies using NMR, UV-visible and FTIR spectroscopy. In particular improved stopped-flow apparatus is required, which will permit research under very high purity conditions at various desired reaction temperatures.

III - Overview of important monomer system

It is possible and convenient to subdivide various important monomer systems according to the following classifications.

A - Nonpolar

Hydrocarbons such as the diene, styrene, 1-alkenes and cycloalkenes

B - Polar

1 - "Electron poor" monomers such as the acrylates, methacryclates and nitriles

2 - "Electron rich" monomers such as the vinyl alkyl ether and N-vinyl carbazole

3 - Cyclic monomer, eq. epoxides, lactones, oxazolines, lactides, lactanes, cylic ethers and cyclosiloxanes.

It is noteworthy that much progress has been achieved in recent years towards the preparation of polymers of controlled architecture, molecular weight, narrow molecular weight distribution and functional end groups. The principles that have made this possible have already been outlined in Section I. Most of these developments are related to the discovery of new initiator/catalyst systems which can provide appropriate stability to the growing species/active center. Emphasis on the nature of the metals and ligands which have been identified suggests that continued studies of organometallic and coordination chemistry are important for future developments.

IV - Architectural considerations

Improved stability of the active centers have enabled significant advancements in the preparation of controlled structures. These in turn have permitted the development of novel rheological, morphological and physical property behavior. Some of the most important architectures are outlined below.

. Macromonomer (telechelics)
. Block, graft and ion-containing copolymers
. Star-like controlled branching
. macrocyclics

Macromonomers or telechelics are functionalized oligomers and polymers which can be utilized in subsequent reactions to produce well defined multiphase graft or segmented (multiblock) advanced materials. Then development has only been possible through improved understanding and stability of active centers : some important applications of these materials are considered in Section V.

V - Applications of advanced materials

One may consider several areas where either bulk or surface enhancement has been already improved significantly in recent years through the utilization of multi-phase advanced materials. These are outlined below. Essentially, these improvements have been possible because the controlled polymerization methods developed have permitted preparation of unique homopolymers and especially well-defined morphologies and physical behavior.

* Bulk physical behavior
  . Unique homopolymer
  . Multiphase systems
    + Toughered thermoplastics and networks
    + Thermoplastic elastomers
    + Transparent toughered coatings
    + Controlled transport of gases and liquids

* Surface enhancement
    - Permeability control
    - Biomaterials
    - Plasma etch resistance
    - Adhesion and release phenomena

III - CATALYSIS OF POLYCONDENSATION REACTIONS

Chairman : B. SILLION

NOVEL POLYCONDENSATION SYSTEM FOR THE SYNTHESIS OF POLYAMIDES AND POLYESTERS

N. Ogata
Department of Chemistry, Sophia University
7-1 Kioi-Cho, Shiyoda-Ku, Tokyo 102
Japan

ABSTRACT. Novel polycondensation reaction has been developed for the synthesis of polyamides and polyesters under mild conditions. Phosphorylation reaction using triphenyl phosphine or triphenyl phosphine dichloride could initiate a direct polycondensation reaction under such mild conditions as ambient temperature to form either polyamides or polyesters.

1. INTRODUCTION

Polycondensation reactions are usually carried out at elevated temperatures above 250°C to form either polyamides or polyesters having high molecular weights simply because the polycondensation reactions essentially accompany with an equilibrium reaction so that one has to shift the equilibrium toward polymer formation side by eliminating byproducts such as water out of the reaction phase.

On the other hand, polycondensation reactions in living animals take place under mild conditions to form a very high molecular weight proteins from α-amino acids, which belong to families of condensation-type polymers. Apparently the polycondensation reactions in living animals are greatly enhanced by enzyms which induce a phosphorylation reaction of α-amino acids so that the reactivity of functional groups such as carboxylic acid or amino groups is enough enhanced to initiate the polycondensation reaction under mild conditions.

Recently, phosphorylating reactions using various phosphorus compounds have been developed for the direct polycondensation reactions under mild conditions to produce either polyamides or polyesters having high molecular weights. It was found in our laboratory (1- 9) that triphenyl phosphine or triphenyl phosphine dichloride could initiate the direct polycondensation under such mild conditions as

M. Fontanille and A. Guyot (eds.), Recent Advances in Mechanistic and Synthetic Aspects of Polymerization, 179–189.

ambient temperature to produce either polyamides or polyesters having high molecular weights. These phosphorus compounds were required to use in a stoichiometric amount to monomers so that the recycling systems of these phosphorus compounds had to be established from view points of industrial applications.

When the direct polycondensation reactions were carried out in the presence of triphenyl phosphine dichloride, the recovered triphenyl phosphine oxide, which was formed as a result of the direct polycondensation reaction, was easily able to be reconverted again to the reactive triphenyl phosphine by treating with either phosgene or oxalyl chloride gas. Also, triphenyl phosphine moiety could be attached on polystyrene so that the recycling system of the phosphorylating agent became possible with respect to the industrial application of the direct polycondensation method.

This paper deals with the direct polycondensation reactions to form either polyamides or polyesters under mild conditions by using various types of phosphorylating agents with respect to the recycling system of the phosphorylating agents.

2. EXPERIMENTAL

2.1 Polycondensation by using triphenyl phosphine

A typical procedure by using triphenyl phosphine as an initiator was as follows: a give amount of p-aminobenzoic acid (4-ABA) and triphenyl phosphine (Ph_3P) was dissolved in 10 cm^3 of solvent, and a given amount of hexachloroethane (C_2Cl_6) was added to the solution with stirring at room temperature. As soon as the hexachloroethane was added, the reaction solution became yellow and an exothermic reaction took place. Then, the reaction solution became heterogeneous with the separation of a solid polymer, and at times the entire solution solidified within 2 min. Excess methanol was added to remove the resulting polymer which was isolated by filtration. The polymer was repeatedly washed with water and methanol, and dried under reduced pressure.

2.2 Polycondensation by using polymeric triphenyl phosphine

A polymeric initiator containing triphenyl phosphine moiety was synthesized by a radical polymerization of diphenyl styryl phosphine. Copolymers from diphenyl styryl phosphine and 4-vinyl pyridine were also synthesized.

$CH_2=CH-C_6H_4-PPh_2 \longrightarrow \{CH_2-CH(C_6H_4-PPh_2)\}$, $CH_2=CH-C_6H_4-PPh_2$ + $CH_2=CH-C_5H_4N$

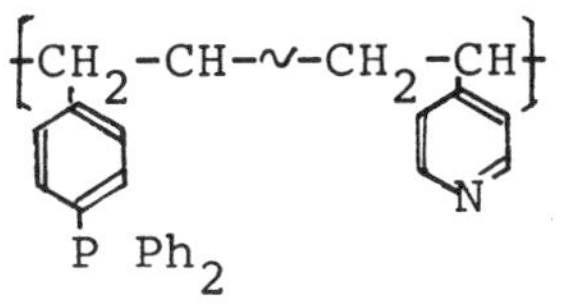

$$\left[CH_2-CH-\sim-CH_2-CH \right]$$

P Ph_2 N

2.3 Polyconensation by using triphenyl phosphine dichloride ($TPPCl_2$)

It is known that triphenyl phosphine dichloride ($TPPCl_2$) is quantitatively formed when triphenyl phosphine oxide is treated with phosgene or oxalyl chloride. The direct polycondensation using triphenyl phosphine resulted in the formation of triphenyl phosphine oxide which had to be reconverted to the initial triphenyl phosphine from industrial point of view for the practical application.

$$Ph_3P{=}O + COCl_2 \longrightarrow Ph_3PCl_2 + CO_2$$

$$Ph_3P{=}O + (COCl)_2 \longrightarrow Ph_3PCl_2 + CO + CO_2$$

Therefore, when the direct polycondensation is carried out by using $TPPCl_2$, the recovered triphenyl phosphine oxide can be easily reconverted again to the reactive $TPPCl_2$ by treating with phosgene or oxalyl chloride.

Polycondensation of bisphenol A with terephthalic acid and isophthalic acid was carried out by using $TPPCl_2$ as follows: a 100 cm^3, four-necked flask equipped with a mechanical stirrer, a dropping funnel, a reflux condenser, and a nitrogen inlet was flushed with dry nitrogen gas and then charged with 20 cm^3 of monochlorobenzene and 8.94g (0.032 mol) of tripheny phosphine oxide (TPPO). To this mixture was added dropwise during 10 min with stirring a solution of 4.10g (0.032 mol) oxalyl chloride in 10 cm^3 of monochlorobenzene. Gas evolved out of the solution. After the evolved gas was stopped, a mixture of 1.25g (0.0075 mol) of terephthalic acid (TPA) and 1.25g (0.0075 mol) of isophthalic acid (IPA) was added.

The mixture was heated with stirring at a reflux temperature for 5 min, then cooled to room temperature. A portion of 3.42g (0.015 mol) of bisphenol A (BPA) was added to a complete dissolution, and finally a solution of 6.07g (0.06 mol) triethylamine in 5 cm^3 of pyridine was added with stirring at room temperature. After the solution was heated at a reflux temperature with stirring for 4 hr, the flask was immersed in a water bath and 50 cm^3 of chloroform was added to this solution. The solution was poured into 1 l of methanol and the precipitate was washed with hot methanol and dried for 18 hr at 50^oC in a vacuum.

3. RESULTS AND DISCUSSION

3.1 Synthesis of polyamides

The direct polycondensation of p-aminobenzoic acid (4-ABA) was carried out by using triphenyl phosphine as an initiator. Various reaction conditions were investigated in terms of molar ratios of initiators/monomer, monomer concentrations and reaction temperature.

The solution viscosities and yield of the resulting polyamides varied with the molar ratio of the reagent to the monomer 4-ABA. The polyamide was obtained in a quantititive amount in pyridine when molar ratios of Ph_3P and C_2Cl_6 ranged from 1 to 2. The optimum molar ratio among Ph_3P, C_2Cl_6 and 4-ABA was 1.2:1.5:1.0 as can be seen in Fig.1, and the aromatic polyamide having a solution viscosity as high as 3 was obtained in a quantitative yield in pyridine at room temperature.

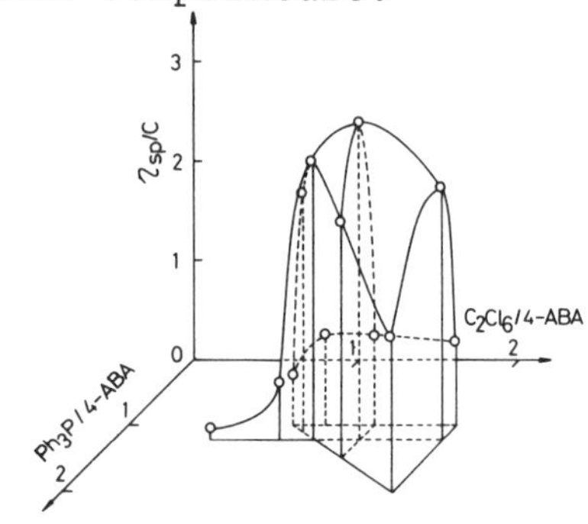

Figure 1. Effect of molar ratios of triphenylphosphine, hexachloroethane, and monomer on the solution viscosities of poly(*p*-benzamide). Monomer concn, 0.4 mol dm^{-3}; temp, room temp; time, 30 min; solvent, pyridine.

The monomer concentration was also found to have a significant effect on the moleuclar weight of the polyamide, as shown in Fig. 2.

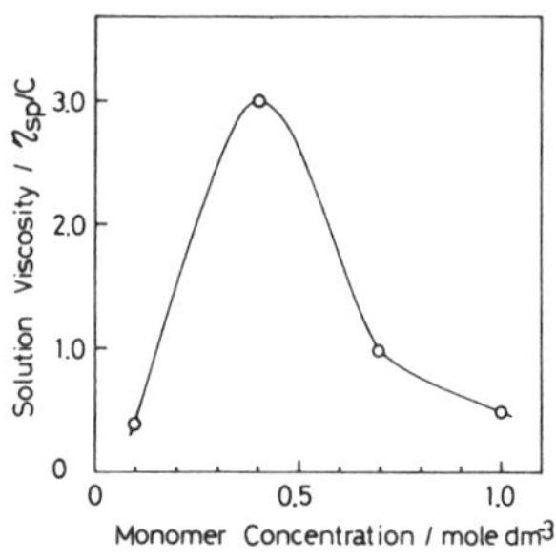

Figure 2. Effect of monomer concentrations on the polycondensation of 4-ABA. Molar ratios of C_2Cl_6–Ph_3P–4-ABA = 1.5 : 1.2 : 1.0; temp, room temp; time, 30 min; solvent, pyridine.

The polyamide having the highest solution viscosity of 3 was obtained at a monomer concentration of 0.4 mol dm^{-3} in pyridine. Subsequently, the solution viscosity of the polyamide decreased with increasing monomer concentrations. The decrease in solution viscosity can be explained by the entanglement of growing polymer chains with each other, with increasing monomer concentration, leading to low reactivity of the polymer end groups.

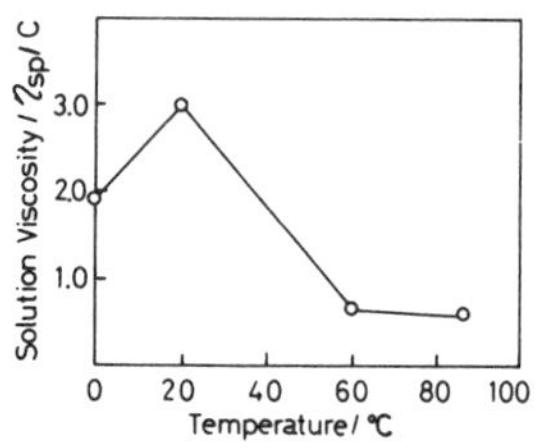

Figure 3. Effect of reaction temperature on the polycondensation of 4-ABA. Monomer concn, 0.4 mol dm^{-3}; molar ratios of C_2Cl_6–Ph_3P–4-ABA = 1.5 : 1.2 : 1.0; time, 30 min; solvent, pyridine.

Solution viscosity of the resulting polyamide was

greatly dependent on reaction temperature as evident from Fig. 3, which indicates that the maximum solution viscosity was attained at 20°C. With increasing reaction temperature, the resulting polyamide became violet, and the solution viscosity of the resulting polyamide decreased, presumably owing to side reactions such as thermal decomposition of reaction intermediates.

The results of the direct polycondensation reaction of 4-ABA , carried out in various solvents,are summarized in Table I, where it is seen that the solvent had an important influence on the yield and solution viscosity of the resulting polyamide and that pyridine had a specific effect on the polycondensation.

The results of the polycondensation of 4-ABA are summarized when various organic bases were used as an acid acceptor. No polyamide was obtained in triethylamine, while pyridine gave a high molecular weight polyamide.

The highest polymer solution viscosity was attained to 5 in a mixed solvent of tetramethylurea and pyridine of 30% content in volume as can be seen in Fig. 4. The effect of these solvents on the polycondensation might be related to both the polarities of solvents and the solubilities of the resulting polyamide, which would enhance the reactivity of end reactive groups.

Table I. Solvent effect on the polycondensation of 4-ABA[a]

Solvent	Dielectric constants	Polymer	
		Yield/%	η_{sp}/c
Triethylamine	2.42	0	—
Quinoline	9.0	88	0.17
Pyridine	13.3	100	2.70
Tetramethylurea	23	96	0.33
N-Methyl-α-pyrrolidone	32.2	100	0.15
Hexamethyl-phosphortriamide	34	0	—

[a] Monomer concn, 0.4 mol dm^{-3}; molar ratios of C_2Cl_6–Ph_3P–4-ABA = 1.5 : 1.2 : 1.0; temp, room temp; time, 30 min.

Table II. Solvent effect on the polycondensation of 4-ABA[a]

Solvent	pK_a[b]	Polymer			
		Yield/%		η_{sp}/c	
		S[c]	M[d]	S[c]	M[d]
2-Chloropyridine	0.49	21		0.09	
Quinoline	4.81	88		0.17	
Pyridine	5.25	100	100	2.99	5.16
3-Picoline	5.63	87	89	0.71	0.69
N,N-Dimethyl-*o*-toluidine	5.86	4		—	
2-Picoline	5.94	89	81	0.26	0.26
4-Picoline	6.03	96	83	0.37	1.27
2,6-Lutidine	6.60	79	100	0.25	0.30
Triethylamine	10.75	0		—	

[a] Monomer concn, 0.4 mol dm^{-3}; molar ratios of C_2Cl_6–Ph_3P–4-ABA = 1.5 : 1.2 : 1.0; temp, room temp; time, 30 min.
[b] Cited from "CRC Handbook of Tables for Organic Compound Identification, 3rd edition."
[c] S, solvent only.
[d] M, mixed solvents containing 70% tetramethylurea.

Various aromatic polyamides could be obtained from aromatic diamines and dicarboxylic acids such as terephthalic acid under the same reaction conditions as the direct polycondensation of 4-ABA. High solution viscosities of more than 1 were easily attained by using the combination of triphenyl phsophine and hexachloroethane as an initiator.

However,it was very difficult to obtain aliphathic polyamides by means of the combination of triphenyl phosphine and hexachloroethane.

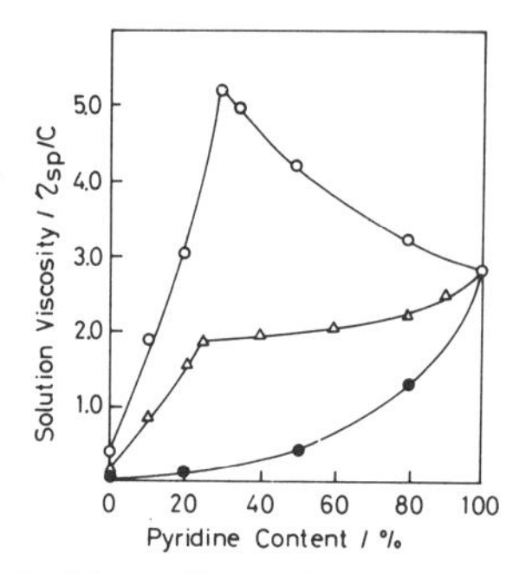

Figure 4. Solvent effect on the polyconcensation of 4-ABA. Monomer concn, 0.4 mol dm^{-3}; molar ratios of C_2Cl_6–Ph_3P–4-ABA = 1.5 : 1.2 : 1.0; temp, room temp; time, 30 min. Mixed solvent: (○), tetramethylurea–pyridine; (△), *N*-methyl-α-pyrrolidone; (●), hexamethylphosphortriamide–pyridine.

Various halides were selected as a partner of triphenyl phosphine in the polycondensation of aliphatic nylon salt from aliphatic diamines and dicarboxylic acids. It was found that carbon tetrabromide CBr_4 was much superior to hexachloroethane as a partner of triphenyl phosphine to initiate the direct polycondensation of nylon salts, as shown in Table III, where various aliphatic nylon salts were used for the direct polycondensation.

Table III. Syntheses of polyamides[a] with triphenylphosphine

Diamine	Dicarboxylic acid	Temp °C	Yield %	η_{sp}/c[b]
	$HOOC(CH_2)_4COOH$	40	0	—
	$HOOC(CH_2)_5COOH$	40	32	0.03
	$HOOC(CH_2)_6COOH$	40	50	0.75
$H_2N(CH_2)_6NH_2$	$HOOC(CH_2)_7COOH$	35	75	0.76
	$HOOC(CH_2)_8COOH$	40	84	0.75
	$HOOC(CH_2)_{10}COOH$	40	64	0.22
$H_2N(CH_2)_2NH_2$	$HOOC(CH_2)_8COOH$	30	0	—
$H_2N(CH_2)_{10}NH_2$	$HOOC(CH_2)_8COOH$	40	81	0.34

[a] Monomer concn, 0.25 mol dm^{-3}; Ph_3P, 0.75 mol dm^{-3}; CBr_4, 0.75 mol dm^{-3}; solvent, pyridine.
[b] 0.1 g/10 cm^3 in H_2SO_4 at 30°C.

It was found that the combination of triphenyl phosphine and halides could initiate the direct polycondensation under mild conditions to produce either aromatic or aliphatic polyamides having sufficient molecular weights in almost quantitative yields. However, the optimum reaction conditions in terms of kinds of halides, molar ratios of the initiators and so on, had to be established in each system of monomers used for the direct polycondensation.

3.2 Synthesis of polyesters

Direct polycondensation for the synthesis of polyesters was carried out by using the combination of triphenyl phosphine and polyhalo compounds under mild conditions. Various bisphenols such as bisphenol A (BPA) or resorcinol (RE) and aliphatic diols were used with dicarboxylic acids such as isophthalic acid (i-PTA) or 2,5-pyridine dicarboxylic acid (2,5-Py) as monomers for the direct polycondensation. p- or m-Hydroxylic benzoic acids (p-HBA or m-HBA) were also used as monomers. Results of the direct polycondensation are summarized in Table IV.

TABLE IV
Effect of Solvents on the Polycondensation of BPA and 2,5-Py[a]

Solvent	Ratio	Temp. (°C)	Yield (%)	η_{sp}/C[b]
Pyridine	1	23	98	0.31
Pyridine–TEA[c]	1:1	23	96	0.20
Pyridine–Quinoline	1:1	23	94	0.24
Pyridine–TMU[d]	1:1	23	96	0.29
Pyridine–NMP[e]	1:1	23	98	0.27
Pyridine–HMPA[f]	1:1	23	93	0.22
Pyridine–β-Picoline	1:1	26	90	0.39
β-Picoline	1	26	89	0.30
β-Picoline–TMU	1:1	26	82	0.23

[a] Molar ratios of Ph_3P/C_2Cl_6/BPA/2,5-Py, 1.2:1.5:0.5:0.5; monomer concentration, 0.5 mol/dm^3; 60 min.
[b] Measured in *m*-cresol at 30°C.
[c] TEA = Triethylamine.
[d] TMU = Tetramethylurea.
[e] NMP = *N*-Methyl-2-pyrrolidone.
[f] HMPA = [illegible]

It is seen in Table IV that various polyesters were obtained by using the combination of Ph_3P and C_2Cl_6 as an initiator under mild conditions. Generally speaking, the direct polycondensation reaction for the synthesis of polyesters requests a longer time of more than 1 hr for the completion of the polycondensation owing to less reactive hydroxyl groups than amino groups.

Synthesis of polyethylene terephthalate (PET) was also carried out by the direct polycondensation of ethylene glycol and terephthalic acid under mild conditions. Monochlorobenzene was used as a solvent and various organic bases were tested for the direct polycondensation. Polymeric bases such as poly(4-vinyl pyridine) were added to the direct polycondensation with an expectation that these polymeric bases might enhanced the polycondensation reaction owing to a matrix effect which might help the adsorption of monomers onto the polymeric bases so as to increase a local concentration of monomers at the reaction phase.

Table V summarizes results of the synthesis of PET, in which it is shown that PET was successfully obtained by using the combination of triphenyl phosphine dichloride and poly(4-vinyl pyridine) as an initiator.

Table V Synthesis of PET by $TPPCl_2$ in the presence of various bases

Acid acceptor		PET	
Kind	$\eta_{sp/c}$	Yield (%)	$\eta_{sp/c}$
2-methyl pyridine	-	45	0.14
4-methyl pyridine	-	41	0.15
2,5-dimethyl pyridine	-	37	0.14
2-bromo pyridine	-	26	0.11
poly(4-vinyl pyridine)	0.15	10	0.13
" "	0.25	35	0.32
" "	0.45	90	0.25
" "	0.53	92	0.35
Imidazole	-	55	0.15
pyrrole	-	65	0.20
piperidine	-	45	0.15

Solvent for the polycondensation: monochlorobenzene

Reaction temperature: room-temperature followed by 130°C

Amount of base: equimolar amount of pyridine to monomer

$TPPCl_2$ is an effective initiator for the synthesis of aromatic polyesters from bisphenols and TPA or IPA and high molecular weights of polyesters were easily attained as shown in Tables VI and VII.

TABLE VI
Polycondensation of Various Bisphenols and IPA, TPA with $TPPCl_2$[a]

Monomer		Solvent	Yield (%)	η_{sp}/C[b]
Acid	Bisphenol			
IPA	BPA	Chlorobenzene	99	0.63
IPA	BPA	Dichlorobenzene	98	1.00
TPA	BPA	Chlorobenzene	99	0.74
TPA	BPA	Dichlorobenzene	99	0.96
TPA	RN[d]	Chlorobenzene	86	0.36[c]
TPA	BPA/RN (50/50)	Chlorobenzene	96	0.60
IPA/TPA (50/50)	BPA	Chlorobenzene	96	1.66
IPA/TPA (50/50)	DDS[e]	Chlorobenzene	99	0.56

[a] Molar ratios of TPPO, $(COCl)_2$ to monomer: 1.07; total monomer concentration: 1.0 mol/dm^3; acid acceptor: $(C_2H_5)_3N$ + pyridine; reflux temperature: 4 h.
[b] Measured in *o*-chlorophenol at 30°C.
[c] Measured in *p*-chlorophenol at 50°C.
[d] RN = Resorcinol.
[e] DDS = 4,4′-Dihydroxydiphenyl sulfone.

TABLE VII
Result of GPC Measurements for Aromatic Polyesters

Composition	Acid acceptor	η_{sp}/C[a]	M_n[b]	M_w[b]	M_w/M_n
TPA, BPA/RN (50/50)	Triethylamine	0.60	11,200	28,100	2.51
IPA/TPA (50/50), BPA	Triethylamine	0.68	13,700	37,500	2.74
	Pyridine (Py)	1.26	10,900	66,300	6.08
	Triethylamine	1.42	37,800	86,400	2.29
	Triethylamine+Py	1.66	18,900	103,200	5.46
m-HBA/*p*-HBA (80/20)	Triethylamine	0.47	7,600	23,500	3.09

[a] Measured in *o*-chlorophenol at 30°C.
[b] Calculated from the poly(styrene) standard calibration curve.

The direct polycondensation reaction with $TPPCl_2$ was very effective for the synthesis of polyamides or polyesters of high molecular weights. The advantage of $TPPCl_2$ for the direct polycondensation is the possible recovery of the reactive $TPPCl_2$ by treating the formed TPPO with phosgene or oxalyl chloride and the recovery and recycling systems thereby established as shown below:

$COCl_2$ or $(COCl)_2$ → $(C_6H_5)_3PCl_2$ → Polyester

CO, CO_2 ← $(C_6H_5)_3PO$ ←

3.3 Polymeric triphenyl phosphine initiator

When triphenyl phosphine moiety is incorporated into polymeric matrix so that the active intiator is immobilized in a resin, it would be possible to produce either polyamides or polyesters through a column of the resin so that a semi-continuous process would be possible, since the regeneration of the initiator would be easy. The idea is schematically drawn as follows:

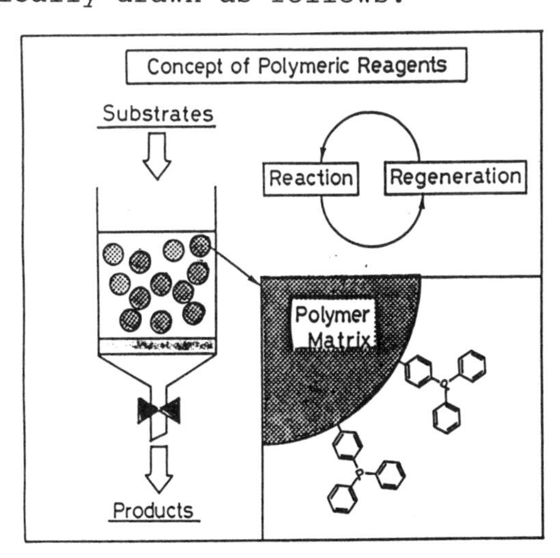

Diphenyl styryl phosphine was synthesized and this monomer was polymerized or copolymerized with 4-vinyl pyridine by means of conventional radical method. Homopolymer and copolymers containing triphenyl phosphine moiety were obtained as shown below:

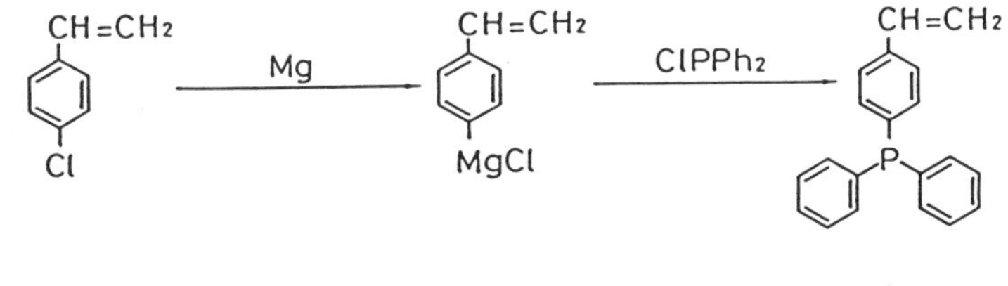

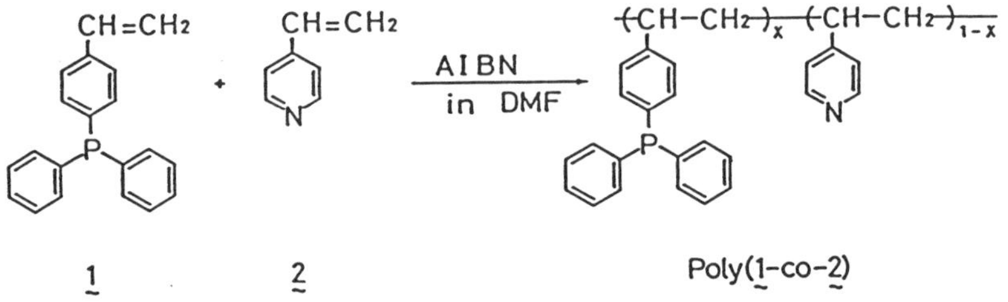

CH=CH2 (Cl) —Mg→ CH=CH2 (MgCl) —ClPPh2→ CH=CH2 (P) 1

1 + 2 —AIBN in DMF→ Poly(1-co-2)

Room-temperature polycondensation of 4-ABA was carried out by using these polymeric initiators and results are summarized in Table VIII. It is seen in this Table that aromatic polyamide was obtained by using these polymeric initiators under mild conditions.

Table VIII Results of room temperature polycondensation of 4-aminobenzoic acid (4-ABA) in the presence of poly(1-co-2) a)

Run No.	Mole fraction of 1 (x) in poly(1-co-2)	Solvent	Concn. of 4-ABA mol/dm^3	$\eta sp/C$ b) 100cm^3/g
1	1.0	Py c)	0.042	0.37
2	1.0	Py	0.083	0.46
3	1.0	Py	0.17	0.51
4	1.0	Py	0.42	0.71
5	1.0	Py	0.83	0.90
6	0.88	NMP d)	0.25	0.78
7	0.79	NMP	0.25	0.75
8	0.43	NMP	0.25	0.66
9	0.35	NMP	0.10	0.25

a) [4-ABA] : [1] : [C_2Cl_6] = 1.0 : 1.2 : 1.5.

b) Measured in conc.H_2SO_4 (0.1g/10cm^3) at 30°C.

c) Pyridine.

d) N-Methyl-2-pyrrolidone.

The activity of the recovered initiator became less and less after the treatment with oxalyl chloride. Possibly, the reason of the decrease in the initiator activity would be related with some amount of the combined polyamide. Therefore, more detailed purification of the revovered polymeric initiator is required for the practical application.

4. CONCLUSION

The combination of triphenyl phosphine and polyhalo compounds could initiate the direct polycondensation of various condensation monomers to produce either polyamides or polyesters under mild conditions.

The disadvantage of using a stoichiometric amount of the triphenylphosphine could be overcome by immobilizing the triphenyl phosphine moiety onto polymeric resins which would make possible to establish a semi-continuous process with a recycling system.

5. REFERENCES

1) G.Wu, H.Tanaka, K.Sanui, and N.Ogata, J.Polym.Sci., Polym. Letters, 19, 343(1981).
2) N.Ogata, K.Sanui, H.Tanaka, and S.Yasuda, Polym.J., 13, 989(1981).
3) G.Wu, H.Tanaka, K.Sanui, and N.Ogata, Polym.J., 14, 571(1982).
4) G.Wu, H.Tanaka, K.Sanui, and N.Ogata, Polym.J., 14, 797(1982).
5) S.Yasuda, G.Wu, H.Tanaka, K.Sanui, and N.Ogata, J.Polym.Sci.,Polym.Chem.Ed., 21, 2609(1983).
6) N.Ogata, K.Sanui, and S.Tan, Polym.J., 16, 569(1984).
7) S.Kitayama, K.Sanui, and N.Ogata, J.Polym.Sci., Polym.Chem.Ed., 22, 2705(1984).
8) N.Ogata, K.Sanui, M.Watanabe, and Y.Kosaka, J.Polym. Sci., Polym.Letters, 24, 65(1986).
9) Y.Kosaka, M.Watanabe, K.Sanui, and N.Ogata, J.Polym. Sci., Polym.Chem.Ed., 24, 1915(1986).

RECENT PROGRESS IN THE USE OF PHASE TRANSFER CATALYSIS FOR POLYCONDENSATION REACTIONS. APPLICATION TO THE PREPARATION OF POLYETHERS, POLYCARBONATES AND POLYESTERS.

O. Mahamat, M. Majdoub, F. Méchin, H. Sleiman and
S. Boileau (1).
Collège de France
11 Place Marcelin Berthelot
75231 PARIS CEDEX 05
FRANCE

ABSTRACT. Phase transfer catalysis has been applied to the synthesis by polycondensation of polyethers, polycarbonates and polyesters, either using a liquid-liquid process (polyethers and polycarbonates) or a solid-liquid process (polyesters). The kinetics and mechanism were discussed, and the influence of several parameters such as the nature of the solvent and the stoichiometry and concentration of the reagents was investigated. It was possible to prepare polycondensates having high enough molecular weights, low polydispersities and well-controlled end groups, starting from dihalogeno compounds and diphenols (liquid-liquid system) or alkaline dicarboxylates (synthesis of polyesters). Further physico-chemical experiments were carried out with the resulting polymers. End-to-end cyclization dynamics in dilute solution was studied for a polycarbonate. Interesting thermal properties were found in the case of polyesters for which an odd-even effect was detected.

1. INTRODUCTION

During the past few years, phase transfer catalysis (PTC) has been widely applied to nucleophilic displacement polycondensation reactions as well as to the chemical modification of polymers. This process allows two compounds located in two different phases to react, thanks to a phase transfer agent (the system being either liquid-liquid or solid-liquid). It requires shorter reaction times and lower temperatures than the corresponding classical reactions (thus limiting the side reactions). Moreover, the process is more simple, and in the case of liquid-liquid systems the use of expensive anhydrous solvents and reagents becomes unnecessary. To begin with, we will give a short review of the recent work concerning polyethers, polycarbonates and polyesters.

M. Fontanille and A. Guyot (eds.), Recent Advances in Mechanistic and Synthetic Aspects of Polymerization, 191–206.

1.1. Polyethers

Among the numerous nucleophilic displacement polycondensations the preparation of polyethers has been most widely studied. They were synthesized from dihalogeno aliphatic, linear or aromatic compounds (or epichlorhydrin (2,3)) and several bisphenols, usually with a liquid-liquid system. The polymers, even when obtained from a polyhalogeno compound (4) were linear and often had well-defined halogenated end groups (5-11) and low polydispersities (5,10). All the authors noticed that the structure and characteristics of their samples were greatly influenced by several parameters such as the nature of the organic solvent (2-6,8-12) and of the catalyst (2,3,6,7,12,13) the structure of the bisphenol (4,6,7) and of the dihalogeno compound (6,8,11) and their initial molar ratio (5,8,10,11). The effect of the reagents and catalyst concentrations (5,7-11,13), of the addition of an alkaline halide (11,13) and of the stirring rate (5,8) were also studied as well as the influence of water in the case of a solid-liquid system (14,15). Moreover polyformals were synthesized from bisphenols and dichloromethane (16-18), and using the same method, it was possible to enhance greatly the molecular weight of a polymer having hydroxyphenyl end groups (19).

Since PTC proved to be an efficient, simple and cheap way of preparing polyethers having well-defined structures, PERCEC (20-29) and several other authors (30-32) applied it to the synthesis of new peculiar polymers such as liquid crystal polyethers (24-29,31), alternating block copolyethers (20,21,28) or ABA triblock copolyethers (22,23). Sometimes they reexamined the effect of the formerly studied parameters on the properties of these new polycondensates.

1.2. Polycarbonates

Moreover, PTC also allows the synthesis of polycarbonates by using either a solid-liquid or a liquid-liquid system. The first synthetic route to be studied was the reaction between 1,4-bis(bromomethyl)benzene and potassium carbonate, using 18-crown-6 ether as catalyst (33). The influence of the nature of the solvent, of the amount of catalyst and reagents, of the reaction time and temperature was investigated. The system was then improved by using dibromo compounds, potassium salts of diols and carbon dioxide (34-36) and a mechanism was proposed.

Another possible way is the transcarbonatation reaction studied by KURAN (37) and still recently used by other authors (38) : PTC is used in the first step to prepare a biscarbonate, such as 1,4-bis-(cyclohexyloxycarbonyloxymethyl)benzene, which can then react with a diol in the presence of a convenient catalyst, to give polymers having quite high molecular weights. Once again different diols and solvents were used.

More recently, several liquid-liquid systems were studied. Among them, a general process consists in using a diol and a bis(chlorocarbonyloxy) compound (bischloroformate) (39). Apart from the nature of both monomers, several parameters proved to affect the results, the most important ones being the amount of sodium hydroxide and the reaction time. At last, phosgenation of several diols was carried out under PTC conditions (40,41), leading to macromolecules with high enough molecular weights.

1.3. Polyesters

The liquid-liquid process involving a diol or a bisphenol and a dichlorocarbonyl compound was successfully applied to the synthesis of polyesters and is now widely used by numerous authors to prepare well-defined polymers, and especially liquid crystal polyesters. The first studies related the influence of several parameters on the kinetics (3,42-45) and on the synthesis of alternating block copolymers (46-48). But the preparation of polyesters is also a very nice example of application of the solid-liquid process. The starting materials are most of the time an alkaline dicarboxylate salt and a dihalogeno compound (11,49-51), whereas the self-condensation of sodium 4-(p-bromoacetylphenyl) butanoate was also investigated (52). The characteristics of the samples varied with the nature of the catalyst (11,49,50) but they had well-defined structures and no side reactions (such as the isomerisation of a double bond in the case of maleic and fumaric acids) occurred (50). More recently, the reaction of potassium 2,6-naphthalene-dicarboxylate with various α,ω-dibromoalkanes and 1,4-dichlorobutene lead to new polyesters having high enough molecular weights, well controlled end groups and interesting thermal properties (11,51).

PTC is thus a powerful and easy way of preparing new polymers by polycondensation. We wish to report our recent results concerning the synthesis and characterization of polyethers (8,11), polycarbonates (39) and polyesters (11,51). To begin with, we will report our studies on the kinetics and mechanism of the preparation of polyethers derived from 2,2-bis(4-hydroxyphenyl)propane (BPA) and 1,4-dichlorobutene (DCB) or α,ω-dihalogenoalkanes, especially 1,6-dibromohexane (DBH). Moreover, the synthesis of new polycarbonates from various bischloroformates and BPA or 2,2-bis(4-hydroxyphenyl)perfluoropropane (BPAF) will be reviewed and discussed. Finally, we will examine the preparation of polyesters derived from the application of solid-liquid PTC, i.e. from a 2,6-naphthalene dicarboxylate and various α,ω-dibromoalkanes $Br(CH_2)_nBr$, with $4 \leqslant n \leqslant 9$. Special atttention was paid to thermal properties of the samples and an odd-even effect was detected.

2. MATERIALS AND METHODS

2.1. Materials purification

Phase transfer catalysts i.e. tetrabutylammonium hydrogen sulfate (TBAH)(Fluka,Merck), dicyclohexyl-18 crown-6 (DCHE)(Aldrich,Fluka) and cryptand [2,2,2] (Merck) were used as received. Pure grade commercial solvents, such as dichloromethane, chloroform, toluene, chlorobenzene,1,2,4-trichlorobenzene, acetonitrile , nitrobenzene, N,N-dimethyl formamide (DMF) and dimethylsulfoxide (DMSO) were also used without any further purification. Tetrahydrofuran (THF) was distilled over sodium ; 1,2-dichlorobenzene (ODCB) was distilled under nitrogen as well as 1-methyl-2 pyrrolidinone (NMP)(Janssen) which was then stored on molecular sieves; 1,1,2,2-tetrachloroethane was passed through a column of alumina before being used and stored in the dark.

BPA(Merck) was recrystallized from toluene whereas BPAF(Atochem) and the dipotassium salt of 2,6-naphthalene dicarboxylic acid(Aldrich) were used as received, as well as α,ω-dihalogenoalkanes (Aldrich and Janssen). DCB was purchased from Aldrich and distilled. Bischloroformates were stored in the dark at low temperature (-30°C) and used as received from SNPE (53).

2.2. Polymer synthesis

2.2.1. Polyethers. As the order in which the reagents are put together seems to influence the results, two different procedures were used : in the first one, BPA (3 mmol) was mixed with 10 ml of aqueous NaOH and 10 ml of the organic solvent and the mixture was mechanically stirred. Then 0.6 mmol of TBAH, and finally the dihalogeno compound were added. The temperature was then raised to 65°C and the mixture stirred for 5 hours. Other experiments were carried out by preparing the sodium bisphenolate (3 mmol of BPA in 10 ml of aqueous NaOH) at 65°C at first. Then the catalyst TBAH (0.6 mmol), the organic solvent (10 ml) and the dihalogeno compound were successively added. The reaction was then conducted in the same way. This second method seems to lead to polymers having higher molecular weights and to give more reproducible results in the case of high concentrations of BPA and NaOH (11).

2.2.2. Polycarbonates (39). Quite similarly, the sodium bisphenolate solution (5 mmol of bisphenol in 20 ml of aqueous NaOH) was prepared at first. After the addition of TBAH, a solution containing 5 mmol of bischloroformate in 20 ml of solvent was introduced, either dropwise or all at once. The heterogeneous mixture was then stirred at 500 rpm for 3 hours (the whole reaction was conducted at 20°C under nitrogen).

In the case of a liquid-liquid procedure, the organic layer was diluted with solvent after cooling and washed with slightly acidic HCl solutions, after the aqueous layer had been discarded. The polymer was then precipitated in methanol, filtered and purified by dissolution and reprecipitation. It was finally dried under high vacuum. Moreover, in the case of polycarbonates (39), the methanolic filtrate was evaporated. The residue was dissolved in dichloromethane and poured into n-hexane. In this way it was sometimes possible to recover large amounts of oligomers.

2.2.3. Polyesters (11,51). Typically, 5.25 mmol of the potassium salt, 5 mmol of dihalogeno compound, 0.525 mmol of catalyst and 4 ml of NMP were mechanically stirred for 3 hours at 110°C. After cooling, water was added to dissolve inorganic salts and precipitate the polymer. The solid was filtered, washed with water, acetone, slightly acidic HCl solutions and methanol. It was then dissolved in a proper solvent (usually 1,1,2,2-tetrachloroethane) and precipitated in light petroleum or n-hexane. After filtration, it was dried under high vacuum.

2.3. Polymer characterization

Composition of the polymers was determined by elemental analysis. Except for polycarbonates the chlorine or bromine content allowed to evaluate $\bar{M}_n$. Moreover, the molecular weights of the polymers were examined by GPC, usually run in THF or toluene at 30°C, using two different detections (UV spectrometry and refractive index). When this was not possible because of a poor solubility (polyesters), the analyses were kindly performed by Dr DEVAUX (54) either in m-cresol or in a mixture of phenol and 1,2,4-trichlorobenzene at 110°C. The structure of the polymers was determined by 200 MHz ^1H and 50 MHz ^{13}C NMR spectroscopy, using a BRUKER AM 200 SY spectrometer.

Finally, thermal properties of the polymers were studied by differential scanning calorimetry (DSC), performed on a Du PONT de NEMOURS 1090 apparatus, at the ESPCI (55).

3. RESULTS AND DISCUSSION

3.1. Polyethers

Some results concerning the preparation of polyethers from BPA and DCB in aqueous alkali-toluene system, at 65°C, using TBAH as catalyst have been previously reported (5,10). The polymers had the following structure :

$$Cl-CH_2-CH=CH-CH_2-[-O-\emptyset-C(CH_3)_2-\emptyset-O-CH_2-CH=CH-CH_2-]_n-Cl$$

The absence of phenolic end groups was confirmed by UV spectroscopy according to the method described by SHCHORI and Mc GRATH (56). The index of polydispersity was remarkably low and the authors also noticed that $\bar{M}_n$ increased on increasing the catalyst concentration as well as the NaOH concentration. Finally, the kinetics of disappearance of DCB in the organic phase was studied by gas chromatography and a rate constant could be determined. All those experiments allowed the authors to propose a mechanism also assumed by JIN and CHANG (7) : polycondensation occurs via the alkylation of a phenolate end group by DCB in the organic phase. The cation of the catalyst Q^+ transports the phenolate to the organic phase, and takes back Cl^- to the aqueous phase continuously. This process governs the kinetics, and it is important to see that $[\sim O^-,Q^+] \ll [Cl \text{ end groups}]$ in the toluene phase $([O^-,Q^+]/[TBAH] \simeq 4\%)$.

According to this mechanism, we tried to improve the properties of the resulting polymer, and especially to enhance its molecular weight by playing on different factors, the main idea being to increase the concentration of active species derived from BPA in the organic phase. A very simple way of achieving such a goal was investigated by MAHAMAT (8) who studied the effect of the rate of stirring. He found that up to 500 rpm, $\bar{M}_n$ increased on increasing this rate; it seemed that over 500 rpm, the value of $\bar{M}_n$ had reached a plateau and no longer increased. He was also able to make sure that the longer the reaction time was, the higher the molecular weight became.

Another means of changing $[O^-,Q^+]$ in toluene is to use various [BPA]/[DCB] ratios. Such a study was started by N'GUYEN (5) and completed by MAHAMAT (8) and SLEIMAN (11). The results are summed up in Table I. It seems that for [BPA]/[DCB] $>$ 3, putting too much BPA has a bad effect on both the kinetics and the structure of the polymer; the mechanism is probably no longer valid.

Table I.
Preparation of polyethers from BPA and DCB in the system : t=5h; Toluene/NaOH 3N (10 ml/10 ml) at 65°C (DCB : 3.07 mmol; TBAH : 0.614 mmol; 500 rpm). Influence of the ratio [BPA]/[DCB].

Run	[BPA]/[DCB]	Yield (%)	$\bar{M}_n$ (elemental analysis)	$\bar{M}_n$ NMR	$\bar{M}_n$ GPC	$\bar{M}_w$ GPC	I_p
1	1	100	2250	3100	3100	5500	1.8
2	2	74	5000	5200	5300	13200	2.5
3	3(a)	70	12500	-	(b)	-	-
4	4(a)(c)	58	8000	6700	2600	3400	1.3

(a) The aqueous phase contains a solid; (b) $\bar{M}_n$ by osmometry of the soluble fraction in toluene : 52000; (c) Presence of phenolic end groups in the polymer.

The same type of results was observed by SLEIMAN (11) when performing the same reactions in ODCB.

Considering the mechanism that we proposed, it is easy to see that $[\sim O^-,Q^+]$ in toluene can be increased by adding an inorganic chloride to the aqueous phase. Such experiments were carried out by SLEIMAN (11) whose results were in good agreement with this idea, as shown in Table II.

Table II
Determination of anions derived from BPA in the organic phase in the absence of DCB (TBAH : 0.614 mmol; toluene/NaOH 3N:(10ml/10ml); 65°C).

Run	BPA 10^3mol	NaCl 10^3mol	[NaOH]/2[BPA]	$[Na^+]/[-OH]$	$[-O^-,Q^+].10^5$eq in organic phase (10ml)	$\bar{M}_n$ elemental analysis
5	3.07	0	4.88	4.88	0.36	2250
6	6.14	0	2.44	2.44	0.44	5000
7	9.21	0	1.62	1.62	0.44	12500
8	3.07	3.07	4.88	5.38	1.47	3300
9	3.07	6.14	4.88	5.88	0.78	1800
10	3.07(a)	0	9.76	9.76	3.37	3400

(a) ODCB/NaOH 6N.

The concentration of anions was determined by UV spectroscopy. Run 9 shows that too much salt lowers the molecular weight, probably because it prevents the transfer of Q^+Cl^- back to the aqueous phase. It is obvious in run 10 that the concentration of anions in ODCB is much higher than in toluene. This illustrates the last effect that we studied, i.e. the nature of the solvent and more specially its dielectric constant, keeping in mind that the solvent used had to be a good solvent for the polymer and a bad one for BPA. The results obtained by SLEIMAN (11) are shown in Table III.

Table III

Preparation of polyethers from BPA and DCB under PTC conditions. Influence of the solvent.
(Solvent/NaOH 6N:10ml/10ml; BPA:3.07mmol;DCB:3.07mmol; TBAH:0.614mmol; t=5h; 500 rpm; 65°C).

Run	Solvent (a)	ε (25°C)	Yield %	$\bar{M}_n$ elemental analysis	$\bar{M}_n$ NMR	$\bar{M}_n$ GPC	$\bar{M}_w$ GPC	I_p
11	toluene	2.38	84	4600	4600	3500	4900	1.4
12	TCB	3.94	83	3300	3200	8400	13700	1.6
13	CB	5.62	90	4900	5200	9500	21900	2.3
14	ODCB	9.93	75	20000	-	31000	53000	1.7
15	NB	34.82	80	11000	9700	19300	48700	2.5

(a) TCB:1,2,4-trichlorobenzene; CB:chlorobenzene;
ODCB:1,2-dichlorobenzene; NB:nitrobenzene

$\bar{M}_n$ increases on increasing ε, except for a strongly polar solvent such as nitrobenzene. However, phenolic end groups appear in the polymers prepared in ODCB and nitrobenzene, and the polydispersity becomes larger. In fact, the polarity of the organic phase must have an influence on both the extraction of active species $\sim O^-,Q^+$ and on their own reactivity.

MAHAMAT (8) tried to study the same effect on the preparation of polyethers from BPA and 1,6-dibromohexane (DBH). He found that it was necessary to operate in a more polar solvent than toluene with concentrated NaOH as shown in Table IV. According to JIN and CHANG (7), the ratio $[\sim O^-,Q^+]/[Q^+]_o$ reaches 80 to 90% in a very polar solvent such as nitrobenzene.

TABLE IV

Preparation of polyethers from BPA and DBH (Solvent/aqueous NaOH : 10 ml/10ml; BPA:3.07 mmol; DBH:3.07 mmol; TBAH:0.614 mmol; 65°C; t=5h).

Run	Solvent	ε 25°C	Yield %	$\bar{M}_n$ elemental analysis	$\bar{M}_n$ osmometry	$\bar{M}_n$ NMR	$\bar{M}_n$ GPC	$\bar{M}_w$ GPC	I_p
16(a)	toluene	2.38	33	1700	-	1800	1200	1300	1.1
17(a)	CB	5.62	76	2700	-	3000	1300	1500	1.1
18(b)	ODCB	9.93	96	14800	7400	-	9400	25700	2.7
19(b)	CB	5.62	86	17000	7800	-	13000	23000	1.8
20(b)	NB	34.82	88	42000	8100	-	11000	22000	2.0
21(b)(c)	NB	34.82	95	80000	8700	-	12000	29000	2.4

(a) NaOH 3N, 300 rpm
(b) NaOH 6N, 500 rpm
(c) BPA : 9.21 mmol

$\bar{M}_n$ increases with the concentration of NaOH and with the stirring rate. In monochlorobenzene using NaOH 6N, no phenolic end groups were found (56). SLEIMAN (11) established that $\bar{M}_n$ was influenced by the ratios [BPA]/[DBH] and [NaBr]/[BPA] in the same way as in the case of DCB, as shown in Table V.

Table V

Preparation of polyethers from BPA and DBH in the system monochlorobenzene/NaOH 6N (10 ml/10 ml) at 65°C. (DBH:3.07 mmol; TBAH:0.614 mmol; t=5h; 500 rpm).

Run	[BPA]/[DBH]	[NaBr]/[BPA]	Yield %	$\bar{M}_n$ elemental analysis	$\bar{M}_n$ NMR	$\bar{M}_n$ GPC	$\bar{M}_w$ GPC	I_p
22	1	0	86	2200	2400	2900	4400	1.5
23	2	0	88	30000	-	5600	8000	1.4
24(57)	1	1	67	71000	-	8600	23000	2.7
25	1	2	87	3800	3400	6300	7800	1.3

The same kind of evolution was observed in the system ODCB/NaOH 6N but some phenolic end groups were present. However, the molecular weights were a little higher.

At last, several other observations were made :

- MAHAMAT (8) compared several α,ω-dibromoalkanes and the corresponding dichloroalkanes. He found that while it was possible to reach excellent yields and high molecular weights starting from dibromo compounds, the dichloroalkanes were very little reactive.

In the case of DBH, SLEIMAN (11) showed that it was possible to hydrolyse quantitatively the CH_2Br end groups into CH_2OH end groups under PTC conditions without affecting the chain. This reaction allows further coupling reactions, e.g. using a bischloroformate or a diisocyanate.

In conclusion, the preparation of polyethers from BPA and dihalogeno compounds under PTC conditions has been thoroughly investigated. The mechanism formerly proposed has been confirmed by numerous experimental facts and it is thus possible to prepare polymers having high molecular weights and well-controlled end groups with excellent yields by choosing an appropriate solvent/aqueous NaOH/NaX system with a convenient [BPA]/[XRX] ratio.

3.2. Polycarbonates

Since liquid-liquid PTC proved to be a reliable means of preparing polyethers we decided to apply it to the synthesis of polycarbonates using a new type of reaction : the dihalogeno compound was in this case a di(chlorocarbonyloxy) compound (bischloroformate ClOC-O-R-O-COCl). Two diphenols, BPA and BPAF were mainly studied by MADJOUB (39) who tried to proceed in the same way as what had been done for polyethers before. Since those studies were totally new, he began by making sure that PTC offered advantages over other methods. Consequently, he prepared the same polymer from the same starting materials (BPAF and diethyleneglycol bischloroformate $ClOC-O-CH_2-CH_2-O-CH_2-CH_2-O-COCl$ ($DEGCl_2$)), but following three different procedures. His results are shown in Table VI.

Table VI

Preparation of polycarbonates from BPAF and $DEGCl_2$
(20°C; BPAF:5 mmol; $DEGCl_2$:5mmol)

Run	Process	Base	Solvent	Yield %	$\bar{M}_n$ GPC	$\bar{M}_w$ GPC	I_p
26(a)	PTC	NaOH 3N	ODCB	66	38900	71750	1.8
27(b)	solution	pyridine	ODCB	78	2350	2950	1.3
28(c)	emulsion	NaOH 1N	THF/H_2O	30	9200	45500	4.9

(a) TBAH:1 mmol; NaOH 3N/ODCB:20ml/20ml
(b) ODCB:20ml; pyridine:20 mmol
(c) THF:20ml; NaOH 1N:20 ml; emulsifier:sodium laurylsulfate (0.7mmol)

Run 26 shows that PTC allows the synthesis of a polycarbonate having the highest molecular weight with a quite low index of polydispersity. The process was consequently worth being studied more thoroughly.

The effect of NaOH concentration was studied in ODCB, both for BPA and BPAF and for a mixture of the two diols (1:1). It appeared that $\bar{M}_n$ increased and then decreased when increasing the concentration of NaOH, as shown in Table VII; in run 31 no polymer was recovered after 3h of reaction.

Table VII
Preparation of polycarbonates from BPA or BPAF and $DEGCl_2$ under PTC conditions (diphenol:5 mmol; $DEGCl_2$:5 mmol; ODCB/aq.NaOH:20ml/20 ml; t=3h; 20°C). Effect of NaOH concentration.

Run	diphenol	[NaOH]	Yield %	$\bar{M}_n$ GPC	$\bar{M}_w$ GPC	I_p	T_g °C
29	BPA	1N	28	3300	6000	1.7	-
30	BPA	3N	71	14000	23600	1.7	60
31	BPA	6N	0	-	-	-	-
32	BPAF	1N	71	15000	37000	2.5	-
26	BPAF	3N	66	39000	71500	1.8	76
33	BPAF	6N	56	15500	24000	1.6	-
34	BPA/BPAF	3N	61	32000	50000	-	58

A second fact becomes probably preponderant when the medium is too strongly basic:presumably, reactions of degradation can occur between a phenolate end group and a carbonate unit. This was suggested by the fact that polymer formation actually takes place even for the system ODCB/NaOH 6N/BPA: good yields of polymer showing high enough molecular weights could be recovered by stopping the reaction very early, e.g. after 15 or 30 minutes. Moreover, MAJDOUB noticed that the way of introducing the bischloroformate had a large influence on the results. When adding dropwise the dihalogeno compound, the resulting polymer showed a high molecular weight but the yield and polydispersity were poor, whereas more polymer with lower polydispersity and molecular weight was recovered when the bischloroformate was added all at once.

Some experiments were also carried out in order to study the effect of the solvent. A comparison between dichloromethane and ODCB was made. Little polymer was recovered from the reactions conducted in dichloromethane, even when using lower concentrations of NaOH. This suggested that the amount of bisphenolate anions is higher in CH_2Cl_2, and this was confirmed by the determination of anions in the organic phase by UV spectroscopy, as shown in Table VIII.

Table VIII
Determination of anions in the organic phase (bisphenol: 5 mmol; TBAH: 1 mmol; solvent/aq NaOH : 20 ml/20 ml).

System	CH_2Cl_2/NaOH 6N	CH_2Cl_2/NaOH 3N	ODCB/NaOH 6N	ODCB/NaOH 3N
$ØO^-Q^+$ mmol.l^{-1}	BPA 88 BPAF -	BPA 88 BPAF 29	BPA - BPAF 0.025	BPA 0.28 BPAF 0.03

The amount of phenolate in the organic solvent is much larger in the case of CH_2Cl_2. Consequently, the degradation reactions must be greatly enhanced in CH_2Cl_2 hence the poor yields and molecular weights observed. In conclusion, it was established that a competition between the chain growth and its degradation occurred when the transfer of bisphenolate in the organic phase was more important, the degradation becoming rapidly preponderant. Thus this type of reaction seems more delicate and sensitive to the different parameters than the preparation of polyethers. However the structure of the polymers was always well-defined :

$$HOCH_2CH_2OCH_2CH_2[OCOOROCOOCH_2CH_2OCH_2CH_2]_nOH$$

with hydroxyl end groups resulting from the hydrolysis of the chloroformate groups.

Several other experiments have been recently carried out :

- the synthesis of a polycarbonate starting from a mixture of BPA and BPAF (1:1) and $DEGCl_2$ lead to a copolymer containing roughly 50% of each moiety.
- Two other bischloroformates were successfully reacted with BPA. Butanediol bischloroformate ($BDCl_2$) and hexanediol bischloroformate ($HDCl_2$) allowed the synthesis of polymers having high molecular weights with excellent yields as shown in Table IX (58).

Table IX

Preparation of polycarbonates from BPA and several bischloroformates under PTC conditions (BPA:5 mmol; bischloroformate:5 mmol; ODCB/aq.NaOH:20ml/20 ml; t=3 h; 20°C).

Run	bischloroformate	[NaOH]	Yield %	$\bar{M}_n$ GPC	$\bar{M}_w$ GPC	Ip
35	$HDCl_2$	3N	95	6800	15500	2.3
36(a)	$HDCl_2$	6N	76	11400	20000	1.8
37	$BDCl_2$	3N	88	16600	39500	2.4

(a) t= 15 minutes.

Recently (59), a physico-chemical study of end-to-end cyclization dynamics in solution was undertaken with a polycarbonate prepared from BPA and $DEGCl_2$. The process used relies on fluorescence spectroscopy measurements and was first applied by WINNIK (60). Both ends of the polymer chain are labelled with a fluorescent probe. In our case, 3-pyrenebutyroylchloride was used. It reacted with the hydroxy end groups to give 100% functionalized polymer. The rate of formation of the resulting intramolecular excimer in highly dilute solutions was then measured to determine the kinetic parameters for end-to-end cyclization dynamics. More experiments will be soon carried out with other polycarbonates.

3.3. Polyesters

Lastly, we wish to report the synthesis of new polyesters as an example of solid-liquid PTC. The polymers were synthesized from potassium 2,6-naphthalene dicarboxylate and various dihalogeno compounds XRX, including DCB and α,ω-dibromoalkanes. The solvent was 1-methyl-2 pyrrolidinone (NMP) and dicyclohexyl-18 crown-6 (DCHE) or cryptand [222] was used as catalyst. Those systems have been studied by SLEIMAN (11) and DORIGO (51). They tried at first to find the best system by playing on different factors such as the nature of the catalyst or the stoichiometry of the reagents. It turned out that the former had no appreciable effect, while the latter was more important: it is better to operate with a slight excess of dipotassium salt.

Attempts were also made to study the effect of the nature of the halogen X : 1,6-dichlorohexane gave poor results compared to 1,6-dibromohexane. This illustrates the fact that Br^- is a better nucleofuge than Cl^- in a polar aprotic solvent such as NMP. The following experiments were thus conducted with DCHE and a ratio [dicarboxylate]/[XRX] = 1.05. All the results are shown in Table X.

Table X
Preparation of polyesters from the dipotassium salt of 2,6-naphthalene dicarboxylic acid (5.25 mmol) and XRX (5 mmol) under PTC conditions (Solvent : NMP (4 ml); Catalyst : DCHE (1.05 mmol); t=3h; 750 rpm at 110°C).

Run	XRX	Yield %	$\bar{M}_n$ elemental analysis	$\bar{M}_n$ NMR	$\bar{M}_n$ GPC	$\bar{M}_w$ GPC	I_p
38	DCB	84	3000	-	3800	8800	2.3
39	$Br(CH_2)_4Br$	97	1800	1300	3800	5100	1.4
40	$Br(CH_2)_5Br$	94	4200	-	3600	8100	2.3
41(a)	$Br(CH_2)_6Br$	(d)	6900	-	6900	14600	2.1
42	$Br(CH_2)_6Br$	87	2600	2300	5300	7700	1.5
43(b)	$Br(CH_2)_6Br$	86	2800	2600	6400	9700	1.5
44(c)	$Br(CH_2)_6Br$	74	1700	1700	3200	4600	1.35
45	$Cl(CH_2)_6Cl$	20	1050	1050	1800	2500	1.4
46	$Br(CH_2)_7Br$	87	26500	-	11000	-	-
47	$Br(CH_2)_8Br$	90	32900	-	11000	-	-
48	$Br(CH_2)_9Br$	87	27800	-	-	-	-

(a) the polymer was not purified by reprecipitation
(b) catalyst : [222]
(c) dicarboxylate : 4.17 mmol; DBH : 5 mmol
(d) not determined

The yields are very good except in the case of 1,6-dichlorohexane. Moreover, NMR and IR spectroscopy showed that the polymers have a well-defined structure X-[R-O-CO-Napht-COO]$_n$-R-X.
This allowed the calculation of $\bar{M}_n$ from NMR spectra and elemental

analysis. However, those data become rather difficult to use when $\bar{M}_n$ is too high and the precision is not so good. DSC analysis was then run with those samples. They did not have thermotropic properties, however two different crystalline phases were detected in all cases. The way in which the main melting temperature T_m was affected by the polymer structure was then investigated and two major effects appeared as shown in Table XI.

Table XI
Melting points of polyesters prepared from the dipotassium salt of 2,6-naphthalene dicarboxylic acid and XRX (main peak).

Run	XRX	$\bar{M}_n$ GPC	T_m °C
38	DCB	3800	217
39	$Br(CH_2)_4Br$	3800	206
40	$Br(CH_2)_5Br$	3600	120
44	$Br(CH_2)_6Br$	3200	179
42	$Br(CH_2)_6Br$	5300	189
43	$Br(CH_2)_6Br$	6400	197
41	$Br(CH_2)_6Br$	6900	206
46	$Br(CH_2)_7Br$	11000	137
47	$Br(CH_2)_8Br$	11000	170

Considering the case of 1,6-dibromohexane it is easy to see that T_m increases on increasing $\bar{M}_n$. Introducing a double bond in the flexible spacer also leads to an increase of T_m (runs 38 and 39). When plotting T_m versus the flexible chain length n an "odd-even" effect appears : the values of T_m are much higher when n is even than when it is odd. Considering only the even values of n, a general decrease is observed when increasing n. Both effects are well known in the case of thermotropic liquid crystal polymers for the different phase transition temperatures. However, the increase of T_m between n=5 and n=7 seems rather curious, although it might be due to a molecular weight effect.

In conclusion, solid-liquid PTC is a very convenient means of synthesizing polyesters containing the 2,6-naphthalene dicarboxylic moiety, with high yields and well-controlled end groups. The polydispersity of the polymers is quite low. Moreover, those polymers, although not thermotropic, show very interesting thermal properties and it seems worth going on studying similar systems.

4. CONCLUSION

Phase transfer catalysis is a very exciting process for the preparation of polyethers, polycarbonates and polyesters with high molecular weights and well-controlled end groups. Having some insight on the kinetics and the mechanism as well as on the nature of the active species, it becomes easier to prepare a polymer having given properties by choosing the good system. Consequently, we wish to apply the PTC process to the synthesis of new polyethers, polyesters and polycarbonates having liquid crystalline thermotropic or lyotropic properties. This might be done by improving the systems that we have already studied or by using new monomers such as 4,4'-dihydroxybiphenyl or 2,5-dihydroxymethylfuran. Moreover, it would be very interesting to develop the dynamic studies of end-to-end cyclization with other polymers in dilute solution, or even in the solid state, which would be particularly interesting with thermotropic liquid crystal polymers.

REFERENCES

(1) Laboratoire de Chimie Macromoléculaire associé au CNRS UA 24.
(2) A.K. BANTHIA, D.C. WEBSTER and J.E. Mc.GRATH, Org. Coatings and Plastics Chem. Prep., **42**, 127 (1980).
(3) A.K. BANTHIA, D. LUNDSFORD, D.C. WEBSTER and J.E. McGRATH, J. Macromol. Sci. Chem. **A15**, 943 (1981).
(4) R. KELLMAN, D.G. GERBI, R.F. WILLIAMS and J.L. MORGAN, Polym. Prep. **21**(2), 164 (1980).
(5) T.D. N'GUYEN, Thèse de doctorat d'Etat, Paris (1981).
(6) G.G. CAMERON and K.S. LAW, Makromol. Chem. Rapid. Commum. **3**, 99 (1982).
(7) J.L. JIN and J.H. CHANG, Polym. Prep. **23**, 156 (1982); 'Phase transfer polycondensation of bisphenolate anions and 1,6-dibromohexane'. Crown ethers and phase transfer catalysis in polymer science (L.J. MATHIAS and C.E. CARRAHER,Jr, editors) Plenum Press p. 91 (1984).
(8) O. MAHAMAT, Thèse de doctorat de 3e cycle, Paris (1983).
(9) N. YAMAZAKI and Y. IMAI, Polym. J. **15**, 603 (1983).
(10) T.D. N'GUYEN and S. BOILEAU, 'Synthesis of polyethers by phase transfer catalyzed polycondensation'Crown ethers and phase transfer catalysis in polymer science (L.J. MATHIAS and C.E. CARRAHER, Jr, editors), Plenum Press, p. 59 (1984).
(11) H. SLEIMAN, Thèse de doctorat de 3e cycle, Paris (1985).
(12) Y. IMAI, M. UEDA and M. II, J. Polym. Sci. Polym. Lett. Ed. **17**, 85 (1979).
(13) N. YAMAZAKI and Y. IMAI, Polym. J. **17**(2), 377 (1985).
(14) D.J. GERBI, R.F. WILLIAMS, R. KELLMAN and J.L. MORGAN, Polym. Prep. **22**, 385 (1981).
(15) D.G. GERBI, G. DIMOTSIS, J.L. MORGAN, R.F. WILLIAMS and R. KELLMAN, J. Polym. Sci. Polym. Lett. Ed. **23**, 551 (1985).
(16) A.S. HAY, F.J. WILLIAMS, H.M. RELLES, B.M. BOULETTE, P.E. DONAHUE and D.S. JOHNSON, J. Polym. Sci. Polym. Lett. Ed. **21**, 449 (1983).
(17) A.S. HAY, F.J. WILLIAMS, H.M. RELLES, J.C. CARNAHAN, G.R. LOUCKS, B.M. BOULETTE, P.E. DONAHUE and D.S. JOHNSON, New monomers and polymers (B.M. CULBERTSON and C.U. PITTMAN,Jr, editors) Plenum Press p. 67 (1984).
(18) A.S. HAY, F.J. WILLIAMS, H.M. RELLES and B.M. BOULETTE, J. Macromol. Sci. Chem. **A21**, 1065 (1984).
(19) V. PERCEC and B.C. AUMAN, Polym. Bull. **10**, 385 (1983).

(20) V. PERCEC, B.C AUMAN and P.L. RINALDI, Polym. Bull. **10**, 391 (1983).
(21) V. PERCEC and B.C. AUMAN, Makromol. Chem. **185**, 617 (1984).
(22) V. PERCEC and H. NAVA, Makromol. Chem. Rapid. Commun. **5**, 319 (1984).
(23) V. PERCEC and B.C. AUMAN, Polym. Bull. **12**, 253 (1984).
(24) T.D. SHAFFER and V. PERCEC, Makromol. Chem. Rapid. Commun., **6**, 97 (1985).
(25) T.D. SHAFFER and V. PERCEC, J. Polym. Sci. Polym. Lett. Ed. **23**, 185 (1985).
(26) T.D. SHAFFER, M. JAMALUDIN and V. PERCEC, J. Polym. Sci. Polym. Chem. Ed. **23**, 2913 (1985).
(27) T.D. SHAFFER, M. JAMALUDIN and V. PERCEC, J. Polym. Sci. Polym. Chem. Ed., **24**, 15 (1986).
(28) T.D. SHAFFER and V. PERCEC, Makromol. Chem. **187**, 111 (1986).
(29) T.D. SHAFFER and V. PERCEC, Makromol. Chem. **187**, 1431 (1986).
(30) A.M. MUÑOZ, F.R. DIAZ and L.H. TAGLE, Polym. Bull., **11**, 493 (1984).
(31) P. KELLER, Makromol. Chem. Rapid Commun. **6**, 255 (1985).
(32) P.H. PARSANIA, P.P. SHAH, K.C. PATEL and R.D. PATEL, J. Macromol. Sci. Chem. **A22** (11), 1495 (1985).
(33) K. SOGA, S. HOSODA and S. IKEDA, J. Polym. Sci. Polym. Lett. Ed. **15**, 611 (1977).
(34) K. SOGA, Y. TOSHIDA, S. HOSODA and S. IKEDA, Makromol. Chem. **178**, 2747 (1977).
(35) K. SOGA, Y. TOSHIDA, S. HOSODA and S. IKEDA, Makromol. Chem. **179**, 2379 (1978).
(36) G. ROKICKI, W. KURAN and J. KIELKIEWICZ, J. Polym. Sci. Polym. Chem. Ed., **20**, 967 (1982).
(37) G.ROKICKI, B. POGORZELSKA-MARCINIAK and W. KURAN, Polym. J. **14**, 847 (1982).
(38) J.M.J. FRECHET, F. BOUCHARD, F.M. HOULIHAN, E. EICHLER, B. KRYCZKA and C.G. WILLSON, Makromol. Chem. Rapid Commun., **7**, 121 (1986).
(39) M. MAJDOUB, Thèse de doctorat de 3e cycle, Paris (1985).
(40) J.A. MYKROYANNIDIS, Eur. Polym. J., **21**, 895 (1985).
(41) J.A. MYKROYANNIDIS, Eur. Polym. J., **21**, 1031 (1985).
(42) Z.K. BRZOZOWSKI, J. KIELKIEWICZ and Z. GOCLAWSKI, Angew. Makromol. Chem. **44**, 1 (1975).
(43) Z.K. BRZOZOWSKI, J. DUBCZINSKI and J. PETRUS, J. Macromol. Sci. Chem. **A13**, 875 (1979).
(44) Z.K. BRZOZOWSKI, J. DUBCZYNSKI and J. PETRUS, J. Macromol. Sci. Chem., **A13**, 887 (1979).
(45) Y. IMAI and S. TASSAVORI, J. Polym. Sci. Polym. Chem. Ed., **22**, 1319 (1984).
(46) J.S. RIFFLE, A.K. BANTHIA, D.C. WEBSTER and J.E. McGRATH, Org. Coatings and Plastics Chem. Prep., **42**, 122 (1980).
(47) J.S. RIFFLE, R.G. FREELIN, A.K. BANTHIA and J.E. McGRATH, J. Macromol. Sci. Chem., **A15**, 967 (1981).
(48) F.L. KEOHAN, R.G. FREELIN, J.S. RIFFLE, I. YILGÖR and J.E. McGRATH, J. Polym. Sci. Polym. Chem. Ed., **22**, 679 (1984).
(49) G.G. CAMERON and K.S. LAW, Polymer, **22**, 272 (1981).
(50) J. KIELKIEWICZ, W. KURAN and B. POGORZELSKA, Makromol. Chem. Rapid Commun., **2**, 255 (1981).
(51) R. DORIGO, Rapport d'activité de DEA, Paris (1986).

(52) C.G. CAMERON, G.M. BUCHAN and K.S. LAW, Polymer, 22, 558 (1981).
(53) Société Nationale des Poudres et Explosifs, 12 quai Henri IV 75181 Paris Cédex 04 FRANCE.
(54) Laboratoire de Physique et de Chimie des Hauts Polymères de l'Université Catholique de Louvain la Neuve, BELGIQUE.
(55) Laboratoire de Physicochimie Structurale et Macromoléculaire de l'ESPCI, 10 rue Vauquelin 75005 Paris, FRANCE.
(56) E. SHCHORI and J.E. McGRATH, Appl. Polym. Symp., 34, 103 (1978).
(57) M. HAAMDI, Rapport d'activité de DEA, Paris (1985).
(58) F. MECHIN, unpublished results.
(59) S. BOILEAU, F. MECHIN and M.A. WINNIK, unpublished results.
(60) M.A. WINNIK, Acc. Chem. Res., 18, 73 (1985).

SYNTHESIS OF POLYARYLETHERKETONES

John B. Rose
Chemistry Department
University of Surrey
Guildford, Surrey GU2 5XH
England

ABSTRACT. The first reported synthesis of polyaryletherketones, in 1962, employed polyaroylation of phenyl ether with bis-aroyl chlorides under Friedel-Crafts conditions; analagous reactions between dicarboxylic acids and aryl ethers have been under investigation since 1968. Polyether synthesis by reaction of bis-4-halogenobenzoyl compounds with bis-phenoxides provides an alternative route to polyaryletherketones and this was reported first in 1963. Initially, both general routes were ineffective for synthesis of high melting partially crystalline polyaryletherketones of high molecular weight due to the insolubility of these polymers in conventional solvents. Du Pont solved the problem for the aroylation route by using liquid HF as solvent with BF_3 as catalyst, reporting this work in 1969, while ICI disclosed an effective polyether synthesis in 1973 which used phenyl sulphone as solvent at temperatures close to 300°C. The present paper is concerned mainly to describe extensions of this work to explore the structural scope of these reactions, to find alternatives to the dangerous HF/BF_3 technique for polaroylation and to develop alternatives to the high temperature polyether synthesis. Recent commercial developments are noted.

1. INTRODUCTION

A wide range of polyaryletherketones have now been made in the laboratory but attention has been concentrated on those structures which can develop substantial crystallinity, as polymers of this type, e.g. those shown in Table I, show great promise as high performance engineering thermoplastics for extended use in air at temperatures up to about 250°C. Two of these polymers, PEEK and PEK, are now manufactured by ICI for sale in world markets and other companies have indicated their intention to produce similar materials.

The two main routes to polyaryletherketones, the polyaroylation reactions (1) and the polyether syntheses (2), are shown in generalized forms below. For each route dissimilar functional groups can be in different monomers, reactions (1a) and (2a) or can be

M. Fontanille and A. Guyot (eds.), Recent Advances in Mechanistic and Synthetic Aspects of Polymerization, 207–221.

TABLE I: SELECTED POLYARYLETHERKETONES

STRUCTURE	CODE	T_g, °C	T_m, °C
$-Ph-O-Ph-O-Ph-CO-$	PEEK	143	334
$-O-Ph-O-Ph-CO-Ph-CO-Ph-$	PEEKK	155	365
$-Ph-O-Ph-CO-$	PEK	155	365
$-O-Ph-CO-Ph-CO-Ph-$	PEKK	165–175	385

incorporated in a single monomer as in reactions (1b) and (2b). For polyaroylations, (1), which are typical electrophilic aromatic substitutions, the use of acid chlorides, which requires a Friedel-Crafts catalyst, must be distinguished from the reactions with carboxylic acids which require the presence of dehydrating agents and usually more forcing reaction conditions. The polyether syntheses, (2), are nucleophilic aromatic substitutions and at present are known to give crystalline polymers of high molecular weight only when the leaving group is fluorine and the bis-phenol derivative an alkali metal bis-phenoxide or a bis-trimethylsilylated bis-phenol.

(1a)
$$XCO-Ph-[Y-Ph]_m-Ph-COX + Ph-Y-[Ar-Y]_n-Ph \longrightarrow -CO-Ph-[Y-Ph]_m-Ph-CO-Ph-Y-[Ar-Y]_n-Ph- + 2HX$$

(1b)
$$XCO-Ph-[Y-Ph]_m-Y-[Ar-Y]_n-Ph \longrightarrow -CO-Ph-[Y-Ph]_m-Y-[Ar-Y]_n-Ph- + HX$$

X = F, Cl, OH ; Y = —, -O- ; m, n = 0 or 1.

(2a)
$$Z-Ph-CO-[Ar-CO]_m-Ph-Z + MO-[Ar-CO]_n-Ar-OM \longrightarrow -O-Ph-CO-[Ar-CO]_m-Ph-O-[Ar-CO]_n-Ar- + 2MZ$$

(2b)
$$Z-Ph-CO-[Ar-CO]_m-Ar-OM \longrightarrow -O-Ph-CO-[Ar-CO]_m-Ar-$$

Z = Hal, NO_2 or OSO_2Ph; M = Alkali metal or $SiMe_3$; m, n = 0 or 1

The commercial utility of these polymers is based on a combination of properties in which toughness and substantial crystallinity are crucial and to obtain tough polymers high molecular weights are required, the threshold solution viscosity for toughness being considerably higher for crystalline polymers than for the corresponding amorphous (quenched) materials (for details see Table II). Development of experimental techniques to give polymers with Inherent Viscosities, IVs, greater than the threshold value, 0.8, took several years after the initial discovery of the polyaryletherketones had been made and this was a major factor delaying the commercial development of these polymers.

TABLE II: THRESHOLD VISCOSITIES FOR TOUGH PEK SAMPLES[1]

RV*	[3]+	Morphology#	Flex. Test
0.49	0.43	A	F
0.67	0.54	A	PF
0.76	0.60	A	P
0.76	0.60	C	F
1.03	0.78	C	PF
1.10	0.84	C	P

* For 1% solution in 98% H_2SO_4 at 25°

\+ For solutions in 98% H_2SO_4 at 25°

\# A = Amorphous (quenched) film. C = Partially crystalline (slow cooled) film. F = Fail. P = Pass.

2. POLYAROYLATION

2.1 Polyaroylation with Aroyl Halides under Friedel-Crafts Conditions

The first preparation of a completely aromatic polyetherketone, reported by Bonner in 1962,[2] employed this reaction to obtain polymers of low molecular weight, IV ca. = 0.2, from isophthaloyl and terephthaloyl chlorides and phenyl ether, reaction (3), using nitrobenzene as solvent and $AlCl_3$ as catalyst. Goodman et al.[3] obtained similar products by analogous reactions and were the first to report a polymer with the PEK, see Table I, structure which they obtained by reaction (4) with $AlCl_3$ as catalyst and methylene chloride as the solvent. PEK, and other regularly structured polyetherketones like those shown in Table I, develop considerable crystallinity which coupled with their high melting points makes them insoluble in conventional solvents at temperatures below 250°C. This poses a major synthetic problem as it limits the molecular weight that can be obtained before the growing chains crystallise out from the polymerisation system. The problem was solved by Marks[4] who found that liquid HF is a good solvent for the polymers (the polymer solutions are bright yellow in colour indicating protonation) and the BF_3/HF complex an excellent catalyst for reactions such as (3) and (4), giving polymers with IVs well over the threshold value for toughness.

(3)

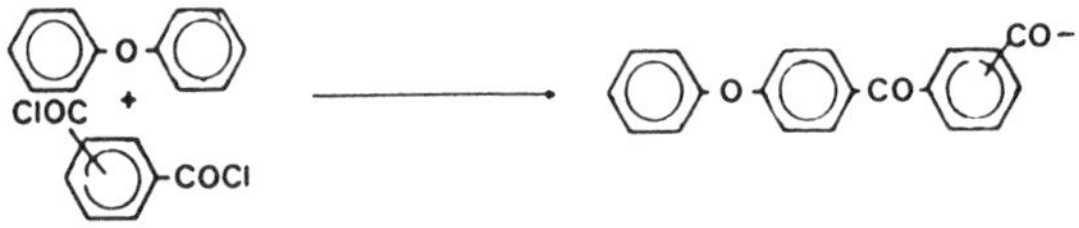

(4)

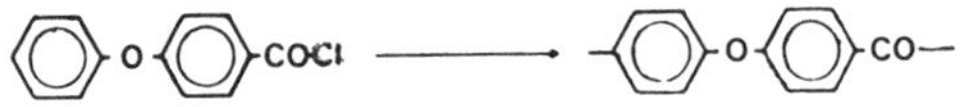

No mechanistic investigations of these reactions have been published but it may be assumed that they proceed conventionally and that aroylation occurs by attack of aryl acylium ions, ArCO+, on reactive phenyl groups as in reaction (5). Thus, suitable co-monomers for reaction with bis-aroyl halides are diphenyl, phenyl ether and 1,4-diphenoxybenzene, but not benzophenone or 1,4-dibenzoylbenzene where electron withdrawal by the carbonyl groups deactivates the phenyl rings to electrophilic attack. Benzene itself is not a suitable co-monomer, because after monoaroylation further reaction is prevented by the deactivating effect of the carbonyl group first introduced.

(5)

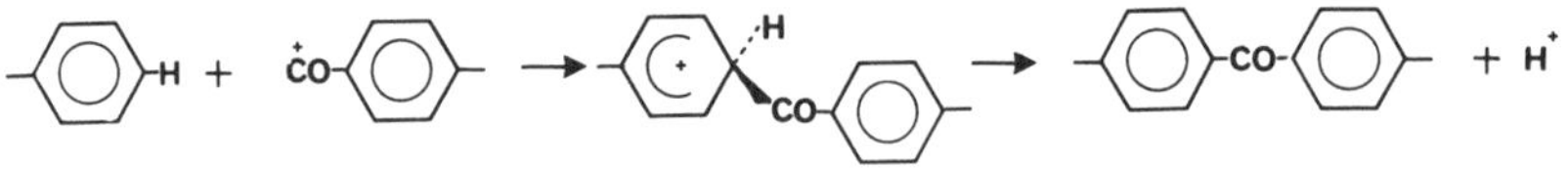

The BF_3/HF process has seen considerable development by Dahl and his co-workers at the Raychem Corporation[5] who showed that the threshold solution viscosity for toughness with partially crystalline PEK specimens (formed by conventional melt fabrication techniques) was IV ≥ 0.8. This group has used the process to make a wide range of polymers some indication of the extent of their work being given in Table III.

It is interesting to note that although the phenylene ring in 1,4-diphenoxybenzene with its two ether substituents appears to be the ring in this compound most activated to electrophilic attack, the polymers formed are reported not to be cross linked as would be expected if the phenylene ring reacted. This is probably because the reaction medium is highly acidic and the monomer completely protonated when delocalisation of the charge deactivates the phenylene ring and one of the phenyl rings, as shown below the equations, so that reaction (6) occurs but (7) does not. When one phenyl group in the monomer has reacted protonation shifts to the carbonyl group and the terminal phenoxyl group becomes reactive but delocalisation of the positive charge continues deactivation of the phenylene ring so that a linear polymer is obtained.

TABLE III: POLYARYLETHERKETONES VIA THE HF/BF_3 PROCESS.[6]

Monomers		Polymer IV	mp (°C)
ClCO–C6H4–COCl	+ C6H5–O–C6H4–O–C6H5	0.90	360
ClCO–C6H4–O–C6H4–COCl	+ C6H5–O–C6H4–O–C6H5	1.62	340
ClCO–C6H4–COCl	+ (C6H5–O–C6H4)$_2$CO	1.31	370
ClCO–C6H4–COCl	+ C6H5–C6H4–O–C6H5	0.43	402
ClCO–C6H4–COCl (meta)	+ C6H5–O–C6H4–O–C6H5	1.36	271
90 T/10 I	+ "	1.44	337
75 T/25 I	+ "	1.51	328
50 T/50 I	+ "	1.45	294

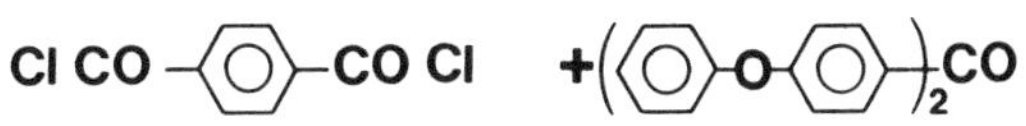

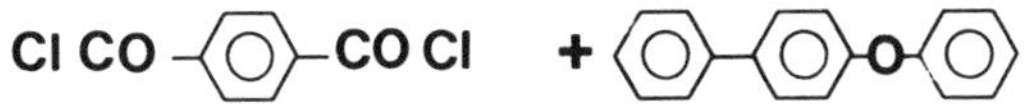

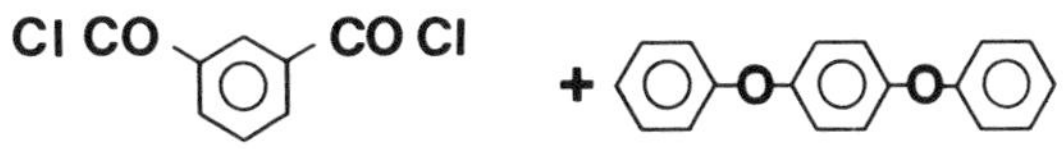

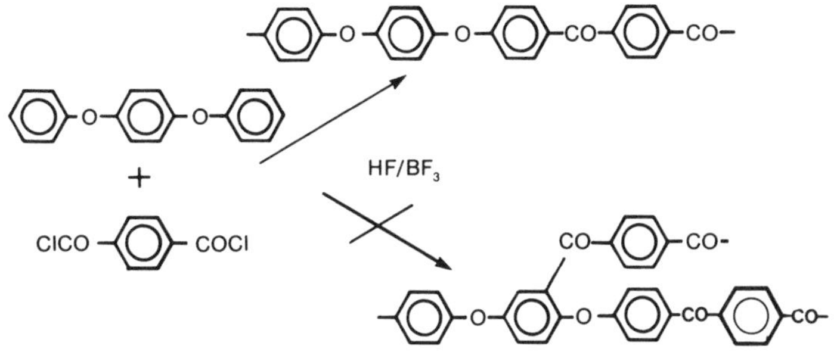

(6)

HF/BF$_3$

ClCO COCl

CO– CO– CO– CO–

(7)

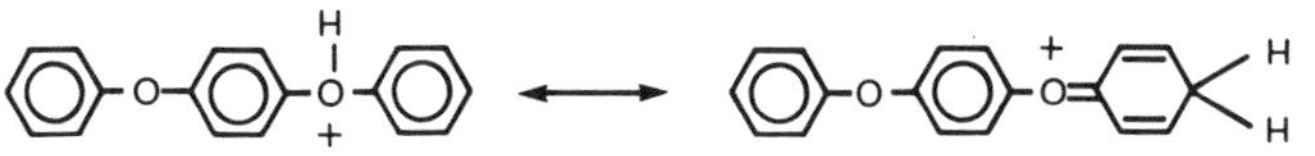

More recently other solvent/catalyst systems have been found which appear to work as well as HF/BF_3 in that they give polymers of comparable molecular weight (see Table IV). Of these, the more interesting ones employ a large excess of $AlCl_3$ as catalyst plus a Lewis Base such as dimethylformamide or lithium chloride to moderate the catalyst's activity and a conventional organic solvent such as 1,2-dichloroethane.[4] The function of the Lewis Base is not understood, but it appears to form a complex with the $AlCl_3$ which acts as a solvent for the polymer/$AlCl_3$ complex during the polymerisation "thereby maintaining the polymer in solution or in a reactive gel state".[7] Trifluoromethanesulphonic acid has also been used as a catalyst/solvent system[8] and polymers of moderate molecular weight have been obtained using the product prepared by dissolving $AlCl_3$ in polyphosphoric acid (when HCl is evolved) as the catalyst/solvent.[9,10]

TABLE IV: CATALYSTS AND SOLVENTS FOR POLYAROYLATION WITH ACID HALIDES

Ph–O–C6H4–COX --(Catalyst, Solvent)--> –[C6H4–O–C6H4–CO]– + HX

X = F or Cl

SOLVENT	CATALYST	IV[a] or RV[b]	REFERENCE
HF	BF_3	2.34	K.J. DAHL Brit.Pat.1,387,303 (1975)
CF_3SO_3H	CF_3SO_3H	1.0	J.B. ROSE Euro.Pat.63874 (1982)
CH_2Cl_2, CH_2ClCH_2Cl	>2 mol $AlCl_3$/LB[c]	2.50	H.GORS and V.JANSON Int.Pat. WO84/03892 (1984)

[a] 0.2 wt.% in H_2SO_4. [b] 1% in H_2SO_4. [c] LEWIS BASE eg. DMF, LiCl, 1-Ethylpyridinium Bromide etc.

2.2 Polyaroylation with Carboxylic Acids

Examples of this reaction are given in Table V. The first reported (in 1968) synthesis of polyaryletherketones by this method was by Iwakura et al.[13] who obtained polymers of moderate molecular weight

TABLE V: POLYAROYLATION WITH CARBOXYLIC ACIDS

REACTANTS	SOLVENT/CATALYST	VISCOSITY	REF.
HO_2C–C6H4–O–C6H5	HO–[P(=O)(OH)–O]$_n$–H	0.53[a]	13
HO_2C–C6H4–CO_2H + C6H5–O–C6H5	CH_3SO_3H/P_2O_5	0.34[a]	11
HO_2C–C6H4–O–C6H4–CO_2H + C6H5–O–C6H4–O–C6H5	CF_3SO_3H	2.68[b]	12

[a] IV for 0.2% solution in H_2SO_4. [b] RV for 1% solution in H_2SO_4.

(IVs up to 0.53) by polycondensation of 4-phenoxybenzoic acid and of the 3-phenoxy derivative in polyphosphoric acid. These workers showed by ^1H nmr that polymer from 4-phenoxybenzoic acid had the all para PEK structure. Polymers have been obtained using methanesulphonic acid as solvent and P_2O_5 as the dehydrating agent, but these were of low molecular weight and probably contained substantial proportions of ortho-substituted repeat units.[11] The most effective system found so far is that described by Colquhoun in 1984[12] where trifluoromethanesulphonic acid is used as both solvent and dehydrating agent and polymers of high molecular weight were obtained from selected monomers. Trifluoromethanesulphonic acid is a "Super Acid" so that the monomers react as protonated species and this limits the scope of the reaction. Thus, 4-phenoxybenzoic acid does not give polymer via reaction (8) due to protonation as indicated below the equations, but the analogous monomer used in reaction (9) gives polymer of high molecular weight because even when protonated it still contains a reactive phenoxyl group. Protonation in this monomer also ensures that the polymer made from it has the all _para_-structure shown in equation (9) for reasons analogous to those discussed in the previous section.

(8)

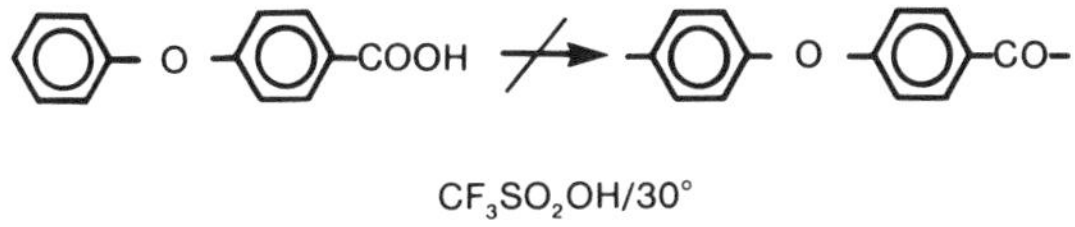

$CF_3SO_2OH/30°$

(9)

T_m 330° ; RV 2.31

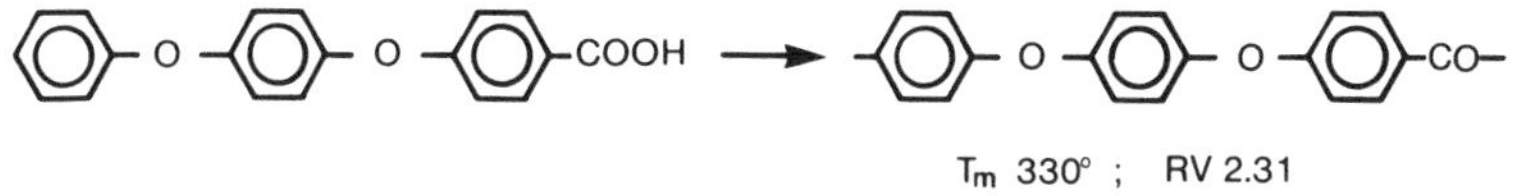

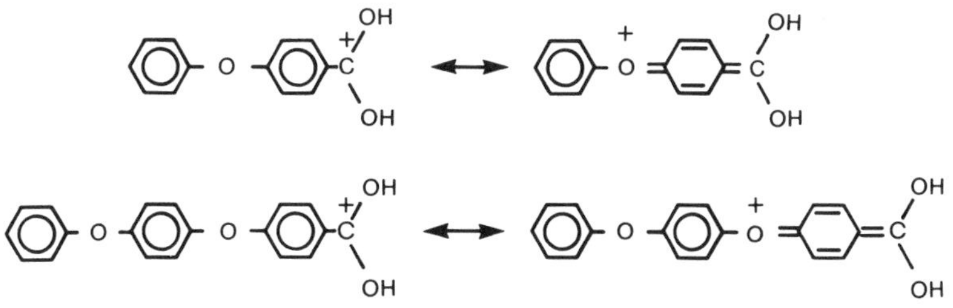

3. POLYETHER SYNTHESIS

3.1. Mechanism and Reaction Conditions

The polyether synthesis was first reported as a general route to polyaryl ethers by Farnham and Johnson in 1963.[14] This process is a nucleophilic aromatic substitution of the type known to proceed via an intermediate complex as indicated in equation (10). In this reaction

(10)

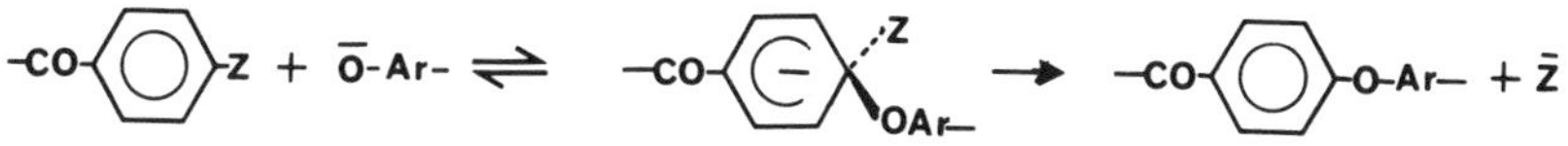

the carbonyl groups play an essential part by activating the leaving groups, Z, to nucleophilic attack via stabilization of the intermediate complex by electron delocalization. Several leaving groups have been exemplified but at present only fluorine has been reported to give polymers with solution viscosities over the threshold value (see Table VI) for toughness.

TABLE VI: LEAVING GROUPS AND REACTION CONDITIONS FOR THE POLYETHER SYNTHESIS

Ar	Z	Solvent	Time/Temp (hrs/C)	RV[a]	Refs
$-C_6H_4-CO-C_6H_4-$	Cl	Ph_2SO_2	1.5-3/335	0.48-0.72	15
"	F	"	1.5-3/335	1.00-1.87	1
"	Ph_2SO_3	"	1/300	0.3	16
$-C_6H_4-CMe_2-C_6H_4-$	NO_2	Sulpholan	35/130	0.3[b]	17

[a] 1% solution in H_2SO_4; [b] 0.2% solution in $CHCl_3$.

Bis-potassium salts appear to be the most useful bis-phenol derivatives, but an interesting variation has been reported[18] in which bis-trimethylsilyl derivatives are employed as indicated in equation (11). A catalyst is required for this reaction of which the most effective found was caesiun fluoride, which is believed to act by the mechanism shown in reactions (12) and (13).[18]

$$(11)\quad F-C_6H_4-CO-C_6H_4-F \;+\; Me_3SiO-Ar-OSiMe_3 \longrightarrow -O-Ar-O-C_6H_4-CO-C_6H_4- \;+\; 2\,Me_3SiF$$

$$(12)\quad -Ar-OSiMe_3 + CsF \rightleftharpoons -Ar-OCs + FSiMe_3$$

$$(13)\quad -Ar-OCs + F-C_6H_4-CO- \rightleftharpoons CsF + -Ar-O-C_6H_4-CO-$$

The polyetherketones first made by the polyether synthesis had structures which prevented the development of substantial crystallinity so that polymers of high molecular weight could be obtained with conventional dipolar aprotic solvents. However, Farnham and Johnson failed to obtain high molecular weight samples of PEK or PEEK by reactions such as (14) or (15) using their preferred solvent, sulpholane, owing to premature crystallisation of the polymers from this solvent during the reactions.[19,20] The problem was solved by ICI[21] who in 1975 reported the use of phenyl sulphone as solvent for preparation of crystalline co-polyketonesulphones via reaction (16). With this solvent the polycondensations could be conducted at temperatures close to the polymers' melting points and products with solution viscosities up to 2.57 obtained. Aryl sulphones appear to be unique solvents for these reactions (see Table

VII) showing far greater resistance to decomposition than conventional dipolar aprotic solvents at the high reaction temperatures required to hold the polymers in solution.[1]

(14)

F-⟨O⟩-CO-⟨O⟩-F
+
KO-⟨O⟩-CO-⟨O⟩-OK

-⟨O⟩-CO-⟨O⟩-O-

(15)

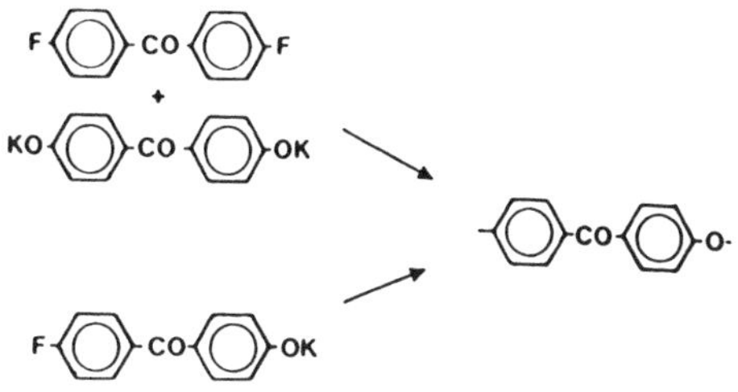

F-⟨O⟩-CO-⟨O⟩-OK

(16)

KO-⟨O⟩-CO-⟨O⟩-OK
+
(1-X)Cl-⟨O⟩-CO-⟨O⟩-Cl
+
X Cl-⟨O⟩-SO_2-⟨O⟩-Cl

$(⟨O⟩-CO-⟨O⟩-O)_{2-X}$ $(⟨O⟩-SO_2-⟨O⟩-O)_X$

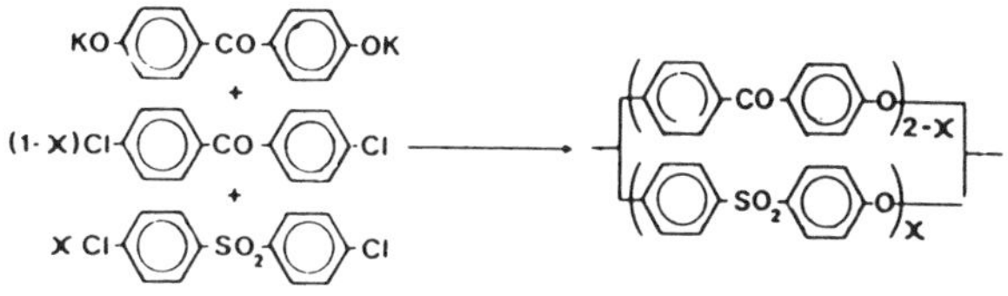

When reactions (14) and (15) were conducted at 335°C using phenyl sulphone as solvent polymers of high molecular weight were obtained without difficulty (see Tables IX and X in the next section). Later on it was found that difficulties associated with prior preparation of the potassium phenoxides employed could be avoided by employing a slight excess of potassium carbonate as the base and adding this to the fluorophenol or mixture of difluoride and bis-phenol employed. This technique increased the effectiveness of the synthesis considerably and was used to obtain high molecular weight samples of PEEK for the first time by reaction (17).[22]

(17)

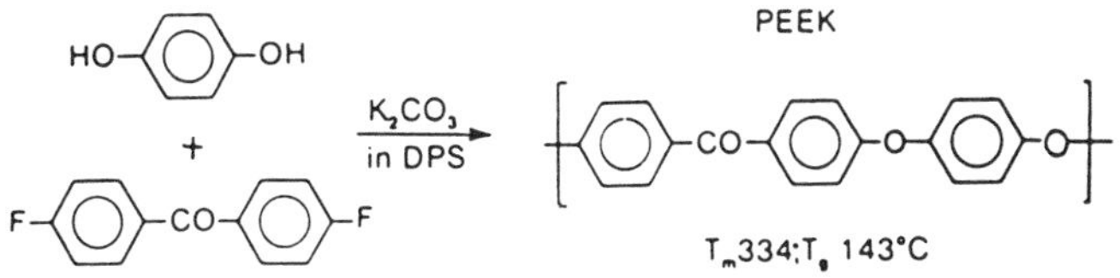

HO-⟨O⟩-OH
+
F-⟨O⟩-CO-⟨O⟩-F

K_2CO_3 in DPS

PEEK

-⟨O⟩-CO-⟨O⟩-O-⟨O⟩-O-

T_m 334; T_g 143°C

TABLE VII: SYNTHESIS OF 80-20 COPOLYETHERKETONE-SULPHONE IN VARIOUS SOLVENTS[7]

Solvent	Polym. Temp.	Polymer RV*	Volatile** products
Hexamethyl phosphoramide	238°C	Insol.	NMe_3
Dimethyl sulphone	250	0.35†	MeSH, Me_2S
Sulpholane	280	0.28†	SO_2, C_4 olefine
N,N-diphenyl acetamide	290	insoluble	CH_3COCH_3
Ditolyl sulphone	"	0.18†	SO_2, PhMe
Methyl phenyl sulphone	"	0.17†	SO_2, PhH
Benzophenone	"	0.14	–
Diphenyl sulphone	"	2.57	–
4 – Phenylsulphonyl biphenyl	"	1.35	–
Dibenzothiophene dioxide	"	1.49	–

* For 1% solutions in H_2SO_4 which had been filtered if necessary to remove solids
** In addition to water
† Solution deeply coloured

More recently, non crystalline high molecular weight derivatives of polyetherketones have been prepared by polyether synthesis in conventional solvents and then converted to the crystalline polyether ketones. Thus, the reaction sequence (18) has been employed to obtain high molecular weight samples of PEK and copolymers containing PEK and PEEK repeat units were obtained by similar methods.[23]

(18) HO–C₆H₄–C(OCH₂CH₂O)–C₆H₄–OH + F–C₆H₄–CO–C₆H₄–F, K_2CO_3, dmac, 150°C → [–C₆H₄–C(OCH₂CH₂O)–C₆H₄–O–C₆H₄–CO–C₆H₄–O–]

I: T_g = 155°C., RV* = 0.8

I $\xrightarrow{H^{\oplus}}$ [–C₆H₄–CO–C₆H₄–O–]

3.2 Scope of the Polyether Synthesis and Occurrence of Side Reactions

The preferred variant of the polyether synthesis using phenyl sulphone as solvent has been widely applied. Some examples using _bis_-4-fluorbenzoyl compounds with _bis_-phenoxides or _bis_-phenols and potassium carbonate are listed in Table VIII.

Polymers made from _bis_-phenoxides and difluorides were usually free from gel and gave pale yellow solutions in sulphuric acid. The molecular weights of these polymers could be controlled by addition of an excess of one monomer in the usual way, and samples of PEK prepared

TABLE VIII: POLYARYLETHERKETONES FROM BIS-4-FLUOROBENZOYL COMPOUNDS AND BIS-PHENOLS[1]

DBP = Hal-⟨◯⟩-CO-⟨◯⟩-Hal DBB = Hal-⟨◯⟩-CO-⟨◯⟩-CO-⟨◯⟩-Hal

Hal-Φ-CO-Φ-Hal Φ ≡ -⟨◯⟩- Hal-Φ-CO-Φ-CO-Φ-Hal

Polymer	Dihalide	Bis phenol	Polymer repeat unit	Tm (°C)	Tg (°C)
III	DBP	HO–Φ–OH	–O–Φ–O–Φ–CO–Φ–	335*	144
IV	DBP	HO–Φ–OH + HO–Φ–CO–Φ–OH	[–O–Φ–O–Φ–CO–Φ–] [–Φ–O–CO–Φ–]	345	154
I	DBP	HO–Φ–CO–Φ–OH	–O–Φ–CO–Φ–	367+	154
V	DBB	HO–Φ–OH	–O–Φ–O–Φ–CO–Φ–CO–Φ–	358	154
VI	DBP	HO–Φ–CO–Φ–CO–Φ–OH	–O–Φ–CO–Φ–CO–Φ–O–Φ–CO–Φ–	383	–
VII	DBB	HO–Φ–CO–Φ–CO–Φ–OH	–O–Φ–CO–Φ–CO–Φ–	384	–
VIII	DBP	HO–Φ–Φ–OH	–O–Φ–Φ–O–Φ–CO–Φ–	416	–
IX	DBP	HO–Φ–OH + HO–Φ–Φ–OH	[–O–Φ–O–Φ–CO–O–] [–O–Φ–Φ–O–Φ–CO–Φ–]	341	160

*Farnham et al. report Tm = 350 +Dahl reports Tm= 365

by reaction (14) showed a threshold solution viscosity for toughness (measured by a creasing test on film) at RV = 1.10 (see Table IX) virtually the same as that recorded by Dahl and his co-workers for PEK made by polyaroylation using the BF_3/HF technique. Samples of PEK were also prepared from potassium-4-fluorbenzoylphenoxide, reaction (15), but these were found to contain ca. 2% gel and to give deep red solutions in sulphuric acid. A measure of this colour was obtained by determining the light absorbance at 420 nm, which appears as a shoulder on the side of the main protonated carbonyl peek (Figure 1) when it was seen that the polymers from the fluorophenoxide (Table X) showed absorbance greater than 1% whereas PEK samples from the difluoride/bis-phenoxide system had absorbance usually below 0.25%.

An important difference between the two series of polymers was that those from the fluorphenoxide showed a threshold viscosity for toughness of RV = 1.9, considerably greater than that required by polymers from the two monomer system. This difference appears due to some structural difference between the two families of PEKs and probably involves chain branching, which is a likely precursor to gel formation. This was confirmed by introducing branches deliberately into polymers made from difluoride and the bis-phenoxide by adding small quantities of 2,4,4'-trifluorobenzophenone to the polycondensation recipe when polymers showing toughness versus solution viscosity behaviour similar to that of the fluorophenoxide

TABLE IX: PEK FROM BIS-4-FLUOROPHENYL KETONE AND THE BIS-PHENOXIDE[1]

(14) F-⟨◯⟩-CO-⟨◯⟩-F + KO-⟨◯⟩-CO-⟨◯⟩-OK $\xrightarrow[\text{at } 335^{\circ}]{Ph_2SO_2}$ [⟨◯⟩-CO-⟨◯⟩-O] + KF

Mole % excess monomer used.			Polymer produced	
$(KO-C_6H_4)_2CO$	$(F-C_6H_4)_2CO$	RV*	Absorbance*	Flex. Test
1.0	–	0.57	1.02	–
–	–	1.87	0.13	P
–	–	1.82	0.25	P
–	1.0	1.60	0.21	–
–	–	1.49	–	P
–	–	1.23	0.12	P
–	–	1.16	–	P
–	–	1.03	0.25	P
–	–	1.03	0.12	PF
–	–	1.00	0.14	F
–	1.0	0.93	0.18	F
–	2.5	0.67	0.20	F
–	3.3	0.55	0.23	–

* For 1% solutions in H_2SO_4 at 25°

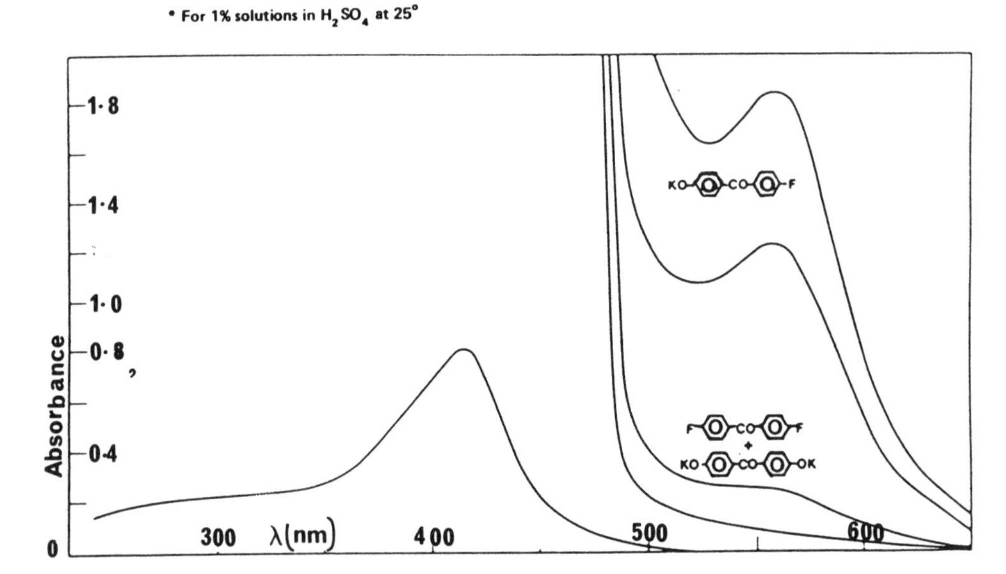

Figure 1. Spectra of PEK in 98% sulphuric acid

TABLE X: PEK FROM POTASSIUM-4-FLUOROBENZOYLPHENOXIDE[1]

(15) $F-C_6H_4-CO-C_6H_4-OK \xrightarrow[\text{at } 335^\circ]{Ph_2SO_2} [-C_6H_4-CO-C_6H_4-O-] + KF$

RV*	% Gel	Absorbance*	Flex. Test
6.0	2.97	–	P
4.66	1.97	–	P
2.17	1.08	0.98	P
2.09	1.12	1.22	P
1.95	1.16	–	P
1.82	0.20	1.55	PF
1.67	0.04	1.63	F
1.40	1.64	> 2	F
1.30	0.08	2	F
1.29	0.92	1.0	F
1.07	1.56	> 2	F
1.04	1.16	1.85	F
1.02	0.72	> 2	F

*For 1% solutions in H_2SO_4 at 25° filtered to remove gel

polymers were obtained (see Table XI), but which gave yellow solutions in sulphuric acid with low absorbence. A mechanism, reaction sequence (19), which involves proton abstraction from the difluoride, (19A), to give a carbanion which then reacts further either to give a trifunctional branching co-monomer, (19B), or adds across a carbonyl group, reaction (19C), to give a triphenyl carbonyl derivative which when incorporated in the polymer would lead to the red colours observed in the sulphuric acid solutions. Side reactions of this type are more likely to occur in the fluorphenoxide polycondensations

TABLE XI: PEK FROM DIFLUORIDE AND THE BIS-PHENOXIDE PLUS 2,4,4'-TRIFLUOROBENZOPHENONE (TFBP)[1]

Mole % TFBP added	Polymer properties				
	% Gel	RV*	Absorbance*	Colour	Flex test
2	3.5	3.58	-	Yellow	PF
1	0	3.69	-	"	P
1	0	1.29	0.20	"	F

* For filtered 1% solutions in sulphuric acid

because here the reactivity of fluorine is reduced compared with that in the difluoride by the presence of a phenoxide substituent in the adjacent ring (see ref. 24 for this effect in the analogous sulphone compounds). Thus, side reactions of the phenoxide functional groups are more likely to occur in the fluorophenoxide than in the difluoride/bis-phenoxide system.

(19)

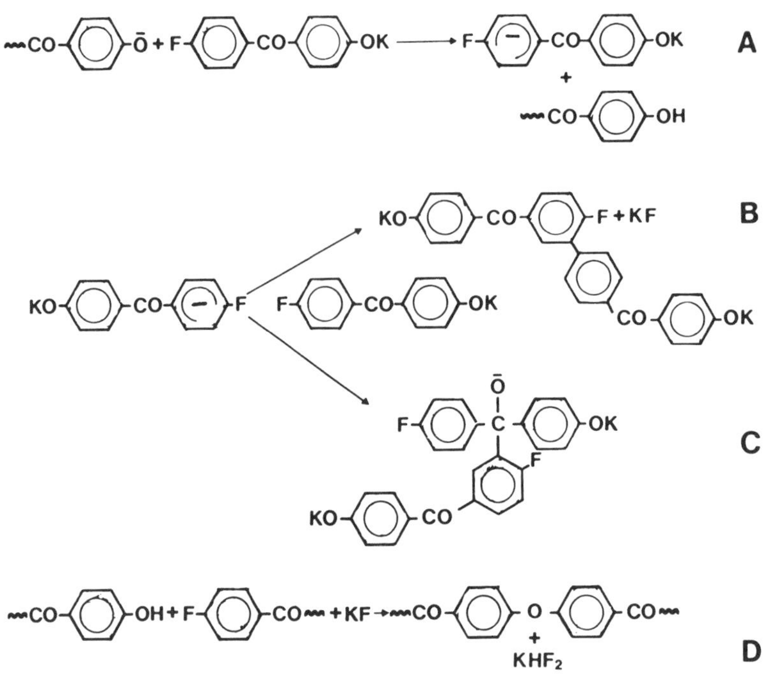

A

B

C

D

COMMERCIAL PRODUCTION OF POLYARYLETHERKETONES

Small scale production of PEK using the HF/BF_3 polyaroylation process was started by Raychem Corporation in 1972. The polymer produced, which was given the trade name "Stilan", was used for in-house manufacture of electrical equipment, but production was discontinued in 1976. ICI started production of "VICTREX" PEEK by polyether synthesis in 1980 and recently announced construction of a new 1000 tonne/year plant; production of PEK was started in 1986. BASF and Hoechst have indicated their intention to start production of PEK this year, but have not disclosed which process will be employed. Union Carbide's engineering plastics business has been brought by AMOCO who also intend to produce a polyaryletherketone in 1987 but have not disclosed the structure of their product, although it is likely to be produced by polyether synthesis. All these developments are summarised in Table XII.

TABLE XII: COMMERCIAL POLYARYLETHERKETONES

STRUCTURE	PRODUCED BY	TRADE NAME	PROD.STARTS
[-C6H4-CO-C6H4-O-]	RAYCHEM.	STILAN	1972 (STOPPED '76)
[-C6H4-CO-C6H4-O-C6H4-O-]	ICI	VICTREX PEEK	1980
[-C6H4-CO-C6H4-O-]	ICI	VICTREX PEK	1986
..	BASF	ULTRAPEK	1987
..	HOECHST	HOSTATEC	1987
?	AMOCO	KADEL	1987

REFERENCES

1. J.B. Rose et al. Polymer, **22**, 1097 (1981).
2. W.H. Bonner, USP 3, 065,205 (1962).
3. I. Goodman et al. BP 971,227 (1964).
4. B.M. Marks, USP 3,442,857 (1969).
5. K.J. Dahl, BP 1,387,303 (1975).
6. K.J. Dahl and V. Jansons, BP 1,471,171 (1977).
7. H. Gors and V. Janson, Int. Pat. WO84/03892 (1984).
8. J.B. Rose, Euro. Pat. EP 63874 (1982).
9. Y. Iwakura et al. J. Polym. Sci. Polym. Lett. Ed., **15**, 283 (1977).
10. Y. Iwakura et al. J. Polym. Sci. Polym. Chem. Ed., **20**, 1965 (1982).
11. M. Ueda and T. Kano, Makromol.Chem. Rapid Commun., **5**, 833 (1986).
12. H. Colquhoun, Polymer Preprints, **2**, 17 (1984).
13. Y. Iwakura et al. J. Polym. Sci. Al., **6**, 3345 (1968).
14. R.A. Farnham and W.F. Johnson, BP 1,078,234 (1965).
15. L.R.J. Hoy, Ph.D. Thesis, Surrey, (1976).
16. BASF, German Patent Application, 24 50 789 (1976).
17. W. Schmidt, E. Radlmann and G. Nischk, BP 1,238,124 (1971).
18. G. Bier and H.R. Kricheldorf, Polymer, **25** 1151 (1984).
19. A.G. Farnham et al. J. Polym. Sci. Al., **5**, 2375 (1967).
20. R.A. Clendinning, BP 1,177,183 (1970).
21. J.B. Rose, BP 1,414,421 (1975).
22. J.B. Rose and P.A. Staniland, USP 4,320,224 (1982).
23. D.R. Kelsey, EP 0,148,633 (1985).
24. T.E. Attwood, A.B. Newton and J.B. Rose, Br.Polym.J. **4**, 391 (1972).

ELECTROCHEMICAL REDUCTIVE POLYCONDENSATION OF DIBROMOAROMATIC DERIVATIVES INTO POLY (1,4 - PHENYLENE)

Jean-François FAUVARQUE[*], Abdelouafi DIGUA[**] and Michel - Alain PETIT[**]

[*]Division Energétique, Laboratoires de Marcousis (CGE), F - 91460 MARCOUSSIS
[**]Laboratoires de Recherches sur les Macromolécules, UA CNRS 5O2, Université Paris - Nord, F-93430 VILLETANEUSE.

ABSTRACT. Poly (1,4 - phenylene), (PPP), can be prepared by electrochemical reductive polycondensation of 4,4' - Dibromo biphenyl under controlled potential conditions in polar aprotic medium in presence of Nickel complexes as catalysts. PPP can be prepared in powder form or electrodeposited as thin films over conductive electrodes. The number of phenyl rings in the chain has been checked from infra-red spectra. Physical properties are consistent with the values obtained. During exhaustive electrolysis the highest molecular weight material is formed at the very beginning of the process. This rules out a mechanism of polycondensation which should give the highest molecular weight material at the very end of the process. This suggests a chain mechanism through reactive intermediates.

Introduction.

In the past few years, there has been an impressive number of reports on the electrochemical preparations and properties of conducting polymers. Such an interest may be explained by the potential applications of these materials. However the litterature is still evasive about the mechanism of the electropolymerization of the aromatic monomers (1).

Anodic polymerizations of monomers such as pyrrole or thiophene are probably initiated by a cation radical, but the following steps of propagation remain unclear ; chain polymerization (2) or oxidative coupling polycondensation (3,4) have been proposed. The presence or absence of oligomers in solution before nucleation has received little attention (5) though this species could be more reactive than the monomer. In our opinion, there is a need for further studies of the electrodeposition of such materials. Obviously a better knowledge of the mechanism would help in defining the experimental conditions for getting materials of improved quality.

M. Fontanille and A. Guyot (eds.), Recent Advances in Mechanistic and Synthetic Aspects of Polymerization, 223–235.

Preparative procedure.

Recently, we reported (6) on the electrochemical synthesis of a highly regular poly (1,4- phenylene) , (PPP), by catalytic electroreduction of 4,4' - dibromo biphenyl. For obtaining large amounts of PPP in powder form, exhaustive electrolyses were performed in a jacketed three electrode cell with two compartments, separated by a glass-sintered disk. A mercury pool with its continuously stirred surface was polarized against an Ag / Ag^{+} reference electrode, while a piece of lithium was generally used as a sacrificial counter-electrode. The typical electrolysis was carried out as follows. To a mixture of 30 mmol. of lithium perchlorate electrolyte in 100 ml of solvent (THF - HMPA, 30% by vol.) were added under argon 10 mmol. of monomer and 1 mmol. of catalyst, preferably Ni Cl_2 (dppe), [dppe : 1,2 bis-(diphenyl phosphino) ethane]. The cathode was polarized at the working potential ($\approx$ -2.5 volt). The initially red solution began to darken as the formation of zerovalent nickel proceeded, and then a yellow suspension was formed whose amount increased as the electrolysis proceeded. Meanwhile, the high initial current decreased to the stable electrolysis level about 10 mA per sq. cm of mercury cathode at 35 ° C, usual electrolysis temperature. When about 90% of the monomer had disappeared, the current slowly fell down towards zero. The electrolysis was stopped when the current intensity dropped at 5 - 10 % of the electrolysis level. At this point, the amount of charge passed through the cell agreed with two Faraday equivalents for each Nickel (II) and aromatic dibromide (2 100 C). The yellow suspension was set apart and stirred in 100 ml of 6M sulfuric acid. The bromide ion content of the solution was checked by potentiometric titration, and, when the electrolysis was not exhaustive, the amounts of remaining dibromo biphenyl and of monobromo biphenyl checked by GC analysis led to an excellent bromine balance at any time. The light yellow powder was then washed with water and acetone. Yields were in the range of ≥80%.

Elemental analysis indicated a composition in agreement with polyphenylene (keeping in mind the difficulty to get complete combustion of carbon from these materials)

$(C_6H_4)_n$	Calc.	C 94.74	H 5.26	
	Found	C 91.43	H 5.35	Br 0.6

For obtaining well defined thin films over conductive electrodes, the procedure is modified as follows : The cathode is made as a plaque of chosen material (gold, vitreous carbon, ITO glass...), put in the cathodic solution and polarized at -2.5 volt. The molar ratio of catalyst over monomer is raised to 0.25. The temperature is room temperature and the amount of current is limited to 0.2 C per sq. cm. Under these conditions PPP films of about one micron thickness are obtained, whose optical spectra do not differ from those of the best powder products obtained in the previous procedure.

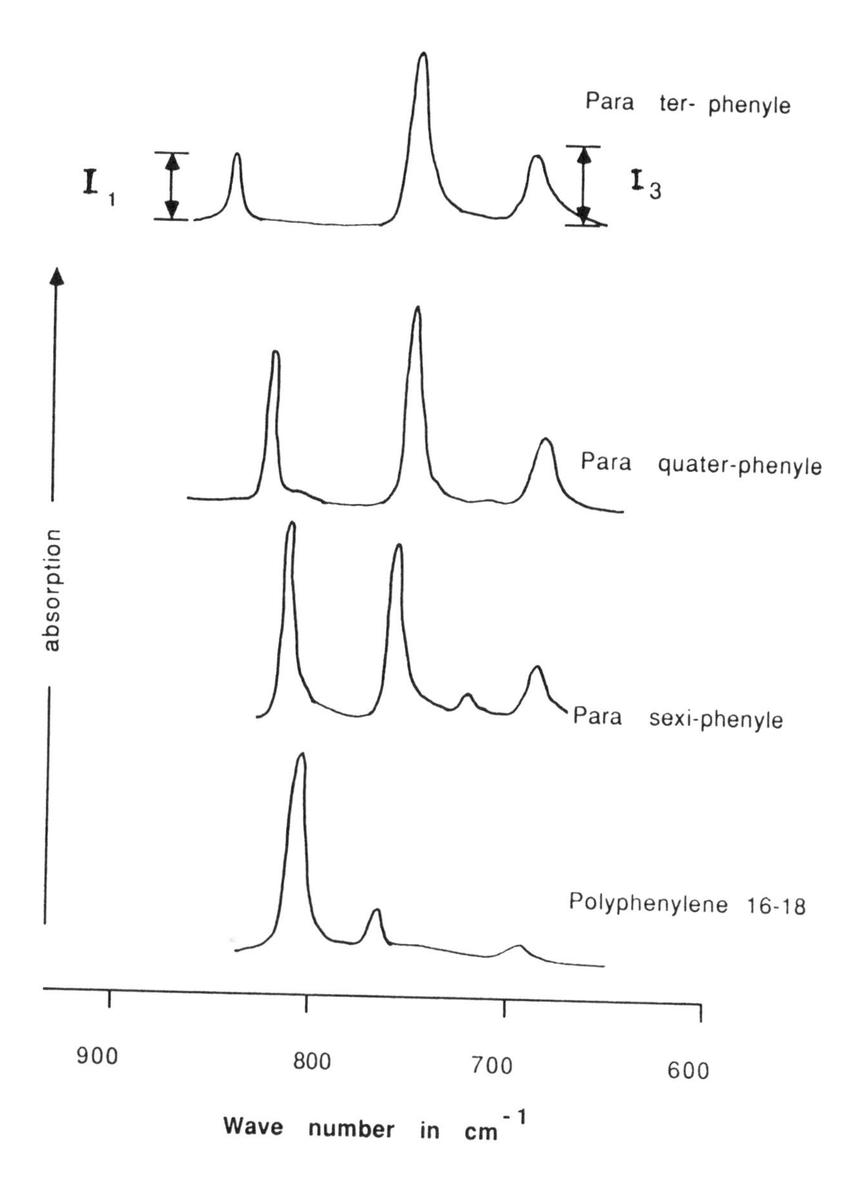

Infra-red determination of the number of phenyle units

FIGURE 1

Characterization :

By refluxing the polymer in trichlorobenzene, no more than 5 - 10% of the solid could be solubilized. Likewise, heating under vacuum at 250 °C could not induce a weight loss higher than 10%. These properties indicate that the polymer is of sufficient molecular weight for being insoluble. Therefore, classical characterization techniques in solution cannot be applied.

We were then conducted to study the relation between IR spectrometric measurements and the average chain length (*n*). Figure 1 shows the three bands observed in the 810 - 650 cm^{-1} range. The first band (intensity I_1) close to 800 cm^{-1} and the 760 cm^{-1} band are related to C – H out of plane deformation modes in para-substituted and mono-substituted aromatics, respectively. A ring puckering deformation appears close to 690 cm^{-1} (intensity I_3) which is distinctive for mono-substituted phenyl rings. The effect of *n* on these bands is shown for some oligo (1,4 - phenylene) (terphenyl, quaterphenyl and sexiphenyl).

As expected, it was found that the 800 cm^{-1} band shifted towards lower frequency with increasing chain length. Taking into consideration the number of C – H bonds involved in this absorption, I_1 was divided by a scaling factor according to Equation (1).

$$i_1 = I_1 / 4 (n - 2) \qquad (1)$$

It was also found that the ratio i_1 / I_3 is always close to 0.25 (p-terphenyl : 0.25, p-quaterphenyl : 0.26, p-sexiphenyl : 0.24). We assume that this relation prevails even for longer polymer chains. Thus, a good estimation of the *n* value of the polymer can be obtained from Equation (2).

$$n \approx (I_1 / I_3) + 2 \qquad (2)$$

Typical values for *n* are in the range of 10 to 14 after exhaustive electrolysis and superior to 16 for powders or films obtained at the beginning of the electrolysis. The precision of the method becomes bad with increasing *n* since the intensity I_3 becomes too low for being measured with precision.

Other characteristics have been reported in (7). When doped by AsF_5 the polymers show an electrical conductivity in the range of 5 to 50 S . cm^{-1}. In THF medium PPP films can be electrochemically reversibly reduced at -1.9 V versus Ag / Ag^+, in acetonitrile it can be reversibly oxidized at +0.6 Volt.

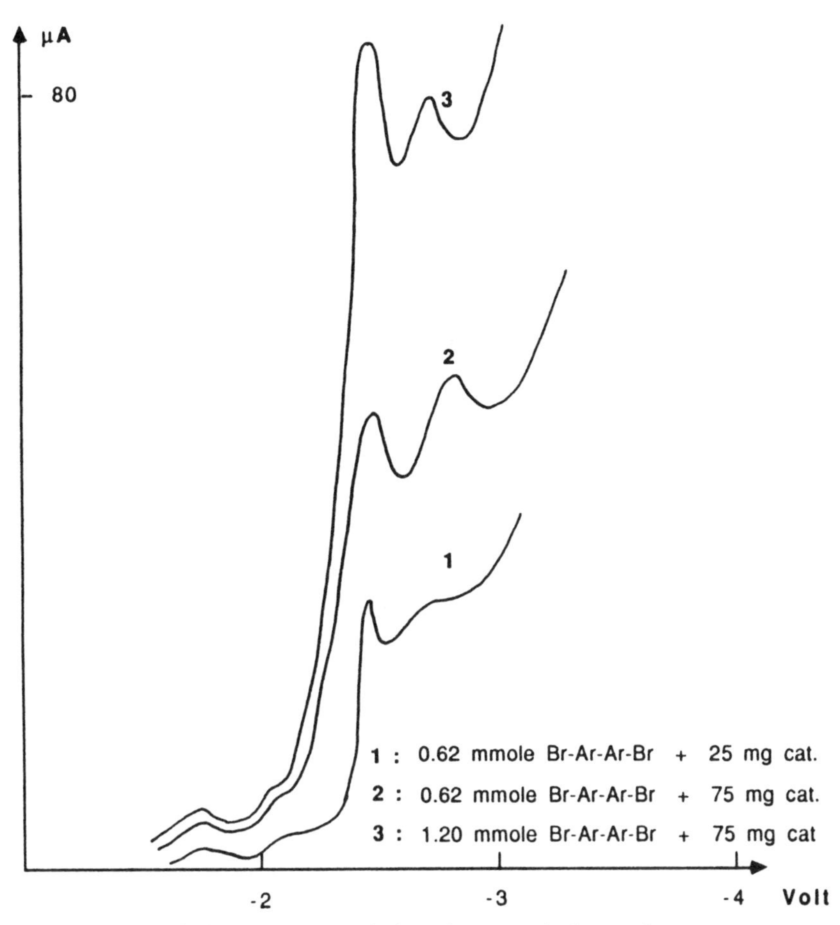

Voltamperogram of the electrosynthetic medium

at a gold microelectrode

FIGURE 2

Mechanism of the polymerization :

Electrochemical coupling of aromatic halides :

The formation of the carbon to carbon bond between two aromatic rings is based on an electrochemical coupling reaction catalyzed by Nickel complexes (8). The following electrochemical steps have been already described by previous reports on arylhalide coupling in the presence of Ni Cl_2 dppe (9) or Ni Cl_2 $[P(C_6H_5)_3]_2$ (8,10). It has been shown that bivalent Nickel complexed with aromatic phosphines can be reduced to zerovalent Nickel complexes. If the reduction is carried out in the presence of aromatic halides, these halides oxidatively add to zerovalent Nickel forming aryl-nickel complexes at the reduction potential (about - 1.8 V) of the bivalent Nickel (equation 3).

$$\mathrm{Ni\,Cl_2\,L_2} + \mathrm{Ar-X} + 2\,e^- \longrightarrow \mathrm{Ar\,Ni\,X\,L_2} + 2\,\mathrm{Cl^-} \quad (3)$$

$$L = P(C_6H_5)_3 \quad \text{or} \quad L_2 = \text{dppe}$$

The aryl-nickel complexes are themselves electroactive (at - 2.5 V) and their reduction yields a mixture of biaryl Ar – Ar and Ar – H (depending on the proticity of the medium) (equation 4).

$$2\,\mathrm{Ar\,Ni\,Cl\,L_2} + 2\,e^- \longrightarrow \mathrm{Ar-Ar} + 2\,\mathrm{Cl^-} + 2\,\mathrm{Ni^\circ\,L_2} \quad (4)$$

During that process, the zerovalent nickel catalyst is regenerated in a very reactive, coordinatively unsaturated, form. Several studies of the coupling mechanism have been reported (8, 10, 11). Clearly, the structure of the various intermediates remains uncertain. It has been suggested (8) the formation of an active aryl-nickel anion $[\mathrm{Ar\,Ni\,L_2}]^-$ which can interact with Ar X to yield biaryl by some kind of aromatic nucleophilic substitution perhaps by successive oxidative addition and reductive elimination. It is the adaptation of these reactions to dihalogeno aromatics which yields PPP. For that objective, 4,4' - dibromo biphenyl and Ni Cl_2 dppe have proven to be the best starting material and catalyst precursor (6).

Electrochemical characteristic of the polycondensation reaction.

The figure 2 shows voltamperograms recorded at platinum, gold or vitreous carbon eletrodes put in the synthetic medium of a standard procedure with a scan rate of 100 mV / second. Have been identified the peaks at -1.6 V (Ni(II) to Ni(I) reduction); -2.1 V (Ni(II) to Ni° reduction) and - 2.45 V (catalytic reduction of Br-Ar-Ar-Br to PPP at the reduction potential of Ar Ni X). The presence of this wave shows that the chemical

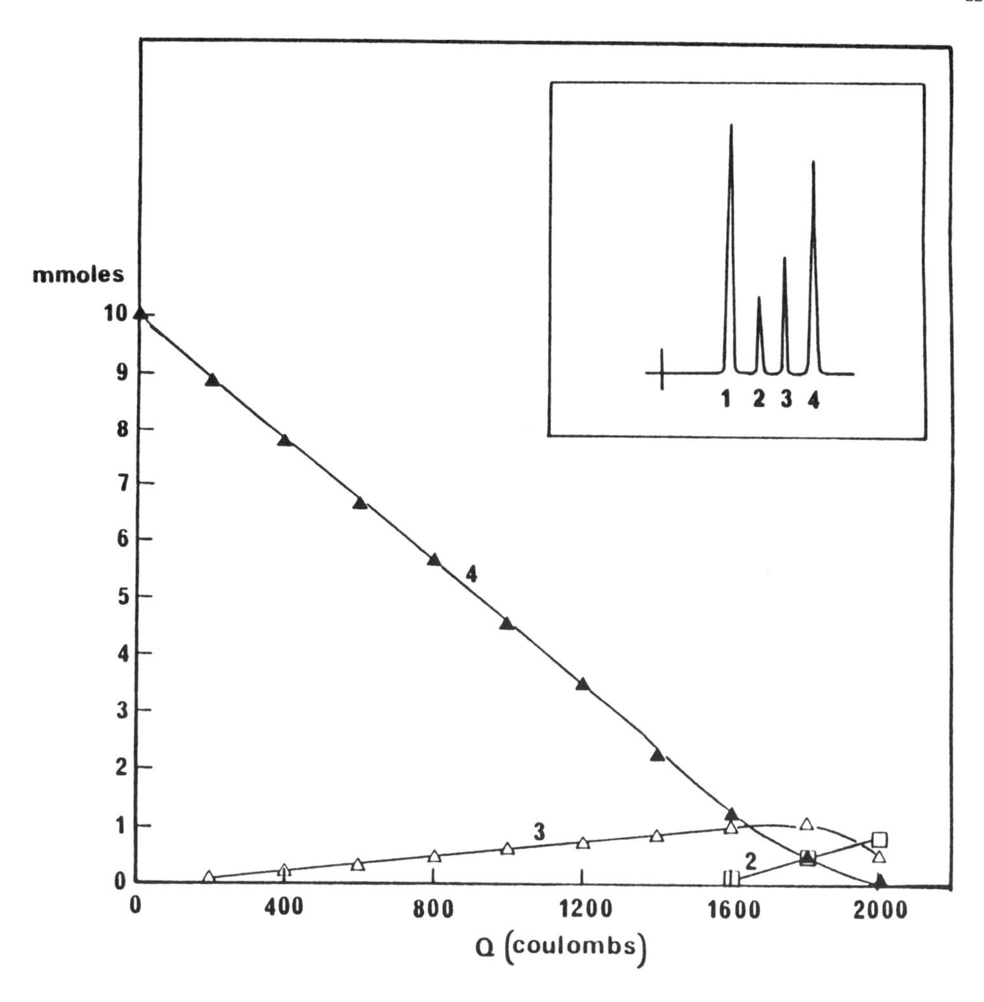

Insert: Typical Chromatogramm.

Anisole **1**, Ar-Ar **2**, Br-Ar-Ar **3**, Br-Ar-Ar-Br **4**

Evolution of a standard exhaustive electrolysis medium

FIGURE 3

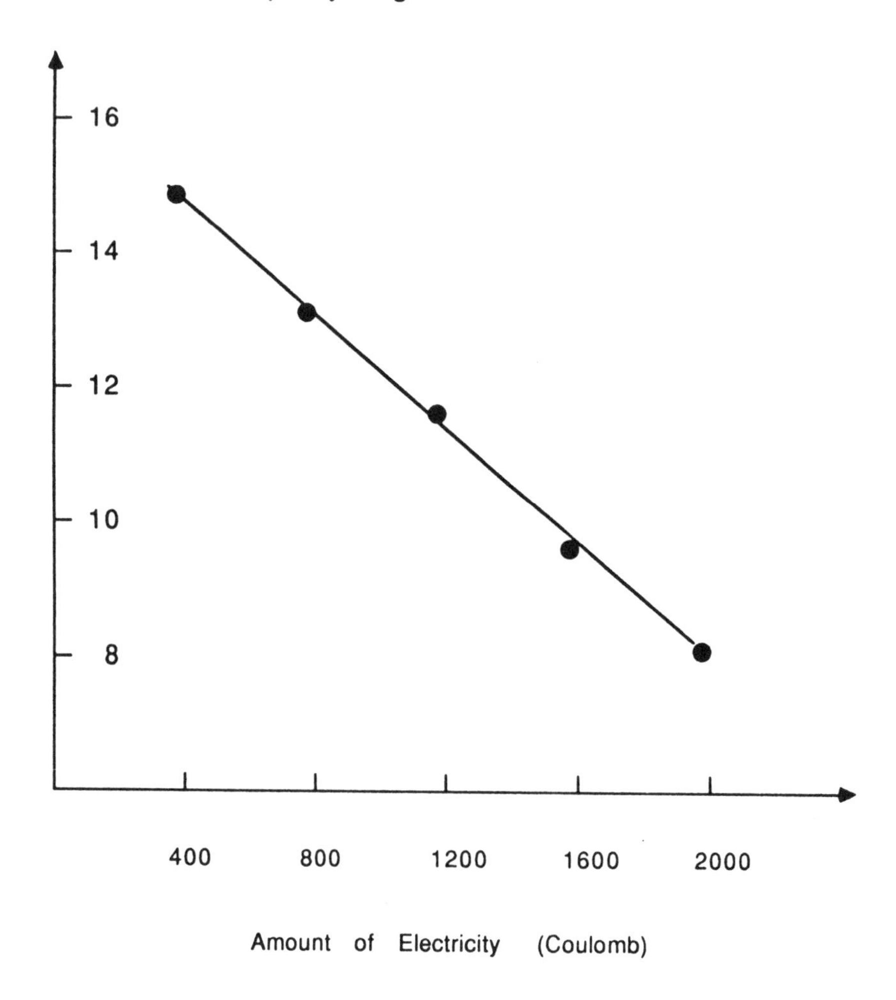

Average number of Phenyl units in electrosynthetized PPP as a function of the charge passed

FIGURE 4

reaction of oxidative addition is rapid on the electrochemical time scale (3.5 seconds between Ni° formation and Ar Ni X reduction). Direct reduction or Br-Ar-Ar-Br does not occur before -2.8 V.

When the molar ratio of the catalyst to monomeric dibromide is inferior to 0.2 (small amounts of catalyst), the height of the wave at -2.45 Volt is proportional both to the catalyst and to the monomeric dibromide concentrations. For higher ratios the reaction rate is limited by the dibromide concentration. Operating at different sweep rates (10 to 200 mV / second) shows a peak height proportional to the square root of the sweep rate. This is indicative of an electrochemical reaction rate limited by the diffusion of the reactants to the working electrode. Thus, the chemical reactions following the aryl-nickel electrochemical reduction are very rapid (coupling and oxidative addition).

Evolution of the medium during exhaustive electrolysis.

During a standard procedure of exhaustive electrolysis, the evolution of the medium was continously checked by quantitative HPLC analysis.

Every 200 C, samples of 100 µl were withdrawn from the electrolyzed solution and diluted (x 2000) with a mixture of acetonitrile and water (65 / 35 by vol.) containing anisol as an internal standard. 10 µl of the diluted solution were injected into an HPLC system equipped with a detector set at 254 nm. A stainless steel (RP8 10µm) column (250 x 4mm) was used. The mobile phase was a mixture of acetonitrile and water (65 / 35 by vol.) pumped at a flow rate of 1 ml/min. Area measurements of the peaks of the recorded detector output signal allowed to calculate the amounts of Br-Ar-Ar-Br, Br-Ar-Ar, and Ar-Ar present in the electrolyzed solution. The following capacity factors K' were found : anisol 1.5, Ar-Ar 2.6, Br-Ar-Ar 3.4, Br-Ar-Ar-Br 4.5.

The results are reported on the figure 3.

Every 400C, samples of 500 µl were taken from the electrolytic medium and filtrated. The yellow deposits were washed successively with THF; water, H Cl 6 M, water, acetone and then dried. IR absorption spectra of these powders were obtained using a Perkin Elmer 580 spectrometer from potassium bromide pellets. As previously described, the mean number of phenyl units in the polymer can be determined from the ratio of the absorbance at 800 and 690 cm^{-1}.

The results are reported on figure 4.

A few electrolyses were stopped at mid-electrolysis. The solution was filtrated, giving the polymer (A) wich was washed only with dry THF, i. e. under mild conditions which are not destructive for any Ar Ni Br dppe functional group eventually present in the material. A part of polymer (A) was treated by boiling H Cl 6M and the chemical composition of the resulting powder (polymer B) compared with that of polymer (A). Polymer (C) was obtained after exhaustive electrolysis as described in the standard procedure.

$(C_6H_4)_n$	Calc.	C 94.74	H 5.26		
Polymer (A)	Found	C 78.5	H 4.9	Br 5.1	Ni 1.6
Corresponding	formula : $C_6\,Br_{0.06}\,Ni_{0.025}\,H_{4.5}$				
Polymer (B)	Found	C 85.7	H 5.2	Br 5.3	Ni 0.4
Corresponding	formula : $C_6\,Br_{0.06}\,Ni_{0.006}\,H_{4.3}$				
Polymer (C)	Found	C 91.43	H 5.35	Br 0.6	
Corresponding	formula : $C_6\,Br_{0.006}\,H_{4.2}$				

Discussion :

The results in figure 3 show an excellent correspondance between the amount of electricity passed through the cell and the number of Ar-Br bonds broken as calculated from analysis of the soluble products. Therefore there are no significant amounts of Ar-Br bonds in the solid materials. This point is corroborated by the elemental analysis results.

The results in figure 4 show that there is a decrease of the average number of phenylene units in the precipitated polymer as the electrolysis proceeds; the largest value (about 16) beeing obtained at the very begining of the electrolysis. Therefore an usual polycondensation process is ruled out since, in such a process, the average molecular weight is known to increase with time, and more rapidly at the end of the reaction.

As expected, the untreated polymer (A) yields unsatisfactory elemental analysis, probably due to the presence of some electrolyte or phosphine ligands. Noteworthy is the slight amount of nickel contained in the material. After thorough washings of polymer (A), the elemental analysis of the material becomes more satisfactory. The main point is that the acidic treatment does not change the bromine content thereby indicating the lack of Ar Ni Br terminal groups in the precipitated polymer. After exaustive electrolysis almost Ar-Br bond is cleaved, even the few ones remaining in the precipitated polymer (A), it can thus be assumed that Ar-Br bond in precipitated polymer remain accessible for a chemical or an electrochemical reaction.

Suggested Mechanism.

Since Ar Ni Br terminal functions are not found in the precipitated polymer, the only nickel complexes which have to be considered in the polymerization mechanism are exclusively the soluble species such as Br-Ar-Ar-Ni Br dppe and H-Ar-Ar-Ni Br dppe; but other nickel species may be transitively present in the diffusion layer of the cathode.

The initial formation of high molecular weight polymer suggests that reactive intermediates are involved as in an addition polymerization. Such intermediates may be radical or anionic species and regenerated after each propagation step. Although we have no absolute evidence, we are in favor of anionic species (8) which are probably more stable than radical species which could rapidly abstract hydrogen from good hydrogen donor solvents such as THF or HMPA. Thus reactive intermediates will be written in an anionic form in the following suggested mechanism operating in the diffusion layer of the cathode.

Initiation

(1) Electrochemical reduction :

$$\underset{\mathbf{C_1}}{\text{Br-Ar-Ar-Ni Br dppe}} + 2\,e^- \longrightarrow \underset{\mathbf{A_1}}{[\text{Br-Ar-Ar-Ni}^\circ\ \text{dppe}]^-} + Br^-$$

Propagation

(2) Oxidative addition :

$$\underset{\mathbf{A_n}}{[\ \text{Br}\ (\text{Ar-Ar})_n\ \text{Ni}^\circ\ \text{dppe}\ ]^-} + \text{Br-Ar-Ar-Br} \longrightarrow \underset{\mathbf{B_{n+1}}}{\text{Br- Ar-Ar-Ni(II)} / \text{Br-(Ar-Ar)}_n} + Br^-$$

(3) Reductive elimination and zerovalent nickel insertion :

$$\mathbf{B_{n+1}} \longrightarrow \text{Br}\ (\text{Ar-Ar})_{n+1}\text{Br...Ni}^\circ\ \text{dppe} \longrightarrow \underset{\mathbf{C_{n+1}}}{\text{Br-(Ar-Ar)}_{n+1}\text{Ni Br dppe}}$$

(4) Electrochemical reduction :

$$\mathbf{C_{n+1}} + 2\,e^- \longrightarrow Br^- + \mathbf{A_{n+1}}$$

Unstable diaryl nickel complexes such as $\mathbf{B_{n+1}}$ have been already postulated by many authors in nickel catalyzed biaryl syntheses (12, 13, 14), the formal oxidation state of nickel being still controversial. Diaryl nickel complexes are known to undergo rapid reductive elimination which may even be facilitated if a bidentate ligand such as dppe is used (step 3).

We consider that Ni° dppe generated by the reductive elimination reaction remains associated to the growing chain and inserts rapidly in the proximate Ar-Br bond giving the $\mathbf{C}_{n+1}$ aryl nickel complex. This complex is then immediately reduced into $\mathbf{A}_{n+1}$ which reenters the reaction cycle at step 2. The rapid propagation proceeds as long as the nickel complexes remain in the vicinity of the electrode, the rate being limited by Br-Ar-Ar-Br diffusion.

Termination :

A termination step occurs if zerovalent nickel Ni° dppe produced at step 3 leaves the electrode. In that case the PPP chain is deactivated and precipitates in the solution. The free Ni° dppe may add dibromo biphenyl monomer or another Ar-Br bond and then the initiation step may again occur.

Another termination step is an hydrogen abstraction from solvent or residual water molecule by an anionic species. Such a reaction competes with the propagation step 2 and becomes predominant at the very end of the electrolysis when the concentration of monomer becomes very low.

A termination step occurs also if step 2 involves Br-Ar-Ar instead of Br-Ar-Ar-Br. The HPLC analysis indicates that the frequency of such a coupling should increase as the electrolysis proceeds since the concentration ratio of Br-Ar-Ar to that of Br-Ar-Ar-Br increase. This may well explain the diminution of the mean molecular weight during electrolysis.

Conclusion :

Although the exact nature of the intermediates are not definitely established, the proposed mechanism agrees well with most of the experimental facts:

- Nickel complexes must intervene in the polymerization process.
- Nickel is not found in the insoluble products.
- Carbon-bromine bonds may be found in the insoluble products but are cleaved before the end of the electrolysis.
- Mean molecular weight decreases as the electrolysis proceeds suggesting a polyaddition mechanism with regeneration of active intermediates in the propagation steps.

The proposed mechanism is also compatible with other works on chemical and electrochemical properties of nickel complexes. Although it should apply strictly for that particular reaction, we think that some features of this mechanism may also be relevant for other electropolymerizations.

Literature :

1) A. F. DIAZ, J. BARGON;
in Handbook of Conducting Polymers, pp 81 ff., Terje A. SKOTHEIM Editor, Marcel Dekker Inc., New York 1986.

2) J. PREJZA, I. LUNDSTOM, T. SKOTHEIM;
J. Electrochem. Soc., 1982, **129**, 1685.

3) S. ASAVAPIRIYANONT, G. K. CHANDLER, G. A. GUNAWARDENA, D. PLETCHER;
J. Electroanal. Chem., 1984, **177**,229.

4) E. M. GENIES, G. BIDAN, A. F. DIAZ;
J. Electroanal. Chem., 1983, **149**, 101.

5) F. CHAO, M. COSTA, P. LANG, E. LHERITIER, F. GARNIER, G. TOURILLON;
Annales de physique, Suppl. 1, 1986,**11**, 21.

6) J. F. FAUVARQUE, A. DIGUA, M. A. PETIT, J. SAVARD;
Makromol. Chem., 1985,**186**, 2415.

7) G. FROYER, F. MAURICE, J. Y. GOBLOT, J. F. FAUVARQUE, M. A. PETIT, A. DIGUA;
Mol. Cryst. Liq. Cryst. 1985, **118**, 267; and results submitted for publication at Synth. Metals.

8) M. TROUPEL, Y. ROLLIN, S. SIBILLE, J. F. FAUVARQUE, J. PERICHON;
J. Organomet. Chem., 1980, **202**, 325.

9) O. SOCK, M. TROUPEL, J. PERICHON, C. CHEVROT, A. JUTAND;
J. Electroanal. Chem., 1985, **183**, 237.

10) G. SCHIAVON, G. BONTEMPELLI, M. DE NOBILI, B. CORAIN;
Inorg. Chim. Acta, 1980, **42**, 211.

11) G. BONTEMPELLI, M. FIORANI;
Ann. di Cimica, 1985, **75**, 303.

12) M. F. SEMMELHACK, P. M. HELQUIST, L. D. JONES;
J. Amer. Chem. Soc., 1971, **93**,5908

13) T. T. TSOU, J. K. KOCHI;
J. Amer. Chem. Soc., 1979, **101**, 7547.

14) I. COLON, D. R. KELSEY;
J. Org. Chem., 1986, **51**, 2627.

POLYMERIZATION OF REACTIVE TELECHELIC OLIGOMERS. FORMATION OF THERMOSTABLE NETWORKS.

Bernard SILLION
CEMOTA
BP n° 3
69390 VERNAISON
France

ABSTRACT.

The rigid structure of heat resistant polymers is a drawback for structural applications because these high molecular weight heterocyclic polymers do not melt or exhibit very high Tg.
This consideration explain the development of a new type of low molecular weight telechelic aromatic oligomers which crosslink by addition polymerization when they are heated.
Three main points have to be considered:
- the melting point Mp or initial Tg of the resine
- the reaction temperature
- the Tg of the final crosslinked network
The chemical structure and the molecular weight of the central moiety will govern the initial Tg or Mp and also affect the reaction temperature.
We will briefly discuss the state of the art concerning the reaction of telechelic oligomers containing nadimide, acetylene and phthalonitrile with special emphasis on phthalonitrile resins.
As a conclusion, the telechelic oligomers approach seems a promising solution to the problem of the processability of heat resistant polymers.
However, it is important to point out the lack of information concerning the polymerization mechanism, the structure and the mechanism of degradation of these thermostable networks.

1. INTRODUCTION

In order to get a heat resistant macromolecule, two main requirements have to be met[1]. First, the retention of the mechanical properties when the temperature increases, depends on the melting point (Tm) in the case of the cristalline polymers and on the glass transition (Tg) in the case of the amorphous polymer, according to the Van Krevelen equation[1] (Fig. 1).

The second requirement is the chemical stability -i.e. thermal stability, oxydation resistance, solvant resistance- against an agressive environment at high temperature.

These both conditions are fullfilled in the case of the aromatic and heterocyclic polymers which exhibit high Tm or Tg, due to their rigid structure and high bonding energy due to their aromaticity (Fig. 2).

M. Fontanille and A. Guyot (eds.), Recent Advances in Mechanistic and Synthetic Aspects of Polymerization, 237–254.

FIGURE 1

EQUATION DE VAN KREVELEN

$$T_{m.v} = aTg\left[1 + x_c \frac{Tm - Tg}{Tg}\right]$$

$$= aTg + \left[1 + 1/2\, x_c\right]$$

a = 0.96 for 200 hours

a = 0.91 for 1000 days

FIGURE 2

HEAT RESISTANT PLASTICS

How TO make them

CHEMICAL REQUIREMENTS

- Oxidation process
- Pyrolytic process

CONCLUSION : High bonding energy
Aromatic carbocyclic and heterocyclic structures

FIGURE 3

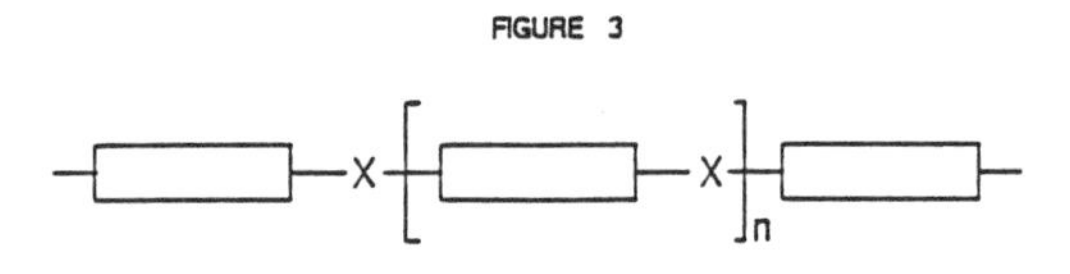

Flexible Links X =

$-C(CF_3)_2-$: $-C(CH_3)_2-$: O : CO : SO_2 : Siloxanes

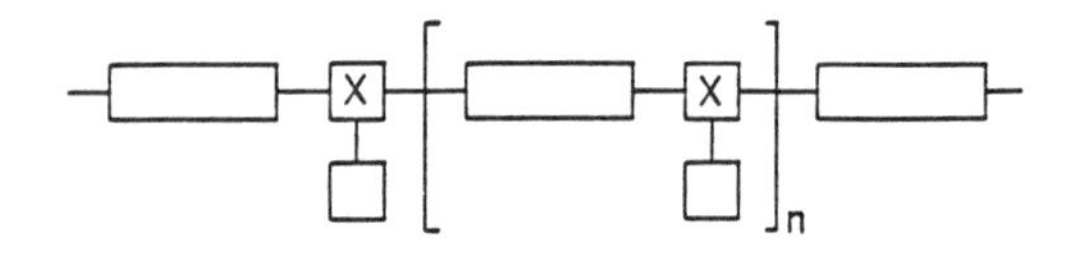

Side Groups or Asymetric Structures

However, the main drawback for the development of the heat resistant polymers is the low processability. That is specially true for the structural applications (adhesives and laminates) since the high molecular weight heterocyclic polymers do not melt and are not soluble, so that it is impossible to ensure a good wettability of the inorganic component.

The improvement of the processability could come from the introduction of, either flexible linkages (O, SO_2, $C(CF_3)_2$, ...siloxanes...) between the heterocycle, or bulky branching on the main chain (Fig. 3).

Although these amorphous thermoplastic polymers show lower Tg than the rigid ones, they need high temperature and high pressure for the processing.

The trend is to develop heat resistant resins with a processability as easy as for the epoxies, but with better properties- i.e. high stability, better electrical properties and low moisture absorption-.

These resins are basically a new type of telechelic aromatic or heterocyclic oligomers which crosslink by an addition polymerization process when they are heated without evolution of volatils.

2. GENERAL FEATURES OF THE THERMOSTABLE TELECHELIC OLIGOMERS

The general formula is given on Fig. 4.

The block A is an oligomer with an average molecular weight ranging from 1000 to 2000 $g.mole^{-1}$. The main chemical structure used for the synthesis of the block A is aromatic ether sulfone, aromatic esters, pyridine ether and heterocyclic compound like quinoxaline imide [(3)]. The chemical structure and the molecular weight of the central moity A will govern initial Tg or Tm.

The telechelic groups Z react by thermal activation leading to an addition polymerization without volatil evolution, this point is important for the structural applications.

The range of the thermal reaction temperature for the different reactive groups is given in (Fig. 5).

As it can be reported, the polymerization temperature appears to be very different for the same type of reactive group. Actually the polymerization temperature is strongly dependent of the physical state of the resin. The thermal polymerization starts, in fact, when the resin melt. This point was clearly demonstrated in the case of maleimide [(4)], the resins with low Mp polymerize at 180°C and resins with high Mp react only at the Mp, the same observation was made with acetylenic resins.

Three main points have to be considered, (Fig. 6):
. The melting point Mp or the glass transition Tg of the resine
. The polymerization temperature
. The Tg of the final network

FIGURE 4

TELECHELIC OLIGOMERS

Z —[A]— Z

1 000 < Mn < 2 000

A : Aromatic or Heterocyclic Block

Z : Latent Crosslinking Agent for Addition Reaction Curing

CYANAMIDE	$-NH-C\equiv N$
CYANATE	$-O-C\equiv N$
NITRILE	$-C\equiv N$
MALEIMIDE	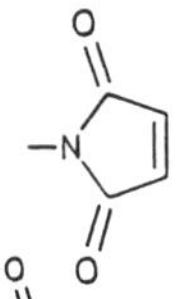 [structure]
NADIMIDE	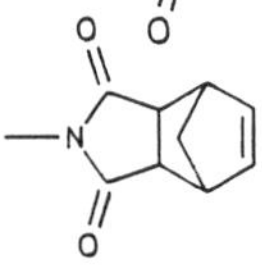 [structure]
ACETYLENE	$-C\equiv CH$
PHTALONITRILE	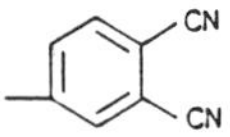 [structure]

FIGURE 5

REACTION TEMPERATURE OF REACTIVE END-GROUPS

FUNCTION	CATALYST	TEMPERATURE (°C)
NITRILE	ATPS	350
CYANATE	ALKYLPHENOL	177
CYANAMIDE	—	150–200
MALEIMIDE	—	177–286
NADIMIDE	—	250–275
ACETYLENE	—	130–140, 200
PHTALONITRILE	REDOX	220

FIGURE 6

TELECHELIC THERMOSTABLE POLYMERS

MAIN PARAMETERS FOR THE PROCESSABILITY

Melting Point, Mp, or Glass Transition, Tg, of the Resin

Polymerization Temperature

Glass Transition Temperature of the Final Network

The difference between the Mp or Tg of the resin and the reaction temperature determine the "processability window" of the considered resin ; At first it seems rather interesting to use an oligomer with the lowest Mp or Tg. However if the initial Tg is too low, the Tg of the final network will also be to low for high temperature applications.

It is also very important to take into account that the final Tg is only reached after a postcure at high temperature because the total extend of the reaction cannot occur in the glassy state.

3. REACTION OF THE DIFFERENTS TELECHELIC GROUPS:

The nitrile groups polymerize by trimerization leading to a triazine cycle (Fig. 7-1), but they are very stable and need a strong acidic catalyst in order to react at 350°C [5]. The cyanates which are trimerized at 180°C with a phenol as a catalyst[6] giving a tris-aryloxytriazine network are more reactive (Fig. 7-2).

The dicyanate prepared from bis phenol A is mainly used in formulation with maleimide (Resine B.T. sold by Mitsubishi) or with epoxies. But semi interpenetred network with polysulfone and other thermoplastics are also studied[7].

The trimerization of the aryl-cyanamides is an old reaction, leading to triaryl melamine. Russian workers have prepared bis-cyanamides (Fig. 7-3) with aromatic[8] or imidic[9] central group. The crosslink reaction give a network with melanine structures. Another way to get that type of heat resistant network is to copolymerize bis-cyanamide with bis-N-sulfonyl-cyanamid [10].

The bis-maleimides[11] where introduced on the market in the early 70's. They can react by nucleophilic addition with, for example, bisthiol or bis-amine[12] or by radical polymerization initiated by thermal activation[4] (Fig. 8).

The chemistry for the bis-maleimide (BMI) is given on Fig. 9. Maleic anhydride reacts with a diamine, then the bis-maleamic acid is cyclised using sodium acetate and acetic anhydride. That type of BMI allows an interesting relationship between the DP of the initial resin and the Tg of the final network. Starting from DP = 1, the Tg is higher than 330°C, with resin having a DP = 2, the Tg is 239°C and if the DP = 3, the final Tg of the network is 185°C[13] (Fig. 9).

The BMI's resin exhibit a processability very similar to the epoxies, but a postcure at 260°C is needed.

Many BMI resins are commercially available from Rhône-Poulenc (Kerimid 353, 601, FE 70003..) Mitsui Toatsu, Mitsubishi (resine BT) Ferro, CIBA (XU 292), Narmco (5245, 5250), Hysol (9655, 9102), Technochemie (H795)...

However, the thermal stability of the BMI is limited at about 200°C.

The nadimide resins seem now more promising for structural applications. The first product was patented by Lubowitz in 1971[14]. The reaction of a mixture of benzophenone tetracarboxylic dianhydride, BTDA, nadic anhydride, NA, and

FIGURE 7

CROSSLINK BY TRIMERIZATION OF NITRILES

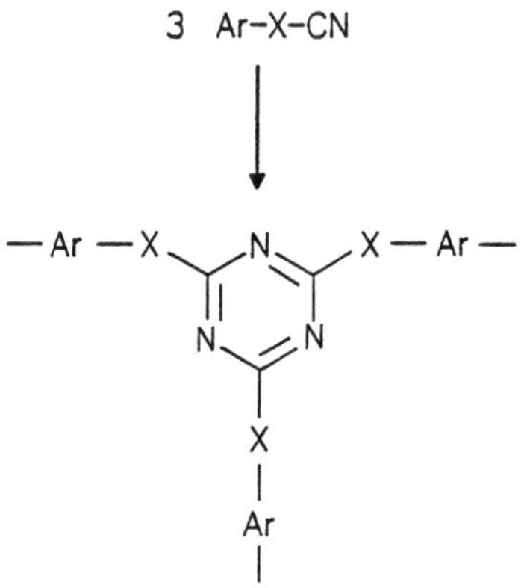

NITRILE	X = nil	⟶	TRIAZINE	1
CYANATE	X = O	⟶	ARYLCYANURATE	2
CYANAMIDE	X = NH	⟶	ARYLMELAMINE	3

FIGURE 8

BIS-MALEIMIDES

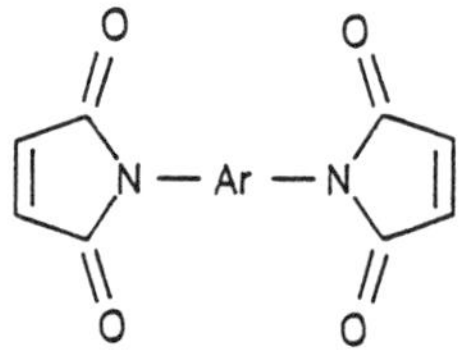

1) Nucleophilic Addition Polymerization

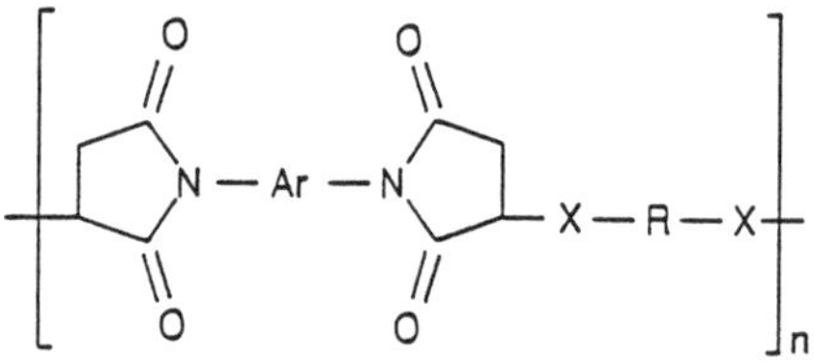

X = S, NH ⟶ Linear Extended Chain

2) Thermal Radicalar Activation

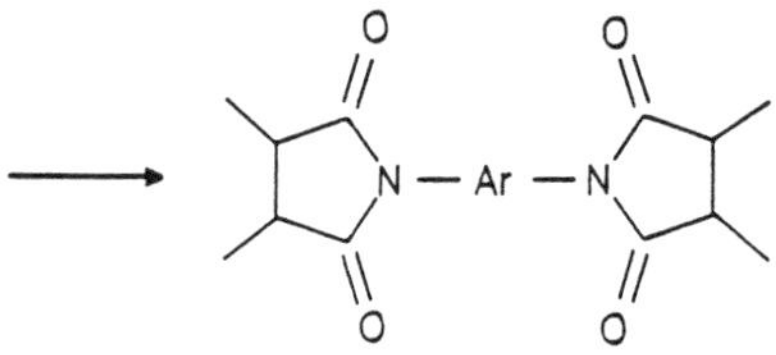

FIGURE 9

BIS-MALEIMIDES

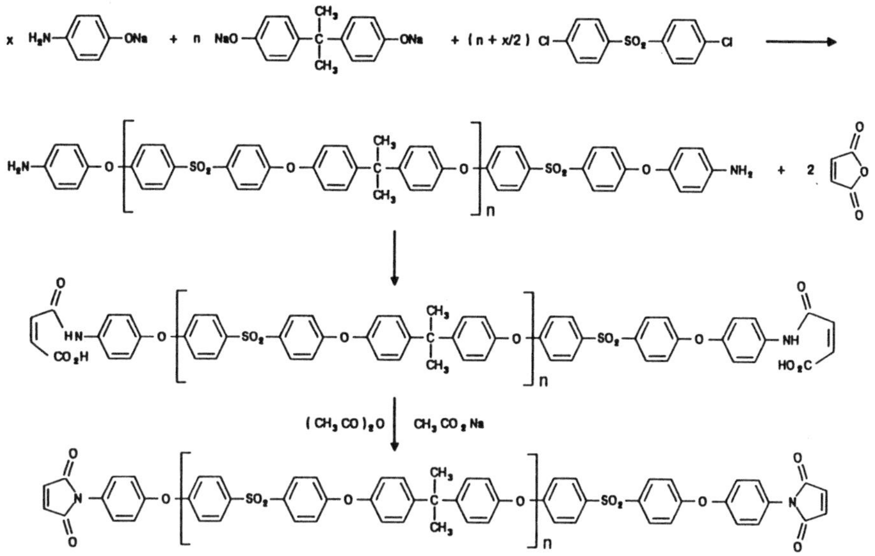

x H_2N–C6H4–ONa + n NaO–C6H4–C(CH_3)2–C6H4–ONa + (n + x/2) Cl–C6H4–SO_2–C6H4–Cl →

H_2N–C6H4–O–[C6H4–SO_2–C6H4–O–C6H4–C(CH_3)2–C6H4–O]n–C6H4–SO_2–C6H4–O–C6H4–NH_2 + 2 (maleic anhydride)

↓

HO2C–CH=CH–CO–HN–C6H4–O–[C6H4–SO_2–C6H4–O–C6H4–C(CH_3)2–C6H4–O]n–C6H4–SO_2–C6H4–O–C6H4–NH–CO–CH=CH–CO_2H

↓ $(CH_3CO)_2O$ CH_3CO_2Na

(maleimide)N–C6H4–O–[C6H4–SO_2–C6H4–O–C6H4–C(CH_3)2–C6H4–O]n–C6H4–SO_2–C6H4–O–C6H4–N(maleimide)

FIGURE 10

NADIMIDE-TERMINATED OLIGOMERS

P 13 N TYPE

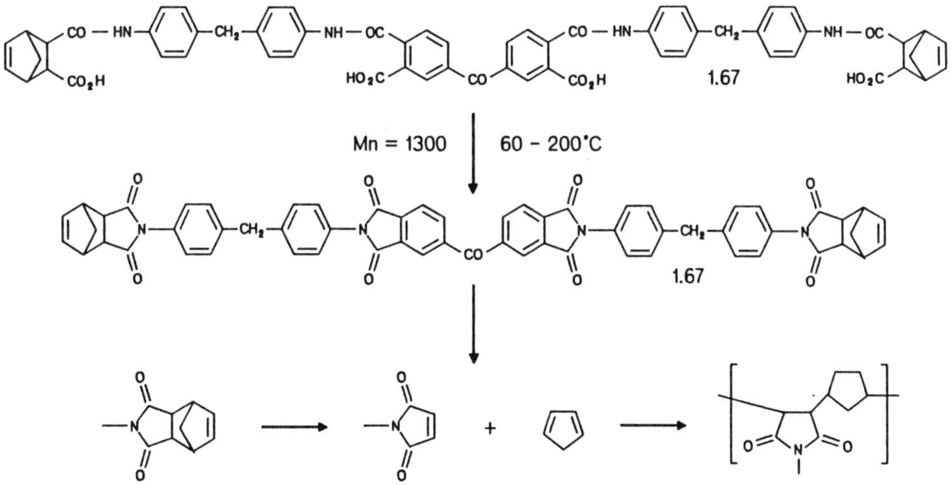

CO — HN–C6H4–CH_2–C6H4–NH — OC, CO_2H, HO_2C, CO, CO — HN–C6H4–CH_2–C6H4–NH — OC, CO_2H, 1.67, HO_2C

Mn = 1300 ↓ 60 - 200°C

(nadimide)N–C6H4–CH_2–C6H4–N(imide)–CO–(imide)N–C6H4–CH_2–C6H4–N(nadimide) 1.67

↓

–N(nadimide) → –N(maleimide) + cyclopentadiene → [polymer]

methylene-4,4'-dianiline, MDA, is carried out in N-methylpyrrolidone solution, in order to obtain an oligomer with theoritical molecular weight limited at 1300 g.mole^{-1} (Fig. 10).

However, the cyclised resin is not soluble and does not melt before the thermal reaction at 275-300°C.

The second drawback is the retro-Diels reaction occuring at the reaction temperature and the resulting evolution of cyclopentadiene[(15)].

A first improvement was the utilisation of the intermediate amic acid in solution in the NMP, but this solution of amic acid is not stable and their shelf life is very short.

In 1975, Serafini[(16)] developped a new concept: the polymerization of monomeric reactant, PMR (Fig. 11).

That type of product is prepared by dissolution of the preceding starting material in methanol, so the dianhydride is transformed in monomethylester of orthodicarboxylic acid. The molecular ratio is calculated in order to obtain specific theoretical molecular weight (for example a resin with theoretical M = 1500 will be noted PMR 15). After the methanol removal, the differential scanning calorimetric curve shows the melting of the reactant mixture, then the condensation and imidization between 100 and 150°C and finally the polymerization which accurs at about 300°C.

If we consider the thermal resistance of the crosslinked nadimides (Fig. 12) exposed at 288°C in air, the loss of weight is less than 2% after 4000 h[(17)].

The polymerization mechanism of the nadimide was studied with the simplest model: the N-phenylnadimide. First, the condensation of the N-phenylmaleimide with the cyclopentadiene gives only the endo isomer at temperature below 200°C (Fig. 13), then by heating at 200°C the exo isomer is partially formed, followed at 275°C by the retro Diels Alder reaction which occurs in competition with the polymerization. According to NMR determination (^{1}H at 270 MHz and ^{13}C) the obtained polymer is saturated. So the structure is non consistent with an alternative copolymer between N-phenylmaleimide and cyclopentadiene. The copolymer contains some endo and exo units coming from the polymerization of the corresponding nadimide and some units resulting of the copolymerization between the N-phenyl maleimide and the cyclopentadiene - N-phenylmaleimide adduct[(18)].

3.1. Acetylenic resins

Two types of acetylenic resins have to be considered[(19)(20)] (Fig. 14).

The resins containing ether sulfone, ester or pyridylether have a low Tg (below 100°) and the polymerization in the liquid state, starts at about 130°C. On the other hand, the resins containing heterocycle like imide or quinoxaline are more rigid and polymerize above Tg at about 200°C. The processability of these resins is given on Table 1.

As it should be expected, the heterocyclic resins are more difficult to process but exhibit a better retention of their mechanical properties at high temperature.

FIGURE 11

NADIMIDE-TERMINATED OLIGOMERS

PMR 15

SOLUTION IN ALCOHOL OF :

BTDE	CH_3O_2C, HO_2C – benzene – CO – benzene – CO_2CH_3, CO_2H	2.09	Mn 1500
MDA	H_2N – benzene – CH_2 – benzene – NH_2	3.09	
NE	norbornene with CO_2CH_3 and CO_2H	2	

B-STAGED OLIGOMERS : Mn = 1500

FIGURE 12

THERMO-OXIDATIVE WEIGHT LOSS OF PMR-11 AND PMR-15 POLYIMIDE RESINS

Isothermal Exposure in Flowing Air (100 cc / min at 550°F)

WEIGHT LOSS (%)

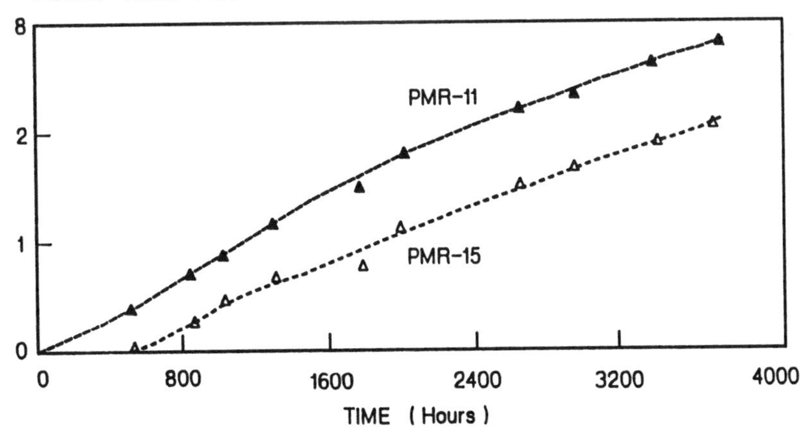

FIGURE 13

POLYMERIZATION OF N-PHENYLNADIMIDE

CYCLOPENTADIENE + N-PHENYLMALEIMIDE

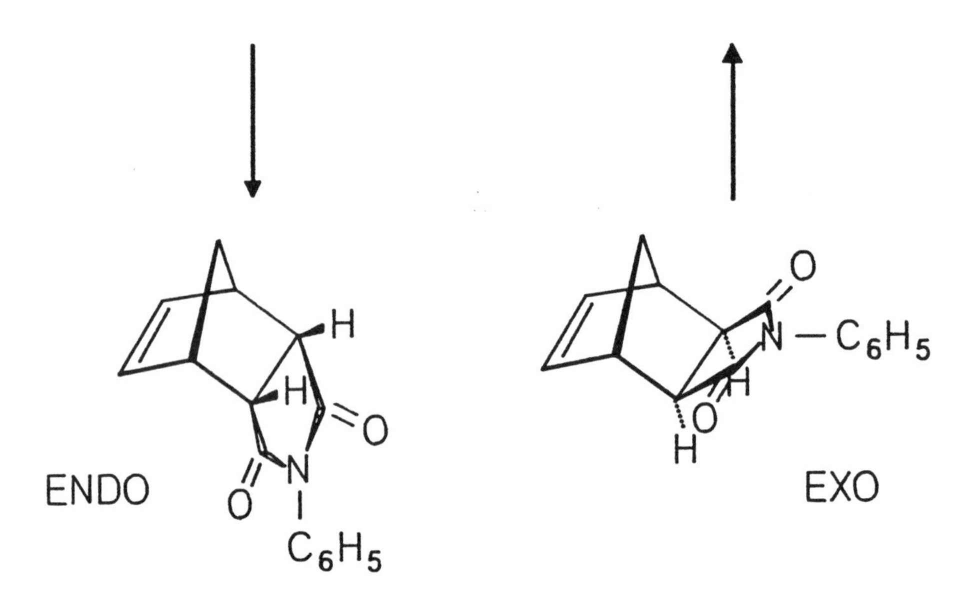

THERMAL POLYMERIZATION AT 285°C

⟶ Saturated Copolymers

- Some units coming from endo and exo nadimides
- Some units coming from copolymerization of N-phenylmaleimide with cyclopentadiene nadimide adduct

FIGURE 14
ACETYLENE-TERMINATED RESINS CURRENTLY STUDIED

SULFONES

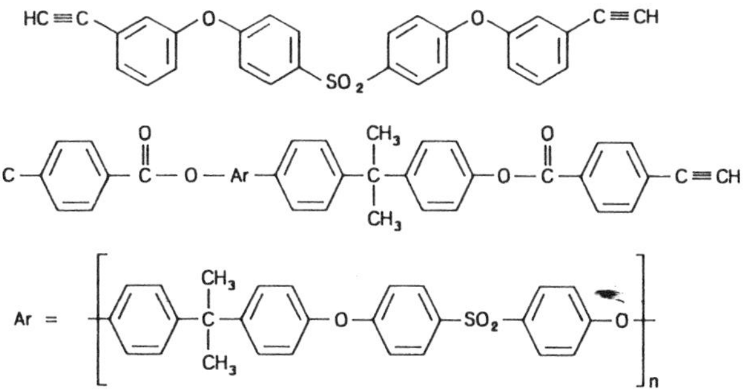

HC≡C — O — SO_2 — O — C≡CH

HC≡C — C(=O) — O — Ar — C(CH_3)(CH_3) — O — C(=O) — C≡CH

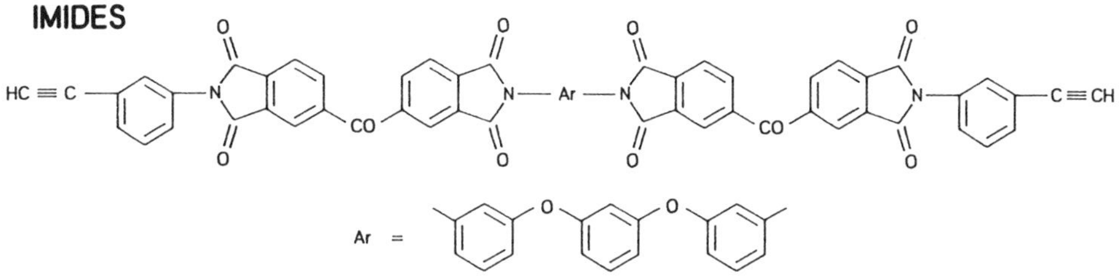

Ar = [C(CH_3)(CH_3) — O — SO_2 — O]$_n$

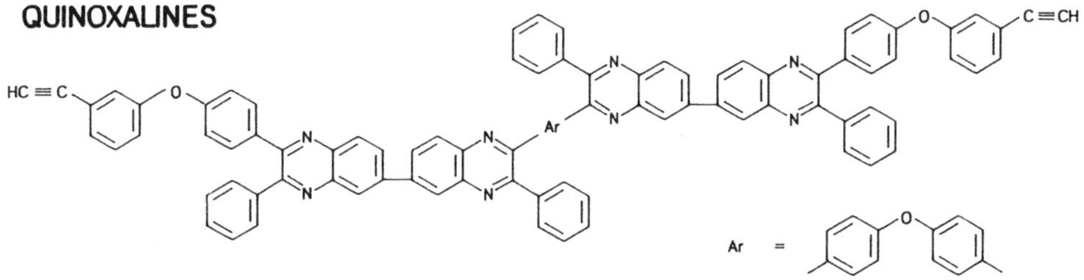

IMIDES

HC≡C — N(O)(O) — CO — N(O)(O) — Ar — N(O)(O) — CO — N(O)(O) — C≡CH

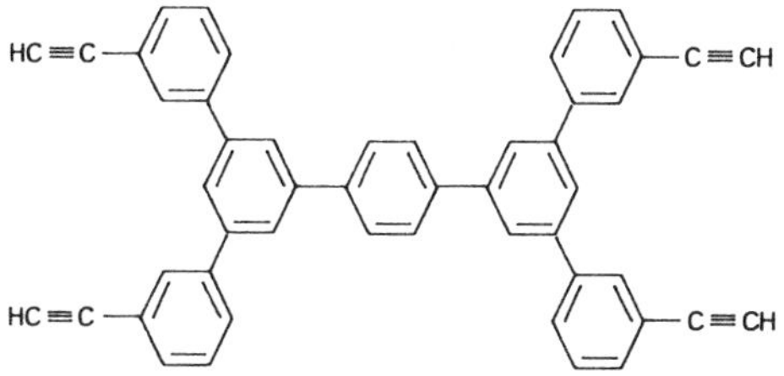

Ar = O O

QUINOXALINES

HC≡C — O — N N — N N — Ar — N N — N N — O — C≡CH

Ar = O

PHENYLENES

HC≡C — HC≡C — C≡CH — C≡CH

TABLE 1

Processability of Acetylenic Resins

Resins with Low Tg (Tg < 100°C)

- Prepreg without solvent
- Laminating conditions :
  - . Temperature 170 – 200°C
  - . Pressure 0.7 MPa
  - . Cooling
  - . Postcure in air at 300°C
- Withstanding of the mechanical properties : 1000 hours at 232°C

Resins with High Tg, heterocyclic resins

- Prepreg with solvent
- Laminating conditions :
  - . Temperature 250°C / 2 hours
    316°C / 16 hours
  - . Pressure 1.4 MPa
- Withstanding of the mechanical properties : 500 hours at 316°C

FIGURE 15

SYNTHESIS OF ACETYLENE COMPOUNDS

Z–C_6H_4–CO—CH_3 $\xrightarrow[HCON(CH_3)_3]{POCl_3}$ Z–C_6H_4–C(Cl)=CH—CHO $\xrightarrow{Base}$ Z–C_6H_4–C≡CH

Z–C_6H_4–X + HC≡C—Y $\xrightarrow[P(Ph)_3,\ N(Et)_3]{PdCl_2,\ CuI}$ Z–C_6H_4–C≡C—Y $\xrightarrow{Base}$ Z–C_6H_4–C≡CH

Y = – Si (CH_3)$_3$

Y = – C (CH_3)$_2$ – OH

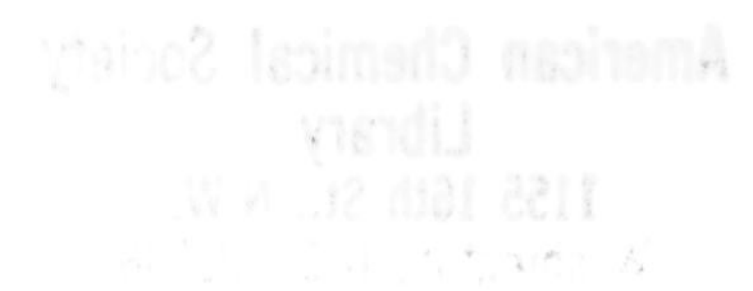

The main problem encountered was the synthesis of the acetylenic group (Fig 15).

Two synthetic pathways can be used. The first is the transformation of a methylarylketone by a Vilsmeir reactive leading to a chloro-cinnamaldehyde which reacts with alkalis and gives the acetylenic compounds[21]. The second pathway is more general: an iodo or bromo derivative is substituted by a protected acetylene, then the protective group is removed by a basic treatment[22].

As for the nadimides, there is a lack of information about the polymerization mechanism and the structure of the final network.

The polymerization enthalpy determined by DSC seems independant of the chemical structure[19][23]. At the beginning of the polymerization some paramagnetic centers appear and their concentration increase as the polymerization extend[24].

The polymerization model suggested by Lee[25] according to kinetical, spectroscopic and paramagnetic resonance data is given on Fig. 16.

At the first time of the polymerization, only one fonction react giving a stable macroradical cluster which crosslinks at higher temperature.

3.2. Phthalocyanine network

The cyclotetramerization of four phthalonitrile molecules into a phthalocyanine macrocycle involves a reduction reaction which requires 2 electrons in addition to the 16 electrons from the 8 nitrile functional groups.

For metallic phthalocyanines, the 2 extra electrons are provided by a metal or a metallic salt[26]. Concerning the metal-free phthalocyanine, the older synthesis consisted in generating a phthalocyanine of a alkaline metal which was then hydrolysed[27]. Another method is based on the preparation of an intermediate isoindolenine which is thermally transformed into phthalocyanine[28].

The high stability of the aromatic phthalocyanine ring leads to many attempts to include it in thermostable resins.

Two methods are known: the use of functionalized phthalocyanine to prepare, for example, imide-phthalocyanine copolycondensate[29], or the direct poly-tetramerization of bisphthalonitrile oligomers[30]. Polymerization can be achieved with the use of a metal or a metallic salt but with problems of heterogeneity of the final material. Polymerization by simple heating is slow and leads to a not very definite composition of the polymer[31]. The use of reducing coreactants such as 1,2,3,6-tetrahydropyridine or hydroquinone[32] is effective for the conversion of phthalonitrile into phthalocyanine, but because of their volatility at the reaction temperature the reaction must be conducted under pressure. Moreover, the oxidation by-products are inevitably trapped in the final polymer.

We previously developed a novel route to metal-free phthalocyanine applicable to polycondensation which involved the fixation on the final phthalocyanine polymer of the molecule which supported the reducing group[33] (Fig. 17).

FIGURE 16

ACETYLENE-TERMINATED POLY (ETHER-SULFONES)

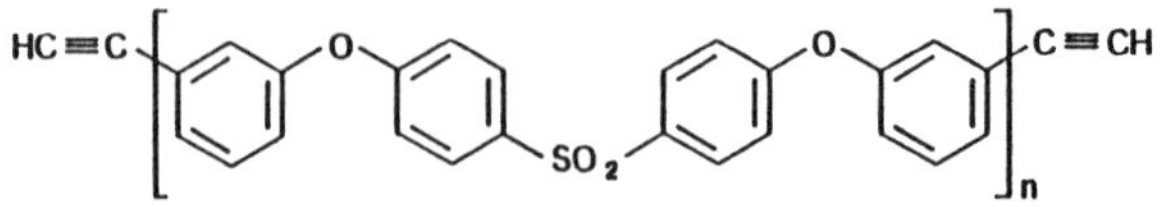

HC≡C–[–O– –SO$_2$– –O–]$_n$–C≡CH

n = 1.23 (77% of n = 1 + Dimer and Trimer)

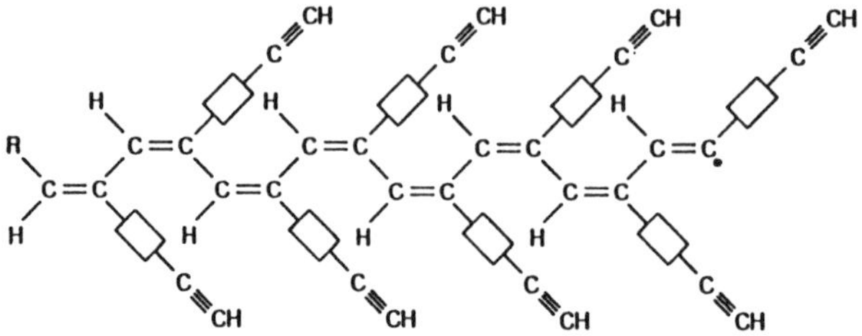

FIGURE 17

PHTHALOCYANINES

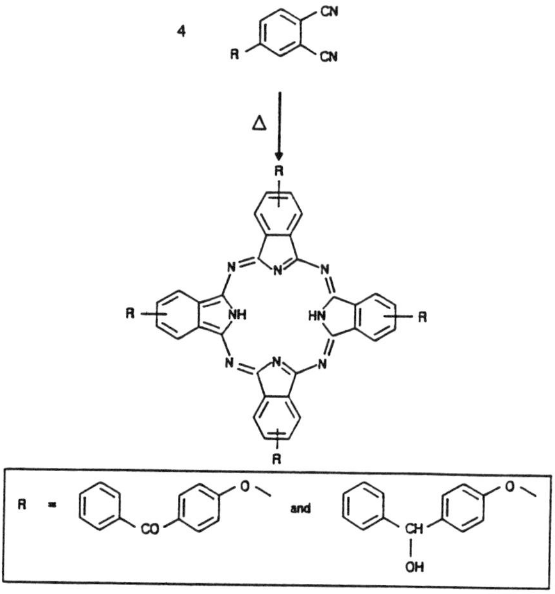

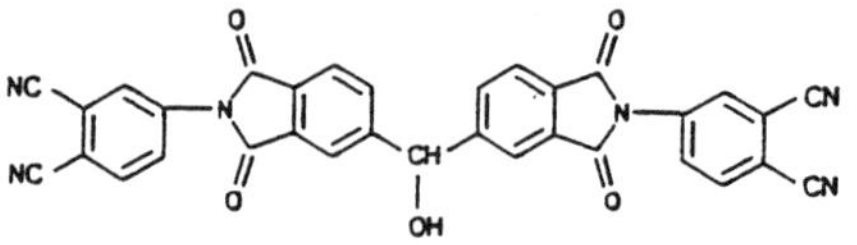

4

Δ

R = and

CO

CH

OH

NC

NC

CN

CN

We have shown that one mole of hydroxybenzyl phthalonitrile reacts with three moles of benzoylphtalonitrile, leading to tetrakis benzoylphthalocyanine.

In this case, the benzhydrol group provides the system with two electrons. We also observed that the reductive capacity of the benzhydrol group is enhanced by electron withdrawing substituant. So, when the benzhydrol group is linked at two imide groups, the tetramerization starts at 220°C, giving a very heat resistant network.

4. NEW REACTIVE GROUPS

4.1. Benzocyclobutene

The biphenylene strained ring is thermally opened, giving a very reactive di-radical system. The biphenylene crosslinking group has been introduced in polyquinoline[34], polyquinoxaline imide oligomers[35] and polyquinoxaline[36].

A catalysed crosslink occurs at about 400°C. Recently, Arnold has been investigating new system based on electrocyclic ring opening of benzocyclobutene[37]. By heating at 200°C, the benzocyclobutene enter in equilibrium with the open form o.xylylene, which can engage in a Diels Alder reaction. It can also react like a diradical at 230-250°C.

The crosslink mechanism is not clearly demonstrated, however two mains pathways are possible cycloaddition and linear addition (Fig. 18).

The Diels Alder reactions with benzocyclobutene where investigated either with A.B monomer carrying a phenylacetylene substituent as a dienophile, or by reaction of bis-benzcyclobutene imide with bis-maleimide. Both systems melt and polymerize with an exotherm maximum at about 260°.

The loss of weight at 343°C after 200 h in air was about 3%.

4.2. Aryl azoethynylarene[38]

The coupling reaction of an arenediazomium chloride with a silver arylacetylide give arylazoethynyarenes. That compound undergoes an head to tail thermal dimerization to give a tetraazapentalene stable up to 480°C; the arylazoethynylarene also reacts by double dipole addition with maleimide (Fig. 19).

5. CONCLUSION

The telechelic oligomer approach seems a promizing solution to the problem of the processability of heat resistant polymers.

It is very interesting to see that new reactive fonctional groups are patented or published, every year. They offer better processability window, or better stability.

However, it is important to point out the lack of information concerning the polymerization mechanism, the network structures and the mechanism of degradation of these thermostable networks.

FIGURE 18

BENZOCYCLOBUTENE

CROSSLINKING REACTION MECHANISM AT 230–250°C

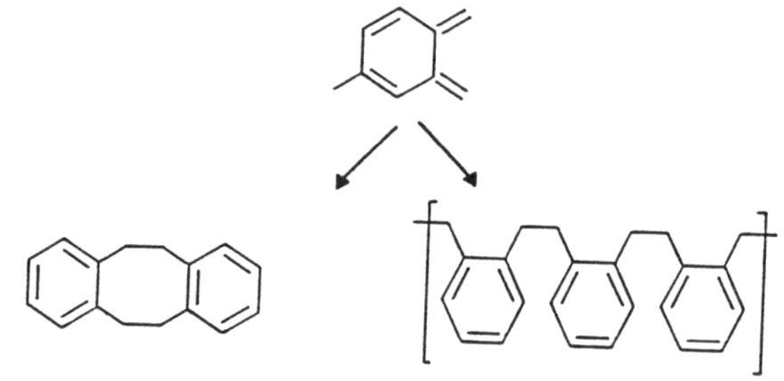

CYCLOBUTENE-TERMINATED IMIDE RESINS

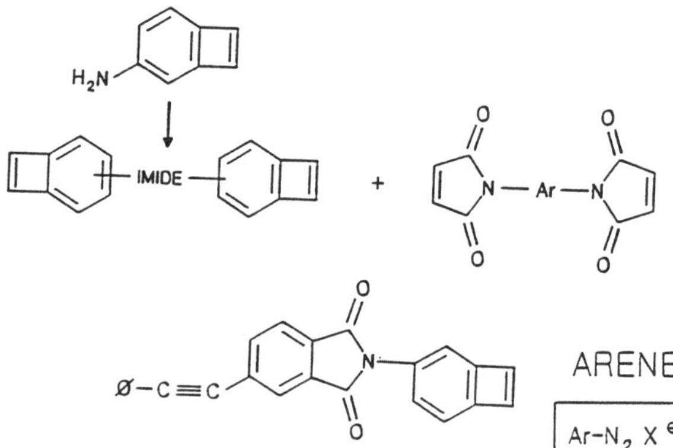

H_2N

IMIDE

+

N — Ar — N

Ø–C≡C

FIGURE 19

ARENE - AZOETHYNYLARENE

$Ar{-}N_2\,X^{\ominus}$ + Ag — C≡C — Ar ⟶
AgX + Ar — N=N — C≡C — Ar

1) Thermal Dimerization

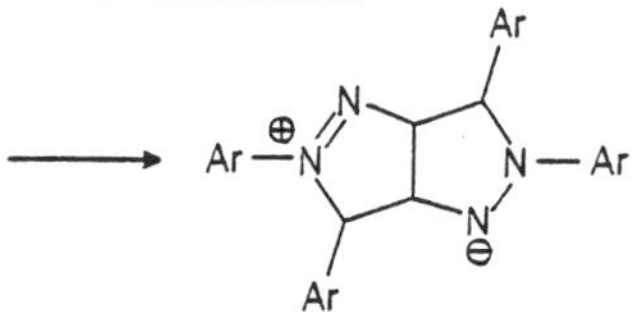

Stable up to 480°C

2) Dipolar Addition with Maleimides

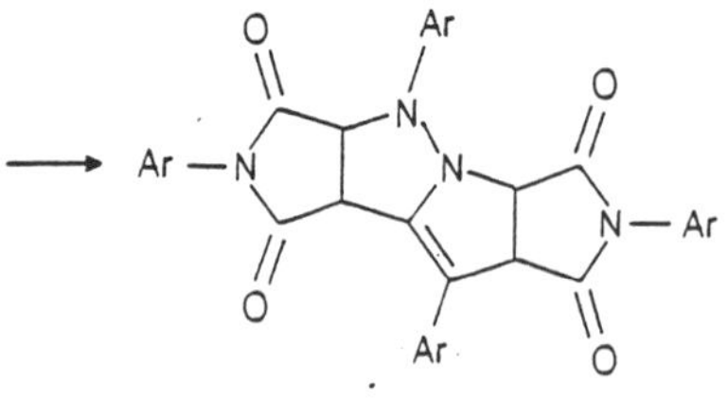

BIBLIOGRAPHY

1) P.E. Cassidy, *Thermally Stable Polymers*, Marcel Dekker, 1980
J.P. Critchley, G.J. Knight, *Heat Resistant Polymers*, Plenum Press, 1983

2) D.W. Van Krevelen, *Chimia*, **35**, (10), 393, 1981

3) P.M. Hergenrother, *A.C.S. Symposium Series*, **282**, 1, 1985

4) I.K. Varna, G.M. Fohlen, J.A. Parker, *J. Polym. Sci. Polym., Chem. Ed.*, **20**, 283, 1982

5) Li Chen Hsu, W.H. Phillips, *A.C.S. Symposium Series*, **195**, 285, 1982

6) D.A. Shimps, *Polym. Mat. Sci.*, **54**, 107, 1986

7) D.W. Wertz, D.C. Prevorsek, *Plast. Eng.*, April 1984, 31

8) V.A. Pandratov, G.E. Shurukov, D.F. Kutenov, E.N. Godovacher, *Plast. Massy*, **10**, 59, 1984

9) G.E. Shukurov, V.A. Pandratov, V.V. Korshak, *Plast. Massy*, **9**, 5, 1985

10) R.J. Kray, *Soc. Adv. Mat., Proc. Eng. Ser.*, **20**, 227, 1975

11) D. Landman ; *Dev. Reinf. Plast.*, **5**, 39, 1986

12) J.E. White ; *Ind. Eng. Chem. Prod. Res. Dev.*, **25**, 395, 1986

13) G.T. Kwiatkowski, L.M. Robeson, G.L. Brode, A.W. Bedwin, *J. Polym. Sci. ; Polym. Chem. Ed.*, **13**, 961, 1975

14) H.R. Lubowitz ; U.S. Pat. 3.528.950 - *Polym. Preprint*, **12**, (1), 329, 1971

15) R.W. Lauver, *J. Polym. Sci., Polym. Chem. Ed.*, **17**, (2), 529, 1979

16) T.T. Serafini, P. Delvigs, G.R. Lightsey ; U.S. Pat. 3.745.149
T.T. Serafini, *A.C.S. Symposium Series*, **132**, 15, 1980

17) D.A. Scola, *22nd National SAMPE Symposium*, **22**, 238, 1977

18) N.G. Gaylord, M. Martan, *Polym. Prepr.*, **22**, 11, 1981
A.C. Wong, W. Ritchey, *Macromolecules*, **14**, 825, 1981
A.C. Wong, A. Garroway, W. Ritchey, *Macromolecules*, **14**, 832, 1981

19) C.Y.C. Lee, *Dev. Reinf. Plast.*, **5**, 121, 1986

20) P.M. Hergenrother, *Encyclopedia of Polym. Sci.*, **1**, 61, 1985

21) G. Kobrich, P. Buck, *Chemistry of acetylene*, Marcel Dekker, 1969, Chap. 2

22) E.T. Sabourin, *Am. Chem. Soc. Div. Petr. Chem. Prepr.*, **24**, 233, 1979

23) A. Dussart-Lermusiaux, M. Senneron, M. Bartholin, B. Sillion, *Colloque GFP, Pau*, 1986, p.

24) T.C. Sandreczki, Y.C. Lee ; *Am. Chem. Soc. Polym. Preprint*, **23**, (2), 185, 1982

25) J.M. Pickard, E.C. Jones, I.J. Goldfarb,
Am. Chem. Soc., Polym. Prepr., **20**, (2), 375, 1979
L. R. Denny, I.J. Goldfarb, C.Y.C. Lee,
31st International SAMPE Symposium, april 7-10, 153, 1986

26) F.H. Moser and A.L. Thomas
The Phthalocyanines, CRC Press Inc., Boca Raton, FL, 1983 ; **II**, Chapters 1 and 2

27) P.A. Barrett, D.A. Frye and R.P. Linstead, *J. Chem. Soc.*, 1157, 1938

28) P.J. Brach, S.J. Grammatica, O.A. Ossanna and L. Weinberger,
J. Heter. Chem., 7, 1403, 1970

29) B.N. Achar, G.M. Fohlen and J.A. Parker,
J. Polym. Sci. Polym. Chem. Ed., **22**, 319, 1984 and 23, 801 1985.

30) B.N. Achar, G.M. Fohlen and J.A. Parker,
J. Polym. Sci. Polym., Chem. Ed., **21**, 111, 1983 and **29**, 353, 1984.
B.N. Achar, G.M. Fohlen and J.A. Parker, *J. Polym. Mater.*, **2**, 16, 1985

31) N.P. Marullo and A.W. Snow, *A.C.S. Symp. Ser.*, 325, 1982

32) A.W. Snow, J.R. Griffith and N.P. Marullo, *Macromolecules*, **17**, 1614, 1984

33) J. Malinge, G. Rabilloud and B. Sillion, *Fr. Dem. 2,568,257*, 1986
T. Pascal, J. Malinge, B. Sillion, P. Claudy, J.M. Letoffé, *J. Polym. Sci.*, in Press

34) J.K. Stille, J. Garapon ; *Macromolecules*, **10**, 627, 1977

35) F.W. Harris, K. Sridhar, S. Das ; *Polym. Preprint*, **25**, (1), 110, 1984

36) W. Vancraeynest, J.K. Stille ; *Macromolecules*, **13**, 1367, 1980

37) F.E. Arnold, Loon Seng Tam ; *31st International SAMPE Symposium*, April 7-10, 1986, 968

38) S.J. Huang, V. Paneccasio, D. Wilson ; *A.C.S. Polymer Preprints*, 27, (1), 125, 1986

REPORT OF DISCUSSIONS AND RECOMMENDATIONS

This section of the meeting showed that new and improved polycondensation techniques are still being discovered and offer the facility to produce novel polymeric structures of value as fibres, plastics, composites and functionalized polymers.

Polyaryletherketones

Synthesis of polyaryletherketones by polyaroylation and by polyether synthesis and especially the solvent systems required to keep these insoluble crystalline polymers in solution until polymers of high molecular weight are obtained : future developments are likely to be concerned with new catalyst/solvent systems, e.g. trifluoromethanesulphonic acid or excess $AlCl_3$ plus a Lewis base in 1,2-dichloroethane for polyaroylation and the use of alkali metal carbonates plus the bis phenol rather than alkali metal bis phenoxides, to react with the bis-fluorobenzoyl compounds for the polyether synthesis. The use of polyether synthesis to make high molecular weight amorphous derivatives of the polyetherketones which are soluble and can then be converted to the insoluble crystalline polyetherketones appears likely to develop further ; in discussion the preparation of polyketamines for further conversions to the polyetherketones was mentioned. Methods for improving toughness by incorporating rubbers, e.g. polyisobutylene with phenolic ends made by living cationic polymerization or polysiloxanes via techniques similar to those used to make polyethersulphone-polysiloxane block copolymers were considered to be worth further investigation.

Synthetic method for specialty polymers

Our approach for the development of specialty for structural application is based on the synthesis of telechelic oligomers which react by addition mechanisms. The problems with the oligomeric group is to find a good relationship between the initial Tg and the mechanical properties of the final network.

M. Fontanille and A. Guyot (eds.), Recent Advances in Mechanistic and Synthetic Aspects of Polymerization, 255–258.

On the other hand, many difficulties appear if we consider the telechelic reaction groups. From the point of view of thermal stability, the more promising groups are now nadimides, acetylene and phtalonitrile, but no precise information are available concerning the mechanism of polymerization, the stability of the networks and their thermal behavior.

Studies on model compound taking into account the rheological aspect have to be performed in order to improve the knowledge of the polymerizations in the solid state.

Use of phase transfer catalysis for polycondensation reactions

It has been well demonstrated that PTC is a useful tool for the preparation of polyethers, polycarbonates, polyesters, polysulfides, etc... The advantages are numerous such as simplicity of the process, shortening of reaction times, lowering of temperatures, increased yields, sometimes suppression of side reactions if compared with the classical processes. Moreover, it is not necessary to use a rigorous stoichiometry of the reagents. If suitable conditions are used, PTC is an excellent method for preparing well-defined telechelic oligomers having a low polydispersity which can be further used to make liquid crystal polymers, complexing polymers, multiblock copolymers, etc... Anyway, more work on the mechanism and the kinetics of such a process is required for optimizing the reactions in some cases. Moreover, the problem of an accurate determination of the molecular weights by GPC was raised. The use of PTC for preparing telechelic oligomers that cannot be otained by classical methods was the major advantage some participants emphasized.

Novel polycondensation system for the synthesis of polyamides and polyesters

Novel polycondensation can be achieved by the phosphenylation reaction using the combination of triphenylphosphine and halides for the synthesis of ether polyamide or polyesters under mild conditions. High molecular weight polyamide or polyester were easily obtained in a good yield. However, a stoichiometric amount of triphenylphosphine and halides is requested for the initiation of the direct polycondensation, in other words 1 mole of triphenylphosphine and halide is necessary to initiate the polycondensation reaction of 1 mole of monomer. Therefore, a good recycling system for the recovery of the used triphenylphosphine must be established

Electrochemical reductive coupling

In an unexpected way, electrochemical reductive coupling of aromatic dibromide catalyzed by nickel complexes afforded high polymers even at the beginning of the electrolyses.

This shows that electrochemical processes can be assigned for obtaining functionalized polymers or specialty polymers, even insoluble, when a rapid polycondensation process occurs at the surface of the electrode. The effectiveness of nickel catalysts was responsible of the result.

New reactions, new catalysts can be discovered. Better control of molecular weight or of product distributions are needed.

Processability of polymers obtained by that way may also be an important quality, which is not yet achieved in the known examples of such electrochemical polycondensation reactions.

POLYCONDENSATION PANEL

Synthetic aspect

- Research on more efficient method at low temperature leading to high molecular weight soluble polycondensation. Including :

+ new solvents or mixtures
+ new catalytic systems
+ modification of the electrophilic or nucleophilic reagents

- Development of new catalyst for bulk polycondensation

- Synthesis of typical polymers

+ reactions on insoluble polymers leaving reactive sites at the surface
+ research on template polycondensation
+ research on biomimetic reactions
+ research on oxydative or reductive coupling
+ development of polycondensation method in connection with electronic conductivity along the chain.
+ preparation of well defined telechelic oligomers

Characterization aspect

+ studies on side reactions occuring during polycondensation
+ studies on the mechanism of telechelic oligomers polymerization
+ studies on interactions occuring between solvents and high molecular weight polycondensate.

IV - POLYMERIZATIONS IN ORGANIZED MEDIUM

Chairmen : K.H. REICHERT and S.L. REGEN

POLYMERIZATION IN INCLUSION COMPOUNDS

Mario Farina
Department of Organic and Industrial Chemistry
University of Milan
Via Venezian 21, I-20133 Milano, Italy

ABSTRACT. After a brief description of the main features of the polymerization in inclusion compounds, a few recent results are described. They concern the regio- and stereoisomeric control in the polymerization of isoprene and pentadiene, the achievement of the first radical polymerization of propylene and the 2D-NMR investigation of hemiisotactic polypropylene.

Polymerization in organized media differ from the most common methods producing regio- or stereoregular polymers in that the growing chain ends do not take part in a catalytic complex nor are they subjected to the influence of a counterion; the structural features of the resulting polymers derive from the ordered arrangement in space of the surrounding molecules, even if these are not formally involved in the reaction. This definition suits the polymerizations which occur inside the channels of crystalline inclusion compounds. The lattice formed by the host molecules delimits the space around the growing chain end, thus imposing a definite reaction path.

The structure of crystalline inclusion compounds and their application to polymer chemistry have been reviewed several times in recent years (1-9); however, I think it useful to outline the most remarkable aspects of this subject for the sake of comparison with other classical or non-classical polymerization methods.

PROPERTIES OF INCLUSION COMPOUNDS

Inclusion compounds, or clathrates, are crystalline adducts containing two (or sometimes more) components: the host which forms the frame of the crystal, and the guest (or the guests) which is placed in the

M. Fontanille and A. Guyot (eds.), Recent Advances in Mechanistic and Synthetic Aspects of Polymerization, 261–280.

cavities existing in the lattice. Particularly important for our purposes are those clathrates which contain very long and narrow cavities: such channel-like or tubulate clathrates are very well suited to include linear macromolecules and to act as reaction site for polymerization.

Even more important is the ability of the host to accomodate the monomer molecules in a favorable disposition: the reactive groups should be close enough to arrive at the reaction distance without a strong disturbance of the crystal structure. It should be pointed out that the forces that stabilize the clathrate are rather weak and nonspecific. A certain mobility exists inside the channels, but - all other things remaining unchanged - those reactions that require the least motion of the involved atoms will be favored.

Formation of inclusion compounds depends on the structural relationships between host and guest, on temperature and, in the case of volatile components, on pressure. According to common definitions (10), the host-guest interactions in true clathrates are mainly or exclusively steric in origin: they depend on the shape and dimension, and not on the chemical nature of the constituents. When the opposite situation occurs, the term "complex" should be used; of course intermediate cases exist, such as the so-called "clathratocomplexes" and "coordinatoclathrates".

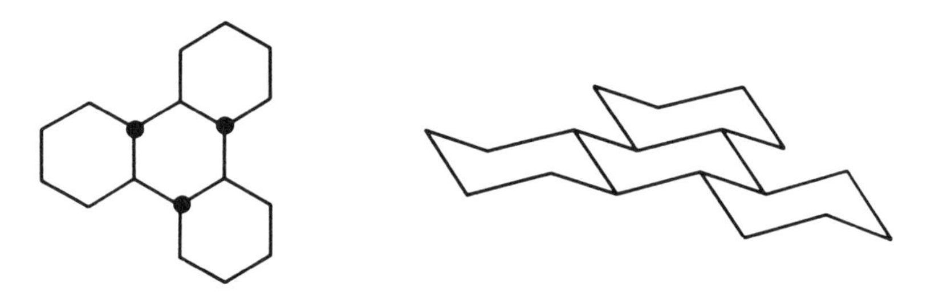

I

Hosts are often high symmetry molecules with trigonal or hexagonal shape and show a marked tendency to polymorphism. As typical hosts of this type I mention perhydrotriphenylene (PHTP, I) (11,12), to which I have dedicated most of my research, some substituted cyclotri-phosphazenes (TPP) (13) and α-cyclodextrin (14,15). Even simple molecules such as urea and thiourea form hexagonal structures stabilized by hydrogen bonds (16,17). Addition of extraneous molecules gives rise to new, often more symmetric, crystal structures, with a better filling up of space and with more favorable intermolecular contacts. The guest molecules are superimposed inside the channels existing in the crystal.

Generally the guest molecules do not occupy fixed positions in the lattice and give rise to characteristic diffused streaks in the X-ray photographs.

Other hosts able to form channel inclusion compounds and used in polymer chemistry are β-cyclodextrin (14,15), deoxycholic acid (DCA) and apocholic acid (ACA) (3,18,19). These are natural or semisynthetic chiral compounds directly obtained in a pure enantiomeric form.

These hosts impart different properties to the corresponding clathrates: PHTP, due to its hydrocarbon nature, forms flexible channels which closely fit the shape of the guests. As a consequence, PHTP has little selection power with respect to the guest structure, but imposes a strong control on the reaction. On the other hand, urea and thiourea form rigid channels with fixed dimension. For this reason, these compounds show a high selectivity and are used in separation processes on the industrial scale. Both hosts are suited for the synthesis of regular polymers. DCA and ACA behave in a different way. Their channels are stiff and wide, thus permitting some orientational and conformational freedom, especially with small guests.

When discussing the stability of true clathrates, we should bear in mind their nature of "phase rule compounds" (20) i.e., of compounds which exist only in the solid phase and disappear when a phase transition occurs.* Stability should be considered with regard to a particular transition: solid-solid, solid-liquid or solid-vapor. Little is known about the first reaction, and only partial studies are reported for the others. A rationale of the thermodynamic behavior was recently proposed by me and my coworkers (12,21). According to these studies, the highest stability is observed when host and guest form an ideal solution in the liquid phase. Stability decreases when repulsive interactions are present in solution, especially when a miscibility gap appears in the liquid. In terms of regular solution theory this phenomenon occurs when the solute-solvent interaction parameter W exceeds 2 RT.

The melting or decomposition points of some inclusion compounds of PHTP and urea with n-alkanes are reported in Table I. It should be noted that the melting points of the pure hosts are very close to each other and the inclusion energy of a given guest in PHTP and in urea are comparable (in case, it is higher in urea); nevertheless, the PHTP clathrates melt 25-80 K above the corresponding ones in urea.

---

* I refer here to inclusion compounds whose crystal structure differs from that of the pure host; in other words, to systems in which the "empty cavities" are thermodynamically unstable.

Table I. Melting or decomposition point of PHTP and urea clathrates with C_n normal alkanes

n	PHTP[a)]	urea[b)]
7	120°C	42°C
8	124	-
10	131	81
12	136	91
16	145	108
20	151	118
24	153	125
28	156	131
36	161	-
polyethylene	178	146
(pure host)	(125)	(132)

a) congruent melting
b) incongruent melting

Figure 1 shows the temperature-composition phase diagrams calculated for a hypothetical 3:1 adduct, by using the same thermodynamic parameters, intermediate between those of PHTP and urea, the only difference being the value of W: in 1a) W = 0 (ideal solution), not too far from the value actually measured for the PHTP-heptane system; in 1b) W = 2000 R, corresponding to an almost complete immiscibility of the two components in the liquid phase. The field of existence of the adduct C is dramatically reduced in the latter case, as experimentally observed by DSC or by tensiometric measurements.

These remarks have a direct relevance to our subject. As monomer inclusion is a preliminary condition for polymerization (there is ample evidence that reaction occurs inside the crystals and not on their surface), the highest temperature at which polymerization can be carried out is determined by the stability of the host-monomer adduct. The butadiene-urea clathrate decomposes around - 30°C and polymerization can be performed only at low temperatures. This caution is not necessary when using PHTP as a host due to the higher affinity of the two partners. Typical polymerization temperatures range from -100° to - 30°C for urea and thiourea, from -80° to + 100°C for PHTP and arrive to + 150°C for DCA and ACA.

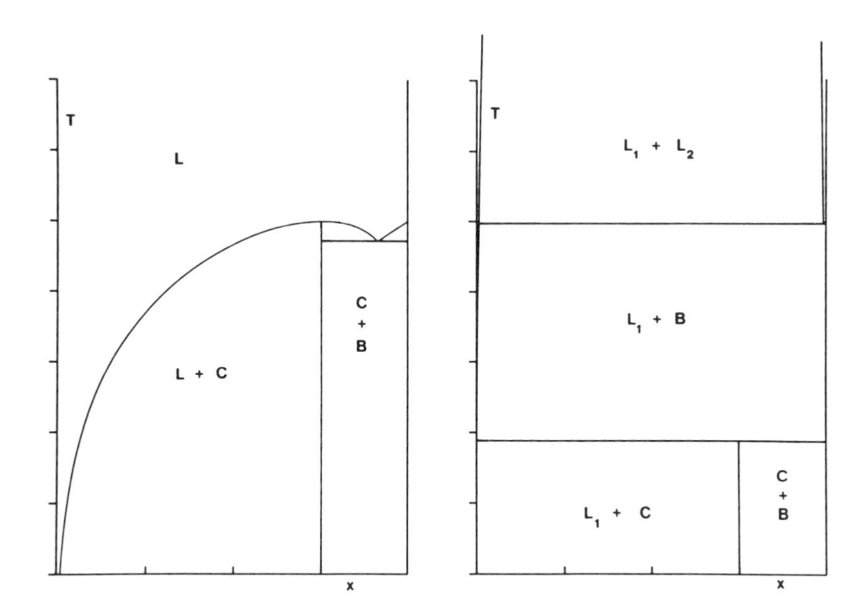

Figure 1. Calculated phase diagrams for hypothetical AB_3 inclusion compounds having different W values. In a) W = 0; in b) W = 2000 R.

The incompatibility of host and guest causes a further difficulty. The time required for clathrate formation, which is of the order of seconds or minutes in the case of PHTP and hydrocarbons, becomes very long, days and even weeks, if the difference in chemical composition is large. For this reason, the reacting system should be chosen very carefully, taking into account the many different factors of structural, thermodynamic and kinetic nature.

A further remark on stability and phase diagrams concerns the polymerization of very volatile monomers. In such cases it is not possible to work in open flasks: the vapor (or decomposition) pressure of the guest in the inclusion compound is lower than that of the pure liquid guest, but can exceed atmospheric pressure. We met with this phenomenon when we tried to polymerize ethylene and propylene in PHTP. Vapor pressure measurements of the propylene adduct showed that the decomposition point is well above room temperature (22). Polymerization was carried out in thick glass vials or in steel tubes at temperatures ranging from -10° to + 30°C. At room temperature (20°C), pressure amounts to 1020 kPa, if an excess monomer is used, or to 325 kPa, if the monomer is used in defect with respect to PHTP. The relatively high thermal stability permits a rapid formation and an easy handling of the adduct. Ethylene, too, was polymerized at room temperature and at high pressure in steel tubes.

It is interesting to note that the use of inclusion compounds considerably increases the field of feasibility of solid state polymerizations. In the just mentioned case of propylene, polymerization in clathrates, a reaction occurring under the control of the crystal lattice, takes place more than 200 K above the melting point of the monomer.

MECHANISM OF INCLUSION POLYMERIZATION

In the narrowest sense, polymerization in inclusion compounds - generally referred to as inclusion, channel or canal polymerization - consists in the conversion of a host-monomer clathrate into a host-polymer one. The real occurrence of this process has been demonstrated both by X-ray (23,24) and by DSC (25,26) investigations. In a few cases, however, the formed polymer lacks the steric requirements for inclusion and the result of polymerization is a mixture of host and polymer (27).

Polymerization generally starts by subjecting the inclusion compound to irradiation with β, γ or X-rays and proceeds by a radical mechanism. Radicals derive both from host and guest, as demonstrated by isotopic labeling. Cleaner and more easily interpreted results were obtained by irradiation of the pure host and subsequent inclusion of the monomer. This method of polymerization in a preirradiated matrix is particularly feasible with PHTP, due to the easy formation of the inclusion compound. In this case radicals derive from the host only: being trapped in the crystal lattice they remain dormant up to several months and become "alive" after monomer inclusion. One of the main advantages of this method is the fact that monomers are not subjected to irradiation.

Recently, a new technique was set up, consisting in the coinclusion (in DCA) of a monomer and a free radical initiator, such as di-ter-butyl peroxide (28).

The radical nature of polymerization was demonstrated many years ago by ESR spectroscopy in the butadiene-urea system (29); its living character was deduced from the linear dependence of molecular mass on polymer yield in a series of strictly controlled reactions carried out in preirradiated PHTP (30) and confirmed by the synthesis of block copolymers (31). Chain transfer is limited by steric constraints and termination is very improbable, at least at low radiation doses. A recent ESR study carried out by including two monomers one after the other in preirradiated PHTP gave a simultaneous demonstration of the living and radical character of the polymerization (32).

The most common monomers used in inclusion polymerization are conjugated diolefins: butadiene, isoprene, pentadiene, 2,3-, 2,4-, 4,4-dimethylbutadiene, 2,3-dichlorobutadiene, 2,4-hexadienes, etc. (included in urea, thiourea, PHTP, DCA, ACA, TPP) (1-9, 23, 25, 30, 33-42). Very active monoolefinic monomers were also used: to vinyl chloride and acrylonitrile (in urea and TPP) (34,42) and to vinylidene chloride (in thiourea) (33), several bulkier monomers were recently added, such as substituted styrenes, methyl methacrylate etc. (in TPP) (43,44). Very recent acquisitions are the polymerization of little reactive monoolefins, such as propylene (in PHTP) (22), and of diacetylene (both in DCA and PHTP) (45,46). The polymer obtained in the latter case has a potential application in the field of electroconductive organic materials.

Random copolymers are obtained when two monomers are included in the same channel, a more common phenomenon than expected in view of the selectivity often observed in clathration; known examples occur with thiourea (33,47), PHTP (35,48), DCA (3) and TPP (44).

Copolymerization may be regarded as an unconventional method of investigation of the structure of ternary inclusion compounds (5). If we admit that interchange between included monomers is slow, the sequence of monomer units in the copolymer represents a permanent image (such as a photographic picture or a magnetic tape) of the guest arrangement in the channel.

When two isomorphic monomers such as 2-methylpentadiene and 4-methylpentadiene are included in PHTP, the distribution of guests is completely at random or Bernoullian; the product $r_1 \cdot r_2$ deduced from NMR sequence analysis equals 1. For monomers largely different in shape such as butadiene and 2,3-dimethylbutadiene, the $r_1 \cdot r_2$ product is greater than 1, indicating a tendency toward formation of blocks of like units.

The production of block copolymers is the consequence and at the same time, the proof of the living character of polymerization in PHTP (31). This process consists of two steps: first, one of the two monomers is included in the preirradiated host, and polymerization allowed for the time required for the formation of the first block (from a few minutes to some hours, depending on the monomer reactivity and on the radiation dose); secondly, the unreacted monomer is eliminated under high vacuum and a second monomer included until completion of the reaction. The NMR spectrum corresponds to that of a mixture of the two homopolymers, but solution properties indicate the presence of two blocks in the same macromolecule. Further evidence of the occurrence of this process is shown by the ESR spectrum, which reveals a change in the structure of the active chain end (32).

Inclusion polymerization has always attracted much attention due to its ability to produce stereoregular polymers. In 1960 Brown and White succeeded in obtaining 1,4-trans crystalline polymers from 2,3-dimethylbutadiene (included in thiourea) (33) and butadiene (in urea) (34). Later, trans-pentadiene and trans-2-methylpentadiene included in PHTP were converted into highly isotactic (96-98 % expressed as isotactic dyads), crystalline, 1,4-trans polymers (25,35,49). Variable degrees of stereoregularity (from low to moderate) were observed with other hosts and other monomers, especially with vinyl monomers.

Several aspects of the topochemical control are known, but until now a comprehensive and general explanation of the reaction is lacking. The absence of branching and the tendency toward 1,4 (and not 1,2) addition with diene monomers are attributed to the geometric constraints existing in clathrates. The prevalence of 1,4-trans over 1,4-cis configuration in polydienes is ascribed to a preference for the transoid conformation, which is more suited to enter the channels, in the included monomers. In DCA and ACA, where these rules are less strict, an increasing tendency toward the formation of 1,4-trans polymers is observed when passing from linear (e.g. butadiene) to bulky monomers (2,3-dimethylbutadiene). This fact is connected with the larger space available inside the rigid cavities of these hosts.

Only head-to-tail junctions are observed in polydienes obtained in PHTP when the monomer contains a substituent in the terminal position (1 or 4). Two converging factors are held responsible for this: the higher electron delocalization existing in the $-CH_2-CH=CH-CH^{\bullet}(CH_3)$ radical with respect to $-CH(CH_3)-CH=CH-CH_2^{\bullet}$, which favors formation of a secondary and not of a primary chain end, in accordance with ESR spectra (32), and the excessive distance between reacting atoms required for the formation of a head-to-head junction (35). In the latter instance, two methyl groups face each other inside the channel and inhibit the approach of the radical to the unsaturated carbon of the next monomer:

$$-CH_2-CH=CH-CH^{\bullet}(CH_3) + CH_3-CH=CH-CH=CH_2 \longrightarrow \text{no reaction}$$

The same argument had already been used to explain the non reactivity of trans-trans-hexadiene in inclusion polymerization.

Regioregularity is much lower in the case of isoprene, when the topochemical factor is suppressed. This reaction will be dealt with in more detail in the next chapter.

More difficult is the explanation of stereoregularity. A fortuitous coincidence was observed between the repeat period of PHTP molecules in the crystal and that of 1,4-trans diene units, both approaching 0.48 nm. As a consequence, all propagation steps in an isotactic polymerization should be isoenergetic, the reactants being placed in crystallographic equivalent positions. However, this hypothesis cannot be extended to other hosts and does not take into account other peculiarities of the process.

PHTP, DCA and ACA are chiral substances existing in enantiomeric forms. Optically active hosts are thus available, and this allows the carrying out of asymmetric polymerizations in the solid state. The first result was obtained in 1967 (50) with the pentadiene-PHTP system (optically active PHTP was obtained by a tedious and expensive resolution procedure (51,52)). Optical yield was rather low (≃7%), nevertheless this achievement was very important because it represented the first example of an asymmetric synthesis performed in the solid state and, at the same time, the first example of radical asymmetric polymerization.

Later on, other asymmetric polymerizations were obtained in DCA and ACA, using cis-pentadiene (38,53) and cis- or trans-2-methylpentadiene (41,53) as monomers. Optical activity and optical purity were higher than in PHTP, in spite of the lower stereoregularity.

The synthesis of an optically active isoprene-pentadiene block copolymer obtained in PHTP (54) and a discussion in probabilistic terms of the structure of optically active polypentadiene obtained in DCA (55) led to an interesting conclusion: both experiments are in accordance with a "through space" and not with the more common "through bond" transmission of asymmetric induction. In other terms chirality in polymers is controlled by the shape and chirality of the cavity and not by the presence of chiral initiators derived from the host.

Further evidence of the power of topochemical control existing in inclusion polymerization is found at the conformational level. As already stated, the final point of most inclusion polymerizations is a host-polymer clathrate where the macromolecules are placed in separated, parallel channels, several hundreds nm long. If decomposition of the adduct is performed without melting or dissolving the polymer, this extended-chain morphology is sometimes retained. The peculiar macroconformation of these "native" polymers can be evidenced by SAXS analysis (56) or by DSC (57). In particular their melting point is 10 -15 K higher than that of conventional samples.

Polymerization of conjugated diolefins in PHTP clathrates often produces highly regular polymers. Even when defects occur along the chain they are always of the same type, thus making the structural analysis very simple. This is the case of isoprene, where the only degree of freedom concerns the direct or inverted insertion of the monomer units.

In spite of its very high steric purity (there is no trace of 1,4-cis units, as well as of 1,2, 3,4 or cyclic units) 1,4-trans polyisoprene obtained in PHTP is amorphous (25,35,58). The reason for this (we recall that balata and gutta percha, two well-known 1,4-trans polyisoprenes of natural origin, are crystalline) is the presence of a substantial number of head-to-head and tail-to-tail junctions, in addition to head-to-tail. As previously discussed, the topochemical factor which imposes head-to-tail propagation is lacking in isoprene. This monomer is approximately spherical in shape and can be accomodated in different ways along the channel. Moreover, we cannot exclude that the isoprene molecules can rotate inside the crystal, especially in connection with the disturbance imposed on the structure by the growing chain end. In this respect, we should note that positional or dynamic disorder of guest molecules in PHTP clathrates had already been observed by X-ray analysis with dioxan and cyclohexane (59).

The polymer structure may be interpreted as a succession of direct (D or 1,4) and inverted (I or 4,1) units:

$-CH_2-C(CH_3)=CH-CH_2-$	$-CH_2-CH=C(CH_3)-CH_2-$
C1 C2 C3 C4	C4 C3 C2 C1
D unit	I unit

Head-to-tail successions derive from DD and II sequences, whereas head-to-head correspond to ID and tail-to-tail to DI sequences. ^{13}C NMR analysis gives clear evidence of the presence of regioisomeric defects (Table II): in the saturated CH_2 region, two additional peaks are observed about 1 ppm distant from the main signals. The weak peak at 38.5 ppm from TMS correponds to head-to-head or ID sequences (carbon C1) and that at 28.3 to tail-to-tail or DI sequences (carbon C4). Analysis of the unsaturated region, and especially that of the C2 carbon placed around 135 ppm, gives a more detailed information. In the spectra run at 50.3 MHz, the low-field signal appears to be split in four peaks, attributed to triads of monomeric units. The most intense peak (placed upfield) corresponds to DDD or III triads (head-to-tail, head-to-tail sequences), the lowest to IDI or DID triads (head-to-head, tail-to-tail), and the remaining two to the other mixed sequences IDD (and IID)

TABLE II. ^{13}C NMR spectrum of 1,4-trans polyisoprene

Carbon	Chemical shift ppm from TMS	Sequence
C2'(CH_3)	16.03	-
C4	26.73	ht
C4	28.29	tt
C1	38.49	hh
C1	39.75	ht
C3	124.23	-
C2	134.93	ht,ht
C2	135.06	ht,hh(or tt,ht)
C2	135.11	tt,ht(or ht,hh)
C2	135.24	tt,hh

and DDI (and DII).

A statistical analysis of sequence distribution was carried out in terms of first-order Markov chains of D or I units, with two independent conditional probabilities p_{DI} and p_{ID}. For any sequence distribution two symmetric solutions exist, obtained by exchanging the values of p_{DI} with those of p_{ID}. In order to have the correct solution independent information is required: this can be obtained by ESR analysis of the growing radical included in PHTP. The spectrum reveals the presence of a radical having the highest spin density on carbons C2 and C4 (32). This corresponds to a 1,4 insertion, or, equivalently, to a propagation of D units:

$$-CH_2-C(CH_3)=CH-\dot{C}H_2$$

The presence of an I radical cannot be excluded, but it does not represent the prevailing species. As a consequence, p_{ID} will be greater than p_{DI}.

For a polyisoprene obtained at room temperature, $p_{ID} = 0.785$ and $p_{DI} = 0.162$. We deduce that polymerization proceeds essentially by inserting D units, with the occasional insertion of isolated I units and of pairs of I units with about the same probability (the average length of I sequences is 1.5):

..DDDIDDD.. and ..DDDIIDDD..

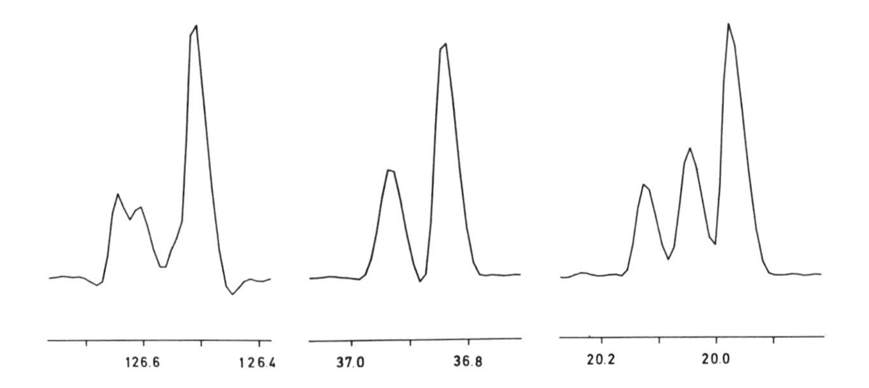

Figure 2. ^{13}C NMR spectrum of 1,4-trans polypentadiene obtained from the cis monomer at +60°C (chemical shift in ppm from TMS). From the left C2, C4 and C4' subspectra.

Polymerization of isoprene in PHTP was examined over a wide temperature range, from -60° to + 70°C. The number of defects increases with temperature, passing from 7 to 16%. A quantitative evaluation of this effect was obtained by assuming that the logarithm of the ratios p_{DI}/p_{DD} and p_{II}/p_{ID} depends linearly on 1/T. In this way we obtained an activation energy for the insertion of a wrong (I) unit comprised between 4 and 4.5 kJ mol^{-1}.

The polymerization of pentadiene offers an example of the way in which stereochemical purity can be controlled (60). In this instance, there are two variables of the process: the configuration, cis or trans, of the starting monomer and the temperature. Both pentadiene stereoisomers are easily included in PHTP and polymerize in the presence of the preirradiated matrix. In both cases the basic polymer structure is the same, head-to-tail 1,4-trans isotactic polypentadiene, but a different degree of tacticity is observed. In the ^{13}C NMR spectrum the sample derived from the trans monomer shows five sharp peaks only, whereas the other polymer presents additional minor peaks related to the presence of steric defects. The most sensitive carbons in this respect are C2 and C4', whose resonances split in three signals, and C4 which appears as a doublet. Details of this spectrum are reported in Figure 2; nomenclature is referred to the following formula:

$$\underset{C1}{-CH_2}-\underset{C2}{CH}=\underset{C3}{CH}-\underset{C4}{CH}(\underset{C4'}{CH_3})-$$

Stereochemical assignments in terms of dyads and triads were made according to already reported studies (61) and to intensity criteria.

Polymerization of both stereoisomers was investigated at -20°, +25° and +60°C. As expected the number of defects increases with temperature and at 60°C even the trans monomer produces a polymer containing a detectable amount of mr and rr triads*, although lower than that observed with the cis monomer at -20°C. Triad distribution in "poly-cis" samples is quite different from that predicted by a Bernoullian process in m/r terms; the lack of data regarding longer sequences and the complexity of factors to be taken into account do not yet permit the formulation of an adeguate scheme for the quantitative interpretation of this reaction.

SYNTHETIC AND SPECTROSCOPIC CONTRIBUTIONS TO THE KNOWLEDGE OF POLYPROPYLENE

In the last part of this article, I wish to outline the contributions inclusion polymerization has made towards the solution of some problems regarding polypropylene: one concerns the reactivity of the monomer and a new method of polymerization; the second the interpretation of ^{1}H and ^{13}C NMR spectra, obtained by using a polypropylene sample having a "hemiisotactic" structure.

Compared to the usual vinyl monomers, propylene should be considered as a poor reactive compound. This is true even when compared with other hydrocarbon monomers such as ethylene or isobutene. The conventional polymerization methods (both ionic and radical) give low molecular mass products and only the use of Ziegler-Natta catalysts yielded polymers with valuable mechanical properties. In particular, radical processes were thought to be ineffective due to the short life-time of the alkyl radicals and the overwhelming presence of

---

* According to the IUPAC rules (62), the terms m and r cannot be used in the case of polypentadiene because the chain segments connecting the stereogenic atoms are not symmetric with respect to the chain observation direction. Their use could however be retained if any reference to their original meaning (m for meso and r for racemo) is avoided and substituted with a conventional definition (m for maintenance and r for reversal of configuration of the successive homologous stereogenic atoms) (63,64). In this sense m and r are used here.

termination and chain transfer reactions.

Inclusion polymerization possesses all the attributes necessary to overcome these drawbacks. Radicals trapped in a solid phase are very stable and side reactions are minimal in comparison with solution or bulk polymerizations. The first step then is the search for suitable conditions for inclusion and hence for polymerization.

The pressure-temperature diagram of the PHTP-propylene clathrate indicates that the field of thermodynamic stability of the adduct extends well above room temperature (22), as discussed in a previous section. Polymerization in PHTP proceeds very slowly: after several days a rubbery polymer is obtained with $\bar{M}_v$ = 8 - 10 000 and with a prevailing syndiotactic structure (65% of *r* dyad content). Pentad concentration was obtained from ^{13}C NMR analysis and follows a first-order Markov distribution with the following conditional probabilities: p_{mr} = 0.437 and p_{rm} = 0.238.

The real existence of a radical mechanism was proved by the synthesis of an isoprene-propylene block copolymer, whose first step, the formation of the polyisoprene block, was unequivocally proved to have a radical character.

The last subject concerns the use of hemiisotactic (hit) polypropylene for a deeper understanding of the NMR spectra. With the term hemitactic we define a polymer in which an ordering rule acts on every second monomer unit only: for instance, the first, third, fifth, ..., monomer units are placed in a isotactic (or syndiotactic) way, whereas the second, fourth, sixth, ..., are arranged at random (65-68).

Hit-polypropylene was obtained by polymerizing trans-2-methyl-pentadiene included in PHTP. The crystalline, insoluble, highly isotactic, 1,4-trans polymer obtained in such a way was treated with tosylhydrazide at 130°C and converted into the desired polymer. During this process, the methyl groups originally bound to tetrahedral carbons keep their configuration, but those placed on the double bonds have almost the same probability to go on one or on the other side of the chain:

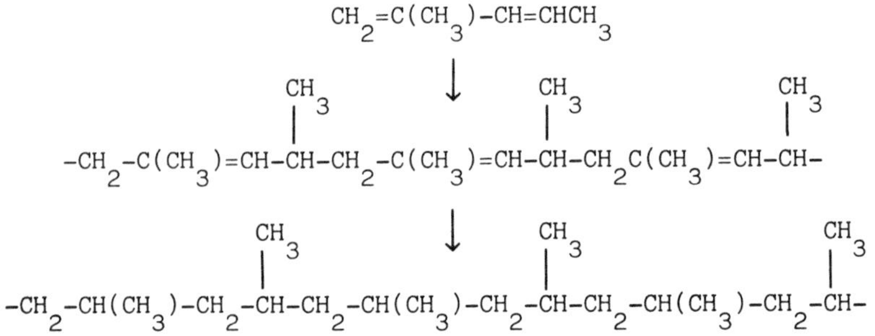

$$CH_2{=}C(CH_3){-}CH{=}CHCH_3$$

$$\downarrow$$

$$-CH_2{-}C(CH_3){=}CH{-}\underset{|}{\overset{CH_3}{CH}}{-}CH_2{-}C(CH_3){=}CH{-}\underset{|}{\overset{CH_3}{CH}}{-}CH_2C(CH_3){=}CH{-}\overset{CH_3}{\underset{|}{CH}}{-}$$

$$\downarrow$$

$$-CH_2{-}CH(CH_3){-}CH_2{-}\overset{CH_3}{\underset{|}{CH}}{-}CH_2{-}CH(CH_3){-}CH_2{-}\overset{CH_3}{\underset{|}{CH}}{-}CH_2{-}CH(CH_3){-}CH_2{-}\overset{CH_3}{\underset{|}{CH}}{-}$$

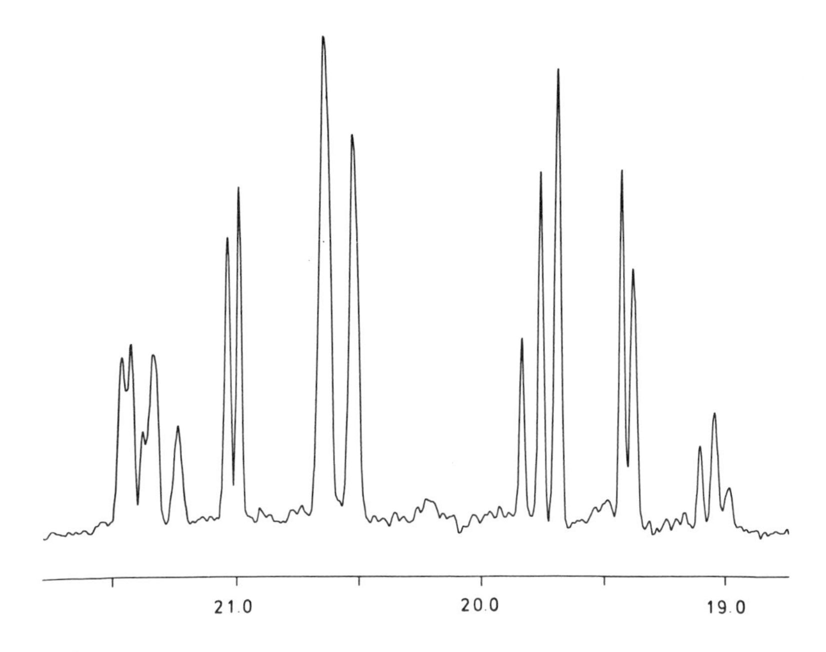

Figure 3. ^{13}C NMR spectrum of the methyl region of hemiisotactic polypropylene.

assigned to undecads mmmmrrmmmm, mmmmrrmmrr and rrmmrrmmrr. This assignment was drawn directly from the selection rules which inhibit the presence of all other sequences. The mechanism of sequence generation is reported in this scheme:

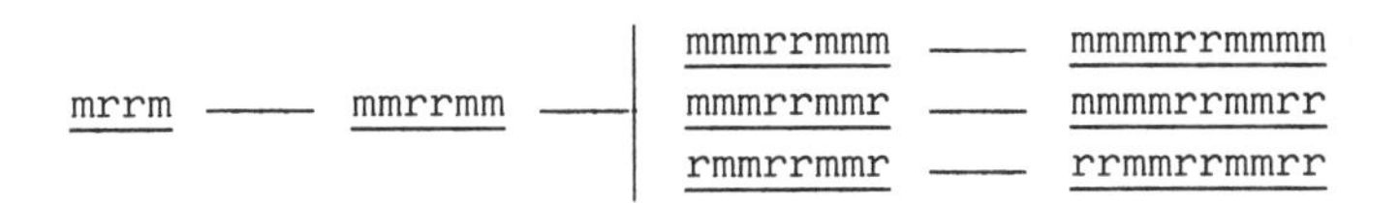

Further undecad peaks are recognized in other spectral regions especially if the spectrum is recorded at varying temperatures, owing to the different temperature coefficient of the chemical shift of different sequences.

2D-NMR spectroscopy was successfully applied to hit-polypropylene (69). The 2D-J resolved ^{1}H-NMR spectrum gives an amount of information that rivals ^{13}C-spectroscopy, particularly in the CH_3 region. As all methyl protons have the same vicinal H-H coupling constant (J = 6.5 Hz), the decoupled spectrum is simply obtained as a section of the 2D spectrum. It contains at least 14 signals showing a sensitivity at the

TABLE III. Number of stereosequences in atactic and hemiisotactic polypropylene

Sequences	atactic	hemisotactic
Dyads	2	2
Triads	3	3
Tetrads	6	4
Pentads	10	7
Hexads	20	8
Heptads	36	14
Octads	72	16
Nonads	136	28
Decads	272	32
Undecads	528	54

(the methyl groups written in brackets in the last formula have an unspecified configuration).

From the point of view of microtacticity, hit-polypropylene is characterized by a succession of non-overlapped mm or rr triads. In other words, the non terminal sections of each sequence must contain an even number of adjacent m or r letters. It follows that isolated dyads (e.g. .. mmrmm..) or more generally odd series of equal letters (...rmmmr...) are forbidden.

As a consequence of these rules the number of the allowed sequences in hit-polymers is considerably lower than in common atactic vinyl polymers (Table III). NMR spectra directly reflect this situation: they are complex enough to contain a great deal of information and at the same time simple enough to permit interpretation at an unprecedently attained level.

Structure of hit-polypropylene was established from ^{13}C NMR spectra recorded under usual conditions (130°C). In the CH_3 region, which is the most sensitive to the stereochemical environment, 7 signals are observed with the theoretically predicted ratio (3:2:1:4:3:2:1), instead of 9 as in atactic samples. The lacking signals correspond to the three forbidden pentads rmrr, mmrm (with the same chemical shift) and rmrm. A more complete analysis was performed by running the spectrum at room temperature (Figure 3). 17 peaks are observed, showing a sensitivity to sequences longer than heptads (which are 14 only). In particular, the three high-field signals all derive from the mrrm pentad and are heptad level or higher, a result which is unimaginable for spectra obtained under the usual decoupling conditions.

Assignment of these peaks was attained by means of 2D-heteronuclear chemical shift correlation spectra. Sequences belonging to the same pentad are generally well resolved, whereas a certain overlap exists among peaks centered on different pentads. Spectral analysis requires consideration of nonads and, in some cases, of undecads. One of the three undecads derived from the mrrm pentad can be detected at high field; other individual sequences identified in the proton spectrum are the nonads mmmmmmrr and mmrrrrmm.

This is one of the most detailed analyses ever performed on a synthetic polymer. It is worthy of note that this result was made possible by the peculiar structure (at the same time ordered and disordered) of hit-polypropylene. In its turn this structure derives from the combined use of two different techniques: a highly selective polymerization process, in particular a polymerization in inclusion compounds, and a non-selective chemical modification of the polymer obtained in such a way.

ACKNOWLEDGEMENTS

This work was partly supported by the Italian National Research Council (CNR) and the Italian Ministry of Education, as a part of the national project on polymers for special uses.

REFERENCES

1) Y. Chatani, Prog.Polym.Sci.Japan, **7**, 149 (1974).
2) M. Farina, in E.B. Mano, ed.: Proceedings of the International Symposium on Macromolecules, Rio de Janeiro, 1974, Elsevier, Amsterdam, 1975, p. 21.
3) K. Takemoto and M. Miyata, J.Macromol.Sci.,Rev.Macromol.Chem., **C18**, 83 (1980).
4) M. Farina, Makromol.Chem.Suppl., **4**, 21 (1981).
5) M. Farina, G. Di Silvestro and P. Sozzani, Mol. Cryst. Liq. Cryst., **93**, 169 (1983).
6) M. Farina 'Inclusion Polymerization' in ref. 7, Vol. 3, Chapt. 10.
7) J.L. Atwood, J.E.D. Davies and D.D. MacNicol, eds., Inclusion Compounds, Academic Press, London, 1983-84, vol. 1-3.
8) M. Farina, Proceedings of the Am. Chem. Soc. PMSE Division, **54**, 288 (1986).
9) M. Farina and G. Di Silvestro, 'Polymerization in clathrates'

Encyclopedia of Polymer Science and Engineering, J. Wiley, New York, in press.
10) E. Weber and H.P. Josel, J.Inclusion Phenomena, **1**, 79 (1983).
11) M. Farina, G. Allegra and G. Natta, J.Am.Chem.Soc., **86**, 516 (1964).
12) M. Farina, 'Inclusion compounds of perhydrotriphenylene' in ref. 7, Vol. 2, Chapt. 3.
13) H. R. Allcock 'Cyclophosphazene inclusion compounds' in ref. 7, Vol. 1, Chapt. 8.
14) F. Cramer, Einschlussverbindingen, Springer Verlag, Heidelberg, 1954.
15) W. Saenger 'Structural aspects of cyclodextrins and their inclusion complexes' in ref. 7, Vol. 2, Chapt. 8.
16) W. Schlenk Jr., Justus Liebigs Ann. Chem. **565**, 204 (1949).
17) K. Takemoto and N. Sonoda, 'Inclusion compounds of urea, thiourea and selenourea', in ref. 7, Vol. 2, Chapt. 2.
18) W.C. Herndon, J. Chem. Ed., **44**, 724 (1967).
19) E. Giglio, 'Inclusion compounds of deoxycholic acid', in ref. 7, Vol. 2, Chapt. 7.
20) J.E. Ricci, The phase rule and heterogeneous equilibrium, Van Nostrand, New York, 1951.
21) M. Farina, G. Di Silvestro and A. Colombo Mol. Cryst. Liq. Cryst., **137**, 265 (1986).
22) G. Di Silvestro, P. Sozzani and M. Farina Am. Chem. Soc. Polym. Preprints, **27(1)**, 92 (1986).
23) Y. Chatani, S. Nakatani and H. Tadokoro Macromolecules, **3**, 481 (1970).
24) A. Colombo and G. Allegra, Macromolecules, **4**, 579 (1971).
25) M. Farina, G. Natta, G. Allegra and M.Löffelholz, J.Polym.Sci., **C16**, 2517 (1967).
26) M. Farina and G. Di Silvestro, Gazz.Chim.Ital., **112**, 91 (1982).
27) Y. Chatani, K. Yoshimori and Y. Tatsuta, Am.Chem.Soc.,Polym. Preprints, **19** (2), 132 (1978).
28) M. Miyata, F. Noma, Y. Osaki, K. Takemoto and M. Kamachi, J. Polym. Sci, Polym. Letters, **24**, 457 (1986).
29) T. Ohmori, T. Ichikawa and M. Iwasaki, Bull.Chem.Soc.Japan, **46**, 1383 (1973).
30) M. Farina, U. Pedretti, M.T. Gramegna and G. Audisio, Macromolecules, **3**, 475 (1970).
31) M. Farina and G. Di Silvestro, J.Chem.Soc.,Chem.Commun., 816 (1976).
32) P. Sozzani, G. Di Silvestro and A. Gervasini, J. Polym. Sci., Polym. Chem. Ed. **24**, 815 (1986).
33) J.F. Brown and D.M. White, J.Am.Chem.Soc., **82**, 5671 (1960).
34) D.M. White, J.Am.Chem.Soc., **82**, 5678 (1960).

35) M. Farina, G. Audisio and M.T. Gramegna, Macromolecules, **5**, 617 (1972).
36) Y. Chatani and S. Nakatani, Macromolecules, **5**, 597 (1972).
37) M. Miyata and K. Takemoto, J.Macromol.Sci.,Chem, **A12**, 637 (1978).
38) G. Audisio and A. Silvani, J.Chem.Soc.,Chem.Commun., 481 (1976).
39) M. Miyata and K. Takemoto, J.Polym.Sci.,Polym.Lett.Ed., **13**, 221 (1975).
40) M. Miyata, Y. Kitahara and K. Takemoto, Polym.J.,**13**, 111 (1981).
41) M. Miyata, Y. Kitahara, Y. Osaki and K. Takemoto, J.Inclusion Phenomena, 2, 391 (1984).
42) J. Finter and G. Wegner, Makromol.Chem., **180**, 1093 (1979).
43) H.R. Allcock, W.T. Ferrar and M.L. Levin, Macromolecules, **15**, 697 (1982).
44) H.R. Allcock and M.L. Levin, Macromolecules, **18**, 1324 (1985).
45) M. Miyata, H. Tsutsumi and K. Takemoto, Preprints of the International Conference on Conductive Polymers, Kyoto, 1986, p. 382.
46) B. Jorgensen, R. Liepins and S. Agnew, Polym. Bull., **16**, 263 (1986).
47) Ch. Schneider, H.H. Greve, H.P. Bohlmann and D. Schuhmann, Preprints of the IUPAC Symposium on Macromolecules, The Hague, 1985, p. 260.
48) P. Sozzani, G. Di Silvestro, M. Grassi and M. Farina, Macromolecules, **17**, 2538 (1984).
49) P. Sozzani, G. Di Silvestro, M. Grassi and M. Farina, Macromolecules, **17**, 2532 (1984).
50) M. Farina, G. Audisio and G. Natta, J.Am.Chem.Soc., **89**, 5071 (1967).
51) M. Farina, G. Audisio, Tetrah. Letters, 1285 (1967).
52) M. Farina and G. Audisio, Tetrahedron, **26**, 1827 (1970).
53) M. Miyata, Y. Kitahara and K. Takemoto, Polym. Bull., **2**, 671 (1980).
54) M. Farina, G. Di Silvestro and P. Sozzani, Makromol.Chem.,Rapid Commun., 2, 51 (1981).
55) G. Audisio, A. Silvani and L. Zetta, Macromolecules, **17**, 29 (1984).
56) Y. Chatani and S. Kuwata, Macromolecules, **8**, 12 (1975).
57) M. Farina and G. Di Silvestro, Makromol.Chem., **183**, 241 (1982).
58) G. Di Silvestro, P. Sozzani and M. Farina, Macromolecules, **20**, 000 (1987).
59) G. Allegra, M. Farina, A. Immirzi, A. Colombo, U. Rossi, R. Broggi and G. Natta, J.Chem.Soc.,Part B, 1020 (1967).
60) unpublished results.
61) L. Zetta, G. Gatti and G. Audisio, Macromolecules, **11**, 763 (1978).
62) IUPAC Macromolecular Division, Macromolecular Nomenclature Commission, Pure Appl. Chem., **53**, 733 (1981).
63) P. Sozzani, cited in ref. 64.
64) M. Farina, Topics Stereochem. **17**, 000 (1987).

65) M. Farina, G. Di Silvestro and P. Sozzani, Macromolecules, **15**, 1451 (1982).
66) M. Farina, G. Di Silvestro, P. Sozzani and B. Savaré, Macromolecules, **18**, 923 (1985).
67) G. Di Silvestro, P. Sozzani, B. Savaré, M. Farina, Macromolecules, **18**, 928 (1985).
68) P. Sozzani and C. Oliva, J. Magn. Reson. **63**, 115 (1985).
69) A. Di Marco and P. Sozzani, unpublished results.

DISCUSSION

The polymerization takes place where the radical is formed. There is no relationship between the dimension of the crystal and the molecular weight of the polymer produced. Then, the distribution is not narrow. Complete conversion may not be achieved or can require several days.

RADICAL POLYMERIZATION ALONG MACROMOLECULAR TEMPLATES

Y.Y. Tan
Laboratory of Polymer Chemistry
State University of Groningen
Nijenborgh 16
9747 AG Groningen
The Netherlands

ABSTRACT. Template polymerizations of synthetic monomers by free radical addition mechanism consist of propagation of chain radicals along template macromolecules. The implications of this special mode of propagation are critically discussed with respect to polymerization rate and structural features of the daughter polymers formed, i.e. their average molar mass, (molar mass distribution) and microstructure.

1. INTRODUCTION

Template polymerization, variously termed matrix or replica polymerization, may be defined as any process in which polymer chains are capable of growing along template macromolecules during at least part of their lifetime. This is rather broadly defined, but it excludes polymerizations proceeding on macroscopic surfaces.

Propagation along a template is made feasible because of cooperative interaction between the growing daughter chain and its parent (template) chain, ending with the formation of a polycomplex.

Any conventional polymerization method, notably addition-, ring-opening-, and condensation polymerization, can be applied to compose a template system by judicial choice of monomer/template combination and reaction conditions. This article concerns mainly template systems that operate according to a radical addition mechanism.

The simplest templates are linear homopolymers carrying groups that act as adsorption sites for monomer molecules and/or monomeric units of propagating chains. Depending on the degree of monomer (M) adsorption by the template (T): $M + -T- \rightleftharpoons -T(M)-$, characterized by the (Langmuir) adsorption equilibrium constant K_M, we can distinguish two extreme types of template polymerization systems [1]: type I with $K_M = \infty$ and type II with $K_M \approx 0$. In type I systems propagation just consists of "zipping-up" monomer arrays, whereas in type II systems propagation along a template molecule ("template propagation") consists of "picking-up" monomer molecules from the surrounding solution. In the latter, template propagation is preceded by complexation of the growing chain, initially formed in solution, with the template after attaining a certain minimum (critical) chain length. The difference in propagation modes of these two types are schematically depicted below:

M. Fontanille and A. Guyot (eds.), Recent Advances in Mechanistic and Synthetic Aspects of Polymerization, 281–292.

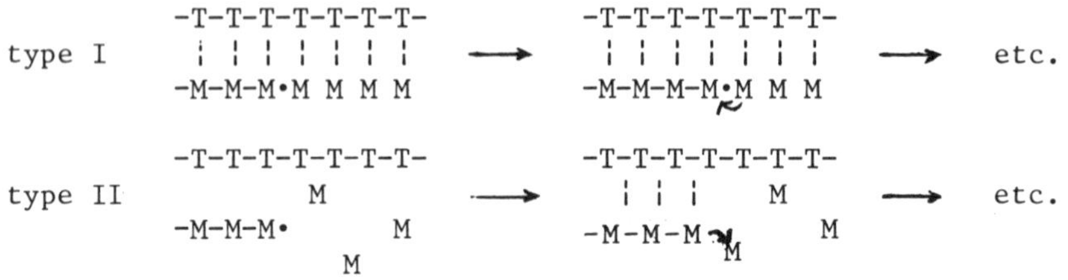

```
         -T-T-T-T-T-T-T-              -T-T-T-T-T-T-T-
type I    ¦ ¦ ¦ ¦ ¦ ¦ ¦      ──→       ¦ ¦ ¦ ¦ ¦ ¦ ¦       ──→   etc.
         -M-M-M•M M M M               -M-M-M-M•M M M

         -T-T-T-T-T-T-T-              -T-T-T-T-T-T-T-
type II           M          ──→       ¦ ¦ ¦    M          ──→   etc.
         -M-M-M•       M              -M-M-M•       M
                     M                      M
```

With respect to conventional polymerization, henceforth called blank polymerization, we may anticipate that such special modes of propagation should affect the kinetics of the polymerization process as well as the molar mass and microstructure of the daughter polymer, these manifestations being called "template effect". In practice, however, interpretation of a template effect encounters many problems as we shall see in the following sections.

2. KINETICS

The most used criterium for claiming a template effect is enhancement (or sometimes reduction) in initial rate of polymerization as compared to a blank polymerization. The latter should preferably be performed in the presence of an appropriate low-molecular analogue of the template. Interpretation of a template effect found in a type I system is in some respects different from that in a type II one.

2.1. Type I systems

Template polymerizations of type I have been invariably performed in (very) dilute solutions, i.e. under conditions far below the critical concentration of coil overlap of template macromolecules, to avoid intertemplate reactions.

In type I systems rate enhancement seems obvious because of the apparent ordered placement of monomer molecules along template macromolecules, provided of course that the adsorbed monomer molecules are immobile and that such parameters as the distance between consecutive monomer molecules and their mutual steric orientation are optimal. Conversely, an experimentally found rate enhancement does not need to be related exclusively to ordering of adsorbed monomer molecules, however. Other factors that may also induce rate enhancement have to be considered.

In the first place, one should take account of the fact that monomer adsorption creates automatically a high local concentration on the template so that rate enhancements should at least in part be attributed to it. To correct for this concentration effect, one needs to know the adsorption equilibrium constant, K_M. For systems where the monomer "molecules" are covalently linked to the template sites [2,3], such a correction constitutes no problem since $K_M = \infty$. Actual systems, based on ionic or hydrogen-bond interaction, for which a type I have been claimed possess finite though large K_M-values, however. Unfortunately such values are practically unavailable. Probably the system that comes nearest to the ideal type I consists of the polymerization of p-styrene sulfonate along a rigid oligomeric [2.2.2],4-ionene template with a $\bar{P}_n \approx 10$ in water/2-propanol (3/1 v/v) at pH = 7 and 70°C, studied extensively by Blumstein et al. [4,5]. From a reported value of 0.62 for the fraction of adsorbed ("condensed") monomer on the template at the 1:1 equivalent ratio of monomer- and template concentration, a K_M of 6800 could be calculated. Assuming complete adsorption the investigators estimated the local monomer concentration to be three orders of magnitude higher than it would have been in solution [4].

The importance of local concentration is further demonstrated by the finding that on using ionene templates with various (greater) charge separations the polymerization rate is directly proportional to the charge density and the related local monomer concentration, signifying that the monomeric counterions possess considerable mobility along the template macromolecules [5,6]. As a consequence an interpretation of the results based on the Manning ion condensation theory appeared to be more appropriate than on the concept of stoichiometric site binding [6,7]. It may well be that the ion condensation model is generally applicable to systems in which strong ionic forces play a role.

Tsuchida et al. [8], who studied the system (meth)acrylic acid/ionene ($\bar{P}_n$ = 106) in aqueous solutions at 50°C, ascribed rate enhancements to reduction in the electrostatic repulsion between ionised propagating radicals and ionised monomer on the template on the one hand and to an increase in intrinsic reactivity of the monomer and polymer radical by adsorption on the template on the other hand. Though these reasons may contribute to the overall rate enhancements, the investigators did not consider concentration effects, nor monomer ordering and possible monomer mobility. It should be remarked that adsorption of the reacting species might likewise be detrimental to their original reactivity.

The rate enhancements found on polymerizing acrylic acid along poly-(ethylene imine) in acetone/water (2/1 v/v) at 25°C for which Bamford et al. [9,10] measured a K_M of 200, was attributed merely to ordering of monomer arrays. In this system where rather weak ionic forces are involved, no monomer mobility could be discerned.

The problem of monomer arrangement is an intrigueing one. In addition to the parameters mentioned at the beginning of this section, the conformation of the template is undoubtedly a factor to be considered. Rigid templates possessing an extended conformation should be most appropriate, while coiled conformations of flexible template chains may impair the probability of polymerization of successive monomer molecules. Flexible chains may lead to lower K_M-values owing to loss of entropy on monomer adsorption. In addition there is a greater loss of conformational entropy during polymerization when the template molecules are flexible than when they are rigid.

In this respect the work of Kawai et al. [11] is worth mentioning. These researchers studied the template polymerization of i.a. acrylic acid along poly-L-lysine (PLL) in aqueous solutions and compared the structures of the formed polycomplexes with those formed by direct complexation between PLL and poly(acrylic acid) by means of X-ray diffraction. They found firstly that there is a correlation between the rate of polymerization and the conformation of PLL: the initial rate decreased in the order β-form PLL (pH = 12.0) > blank (pH = 11.7) > α-helix PLL (pH = 11.7) and blank (pH = 5.4) > random coil PLL (pH = 5.4), and secondly that during polymerization a conformational change of PLL had taken place, viz. the random coil changed into the α-helix, and the α-helix into the β-form which is the most stable conformation of PLL. It was suggested by the authors that the highest rate in the presence of the β-form is due to the most favourable placement of consecutive monomer molecules on the PLL. Note that a negative template effect can occur in the random coil and α-helix conformations probably as a consequence of too great a distance between the adjacent monomer molecules on the PLL chains.

Template polymerizations are often accompanied by the formation of insoluble polycomplex particles revealed by gradual turbidity of the solutions. As various investigators found initial rates that are (nearly) first order in initiator concentration [5,8,12], monomolecular termination was suggested to predominate either by entrapment of template-bound polymer radicals [8,12], or by degradative chain transfer to solvent [5]. There is no clarity, let alone consensus about the role of the precipitates in promoting rate enhancement. This is likely to become relevant only in presence of excess monomer, indications of which were found in polymerizations of monomeric acids in the presence of ionene templates [5].

In contrast to the above findings, Endo et al. found a normal half order in initiator concentration for the system N-vinyl-2-oxazolidone/poly(methacrylic acid) at pH = 4 in water at 60°C [13], despite precipitate formation.

Additional kinetic information can be extracted from overall activation energies, E_a, (and entropies), usually obtained at an equimolar ratio of template and monomer. A conclusive analysis is however, hampered by the lack of activation values of the individual reaction steps, so that recourse is taken to acceptable assumptions. In doing so Blumstein came to the conclusion that the found decrease in E_a is due to a decrease in activation energy of the propagation step and an increase in activation energy of chain transfer to solvent, the first being attributed to the close proximity and orientation of the adsorbed monomer molecules if the stoichiometric site binding model were valid [5]. However, these attributions should rather be related to an increase in activation entropy of the propagation step.

2.2. Type II systems

In these and type II-like systems we are not concerned with local monomer concentration on the condition that preferential monomer adsorption is low. It is now generally accepted that in these systems rate enhancement is primarily due to retardation of the termination step. So far known, there is only quantitative evidence from work done by Gons et al. [14] who obtained rate constants for the propagation and termination step, k_p and k_t, for the system methyl methacrylate/it-poly(methylmethacrylate) at 5°C (no preferential adsorption of monomer or solvent) by means of the rotating sector technique. They found a five-fold reduction in k_p which is however more than compensated by an 82-fold reduction in k_t.

Retarded termination may be caused by a strong hindrance of the segmental mobility of the template-associated chain radicals. How the termination mechanism must be envisaged is a matter of conjecture. There are several possibilities: two template-associated radicals encounter one another on the same template macromolecule or in the solution when each of them grow off the end of a template molecule, but also a non-associated chain radical may participate in the termination act ("cross-termination"). The first mentioned possibility might lead to termination by combination only, owing to sterical hindrance of a hydrogen atom that is needed for a disproportionation reaction.

The formation of gels or precipitates of polycomplexes does not seem to affect the (approximately) half-order in initiator concentration of the polymerization rate [14-18], signifying that gel and occlusion effects do not play a role. This is supported by the fact that in presence of excess monomer no accelerations were observed and that the template rate decreased to that of the blank polymerization after attaining a conversion corresponding to full coverage of the template [19]. One must assume that chains are terminated before a precipitate is formed. Probably during the formation of a microparticle the radical end remains protruding and growing into the solution until its termination followed by aggregation of several microparticles to a precipitating macroparticle. In this way no radicals are occluded and the half-order dependence in initiator remains valid.

Besides a half-order dependence in initiator concentration, lower orders have also been found, like in the template systems N-vinyl-pyrrolidone/poly(acrylic acid) in DMF [20], methacrylic acid/poly-2-vinylpyridine in DMF [21], and even in the mentioned system MMA/it-PMMA [14]. This may be explained by a contribution of primary termination. However, after a kinetic analysis Smid et al. [21] came to the conclusion that this may be caused by a prevalent termination between template-bound chain radicals on the supposition that the presence of the template does not disturb the steady state concentration of the radicals in the bulk solution, i.e. when only part of the radicals in the bulk solution complex with the template. If nearly all radicals are capable of complexing with the template it may be kinetically deduced that the template rate is half-order in initiator concentration; this is an alternative explanation of the one in the preceding paragraph.

In some template systems induction periods have been observed that could be ascribed to adventitious impurities introduced by the template [15,19]. According to Ferguson et al. [22] who noticed a lengthening of an induction period on increasing the molar mass of the template in the system acrylic acid/poly-N-vinyl pyrrolidone in aqueous solutions, suggested this induction time to be the time required to start template polymerization.

Recently, Smid et al. [23] showed that induction times arising in the system of methacrylic acid/poly-2-vinylpyridine at 30°C, could be rationalized by a slow build-up of template-bound chain radicals to its steady-state concentration. The concentration in dioxane reached values to such a level that it was detectable by ESR [24]. The slow build-up of these radicals appeared to be due to a relatively slow complexation rate of the oligomeric radicals and a very much retarded termination rate which amounts to three decades less than that of the blank polymerization. The reason for the low complexation rate is as yet unknown [24,25].

We have seen that the propagation step along the template in the system MMA/it-PMMA is reduced. This may signify that the radical chain end is firmly associated with the template site, thus diminishing its activity. Alternatively, the template chain may sterically hinder the approach of a monomer molecule, leading to a lowered activation entropy of the propagation step [14]. One may speculate whether such steric hindrance might induce a stereospecific addition. We shall discuss this more deeply in a following section.

The factors determining the firmness with which a radical chain end is held on the template site are to date unknown. One may express it as an adsorption equilibrium $R_e + -T- \rightleftharpoons -T(R_e)-$, R_e being the radical end, with an equilibrium constant K_R. If K_R is small the radical (including possible preceding monomeric units) will on the average be in a "loose" state though it is still close to the template chain. Such a condition should not affect k_p, and propagation will proceed in a "normal" manner before complexation of the newly made chain part takes place. Such a mode of template propagation will not produce a neatly ordered polycomplex with a high degree of interacting bonds (pairing).

On the other hand, if $K_R = \infty$, one may imagine that k_p could be strongly reduced. This situation may probably be exemplified by the system (meth)acrylic acid/poly(ethylene oxide) in aqueous solution which reveals a negative template effect. The investigators, Papisov et al. [26] believe however, that the possible lowering of k_p may best be interpreted in terms of a medium effect, i.e. the template-bound radicals reside in a poly(ethylene oxide) environment that is quite different from that in water. This explanation was prompted by the fact that rate decrease was also invoked by addition of methanol (or another alcohol) to the aqueous solutions of the monomer. It should be remarked that the rate reduction by the template is amazingly effective, at least under the dilute reaction conditions employed. Its independence of template concentration was accounted for by complete adsorption of all radicals formed in the bulk solution.

The above discussions around k_p is based on the assumption that k_p is generally reduced on adsorption of the radical, but there is no reason why in other cases k_p could not be raised.

3. MOLAR MASS

To establish convincingly a template effect, the observation of rate enhancement (or reduction) may be insufficient. Supporting evidence from the production of the appropriate type of polycomplex or from complementary features of daughter and template macromolecules are desirable, such as molar mass, molar mass distribution and microstructure c.q. tacticity. Unfortunately, characterization of the daughter polymer is normally hampered by the difficulty of its quantitative isolation from the polycomplex. Only partial isolations have been successful probably because of grafting or for hitherto unknown reasons. Sometimes indirect characterization can be achieved without isolation if a solvent could be found to dissociate fully the polycomplex.

A criterium for a template effect is that molar masses of daughter and template polymers run parallel. However, there is no obvious reason for the degrees of polymerizations to be exactly equal, since primary radicals (in type I systems) or active chains initially produced in the bulk solution (in type II systems) can respectively initiate or associate at any place along the template molecule. Moreover, hopping of growing chains from one template to another is conceivable under some circumstances [27].

3.1. Type I systems

In these systems when working with short templates in highly dilute solutions, initiation could start at one end of the template and stop at the other end by primary termination. This has been shown to be feasible in the polymerization of so-called multi(meth)acrylate oligomers, i.e. (meth)acrylate units covalently linked to an oligo"OH"-template by full esterification, synthesized by Kämmerer et al. [2,3]. After polymerization (giving ladder oligomers) and separation of the daughter oligomer by hydrolysis, its chain length, excluding initiator fragments at both chain ends, measures exactly that of the template.

With much longer multimethacrylates this is no longer true since random initiation becomes probable which may even lead to isolated monomeric sequences. Further, if intertemplate reactions are not prevented, ladder polymers longer than the original multimers are produced that may even be branched or cross-linked. The polymerization of a relatively long multi-methacrylate based on polyvinylalcohol ($\bar{P}_v \approx 100$) as template, studied by Połowinski et al. [28] yielded after hydrolysis of the product a daughter poly(methacrylic acid) possessing a chain length about threefold of the template. A determination of its molar mass distribution could give additional information on the process.

Data on type I systems where the monomer is not covalently attached to the template are practically non-existent. They do not indicate copying of the chain length of the template by the daughter polymer or oligomer [13,29].

3.2. Type II systems

In the second section we have made clear that enhanced polymerization rates primarily originate from a lowering of k_t. Because rate and molar mass show identical dependency on the rate constants, k_p and k_t, any change in these constants should affect both rate and molar mass in the same manner. Thus an enhanced rate should be accompanied by an increase in molar mass of the daughter polymer which have actually been observed in most systems. Whether a negative template effect, manifested by a reduction in rate, is actually attended by a lower molar mass is not known.

One of the most popular experiments to demonstrate a template effect is to measure rates and molar masses of the daughter polymers as a function of template concentration at a fixed monomer concentration. Below the critical overlap concentration where the template molecules function as separate "microreactors", the increase in overall polymerization rate is normally accompanied by an increase in the average molar mass of the total polymer formed [1,15]. However, in the system MMA/it-PMMA in DMF at -5°C, it was observed that contrary to increase in overall polymerization rate, the average molar mass of the daughter PMMA formed decreased down to a minimum value before it started to raise [27]. Several possibilities have been advanced to explain this deviation, such as chain transfer to monomer, solvent, or polymer, but so far it has eluded an adequate explanation.

As to the influence of template chain length, results of a couple of template systems (at equivalent ratio of template and monomer) showed a parallel tendency of increase in polymerization rate and average molar mass of the daughter polymer with template length [15,19, 27,30]. Various explanations have been put forth, the simplest being that of a greater chance of continuous growth of the daughter chain radicals along longer template chains [30]. Koetsier et al. [15] who worked with semi-dilute reaction mixtures stated this more precisely by saying that termination is increasingly segmental diffusion controlled, partly because of the high polymer concentrations lending to the solutions a high viscosity that increases with template chain length and partly because of the increasing immobility of the daughter chain radicals which are associated with longer template molecules, inferring that k_t decreases with template length. On the other hand, Shavit et al. [16] who used dilute solutions, ascribed the dependence as due to an increasing tendency of complexation of the radicals in bulk solution with longer template molecules, in other words the longer the templates the more chain radicals are "caught" by them. The workers proposed the complexation between the daughter polymer radical and template to be reversible, its equilibrium constant increasing as the degree of polymerization of the macromolecular components increased. Which of these explanations represent the actual situation is as yet unsettled, though we are prone to support the first one recognizing that for rather short templates the second explanation applies.

Recently, Połowinski [31] found that with fractions of poly-2-vinylpyridine, ranging in molar mass from 24000 to 144000 ($\bar{P}$ = 231 to 1431), the polymerization rates of methacrylic acid in DMF were identical. It should be recalled, however, that we are concerned here with a type II-like system in which there is some preferential adsorption of monomer. The independency means that the concentration of template-bound chain radicals is constant regardless of template length. It implies that the average molar mass of the daughter polymer should neither be affected by template length. This cannot be verified since molar masses were not measured.

4. MICROSTRUCTURE (TACTICITY)

One may anticipate that the template should impose its microstructure on that of the daughter chains. The extent of influence might depend primarily on the involvement of the relevant neighbouring template sites in the transition state of monomer addition. The various possibilities are diagramatically depicted below.

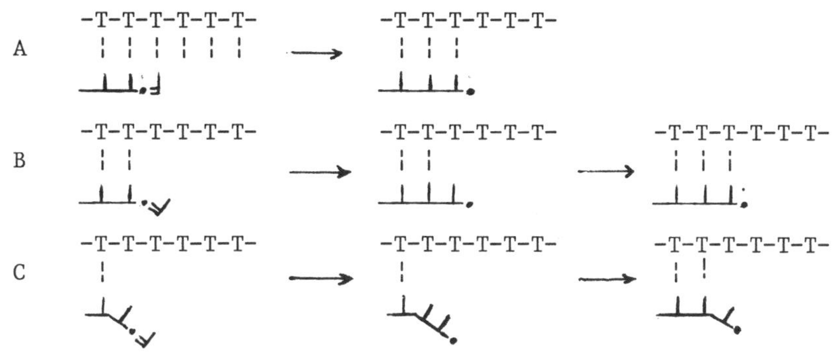

A -T-T-T-T-T-T- ⟶ -T-T-T-T-T-T-

B -T-T-T-T-T-T- ⟶ -T-T-T-T-T-T- ⟶ -T-T-T-T-T-T-

C -T-T-T-T-T-T- ⟶ -T-T-T-T-T-T- ⟶ -T-T-T-T-T-T-

In A the chain end radical as well as the monomer molecule are both adsorbed on adjacent template sites. This undoubtedly occurs in type I propagation but is also feasible in type II-like propagation when addition and adsorption take place concurrently. It remains to be seen however, whether concomittant monomer adsorption is a crucial factor to induce a change in the "normal" stereochemical addition; steric hindrance by the template (site or segment) might suffice in case B. Stereochemical effects will probably be absent when addition proceeds according to C. In C one should not visualize the radical end as to be permanently "free", but as one fluctuating between a "fixed" and a "loose" state, having a low equilibrium constant K_R (see section 2.2).

In order for a propagating chain radical to "follow" the consecutive sites on the template smoothly, each successive addition and the accompanying complexation step should proceed without strain. It is doubtful whether such could be achieved during the whole course of propagation since a high degree of matching of template and daughter chains is required. Smooth "following" may probably be realized by adapting the stereochemical mode of addition (meso or racemic) over a limited number of elementary steps. If this could not be accomplished, the alternative would be variation of the propagation step according to A, B, or C. To ensure process A to take place exclusively, the monomer molecules should be covalently attached to the template. The polymerization of oligomeric multi(meth)acrylates (P up to 6) [2] was expected to give syndiotactic poly(meth)acrylic acid oligomers, but so far no definitive results have been reported.

In the template polymerization of vinylsulfonate along an ionene oligomer as template (type I), Blumstein et al. [30] succeeded in a partial isolation of the daughter polymer. It appeared to possess a microstructure very different from that of a conventional poly(vinylsulfonate) [32]. A strong increase in heterotactic and isotactic triads at the expense of syndiotactic triads was noted (tentative assessment: template polymer rr = 0.23, mr = 0.64, mm = 0.13 vs blank polymer rr = 0.65, mr = 0.35). This was rationalized as follows. The monomeric counterions are aligned along the template in such a manner that isotactic addition would be favoured. However, this is only allowed for 2-3 monomer units. To continue monomer-template pairing a high number of "cross-overs" into syndiotactic additions, leading to heterotactic triads, must occur before isotactic additions could be repeated.

Template systems of type II have led to daughter polymers whose microstructure differ only little from those of corresponding blank polymers. Shavit et al. [16] observed moderate increase in syndiotacticity of the poly(methacrylic acid) formed in the presence of conventional poly-N-vinylpyrrolidone, whereas Ferguson et al. [22] claimed an enhancement in isotacticity of the daughter poly(acrylic acid) on applying the same template. Papisov et al. [26] observed likewise a somewhat better syndiotactic poly(methacrylic acid) induced by poly-(ethylene oxide). On the other hand, Koetsier et al. [15] did not find any difference in tacticity of the poly-N-vinyl-pyrrolidone, obtained in presence and absence of st-poly (methacrylic acid).

A drawback of type II systems is the fact that the daughter polymer chains are not completely created on the template, but also partly in the bulk solution during the initial stages. This means that the template effect on microstructure also depends on the critical chain length of the initially formed chain portion prior to complexation, and on the ultimate length of the daughter polymer. The shorter the first and the longer the latter is, the greater the difference in tacticity between daughter and blank polymer. This has been borne out by some experimental results [16,33].

One may speculate that stereochemical template effects in type II systems mainly proceed according to process B, alternated by C. If there is preferential adsorption of monomer, process A may contribute to the overall process; if there is preferential adsorption of solvent (an example of which is the template system N-vinyl pyrrolidone/poly(methacrylic acid) in DMF [15]), process C may play a substantial role.

A special type II system is the template polymerization of MMA along it-PMMA which leads to the formation of st-PMMA in the early stages of the reaction, studied by Buter et al. [33]. The driving force of this stereospecificity is the well-known stereocomplexation between isotactic and syndiotactic PMMA, made feasible through combination of cooperative Van der Waals forces and stereochemical fitting. It has been rather well established that the stereocomplex is built up of a double helix consisting of an it-PMMA helix lying within a larger helix of st-PMMA, such that the basemolar ratio of it- to st-PMMA is 1 to 2 [34]. How a PMMA chain is able to grow during template polymerization to create a double helix is difficult to envisage unless one assumes the pre-existence of helical segments of it-PMMA. Initially growing atactic PMMA-oligomeric radicals, possessing predominant syndiotactic segments, complex first with these helical segments of it-PMMA template molecules. Further monomer additions may be sterically steered by the template such that these take place preferably in a syndiotactic manner. This would be made possible if the segment of the isotactic template chain just in front of the daughter PMMA radical adopts a helical conformation incited by the formed double helix.

A final note should be made on the influence of template tacticity on the polymerization rate. Different rate enhancements have been found in the polymerization of N-β-methacryl-oxyethyl (MAO) of adenine in presence of it-, st-, and at-poly(MAO of uracil), especially in pyridine solutions [35], and in the polymerization of methacrylic acid in presence of it-, and at-poly-2-vinylpyridine [25]. These results might be interpreted in terms of process A and should at least indicate some stereochemical effect on the daughter polymer. However, in the first example the tacticity of the daughter polymer was similar to that of a blank one. In the second example no characterizations of the polymers were carried out.

5. CONCLUDING REMARKS

Quite a number of template systems, based on radical mechanism, have been examined in the past 20 years, and many interesting features have emerged from these, as well as a great deal of understanding of their underlying mechanisms. Nevertheless, several fundamental problems remain to be solved. To summarize a few: There is no unambiguous explanation for the rate enhancement (or reduction), especially with respect to type I systems. This is mostly hampered by the lack of knowledge concerning the extent of monomer adsorption on the template and the magnitudes of propagation- and termination constants. The determination of rate constants is made impossible owing to the formation of precipitates. Moreover, the assumption of the initiation step to be unaffected by the template, may not always be valid. The difficulty, if not infeasibility in some cases, of isolating the daughter polymer completely from its template, withholds valuable information about its structural features.

In this article only homopolymerization and the use of homopolymers as template have been dealt with. Several copolymerization systems have been studied lately [29,36-39]; they may contribute to a further insight into the mechanism of template polymerization. The application of copolymers and crosslinked polymers as templates is still in its infancy.

Finally, it should be remarked that there is a great diversity in template systems, each having its own peculiarities. This is in fact not surprising since the introduction of a template to a polymerization system brings in a new dimension.

Acknowledgements. The author thanks Professor G. Challa for the simulating discussions on the contents of this manuscript, and Mr. G.O.R. Alberda van Ekenstein and Mr. E. van de Grampel for its critical perusal.

6. REFERENCES

1. G. Challa, Y.Y. Tan, Pure Appl. Chem., **53**, 627 (1981)
2. H. Kämmerer, J. Shukla, N. Önder, G. Schürmann, J. Polym. Sci., Polym. Symp., **C2**, 213 (1967)
3. H. Kämmerer, G. Hegemann, Makromol. Chem., **139**, 17 (1970) and **185**, 635 (1984)
4. A. Blumstein, S.R. Kakivaya, K.R. Shah, D.J. Wilkins, J. Polym. Sci., Polym. Symp., **45**, 75 (1974)
5. A. Blumstein, S.R. Kakivaya, in H.G. Elias, ed., Polymerization of Organized Systems, Gordon and Breach, London, 1977, p. 189
6. A. Blumstein, G. Weill, Macromolecules, **10**, 75 (1977)
7. A. Blumstein, M. Milas, Y. Ozcayir, E. Bellantoni, J. Polym. Sci., Polym. Phys. Ed., **21**, 2159 (1983)
8. E. Tsuchida, Y. Osada, J. Polym. Sci., Polym. Chem. Ed., **13**, 559

(1975)
9. C.H. Bamford, Z. Shikii, Polymer, 9, 596 (1968)
10. C.H. Bamford, 'Template Polymerization', in R.N. Haward, ed., Developments in Polymerization, Applied Science Publ., London, vol. 2, 273 (1979)
11. T. Kawai, A. Fujie, J. Macromol. Sci., Phys., B17, 653 (1980)
12. J. Ferguson, S.A.O. Shah, Eur. Polym. J., 4, 611 (1968)
13. T. Endo, R. Numazawa, M. Okawara, Makromol. Chem., 146, 247 (1971)
14. J. Gons, E.J. Vorenkamp, G. Challa, J. Polym. Sci., Polym. Chem. Ed., 15, 3031 (1977)
15. D.W. Koetsier, Y.Y. Tan, G. Challa, J. Polym. Sci., Polym. Chem. Ed., 18, 1933 (1980)
16. N. Shavit, J. Cohen, in H.G. Elias, ed., Polymerization of Organized Systems, Gordon and Breach, London, 1977 p. 213
17. R. Muramatsu, T. Shimidzu, Bull. Chem. Soc. Japan, 45, 2538 (1972)
18. K. Fujimori, Makromol. Chem., 180, 1743 (1979)
19. T. Bartels, Y.Y. Tan, G. Challa, J. Polym. Sci., Polym. Chem. Ed., 15, 341 (1977)
20. V.S. Rajan, J. Ferguson, Eur. Polym. J., 18, 633 (1980)
21. J. Smid, Y.Y. Tan, G. Challa, Eur. Polym. J., 20, 887 (1984)
22. J. Ferguson, S.Al-Alawi, G. Granmayeh, Eur. Polym. J., 19, 475 (1983)
23. J. Smid, Y.Y. Tan, G. Challa, Eur. Polym. J., 19, 853 (1983)
24. J. Smid, Y.Y. Tan, G. Challa, W.R. Hagen, Eur. Polym. J., 21, 757 (1985)
25. J. Smid, J. Speelman, Y.Y. Tan, G. Challa, Eur. Polym. J., 21, 141 (1985)
26. I.M. Papisov, V.A. Kabanov, Y. Osada, M. Leskano-Brito, J. Richmond, A.N. Gvozdetskii, Vysokomol. Soedin., A14, 2462 (1972); Eng. transl. in Polym. Sci. USSR, 14, 2871
27. J. Gons, L.J.P. Straatman, G. Challa, J. Polym. Sci., Polym. Chem. Ed., 16, 427 (1978)
28. R. Jantas, S. Połowinski, J. Polym. Sci., Part A, Polym. Chem., 24, 1819 (1986)
29. S.R. Kakivaya, A. Blumstein, J. Chem. Soc., Chem. Commun., 459 (1974)
30. J. Gons, E.J. Vorenkamp, G. Challa, J. Polym. Sci., Polym. Chem. Ed., 13, 1699 (1975)
31. S. Połowinski, private communication
32. A. Blumstein, S.R. Kakivaya, R. Blumstein, T. Suzuki, Macromolecules, 8, 435 (1975)
33. R. Buter, Y.Y. Tan, G. Challa, J. Polym. Sci., Part A-1, 10, 1031 (1972)
34. F. Bosscher, G. ten Brinke, G. Challa, Macromolecules, 15, 1442 (1982)
35. M. Akashi, H. Takada, Y. Inaki, K. Takemoto, J. Polym. Sci., Polym. Chem. Ed., 17, 747 (1979)
36. K.F. O'Driscoll, I. Capek, J. Polym. Sci., Polym. Lett. Ed., 19, 401 (1981)
37. S. Połowinski, Eur. Polym. J., 19, 679 (1983)
38. S. Połowinski, J. Polym. Sci., Polym. Chem. Ed., 22, 2887 (1984)
39. Y. Inaki, K. Ebisutani, K. Takemoto, J. Polym. Sci., Part A: Polym. Chem., 24, 3249 (1986)

FORMATION OF POLYMERS OF ß-HYDROXYBUTYRIC ACID IN BACTERIAL CELLS AND A COMPARISON OF THE MORPHOLOGY OF GROWTH WITH THE FORMATION OF POLYETHYLENE IN THE SOLID STATE

D G H Ballard & P A Holmes
ICI New Science Group
The Heath
Runcorn WA7 4QE

P J Senior,*
ICI Agricultural Division
Billingham, TS23 1LB

ABSTRACT. The similarities between the polymerisation of ethylene in the solid state by supported organometallic catalysts and the in vivo synthesis of ß-hydroxybutyric acid polymers in bacteria are explored with particular regard to nascent polymer morphology and molecular weight control.

POLEYTHYLENE FORMATION IN THE SOLID STATE

The surface of silica and alumina freed from physically adsorbed water contain acidic OH groups which will react with transition metal alkyls. These reactions can be followed in the infra-red. It has been found that transition metal alkyl compounds can react with the OH groups in more than one way and the product obtained depends on several factors. For example, $Zr(allyl)_4$ reacts with silica pre-dried at 200°C to give two molecules of propene per metal atom utilizing in the course of this process two OH groups per metal atom. The chemistry of the process is accurately described in figure 1.

We have observed similar reactions with $Zr(CH_2C_6H_5)_4$, $Zr[CH_2Si(CH_3)_3]_4$, $Zr[CH_2OCH_3]_4$ etc. and also silica can be replaced by alumina and other matrices, giving transition metal centres with structures related to (I) in which the organic ligands are $CH_2C_6H_5$, $CH_2Si(CH_3)_3$, etc.

In the presence of ethylene polymerisation occurs by insertion between the metal-alkyl bond to give initially the species (II). This can occur in the gas phase or in the presence of hexane. The polymer subsequently is detached from the metal centre to form a metal hydride bond as shown by III. This occurs in the presence of hydrogen or spontaneously by removal of a hydrogen atom from the carbon atom ß- to

*Now at Delta Biotechnology Limited, Nottingham NG7 1FD

M. Fontanille and A. Guyot (eds.), Recent Advances in Mechanistic and Synthetic Aspects of Polymerization, 293–314.

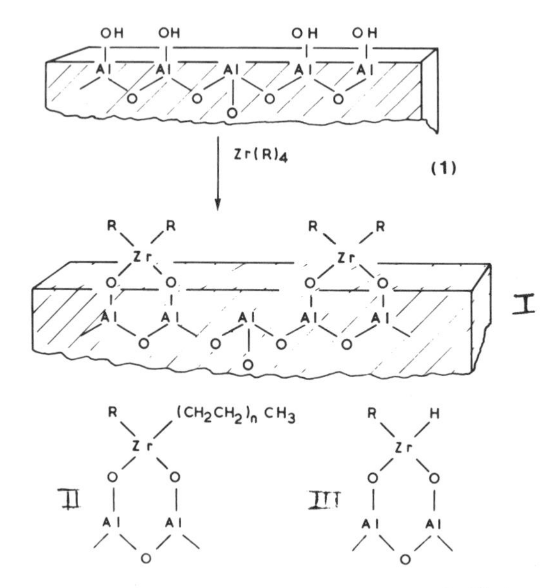

Figure 1. Reaction on Surface of Alumina of Zr Alkyls

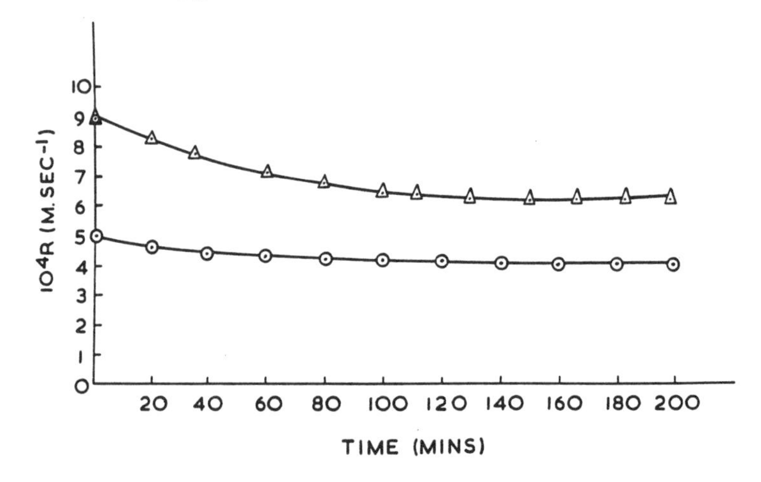

Figure 2. Polymerisation of ethylene by $Zr(benzyl)_4)/Al_2O_3$ in toluene showing rate is independent of conversion. $[C]o = 1.5 \times 10^{-4}$ Zr atoms/ℓ.
Pressure 4 atm. Δ, 40°C; O, 25°C. Equal partial pressures of ethylene and hydrogen.

the metal atom. The polymerisation then continues by insertion between the metal hydrogen bond.

A surprising feature to this polymerisation process is that the rate is independent of conversion and virtually constant for an indefinite period as shown in figure 2. This behaviour is explained by the morphology of the growth in the solid state.

In order to comprehend the chemistry of the process further it is necessary to trace the behaviour of catalyst particles during the polymerization. This is readily deducible from the microscopic studies of the polymer particles produced. Figure 3 is a cross-section of a particle of polyethylene formed using conditions described in figure 2. On closer examination at higher magnifications (figure 4) it is seen that polyethylene is laid down in cylinders of fairly uniform diameter which wind into helices In the example quoted the original alumina catalyst particle has completely disintegrated giving a polymer particle with an increase in diameter of 120 times. Further analysis with the million volt electron microscope shows that these alumina particle fragments are located at the tips of the cylinder as shown in Figure 5. We can therefore visualize the polymerization process as follows.

The initial alumina particle Figure 6 with attached Zr-alkyl groups break up in the early stages of polymerization into particles of 1000 Å or less which form nuclei around which chain growth and crystallization occur. The polymer chains generated fold to give lamellae (polyethylene single crystals) which stack behind the alumina particle with the long axis of polymer chains parallel to the axis of the cylindrical envelope shown in Figure 5. Direct evidence that the orientation of polyethylene chains is along the long axis of the cylinder in Figure 5 was obtained by looking at the electron diffraction pattern of a cylinder in corresponding orientation. Well-defined maxima on the (110) and (200) arcs were perpendicular to the cylinder axis, hence the preferred chain direction is seen to be parallel to this axis. Following an initial degradation of the primary catalyst particle, the catalyst fragments are pushed outwards from the centre at the tips of cylinders which coil to give helices shown in Figure 5. It is evident from this description that the diffusion barrier between catalyst sites and monomer is independent of the amount of polymer formed which accounts for the persistence of high rates of polymerization at high conversion a characteristic of this system.

The physical constraints leading to the overall morphology, however, are not understood. In particular why is such a uniform cylindrical envelope maintained during the growth?

It is evident that the conditions for this type of solid state polymerisation are effectively described by polyethylene. Surprisingly the formation of poly ß-hydroxybutyrate in bacterial cells is very

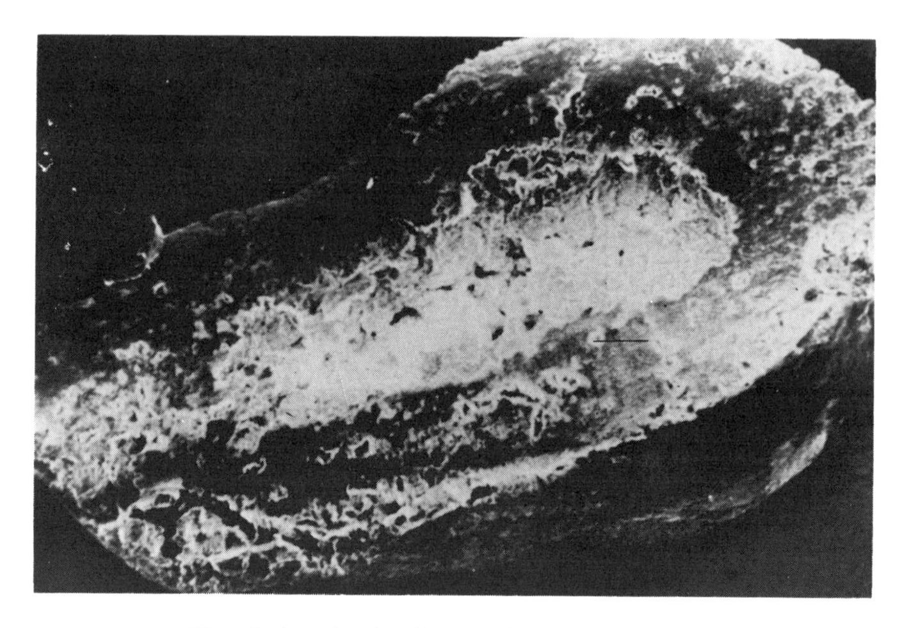

Figure 3. A section through a particle of polyethylene formed in hexane at 70°C from $Zr(CH_2\emptyset)_4/Al_2O_3$ catalysts. Electron micrograph shows porosity of particles. Mangification x 600.

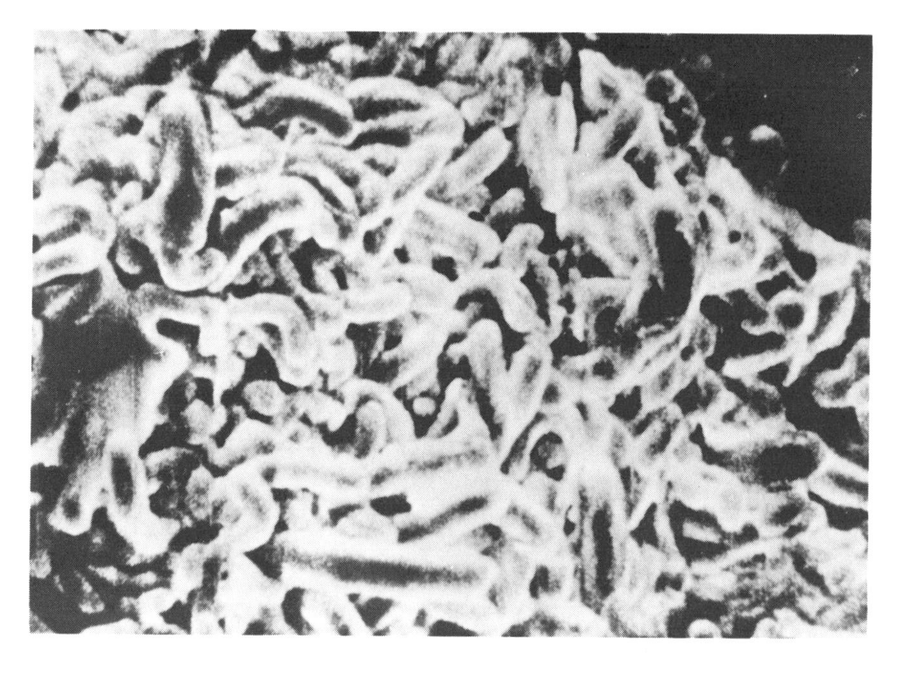

Figure 4. Electron micrograph of polyethylene shown in Figure 3 but now at a magnification of x 10,000.

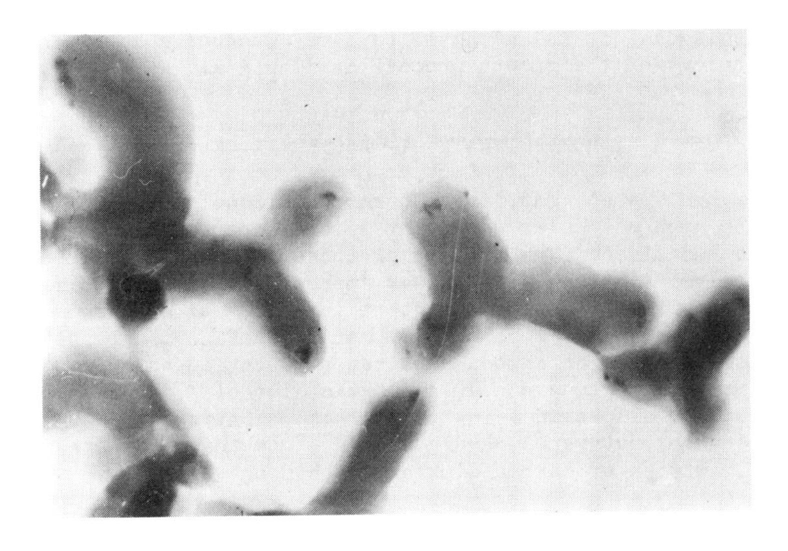

Figure 5. Electron micrograph of polyethylene as in Figure 4 magnification 30,000. Note cluster of catalyst particles at tips of the fragments.

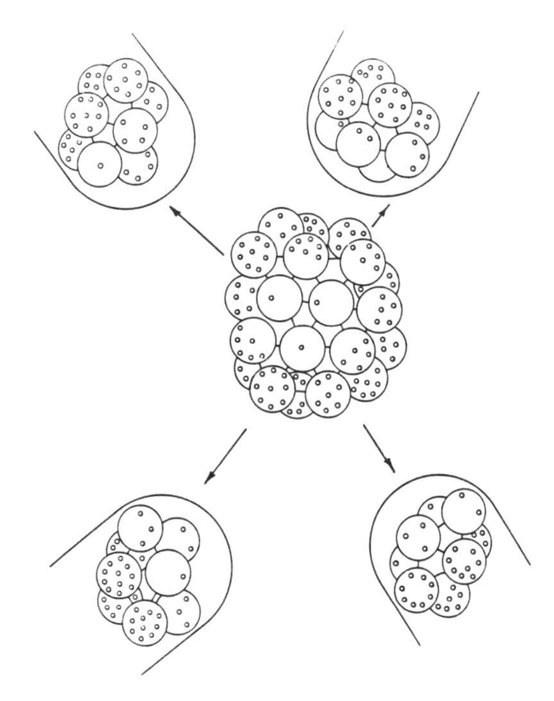

Figure 6. Diagramatic representation of the disintigration during polymerisation of catalyst particles. Primary particles approx. 50Å diameter held together by hydrogen bonds. Particles of catalyst used initially 30 μm. Small circles on primary particles represent zirconium alklyl centres.

similar. In this case it is an enzyme catalyst producing a polyester but the morphology of the growth process is identical.

FORMATION OF POLY(ß-HYDROXYBUTYRATE) IN BACTERIAL CELL

Poly(ß-hydroxybuntyrate), PHB, is a thermoplastic polyester synthesised to high molecular weight by a very wide range of bacteria, cyanobacteria and algae. The ability of these microorganisms to produce the polymer in high yield under appropriate conditions is quite remarkable with 70%-80% PHB content in dried cells being routine[1,2]. Moreover, this level of polymer production can be achieved using organisms growing on carbon substrates as varied as sugars[3] (sucrose, glucose, fructose etc.), aliphatic alcohols[4,5] (methanol, ethanol, butanol), aliphatic acids[6], (acetate, n-butyrate, crotonate, ß-hydroxybutyrate), hydrocarbons[7,8] (methane, ethane) and even gaseous mixtures of carbon dioxide and hydrogen[9].

In all cases the polymer occurs as discrete granules within the cell cytoplasmic space and these become clearly visible in the optical microscope when stained with Sudan Black[3] and as optically empty areas in electron micrographs of ultrathin sections of bacteria as seen in figure 7. Scanning electron microscopy can also reveal PHB granules spilling from lysed cells as in figure 8.

Separation of pure PHB from cell cultures is traditionally achieved by solvent extraction using chlorinated solvents such as chloroform and methylene chloride. However, a more elegant and commercially more attractive non-solvent process, which involves a high temperature heat shock to lyse the cells followed by a series of enzyme and detergent digestive steps to solubilise the non-PHB components, has been developed[10,11] and is being operated by Marlborough Biopolymers Limited of Stockton, England, with a capacity of 500 tonnes per year[12].

PHB itself is a highly crystalline, stereorgular polyester with the structure given below.

$$\left(\overset{\displaystyle CH_3}{\underset{*}{\overset{|}{C}H}}-CH_2-CO-O \right)_n$$

It contains an optically active centre, which is always in the R(-) absolute configuration for the biopolymer although atactic RS polymer can be synthesised chemically from ß-butyrolactone by ring opening polymerisation[13,14]. The properties of PHB are often compared to those of polypropymene because of their similar melting point, glass transition temperature and degree of crystallinity[15]. However, the biopolymer is a relatively weak and brittle material and applications for it would be limited were it not for the fact that the synthesising bacteria are capable of modifying the structure of the polymer, and hence its properties, by introducing other ß-hydroxyalkanoates as

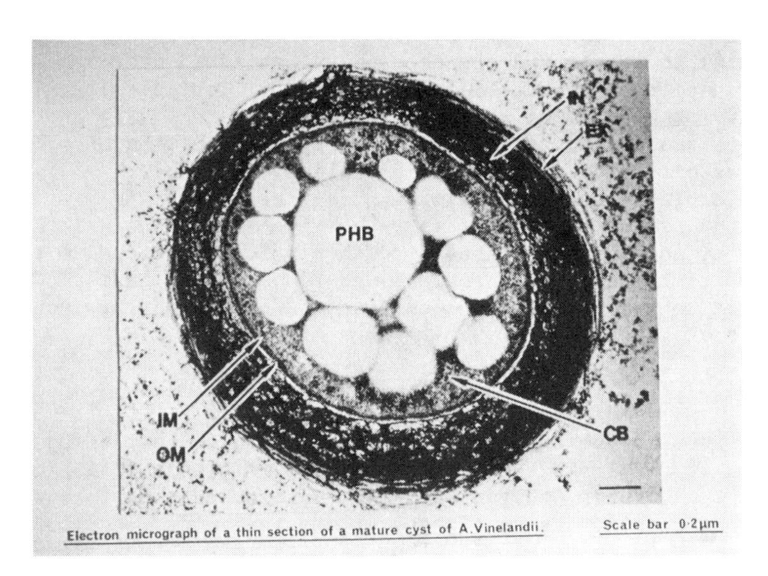

Figure 7. Electron micrograph of ultrathin sections of *A.vinelandii* showing PHB granules as optically-empty areas.

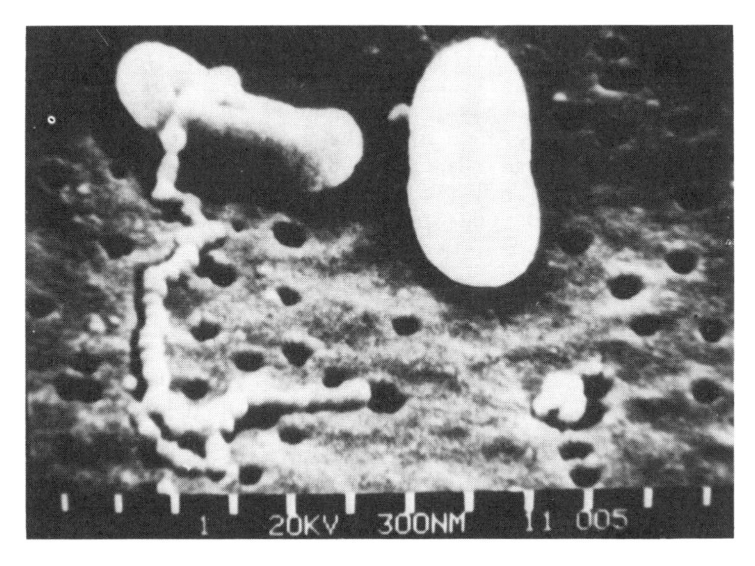

Figure 8. Scanning electron micrograph of partically lysed culture of *A.eutrophus* showing one intact cell and one spilling its contents of PHB granules.

comonomers[16]. Such units with every ß-alkyl substituent from methyl (PHB itself) to linear and branched pentyl units (ß-hydroxyoctanoates) have been detected in bacterial polyesters[17]. Typical mechanical properties for a range of random ß-hydroxybutyrate - ß-hydroxypentanoate copolymers produced by fermentation of *Alcaligenes eutrophus* are given in table 1, and their potential applications have recently been reviewed[18].
The main purpose of this paper is to examine in more detail the mechanism of polymer biosynthesis, with particular regard to the morphology of the nascent granules, and to consider the methods used by bacteria to regulate the composition of the polymer produced and its molecular weight.

BIOSYNTHESIS OF POLY(ß-HYDROXYBUTYRATE)

The first question to examine is why bacteria produce thermoplastic polyesters at all. PHB has been implicated in biochemical processes such as sporulation[19], encystment[20] and gene expression[21], but the most obvious function is as a means of storing reduced carbon without incurring an osmotic penalty[22]. Thus, the primary raison d'etre of bacteria is to grow and reproduce themselves. To do this they require a number of essential elements, notably carbon, oxygen, phosphorus, nitrogen and sulphur. If a PHB synthesising organism has all of these in abundance, then it is counterproductive for it to waste energy in making polyester. If on the other hand, the organism is faced with a plentiful supply of carbon but a complete absence of some other essential nutrient such as nitrogen or phosphorus, the capacity to reproduce is lost and PHB synthesis provides a mechanism for storing the available carbon in a highly reduced and accessible form. Under these circumstances the cell can accumulate up to 90% of its dry weight as polymer[23] without suffering any of the problems of maintaining an osmotic balance with the environment that would result from trying to store large quantities of a soluble carbon source. When the nutrient limitation is lifted, moreover, the endogenous PHB supply is very quickly degraded to useful carbon and energy, allowing growth and reproduction to resume immediately.

It has already been noted that PHB can be produced from a large number of diverse carbon substrates. The common factor in most of these biosynthetic pathways is the production of acetate, or more precisely acetyl coenzyme-A, as the first building block in PHB production. Thus glucose, for example is metabolised via pyruvate to acetyl-CoA. Coenzyme-A (CoA) is a nucleotide derivative of phosphopantetheine which links to acetate by condensation with its thiol group and serves to activate the attached moiety to further biochemical transformation.

A simplified view of the synthesis of PHB from acetyl-CoA is presented in figure 9. The first step, condensation of two molecules of

TABLE 1

Thermal and Mechanical Properties of Random HB-HV Copolymers

HB:HV Ratio	Melting Point (°C)	Glass* Transition (°C)	Youngs Modulus (GPa)	Tensile Strength (MPa)	Notched Izod# Impact Strength (JM^{-1})
100:0	179	10	3.5	40	50
97:3	170	8	2.9	38	60
91:9	162	6	1.9	37	95
86:14	150	4	1.5	35	120
80:20	145	-1	1.2	32	200
75:25	137	-6	0.7	30	400

* Position of E" peak at 5Hz

1mm radius notch

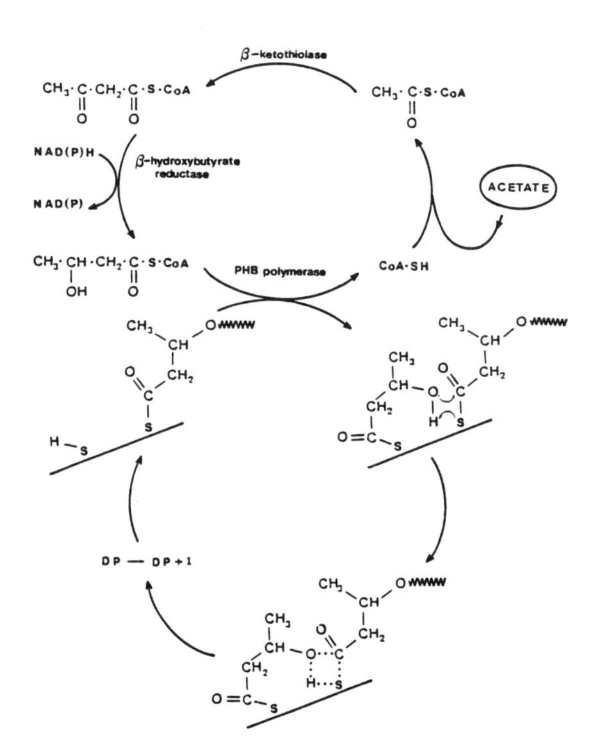

Figure 9 Biosynthesis of PHB from Acetate

acetyl-CoA to form acetoacetyl-CoA is catalysed by the enzyme ß-ketothiolase which has recently been studied in great detail. The gene responsible for its synthesis in *Zoogloea ramigera* has been cloned and transferred successfully to *E.coli* giving overproduction of the enzyme and allowing a full aminoacid sequence to be obtained[24]. Similar studies on the other enzymes in PHB biosynthesis, ß-hydroxybutyrate reductase and PHB polymerase, are planned[25]. The former enzyme is linked to the cofactor nicotinamide adenine dinucleotide (NAD) and effects the reduction of acetoacetyl-CoA to ß-hydroxybutyryl-CoA.

Very little is known about the polymerase enzyme, other than it is inhibited by p-mercuribenzoate and N-ethylmaleimide, which suggests that it is a sulfhydryl enzyme[26]. A model for the action is included in figure 9 with two thiol groups locating the growing polymer chain at one site and a new hydroxybutyrate monomer at the other. Condensation then occurs via a four membered transition state leaving one of the thiol groups available for further monomer. In this model the polymer chain actually flip flops back and forth between the two thiol groups with the free site acting to co-ordinate new ß-hydroxybutyrate units ready for incorporation.

The real situation is, however, almost certainly much more complicated than that described here. For example, the capacity of PHB containing bacteria under nutrient limitation to begin polymer degradation immediately the limitation is lifted, indicates that the depolymerase enzyme is present in significant quantity at the synthesis site throughout. One interpretation is that the polymerase and depolymerase are the same enzyme[27] with some external control factor influencing its direction of operation. Indeed, purified granules of PHB retain their polymerase activity but require the addition of a labile protein activator or trypsin to initiate depolymerase activity[26]. Moreover, although some bacterial depolymerase enzyme systems convert PHB directly to ß-hydroxybutyrate monomer, others stop at the dimer stage and require a separate dimer hydrolase to complete the degradation to monomer. A possible implication is that, in the latter case, the polymerase enzyme also requires ß-hydroxybutyrate dimer as substrate.

Overall, the PHB polymerase system is a complex mixture of soluble and membrane bound proteins that perform both polymerisation and depolymerisation functions[26,29]. In addition they must in some way perform a chain transfer role to control the molecular weight of the polymer formed, but this will be discussed in more detail later.

FORMATION OF COPOLYMERS

Copolymers of ß-hydroxybutyrate (HB) and ß-hydroxyvalerate) (HV) can be prepared by the fermentation of mixtures of acetate/propionate, glucose/propionate or ethanol/propanol using, for example, *Alcaligenes eutrophus* [16]. Biosynthesis of the HV units is by the ß-ketothiolase mediated condensation of one molecule of acetyl-CoA with one molecule of propionyl-CoA to form propionylacetyl-CoA followed by reduction of the ß-carbonyl by ß-hydroxybutyrate reductase to yield ß-hydroxyvaleryl-CoA as illustrated in the top half of figure 10, R=H.

Presumably a similar mechanism operates in the micro organisms of estuarine sediments[17] that produce copolymers containing ß-hydroxyhexanoate, ß-hydroxyheptanoate and ß-hydroxyoctanoate residues. However, *A.eutrophus* grown on a mixture of acetate and n-butyrate does not produce copolymers containing ß-hydroxyhexanoate units, as predicted by the above reaction scheme. In fact it produces pure PHB. The mechanism in this case involves oxidation of the n-butyric acid, as its CoA derivative, to crotonate using the enzyme acyl CoA dehydrogenase followed by addition of water across the double bond catalysed by enoyl-CoA hydratase to give ß-hydroxybutyryl-CoA as shown in the lower half of figure 10. Crotonic acid itself is, of course, similarly converted to PHB via hydration of its CoA derivative.

In the particular case of *A-eutrophus* , n-pentanoic acid is treated by the dehydrogenation-hydration route of figure 10 to produce ß-hydroxyvalerate rather than ß-hydroxyheptanoate via the ß-ketothiolase condensation. This is despite the fact that the enzymes associated with the latter biosynthetic pathways are demonstrably operational since some ß-hydroxybutyrate units are formed from acetate simultaneously with HV from n-pentanoic acid to produce a copolymer of HB and HV as the final storage product. Thus when n-pentanoic acid is the sole carbon source for *A-eutrophus* under conditions of polymer accumulation, the polymer formed is not pure PHV but a 50/50 copolymer of HB and HV because some of the n-pentanoic acid is metabolised to acetate and thence to ß-hydroxybutyrate.

Similar processes are seen in the metabolism of isobutyric acid by *A.eutrophus* . Again we do not see condensation acetate to give ß-hydroxyisohexanoate units. Dehydrogenation to methacryloyl-CoA followed by hydration to ß-hydroxyisobutryl-CoA does happen as for n-butyrate metabolism, but this potential monomer is not incorporated into the polymer, presumably because of the methyl group at the α-carbon. In fact it is further oxidised to the CoA derivative of methylmalonylsemialdehyde, decarboxylated to propionyl-CoA and then processed by ß-ketothiolase to HV.

Thus, microorganisms can produce a range of ß-hydroxyalkanoates but *A.eutrophus* at least has a profound preference or synthesising PHB or HB-HV copolymers and has access to several biosynthetic pathways that allow it to convert diverse substrates to these polymers. Other bacteria, however, have different preferences. *Pseudomonas oleovarans,* for example, has been

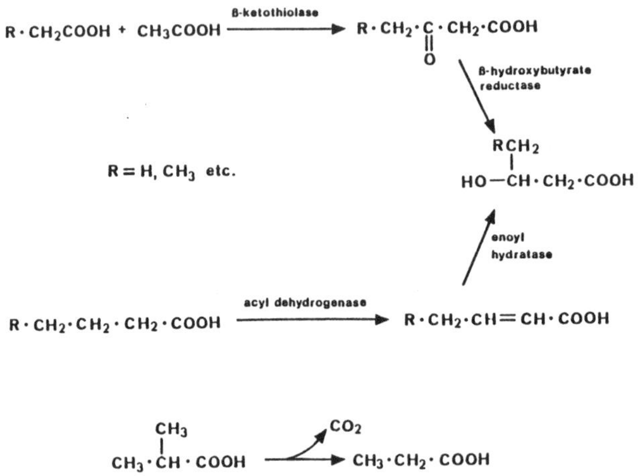

$$R\cdot CH_2COOH + CH_3COOH \xrightarrow{\beta\text{-ketothiolase}} R\cdot CH_2\cdot C(=O)\cdot CH_2\cdot COOH$$

$$R\cdot CH_2\cdot C(=O)\cdot CH_2\cdot COOH \xrightarrow{\beta\text{-hydroxybutyrate reductase}} HO-CH(RCH_2)\cdot CH_2\cdot COOH$$

$R = H, CH_3$ etc.

$$R\cdot CH_2\cdot CH_2\cdot CH_2\cdot COOH \xrightarrow{\text{acyl dehydrogenase}} R\cdot CH_2\cdot CH{=}CH\cdot COOH \xrightarrow{\text{enoyl hydratase}} HO-CH(RCH_2)\cdot CH_2\cdot COOH$$

$$CH_3\cdot CH(CH_3)\cdot COOH \xrightarrow{-CO_2} CH_3\cdot CH_2\cdot COOH$$

Figure 10. Common metabolic pathways in PHB biosynthesis

TABLE 2

Molecular Weight of PHB Extracted from Different Bacterial Strains

Bacteria	Carbon Source	Extraction Method	Weight Average Molecular Weight
Azotobacter vinerandii	Sucrose	Methylene Chloride	1,500,000
Rhizobium meliloti	Glucose	Chloroform	1,000,000
Azotobacter chrooccocum	Glucose	Chloroform	1,000,000
Azotobacter beijerincki	Glucose	1,2 dichloroethane	800,000
Alcaligenes eutrophus	Fructose	Chloroform	600,000
Methylobacterium B3-Bp	Methanol	1,2 dichloroethane	300,000
Pseudomonas AM-1	Methanol	Chloroform	50,000
Bacillus cereus	Glucose	Hypochlorite	10,000
Rhodospirillum rubrum	Glucose	Chloroform	5,000

Note: Results are for typical batch fermentations. The actual values obtained will also depend upon the precise fermentation conditions and extraction method used.

reported to produce pure poly(ß-hydroxyoctanoate) when grown on n-octane[30]. Presumably, microorganisms that specifically produce other homopolymers in the ß-hydroxyalkanoate homologous series exist, but the situation is not clear cut. *Pseudomonas aeruginosa* grown on n-hexadecane, for example, produces PHB rather than the anticipated higher homologue[31].

The composition of poly(ß-hydroxyalkanoate) polymers and copolymers can be determined by gas or liquid chromatography[17,32], infra-red analysis and proton nmr[33]. In addition, carbon-13 nmr can give some information on sequence distribution of comonomers since the carbonyl carbon resonance is sensitive to dyad composition[34]. The HB-HV copolymers produced by *A.eutrophus* have been shown to be random copolymers by this technique[35]. In addition, the crystallinity of these copolymers as a function of HV contents have been studied by X-ray diffraction and FTIR with the surprising result that all the samples are 60-70% crystalline irrespective of composition from 0 to 50 mole% HV[36]. Thus, the HV units are able to enter the PHB lattice with minimal disturbance. Clearly, the polymerase enzyme can incorporate either HB or HV into its growing polymer chain with equal facility.

The structure of PHB granules has been studied in some detail [29,37-41] using a range of techniques including X-ray diffraction and electron microscopy. They are remarkably pure inclusions of apparently crystalline polymer, judging from X-ray[38] and density measurements [42,43] surrounded by a rather ill defined lipid coat containing the polymerase/depolymerase enzymes [34]. Typical granules are 0.2 to 0.5µm in diameter and contain about 2% protein and 0.5% lipid forming the membrane which is around 2nm thick[37]. They appear to be highly swollen *in vivo* [43], containing 40% water, which is surprising since the polymer itself is extremely hydrophobic when extracted. Light scattering[38] indicates that the average molecular weight of PHB granules is about 5 x 10^9 and, since the molecular weight of extracted polymer ranges from 10^3 to 10^6 depending on the organism, (table 2), there must be several thousand polymer chains in each granule.

The polymer chains appear to be organised into a number of microfibrils 10-15nm in diameter which are wound up like a ball of string to form the spherical granules[38,41]. This sub-structure is most clearly seen by freeze fracture electron microscopy[44]. This involves fixing a PHB containing bacterial culture with glutaraldehyde to preserve the internal structure, freezing and fracturing at very low temperature. The freshly exposed surface is then shadowed by evaporation from a heavy metal source located at an acute angle to form a metal replica which can be floated off and examined in the transmission electron microscope. A typical result is shown in figure

Figure 11. Transmission electron,micrograph of a metal replica of the fracture surface obtained from a frozen culture of *A.eutrophus.*

11. Two fractured bacteria can be seen. The polymer granules in one cell have been completely removed by the fracturing process leaving three craters. In the other cell, however, the granules remain but have been pulled out into horns of polymer heavily shadowed by the metal staining technique. Moreover, the fibrillar texture of the polymer can be seen very clearly at the edges of these distorted granules.

The plastic deformation of latex particles of brittle polymers such as polystyrene and polyacrylates during freeze fracture is well established. It is assumed that the energy dissipated during fracture causes massive local temperature rises to bring the polymer above its glass transition temperature. However, the magnitude of the deformation for PHB granules is still surprising. The 0.5µm diameter (0.25µm radius) particles are pulled into horns about 1.5µm long, representing a maximum extension of around 600%. PHB itself is a rather brittle polymer[35] and fails at extensions of less than 10% at all temperatures up to its melting point at 180°C. Thus, the plastic deformation evident in figure 11 is a result of the unique nascent morphology of PHB granules rather than inherent ductility in the polymer.

There remain a number of intriguing questions about the control of PHB biosynthesis. How is the number of granules per bacterium decided and does it change with the level of polymer accumulation? Above all, what determines the molecular weight of the polymer produced? It is instructive to consider a typical nutrient limited PHB fermentation in some detail with particular reference to these questions.

The results given in figures 12 and 13 were obtained for a phosphate limited fed batch fermentation in a $10m^3$ vessel. The fermenter was initially half filled with sterile water containing an excess of all nutrients except phosphorus (phosphate) and carbon (glucose). The levels of these were carefully calculated such that, when an inoculum of bacteria was introduced, it grew to an appropriate concentration and then simultaneously entered phosphate and carbon limitation. From this point the culture was fed continuously over a period of many hours with an appropriate carbon source that is converted with high efficiency to the equivalent poly(ß-hydroxyalkanoate); pure glucose to PHB, mixtures of glucose and propionic acid to HB-HV copolymer with composition determined by the glucose/propionate ratio in the feed.

The initial glucose level is shown in figure 12 and this rapidly decreased after about 36 hours as the cells, growing rapidly under ideal conditions, deplete the available carbon. The PHB content of the cells is shown in figure 13. During their initial logarithmic growth phase the bacteria produce up to 15% polymer but then enter carbon limitation and rapidly degrade the polymer for immediate use.

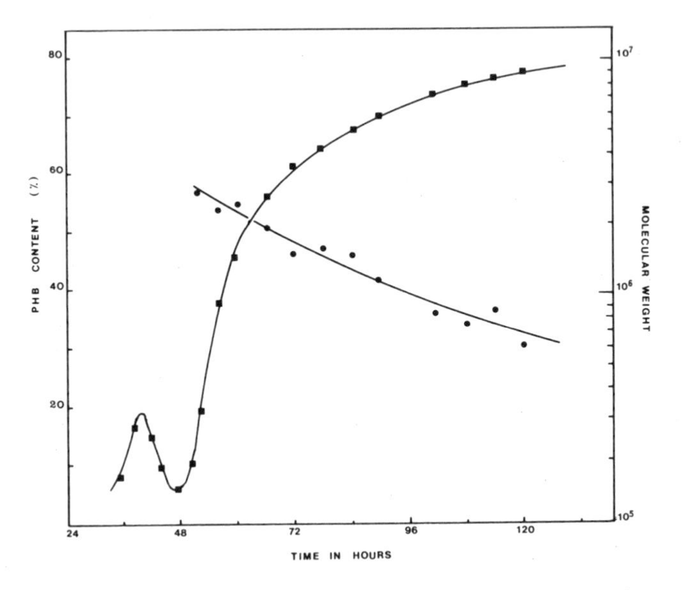

Figure 12. PHB and biomass production during phosphate limited fermentation of *A.eutrophus*.

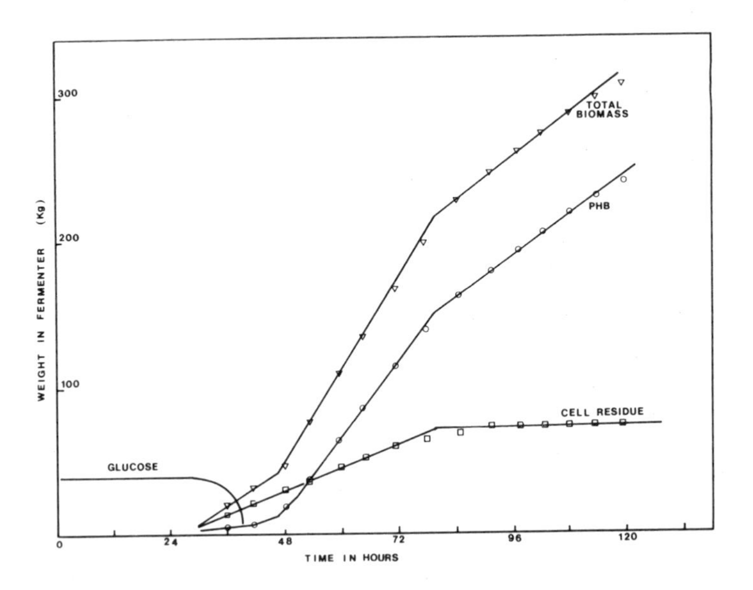

Figure 13. PHB content and polymer molecular weight during phosphate imited fermentation of *A.eutrophus*.

After 44 hours the culture is in carbon and phosphorus limitation, containing less than 5% PHB. The second phase of the fermentation then begins as the carbon feed is switched on. PHB is synthesised very rapidly and the cells contain over 50% polymer within 12 hours. It is also interesting to note that the quantity of non-PHB cell material also increases. Thus, despite the phosphorus limitation, the bacteria are able to shuffle their internal phosphate reserves to allow the synthesis of new cell wall material such that they grow and accumulate PHB simultaneously. This continues for about 36 hours when the synthesis of new non-PHB cell material ceases completely due to lack of phosphorus. Production of PHB, however, continues for a further 36 hours to a polymer content of over 75% on a dry weight basis. At this point the product was harvested because previous experience had shown that polymer synthesis is not maintained beyond about 80% PHB content.

In an attempt to establish why PHB synthesis stops, and to examine granule growth in more detail, a similar fermentation to that described was performed and reported in a separate study[45]. It was done under nitrogen rather than phosphorus limitation and samples were removed every two hours. These were freeze fractured and examined by TEM in the same way as that used to obtain figure 11 except that each photograph contained over 100 individual bacteria. The photographs were then analysed using a Joyce-Loebl Magiscan image analysis system to determine the number and size of polymer granules in each cell. A "cylindrical cell" model was used to interpret the results. Thus, the bacteria were assumed to be randomly oriented cylinders containing spherical granules and the image to represent a fracture plane through the assembly.

The results fitted the theory remarkably well up to a PHB volume fraction of 0.58. The bacteria were found to have an initial average length of 1.75μm with an aspect ratio (length/radius) of 7. The average number of granules per cell remained constant at 12.7±1.0 and these uniformly increased in average radius from 0.12 to 0.25μ. In a repeat experiment, on a much larger scale fermentation, very similar results were obtained except that the average number of granules per bacterium was 8.6±0.6.

The most remarkable feature of the experiments, however, was that there were significant deviations from the cylindrical cell model at PHB fractions exceeding 0.58. The rate of PHB synthesis dropped markedly and the results strongly suggested that, in order to accommodate more polymer, the cells were changing shape to a more spherical geometry.

In addition to this image analysis study, some of the culture samples were monitored for PHB polymerase activity using the method of Merrick and Doudoroff[46]. The level of enzyme activity was found to be high throughout the fermentation with no fall off in the capacity to polymerise ß-hydroxybutyryl-CoA even at very high PHB contents when the rate of polymer synthesis in the bacteria declined. This leads to the conclusion that PHB accumulation ceases because the bacterial cells are physically full and cannot accommodate more polymer within the fixed amount of cell wall material available to them despite the presence of monomer and active polymerase.

The final part of the fermentation experiment described earlier was aimed at investigating the control of PHB molecules weight by the bacteria. Samples of the culture were removed at regular intervals and the polymer isolated by freeze drying the cell mass and extracting it in refluxing chloroform for ten minutes. The cell debris was removed by filtration through glass fibre filter mat (Whatman GF/B) and molecular weight determined using gel permeation chromatography[47] on the diluted chloroform solution (0.1% w/v). The results are given in figure 13 and show a definite reduction in MW from over 2 x 10^6 to 6 x 10^5 during the course of the polymer accumulation.

This appears to be a general phenomenon for *A.eutrophus* and similar results are obtained irrespective of the nutrient limitation (nitrogen, phosphorus or oxygen), carbon substrate (glucose, ethanol, CO_2/H_2) and general fermentation parameters of pH, temperature, essential nutrient concentrations, etc. Moreover, the HB-HV copolymers produced by this organism also have very similar molecular weights, irrespective of composition.

Different organisms, however, do produce polymers of different molecular weights as indicated in table 2. Thus, the methylotroph, 95WA *Pseudomonas AM1,* consistently produces PHB of 50-60,000 molecular weight whereas *Methylobacterium B3-Bp* also grown on methanol gives polymer of much higher molecular weights, typically 250-300,000. The reason for this behaviour is simply not known. Nor do we know how to manipulate the fermentation production of PHB to give polymer of a particular molecular weight significantly different from that produced naturally by the bacterium. Phenylacetic acid is reported to cause a slight reduction in polymer molecular weight[48] but the effect is much smaller than the natural decrease during fermentation that is evident in figure 13.

One must be cautious about drawing conclusions relating to the polymerisation mechanism from molecular weight data, primarily because the MW measured on extracted PHB is not necessarily that of the polymer *in vivo* . Most extraction methods cause some degradation of the polymer, particularly if it is of very high MW, but the technique described here is believed to be extremely mild[38].

However, since polymer of progressively lower molecular weight is being synthesised during the fermentation (figure 13), one would expect the molecular weight distribution (MWD) to broaden with time. Indeed, the MWD increases uniformly from about 1.8 to 2.8 in the final product. The initial value is somewhat less than the figure of 2.0 expected for a purely random polyesterification reaction, especially in view of the instrumental broadening that inevitably leads to high measured values of MWD. Thus, it appears that at any given time polymer of extremely narrow MWD is being produced in the cells. The molecular weight of the polymer being produced, however, decreases with time and level of accumulation such that the average MW decreases and the MWD increases. Moreover, the initial rate of polymerisation must be extremely rapid since, despite repeated attempts, no evidence of low molecular weight PHB was found in the earliest stages of polymer accumulation.

CONCLUSIONS

From the available data it appears that there are three main elements in PHB and HB-HV copolymer biosynthesis in *A.eutrophus*. These are monomer (ß-hydroxybutyryl-CoA and ß-hydroxyvaleryl-CoA) synthesis from the various carbon substrates, using the biosynthetic pathways illustrated in figure 10, polymerisation by an enzyme system located at the surface of the polymer granule according to a mechanism similar to that shown in figure 9 and finally some form of chain transfer reaction. The latter reaction effectively controls the molecular weight of the polymer produced and presumably involves a simple transesterification between the polymer-enzyme thioester bond and that in the monomer-CoA molecule. The precise functioning of the chain transfer process, however, is not known although it may simply be related to the equilibrium concentration of monomer-CoA at the polymerase enzyme. Its effect is to produce polymer of extremely narrow MWD at any particular instant, but the absolute value of the molecular weight decreases with time during a typical nutrient limited fermentation. The process is also independent of parameters such as the nature of the carbon substrate, copolymer ratio, limiting nutrient etc., but does vary from organism to organism such that particular bacteria tend to produce polymer of a characteristic molecular weight.

The growing polymer chains are arranged into 10nm diameter fibrils coiled around to form discrete 0.2 to 0.5µm diameter spherical granules within the cell cytoplasmic space. The polymer itself is crystalline with hydroxybutyrate and hydroxyvalerate being incorporated with equal facility. However, the granules are reported to be highly swollen with up to 40% water[31], despite the hydrophobic nature of these polymers.

The number of granules per bacterium is fixed at the very earliest stages of polymer accumulation and then remains constant with the granules expanding uniformly in diameter to accommodate the PHB produced. There are typically 8-12 granules per bacterium in

A.eutrophus [45] and it would be interesting to see if this varied for other organisms and whether it related to the molecular weight of polymer produced. The bacteria themselves are initially cylindrical and expand with uniform shape to contain the growing granules until the limiting nutrient is completely used up in cell wall formation. At this point, no new non-PHB material is formed and additional polymer is contained by the cell changing shape to a more spherical geometry. In *A.eutrophus* this occurs at PHB volume fraction of 0.58. Polymer synthesis eventually stops when the bacteria are physically full of polymer and the cells lose viability.

Thus PHB biosynthestis has some of the features of an emulsion polymerisation with the individual granules corresponding to latex particles and the membrane lipid to surfactant. However, the polymerisation mechanism is not free radical but a coordination-insertion process bearing more resemblance to a Ziegler type system. Indeed, the fibrillar substructure of PHB granules is reminiscent of the polymer fibrils produced in the zirconium benzyl mediated polymerisation of ethylene[49].

9. REFERENCES

1. A.C. Ward, B.I. Rowley and E.A. Dawes, J.Gen.Microbiol. 102 (1977) 61-68
2. T. Suzuki, T. Yamane and S. Shimizu, Appl. Microbiol. Biotechnol. 24 (1968) 366-374
3. E.A. Dawes and P.J. Senior, Adv. Microbial Phys. 10 (1973) 138-266
4. T. Suzuki, T. Yamane and S. Shimizu, Appl. Microbiol Biotechnol. 23 (1986) 322-9
5. D.J. Leak and H. Dalton, J.Gen.Microbiol. 129 (1983) 3487-3498
6. K. Nicolay, H. Van Gemerden, K.J. Hellingwerf, W.N. Konings and R. Kaptein, J.Bacteriol. 155 (1983) 634-42
7. A.W. Thomson, J.G. O'Neill and J.F. Wilkinson, Arch. Microbiol 109 (1976) 243-246
8. J.A. Asenjo and J.S. Suk, J. Ferment. Technol. 64 (1986) 271-8.
9. G. Gottschalk, Arch. Mikrobiol. 47 (1964) 236-50
10. P.A. Holmes (Imperial Chemical Industries PLC) Eur.Pat. EP46,335 (1982)
11. P.A. Holmes and G.B. Lim (Imperial Chemical Industries PLC) Eur. Pat. EP145,233 (1985)
12. Plastics and Rubber Weekly, (Jan 10, 1987) p12
13. T. Yasuda, T. Aida and S. Inoue, Macromolecules 16 (1983) 1792-96
14. Z. Jedlinski, P. Kurcok, M. Kowalczuk and J. Kasperzyk, Makromol. Chem. 187 (1986) 1651-1656
15. Manufacturing Chemist (October, 1985) p63-65

16. P.A. Holmes, L.F. Wright and S.J. Collins, (Imperial Chemical Industries PLC) Eur. Pat. EP69,497 (1983) and EP52,459 (1981)
17. R.H. Findlay and D.C. White, Appl. Env. Microbiol. 45 (1983) 71-78
18. P.A. Holmes, Phys. Technol. 16 (1985) 32-36
19. D.G. Lundgren and K F Bott, J Bacteriol. 86 (1963) 462-473
20. L.H. Stevenson and M.D. Socolofsky, J. Microbiol. Serol. 39 (1973) 341-350
21. R.N. Reusch and H.L. Sadoff, J. Bacteriol 156 (1983) 778-788
22. J.M. Merrick, "Photosynthetic Bacteria", R.K. Clayton and W.R. Sistrom editors (1978) 199-219, Plenum NY
23. C. Pedros-Alio, J. Mas and R. Guerrero, Arch. Microbiol. 143 (1985) 178-84
24. P.O. Peoples, S. Masamune, C.T. Walsh and A.J. Sinskey, J. Biol. Chem. 262 (1987) 97-102
25. Personal Communication from C.T. Walsh re. MIT Projects 6.07.108 and 6.07.012
26. R. Griebel, Z. Smith and J.M. Merrick, Biochemistry 7 (1968) 3676-3681
27. T. Fukui, Y. Akio, M. Mamoru, H. Shunji, S. Terumi, N. Hiroko and T. Kenkichi, Arch. Microbiol. 110 (1976) 149-156
28. J.M. Merrick and C.I. Yu, Biochemistry 5 (1966) 3563-3568
29. J.M. Merrick, D.G. Lundgren and R.M. Pfister, J. Bacteriol 89 (1965) 234-239
30. M.J. de Smet, G. Eggink, B. Witholt, J. Kingma and H. Wynbert, J. Bacteriol. 154 (1983) 870-878
31. S.K. Layokum, B.O. Solomon and I.A. Fatile, Appl. Microbiol. Biotechnol. 22 (1985) 255-258
32. D.B. Karr, J.K. Waters and D.W. Emerich, Appl. Env. Microbiol. 46 (1983) 1339-1344
33. S. Bloembergen, D.A. Holden, G.K. Hamer, T.L. Bluhm and R.H. Marchessault, Macromolecules 19 (1986) 2865-2871
34. Y. Doi, M. Kunioka, V. Nakamura and K. Soga, Macromolecules 19 (1986) 2860-2864
35. P.A. Holmes, "Developments in Crystalline Polymers II", ed. D.C. Bassett in press (1987) Elsevier London.
36. T.L. Bluhm, G.K. Hamer, R.H. Marchessault, C.A. Fyfe and R.P. Veregin, Macromolecules 19 (1986) 2871-2876
37. D.G. Lundgren, R.M. Pfister and J.M. Merrick, J. Gen. Microbiol. 34 (1964) 441-446
38. D.J. Ellar, D.G. Lundgren, K. Okamura and R.H. Marchessault J. Mol. Biol. 35 (1968) 489-502
39. T.E. Jenson and L.M. Sicko, J. Bacteriol. 106 (1971) 683-686
40. R. Griebel and J.M. Merrick, J. Bacteriol. 108 (1971) 782-789
41. A.W. Robards and W.F. Dunlop, J.Bacteriol. 114 (1973) 1271-1280
42. K.W. Nickerson, Appl. Env. Microbiol 43 (1982) 1208-1209
43. J. Mas, C. Pedros-Alio and R. Guerrero, J. Bacteriol. 164 (1985) 749-756
44. U.B. Sleytr and A.W. Robards, J. Microscopy 110 (1977) 1-25

45. J.F. Stageman and R.S. Kittlety, paper in preparation
46. J.M. Merrick and M. Doudoroff, J. Bacteriol. 88 (1964) 60-
47. P.J. Barham, A. Keller, E.L. Otun and P.A. Holmes, J. Mat. Sci. 19 (1984) 2781-2794
48. M.P. Nuti, M de Bertoldi and A A Lepidi, Can. J. Microbiol. 18 (1972) 1257-1261
49. D.G.H. Ballard, E. Jones, R.J. Wyatt, R.T. Murray and P.A. Robinson, Polymer 15 (1974) 169-174

DISCUSSION

It is difficult to get the appropriate enzymes to work outside of the cell environment so that no attempt was made to isolate them.

Certain types of bacterial cells produce the higher homologs of polyhydroxybutyric acid, and this has been extensively investigated by biochemists.

The molecular weight distribution of the high polymers of polyhydroxybutyric acid is uncertain ; for the low molecular weight fractions, it is narrow.

ADDITIONAL REPORT

D.G.H. BALLARD

The microbial oxidation of aromatics and the use of the latter in the synthesis of aromatic and hydroxylated polymers

It has been demonstrated by ICI (D.G.H. Ballard et al - J.C.S. Chem. Comm. 953, 1983) that substituted cyclo 1:3 hexadienes can be produced economically by the bacterial oxidation of benzene :

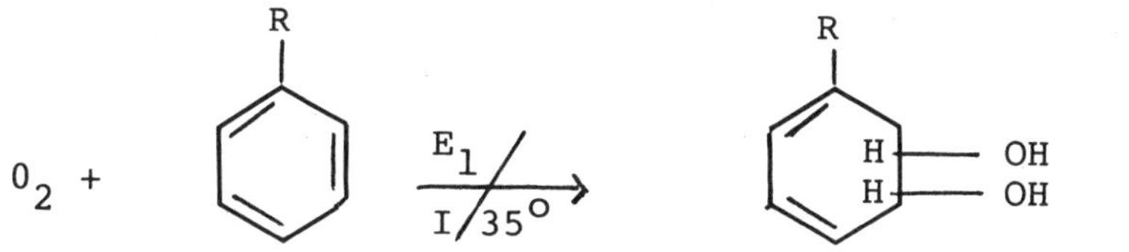

$$O_2 + \text{C}_6\text{H}_5\text{R} \xrightarrow[I/35^o]{E_1} \text{R-cyclohexadiene-}cis\text{-diol (H, OH; H, OH)} \quad (\underline{1})$$

The organisms used are derived from the genius Psuedominium Putida and the enzyme E_1 contains the dioxygenase function. The miroorganisms are modified so that the enzyme which converts the cis-glycol to catechol is inactivated. In these circumstances the only oxidation product produced is the 5:6 cis glycol cyclohexa 1:3 diene (1). Derivatives (2) of 1 (R = H) can be polymerized using free-radical generators and in the absence of solvent, to give a polymer (3) which is soluble in simple organic solvents.

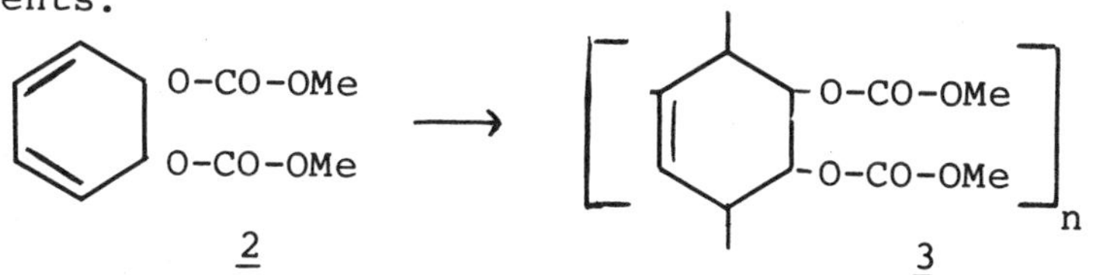

$$\underset{\underline{2}}{\text{cyclohexadiene}(\text{O-CO-OMe})_2} \longrightarrow \underset{\underline{3}}{[\text{cyclohexene}(\text{O-CO-OMe})_2]_n}$$

The soluble precursor polymer 3 can be used to produce coatings or films of polyphenylene which

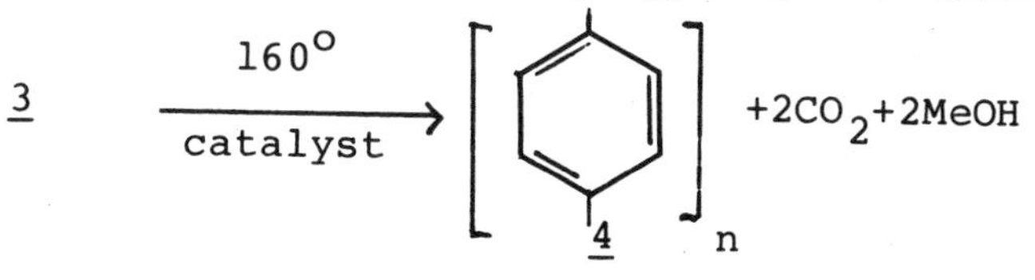

$$\underline{3} \xrightarrow[\text{catalyst}]{160^o} \underset{\underline{4}}{[\text{C}_6\text{H}_4]_n} + 2CO_2 + 2MeOH$$

M. Fontanille and A. Guyot (eds.), Recent Advances in Mechanistic and Synthetic Aspects of Polymerization, 315–316.

contain no metal catalyst residues.

Also, hydroxylation of 3 at the 2:3 position in the presence of strong base produces a highly hydrophilic polymer 5.

$$\underline{3} \xrightarrow[H_2O_2]{KOH} \left[\text{(HO)}_2\text{C}_6\text{H}_6\text{(OH)}_2 \right]_p$$

5

All these processes are being evaluated on a pre-commercial scale and compound 1 (R = H) is generally available.

POLYMERIZED VESICLES

Steven L. Regen
Department of Chemistry, Lehigh University
Bethlehem, Pennsylvania 18015
USA

ABSTRACT: Phospholipid and surfactant vesicles serve as important models for biological membranes, as potential carriers of drugs, and as devices for photochemical energy conversion. While much progress has been made in the construction of polymerized membrane analogs over the past seven years, further synthetic efforts are warranted. In particular, the synthesis of polymerized vesicles which: (a) are potentially biodegradable, (b) are synthesized under extremely mild conditions, and (c) offer improved control over lateral diffusion, stability and permeability properties, are all worthy goals for future studies. This paper briefly reviews one new approach toward a "second generation" of polymerized vesicles that represents a significant step toward achieving all of these goals. Each of these polymers is based on a disulfide backbone.

INTRODUCTION

Phospholipid and surfactant vesicles serve as important models for biological membranes, as potential carriers of drugs, and as devices for photochemical energy conversion. One inherent property of vesicles, however, which limits their usefulness in practical and mechanistic applications, is their limited *in vitro* and *in vivo* stability. In particular, on standing, vesicle dispersions are subject to aggregation and fusion processes which result in the formation of large vesicular structures; in many cases, precipitation from solution is, in fact, observed. This behavior places serious limitations on all applications of vesicles which require their long term use. As drug delivery devices, *in vivo* instability is also of considerable concern. For example, processes such as lipid-exchange with cell membranes, direct removal and net transfer of lipids from liposomes, liposome-liposome fusion and liposome-cell fusion, all contribute to a limited *in vivo* life-time of conventional (nonpolymerized) vesicles. This inherent instability of vesicles, noted above, has been a major driving force for the development of improved synthetic analogs.

M. Fontanille and A. Guyot (eds.), Recent Advances in Mechanistic and Synthetic Aspects of Polymerization, 317–324.

Within a remarkably short span of time, coming between the end of 1980 and the beginning of 1981, we, along with three other research groups, reported the successful synthesis and characterization of highly stable polymerized vesicles.[1-4] Since these initial publications, a large variety of polymerized vesicle structures have been introduced.[5] A pictoral summary of what has been accomplished thus far in the polymerized vesicle area is illustrated below. In brief, polymerized vesicles have been synthesized in which a polymeric backbone runs: (A) through the center of the hydrocarbon bilayer, (B) through the lipid chains of inner and outer monolayers, (C) through the polar head groups of each monolayer, (D) through a monolayer lipid membrane, and (E) adjacent to the inner and outer surface of the bilayer.

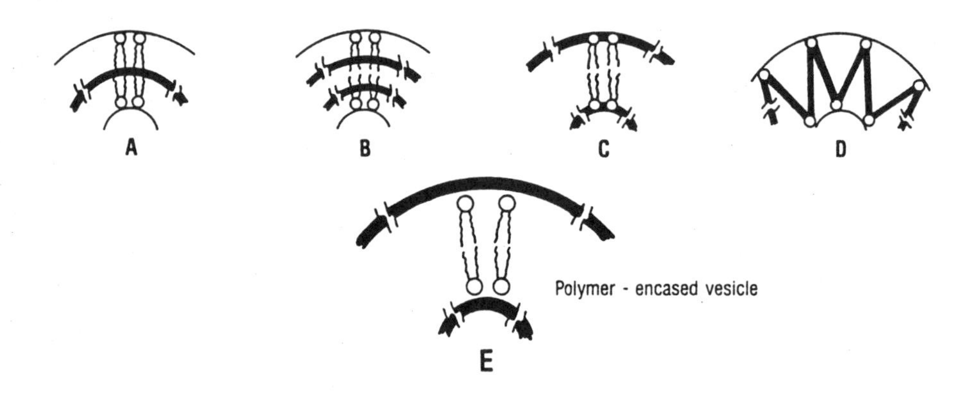

Three types of functional groups which have received the greatest attention in the synthesis of polymerizable phospholipids are the conjugated diacetylene, methacryloyl, and conjugated diene groups. In each of these cases, an all carbon backbone is produced upon polymerization via UV irradiation or thermally induced free radical initiation. In order to maximize their utility for biomechanistic studies (membrane modeling) and for biomedical applications (drug delivery), polymerized vesicles should be prepared under the mildest conditions possible, so that sensitive comembrane and entrapped components can be incorporated without degradation. Both UV irradiation and thermal procedures are now commonly used are not particularly mild, and there is a clear need for the design and synthesis of new types of polymerizable lipids.

In this paper we briefly describe one new class of "second generation" polymerized vesicles which we have devised for use as improved biomembrane models and as potentially biodegradable drug delivery devices. This class is based on lipid membranes which are coupled together under mild conditions via disulfide bond formation. The fact that disulfide moieties are common to many biopolymers implies that such polymers will be biodegradable, although detailed biological studies remain to be carried out.

RESULTS AND DISCUSSION

Oxidative Polymerization. Three thiol-containing phospholipids that have been specifically designed for polymerized vesicle synthesis are 1a, 1b and 1c.[6]

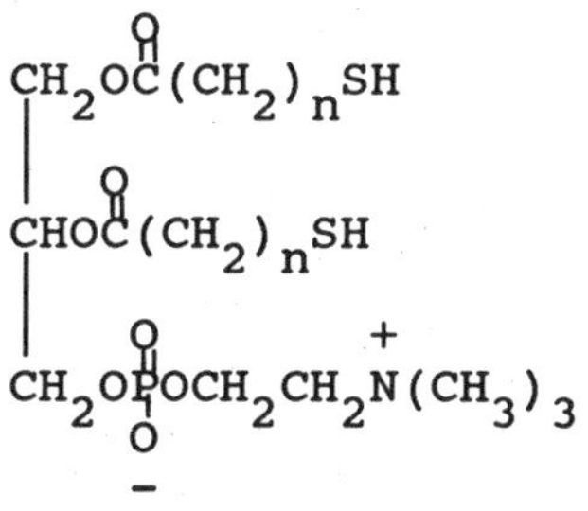

$$\begin{array}{l} CH_2OC(=O)(CH_2)_nSH \\ | \\ CHOC(=O)(CH_2)_nSH \\ | \\ CH_2OP(=O)(O^-)OCH_2CH_2\overset{+}{N}(CH_3)_3 \end{array}$$

1a, n=10

b, n=15

$$\begin{array}{l} CH_2OC(=O)CHSH(CH_2)_{13}CH_3 \\ | \\ CHOC(=O)CHSH(CH_2)_{13}CH_3 \\ | \\ CH_2OP(=O)(O^-)OCH_2CH_2\overset{+}{N}(CH_3)_3 \end{array}$$

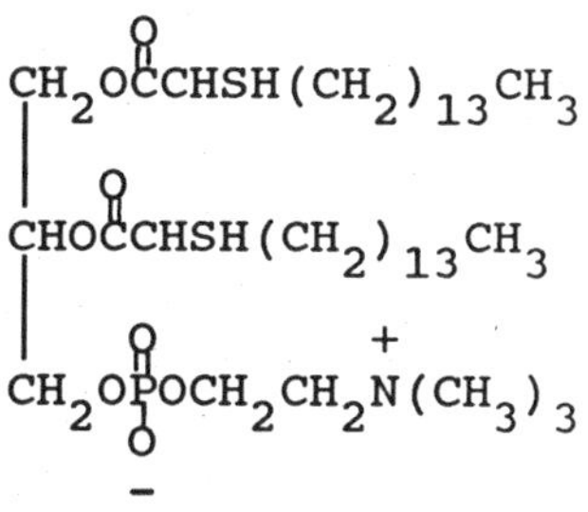

1c

The synthetic routes used for the preparation of 1a, 1b and also 2 (a macrocyclic derivative of 1a) are summarized below. Lipid 1c was prepared using similar procedures, starting from 2-mercapto-hexadecanoic acid.

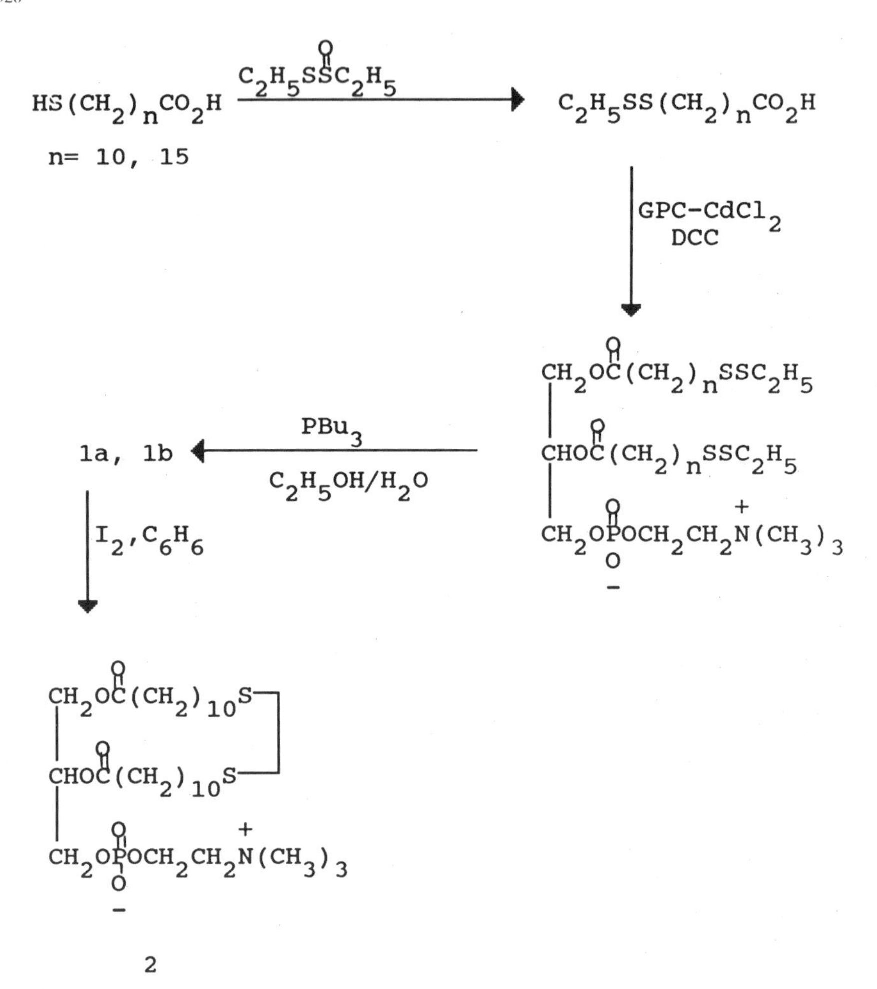

Vesicles derived from lipids 1a, 1b, and 1c readily formed by dispersal in 10 mM borate buffer (pH 8.5) containing 140 mM NaCl and 2 mM NaN_3, with the aid of ultrasonic irradiation and standard procedures. Vesicle polymerization was effectively carried out by oxidation with excess hydrogen peroxide. Quantitative analysis for thiol groups confirmed that oxidation proceeded to greater than 95% in all cases. Thus, analysis of the end-groups indicated that the number average of degree of polymerization was greater than 20. In addition, thin-layer chromatography showed that the polymeric lipids remained at the origin [silica, $CHCl_3/CH_3OH/H_2O$, 65/25/4 (v/v/v)]. Electron micrographs of these polymerized vesicles confirmed the presence of closed microspheres having diameters ranging between 200 and 800 angstroms, and a membrane thickness of ca. 40-50 angstroms, which is consistent with bilayer formation. Significantly, in contrast to their nonpolymerized analogs which precipitate on standing within 48 h, these polymerized vesicles exhibited shelf-lives of greater than 10 days.

Ring Opening Polymerization. One intriguing question that relates to the above vesicle polymerization reactions, is whether the polymers that are produced are thermodynamic or kinetic products. Recently it occurred to us that if vesicles could be prepared from the macrocyclic phospholipid 2, and if a trace amount of a disulfide reducing agent were added, conversion from the monomeric to a polymeric state might proceed, if the latter were thermodynamically favored. Specifically, we envisioned the possibility of a ring opening polymerization process occurring via a thiol--disulfide interchange.[7] Successful ring-opening polymerization of vesicular 2 would not only demonstrate that the polymeric state of such lipids is thermodynamically favored, but it would also provide the mildest synthetic route available to date for the construction of polymerized phosphatidylcholine membranes.

Vesicle dispersions of 2 were prepared using procedures similar to that used for lipids 1a, 1b, and 1c.[8] Dynamic light scattering and transmission electron microscopy established the presence of small vesicles ranging between 240 and 300 angstroms in diameter. Treatment with 5 mol% of dithiothreitol (DTT) for 4 h at 50°C and subsequent analysis of the dispersion by thin layer chromatography showed the complete loss of starting monomer; i.e., all of the lipid remained at the origin. Gel filtration through a Sepahrose 6B column resulted in a 96% vesicle recovery in the void volume. Light scattering showed no significant change in the apparent vesicle diameter upon polymerization. Control experiments carried out by heating vesicles derived from 2 for 4 h at 50°C, in the absence of DTT, showed only starting monomer and no apparent change in vesicle diameter.

Cross-linked Disulfide-based Polymerized Phosphatidylcholine Vesicles. All of the above vesicles are comprised of linear polymers. In principle, cross-linked polymerized vesicles should exhibit stability which is higher than their noncross-linked counterparts. Using the ring-opening polymerization approach described above, we have successfully prepared cross-linked polymerized phosphatidylcholine vesicles based on the use of 1,2-bis[12-(lipolyoxy)dodecanoyl]-*sn*-glycero-3-phosphocholine [designated as DLL]. An efficient synthesis[9] of DLL which we have developed is outlined below.

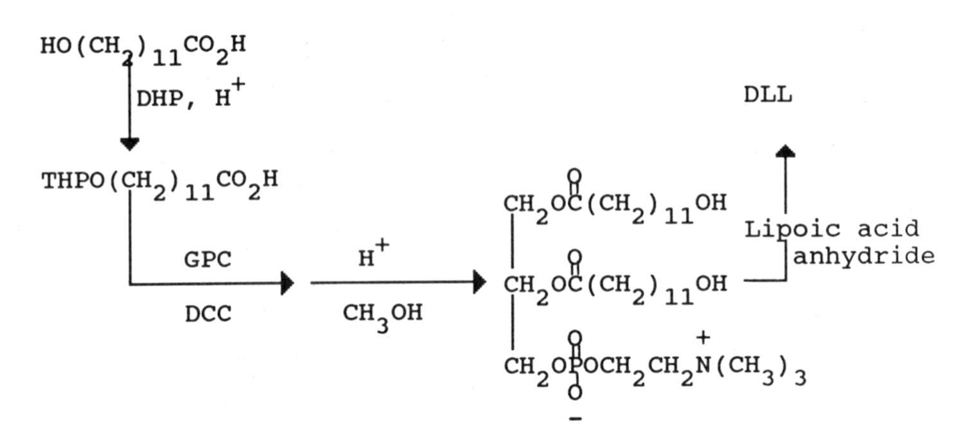

Using standard procedures, injection of an ethanolic solution of DLL into a borate buffer solution at room temperature, produced a vesicle dispersion having an average diameter ranging typically between 270 and 400 angstroms. Analysis of the dispersion by thin layer chromatography indicated negligible polymerization; only trace amounts of lipid remained at the origin, and nearly all of the lipid moved with an R_f which was identical with starting DLL. Subsequent treatment with 10 mol% of DTT resulted in complete polymerization after 4 h at 27 °C. All of the lipid remained at the origin of the TLC plate. Polymerization of these vesicles could also be conveniently monitored by the loss of UV absorbance at 333 nm, which is characteristic of the five-membered ring cyclic disulfide. This decrease in absorbance obeyed clean first-order kinetics and was complete within 4 h. In the absence of DTT, extensive (but incomplete) polymerization was observed after 72 h at 23 °C, and 6 h at 50 °C.

Examination of these dispersions by light scattering and by electron microscopy confirmed the presence of closed vesicles having an average diameter of ca. 320 angstroms; the apparent membrane thickness was estimated to be ca. 75 angstroms, which is consistent with bilayer formation. Gel filtration further confirmed the vesicle state of the polymerized lipid.

Solubility, Stability Toward Detergent, Lateral Diffusion and Permeability Properties. In contrast to liposomes comprised of linear polymers derived from 1a, 1b, 1c and 2, freeze-dried polymerized liposomes of DLL proved insoluble in chloroform and chloroform/methanol (1/1,v/v). This insolubility provides strong indirect evidence for extensive cross-linking. Moreover, unlike the former, which are readily lysed in 0.6% sodium dodecylsulfate (SDS), polymerized liposomes of DLL are completely stable in 1% SDS, even after brief heating at 60°C. Control experiments confirmed that nonpolymerized vesicles of DLL are readily destroyed with 0.05% SDS at room temperature. Preliminary permeability studies carried out with each of the above lipid vesicles, using entraped [^{14}C] sucrose as a permeant, indicate that polymerization decreases membrane permeability in all cases. Qualitatively, polymerized DLL vesicles exhibit the lowest permeability of all of the disulfide-based phospholipid polymers that have been investigated to date. Finally, preliminary photobleaching recovery experiments, using 3,3'-dioctadecyloxacarbocyanine perchlorate as a probe, indicated that lateral diffusion with bilayer membranes of 1a was reduced by approximately one order of magnitude upon polymerization, to a value of 2×10^{-9} cm^2/s.

Further studies which are now in progress are focusing on (1) the synthesis of homologs of 1a, 1b, 1c, 2, and DLL, and (2) a detailed examination of their stability, permeability, lateral diffusion, and biological characteristics.

CONCLUDING REMARKS

This paper has briefly summarized one new approach that we have taken toward the construction of potentially biodegradable vesicles with controllable lateral diffusion, stability, and permeability properties. The efficient synthetic procedures which have been developed for the preparation of 1a, 1b, 1c, 2, and DLL, together with the improved in vitro stability of their corresponding polymerized vesicles make these systems worthy of extensive biochemical and biological exploration and exploitation. Such efforts are now underway in our laboratories.

REFERENCES

1. Regen, S. L.; Czech, B.; Singh, A., J. Am. Chem. Soc., 1980, 102, 6638.

2. Hupfer, H.; Hupfer, B.; Koch, H.; Ringsdorf, H., Angew. Chem. Int. Ed. Engl., 1980, 19, 938.

3. Johnson, D. S.; Songhera, S.; Pons, M.; Chapman, D., Biochim. Biophys. Acta, 1980, 602, 57.

4. O'Brien, D. F.; Whitesides, T. H., J. Polym. Sci., Polym. Lett. Ed., 1981, 19, 95.

5. Reviews: Bader, H.; Dorn, K.; Hashimoto, K.; Hupfer, B.; Petropoulos, J. H.; Ringsdorf, H.; Sunimoto, H. In "Polymeric Membranes", Gordon, M., Ed.; Springer, Verlag: Berlin, 1985, p 1. Fendler, J. H.; Tundo, P. Acc. Chem. Res., 1984, 17, 3.

6. Samuel, N. K. P.; Singh, M. Yamaguchi, K.; Regen, S. L., J. Am. Chem. Soc., 1985, 107, 42.

7. Szajewski, R. P.; Whitesides, G. M. J. Am. Chem. Soc., 1980, 102, 2011.

8. Regen, S. L.; Samuel, N. K. P.; Khurana, J. M., J. Am. Chem. Soc., 1985, 107, 5804.

9. Sadownik, A.; Stefely, J.; Regen, S. L., J. Am. Chem. Soc., 1986, 108, 7789.

DISCUSSION

It is not possible to know how many disulfide bonds between the bilayers are formed.

It is supposed that the disulfide bonds are biodegradable allowing the encapsulated drugs to diffuse out of the liposomes.

Asymmetric vesicles with different surfactants inside and outside of the vesicles were obtained by Mary Roberts at MIT.

Systems with siloxane of fluorocarbon as hydrophobic segment of the surfactant were obtained by Ringsdorf.

POLYMERIZATION OF ACRYLAMIDE IN INVERSE EMULSION

K. H. Reichert
Institut fuer Technische Chemie der Technischen
Universitaet Berlin, 1000 Berlin 12,
Strasse des 17. Juni 135

ABSTRACT. The polymerization of acrylamide was studied in inverse emulsion by using oil and water soluble azoinitiatiors and sorbitan monooleate as nonionic oil soluble emulsifier. Isoparaffin or toluene was used as organic phase. The polymerization was performed in a stirred tank reactor at constant temperature. The kinetic dependencies obtained depend strongly on the system used. Polymerization presumably takes place within the dispersed water droplets. When using oil soluble initiators mass transfer processes must be considered which can be rate determining. Aggregates of the emulsifier could be detected by light scattering in n-heptane. The size of aggregates depends on the water content present in the organic solvent. The solubilization of acrylamide in the organic phase, containing the emulsifier, is very poor at polymerization conditions. The different kinetic expressions found in different systems are discussed.

INTRODUCTION

Synthetic water soluble polymers are frequently produced by polymerization in inverse emulsion. This kind of polymerization is suitable for producing water soluble polymers with high molecular weights at high rates of polymerization. The understanding of polymerization in inverse emulsion is still rather poor although much work has been done in this field since the pioneering work of Vanderhoff in 1962. Most of the work published so far is related to polymerization of acrylamide. Kinetic data of acrylamide polymerization in inverse emulsion is summarized in table 1. The kinetic results differ from each other and seem to depend mainly on the kind of emulsifier and initiator used. The orders of reaction differ in general from those of normal emulsion polymerization indicating that the classical theory of emulsion polymerization should not be applied as such to polymerization in inverse emulsion. In the present case the radical polymerization of acrylamide in inverse emulsion was studied by dispersing an aqueous solution of acrylamide into an organic phase such as isoparaffin or toluene using sorbitan monooleate with a HLB value of 4.3. For initiation, oil as

M. Fontanille and A. Guyot (eds.), Recent Advances in Mechanistic and Synthetic Aspects of Polymerization, 325–334.

oil phase	emulsifier	initiator	kinetic	Ref.
toluene	cetylpyridinechloride	$K_2S_2O_8$	$R \sim c_M^{1,6} c_I^{0,5} c_E^{1}$	Trubitsyna, 1976, 1978
toluene	sentamide 5 (polyethyleneglycolether of monoethhydroxyamide)	$K_2S_2O_8$	$R \sim c_M^{1,7} c_I^{0,9} c_E^{a} c_T^{b}$	Kurenkov, 1978
toluene	Na- and K-p-styrenesulfonate-oligomers and -copolymers	AIBN $(NH_4)_2S_2O_8$	$R \sim c_M^{1,3} c_I^{0,46}$ $R \sim c_M^{1,3} c_I^{0,5}$	Kurenkov, 1982
xylene	Tetronic 1102 (EO-PO blockcopolymer condensed with ethylenediamine)	BPO	$R \sim c_M^{1} c_I^{2} c_E^{1}$	Vanderhoff, 1984
xylene	Tetronic 1102	ADVN KPS	$R \sim c_M^{1} c_I^{0,2} c_E^{0,8}$ $R \sim c_M^{1} c_I^{0,4} c_E^{0,6}$	Visioli, (Vanderhoff), 1984
toluene	sorbitansesquioleate + polyoxyethylenesorbitantrioleate or C_{18}-merkaptoacrylamide oligomer	AIBN ADVN AMBN	$R \sim c_I^{0,1} c_E^{-a}$	Guyot, 1986
toluene	PS-PEO-PS blockpolymer + 2-propanol	AIBN	$R \sim c_M^{1,9} c_I^{0,5}$	Candau, 1981
hexadecane	potassiumoleate + hexanol (octanol)	AIBN/KACV	$R \sim c_M^{1,8} c_I^{0,5}$	Wu, 1983
toluene	Aerosol OT	AIBN $K_2S_2O_8$	$R \sim c_M^{1,1} c_I^{0,1} c_E^{-0,55}$ $R \sim c_M^{1,5} c_I^{0,03}$	Candau, 1985

AIBN = Azoisobutyronitrile ADVN = Azodimethylvaleronitrile AMBN = Azomethylbutyronitrile
BPO = Benzoylperoxide KPS = Potassiumpersulfate KACV = Potassiumazocyanovaleriate

Table 1. Kinetics of acrylamide polymerization in inverse emulsion. A literature survey.

well as water soluble azoinitiators were used. The standard recipe is given in table 2.

	wt.-% related to emulsion	wt.-% related to one phase
Oil phase:		
Isopar M (Esso)	40.52	94.21
Sorbitan monooleate	2.47	5.74
Azodimethyl valeronitrile	0.02	0.05
Water phase:		
Acrylamide	22.74	39.90
Distilled water	34.01	59.70
B_2O_3 (buffer)	0.24	0.30
EDTA (complexation agent)	0.006	0.10

Table 2. Standard recipe of acrylamide polymerization in inverse emulsion with oil soluble initiator

The polymerization was performed in a stirred tank reactor at constant temperature (47 °C) under nitrogen atmosphere. The standard stirring speed of the propeller mixer used was 1000 rpm.

KINETIC RESULTS OF ACRYLAMIDE POLYMERIZATION

If polymerization of acrylamide is started by injection of the initiator into the stirred emulsion purged with nitrogen, inhibition periods of different duration may be observed. Intensive purging with nitrogen usually shortens the length of the inhibition period. Once started the polymerization shows an autocatalytic behaviour like illustrated in figure 1. The initial rate of polymerization can not be reproduced with an acceptable error. This is only the case for the maximum rate of polymerization. The rate of polymerization refers to the water phase. The autocatalytic behaviour is presumably caused by the gel effect of polymerization which can also be responsible for the slight increase of molecular weights with conversion as shown in figure 2. The particle size of the disperse phase is constant with conversion as determined by electron microscopy at conversions larger than 0.2. It depends on stirring speed, temperature and on the kind of oil soluble initiator and organic phase. The same is true for the particle size distribution. Kinetic data of acrylamide polymerization achieved so far for different systems are summarized in table 3. It can be seen that the kind of azoinitiator has a strong effect on the kinetic behaviour of the polymerization. Water soluble initiators lead to kinetic expressions which are typical for radical polymerizations in solution.

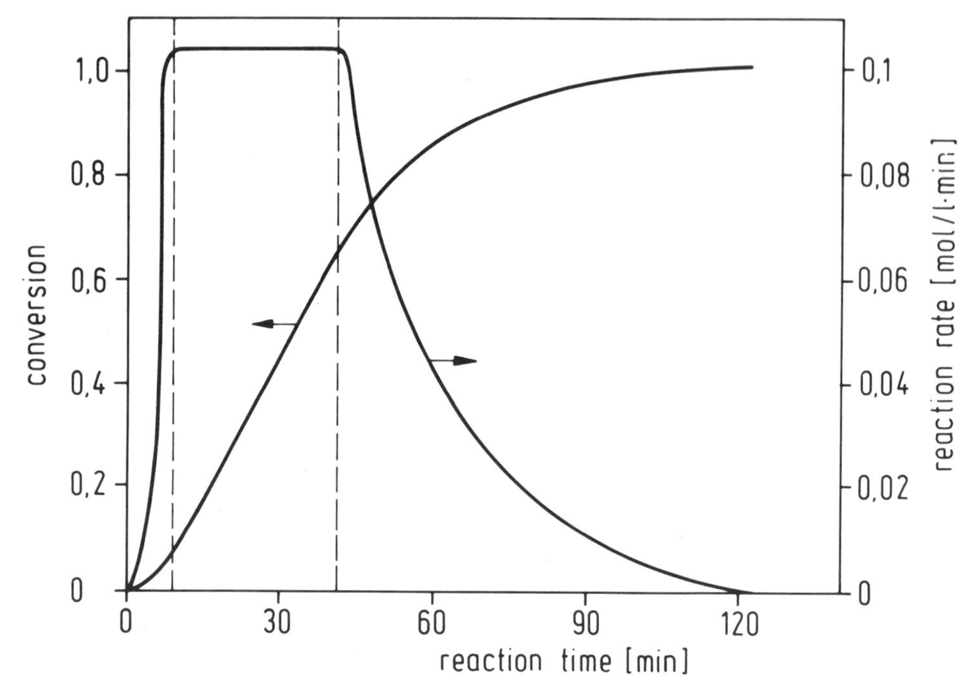

Fig. 1. Time dependence of conversion and reaction rate of acrylamide polymerization (C_{ADVN}= 1.33 mmol/l, 47 °C).

With oil soluble azoinitiators the kinetic equations of acrylamide polymerization are dependent on the nature of the organic phase as can be seen from table 3. While the kinetic expression of acrylamide polymerization in toluene is similar to that of normal emulsion polymerization a completely different kinetic expression is obtained in the case of isoparaffin and n-heptane. Differences are also observed in the case of maximum reaction rate, mean particle diameter and to a smaller extent in the case of average molecular weights at comparable reaction conditions. It is note worthy that in case of polymerization in presence of isoparaffin the overall rate constant is not only a function of temperature but also of interfacial area. This can be seen from figure 3. The linear relationship between rate constant and interfacial area running through the origin of the diagram is indicating that the polymerization is controlled by phase transfer processes or by interface reactions. The activation energy at constant interfacial area (26 kJ/mol) is in the order of magnitude of activation energies of phase transfer processes as well as of chemical reactions with radicals as reactants.

oil soluble azoinitiators		water soluble azoinitiators
in isoparaffin a. n-heptane	in toluene	in isoparaffin
$R_{max} = k \; c_{I,o}^{1} c_{M,o}^{1} c_{E,o}^{a}$	$R_{max} = k \; c_{I,o}^{0,5} c_{M,o}^{1} c_{E,o}^{0,45}$	$R_{max} = k \; c_{I,o}^{0,5} c_{M,o}^{1}$
k = f (RPM, T) a = -0,1 for ADVN a = -0,2 for AIBN E_A = 88,2 kJ/mol with AIBN E_{A_0} = 26 kJ/mol at constant interface and AIBN	with ADVN	with AIBEA E_A = 97,9 kJ/mol with AIBEA
molecular weight (weight average) in g/mol at comparable conditions		
$8,2 \cdot 10^6$	$9,2 \cdot 10^6$	$5 \cdot 10^6$
mean particle diameter of disperse phase in µm at comparable conditions		
1,5	0,2	1,5
maximum rate of reaction in mol/l·min at comparable conditions		
6	20	6

Table 3. Kinetic data of acrylamide polymerization in emulsion with sorbitan monooleate as emulsifier

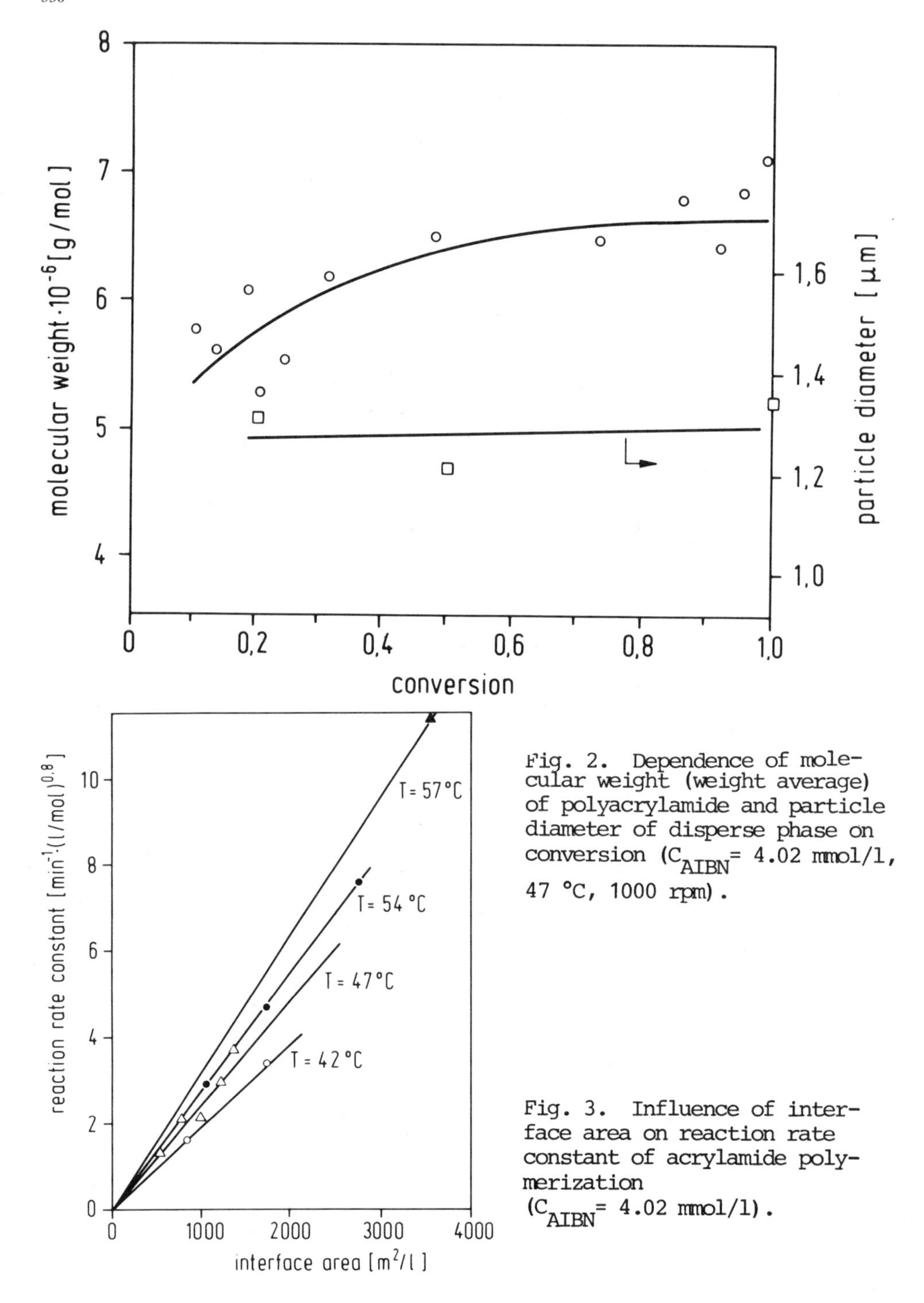

Fig. 2. Dependence of molecular weight (weight average) of polyacrylamide and particle diameter of disperse phase on conversion (C_{AIBN}= 4.02 mmol/l, 47 °C, 1000 rpm).

Fig. 3. Influence of interface area on reaction rate constant of acrylamide polymerization (C_{AIBN}= 4.02 mmol/l).

CHARACTERIZATION OF SURFACTANT IN SOLUTION

Sorbitan monooleate (Disponil 100, Henkel) was purified by dissolving it in ether and storing the solution in a refrigerator for some time. Impurities precipitate and can be filtered off. The surfactant still contains small amounts of byproducts as can be seen by chromatography. The interfacial tension of surfactant solutions against water was measured at room temperature by the method of spinning drop and ring pull out. The results are plotted in figure 4.

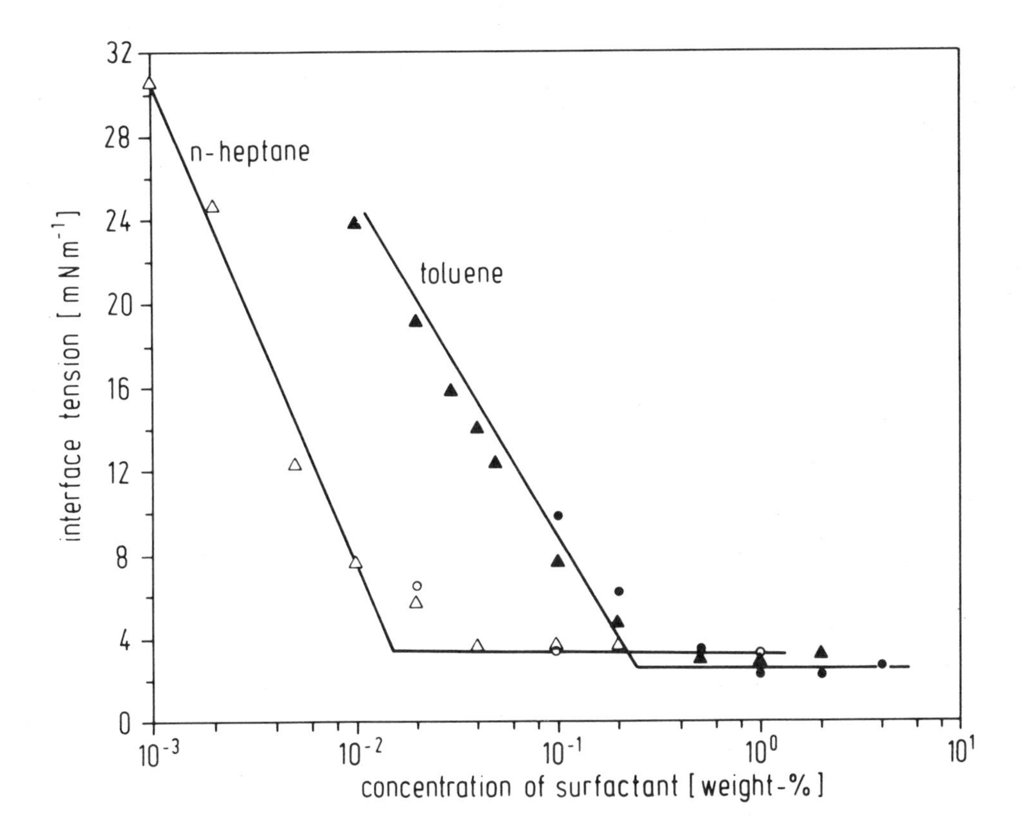

Fig. 4. Effect of sorbitan monooleate concentration on interface tension (oil-water) at room temperature. o spinning drop, Δ ring pull out.

The plotted values are initial values. The interfacial tension is changing with time reaching stationary values at about 2 hours after covering the water phase with the oil phase. As can be seen from figure 4 a constant interfacial tension is reached at surfactant concentrations of approximately 0.02 and 0.2 weight percent indicating the appearance of aggregates of surfactant. Looking at the apparent aggregation number of the surfactant in different solvents as determined by vapour pressure osmometry (see figure 5) one can see that the

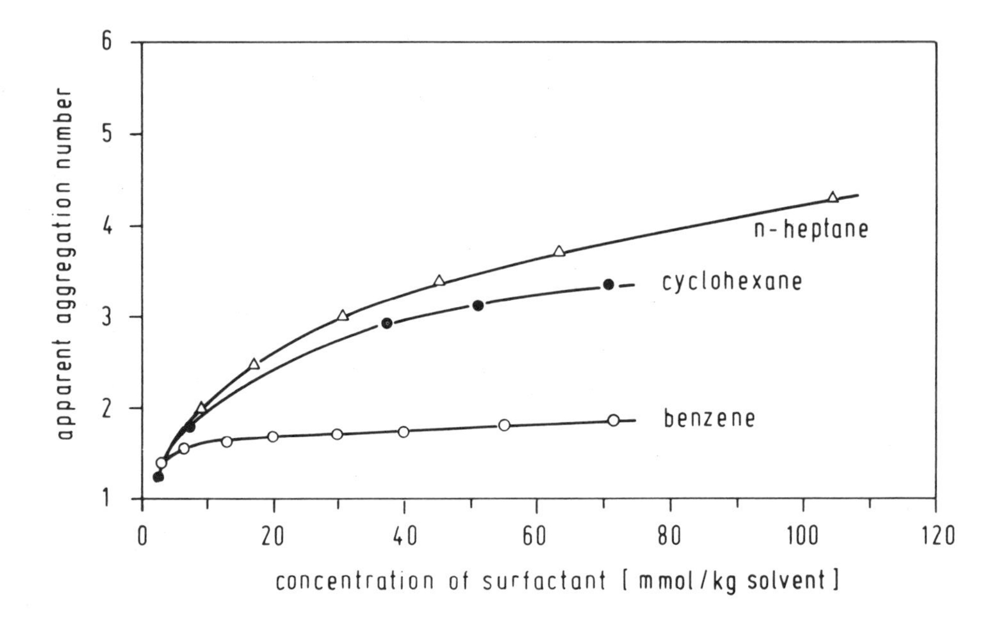

Fig. 5. Relation between aggregation number and concentration of sorbitan monooleate at 40 °C in different solvents (Kitahara et al.).

degree of aggregation is changing with concentration of surfactant reaching larger aggregation numbers in n-heptane and cyclohexane than in benzene. Dynamic light scattering measurements of sorbitan monooleate in benzene show that no scattering of the solution can be detected. This is not necessarily an indication that no aggregates are present but rather due to the fact that the difference of the refractive index increment of the two components is too small. In n-heptane solutions scattering can be observed and the size of the scattering centers are dependent on the concentration of water being present in the surfactant solutions. The size of scattering centers is changing from 5 nm at a molar ratio of surfactant to water of 10 to 280 nm at a ratio of 1 to 0.01. The aggregation number of the 5nm particles is about 90 ± 30. The large difference of this value to that measured by vapour pressure osmometry might be due to the dispersity of the particle size distribution and should be judged in view of the different methods leading to weight and number average molecular weights of the surfactant aggregates. In any case sorbitan monooleate tends to form aggregates in apolar media and the size of the aggregates is dependent on the concentration of water. At present nothing is known on the concentration of the aggregates at given conditions.

DISCUSSION

The polymerization of acrylamide in a disperse system consisting of a continuous oil phase (dilute solution of sorbitan monooleate and azoinitiator in isoparaffin) and a disperse water phase (concentrated aqueous solution of acrylamide) is assumed to take place predominately within the dispersed water droplets. This is likely because the solubilization of acrylamide in isoparaffin by sorbitan monooleate is very poor at polymerization conditions. Furthermore the rate of polymerization referred to volume of emulsion decreases with decreasing water phase volume, but is independent of water phase volume when rate is referred to the water phase.A schematic picture of the course of polymerization is presented in scheme 1.

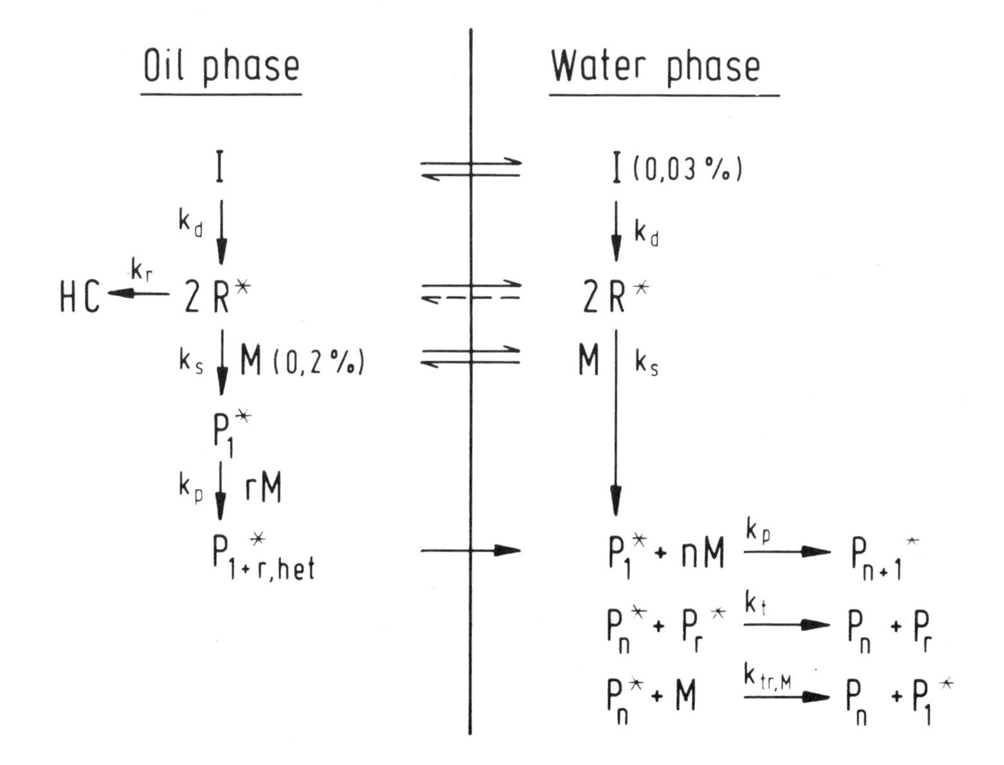

Scheme 1. Kinetic scheme of acrylamide polymerization in dispersion with oil soluble initiators and sorbitane monooleate as surfactant.

The start of polymerization seems to play an important role which is still largely unknown. The rate determining step is either a phase transfer process or an interface reaction with low activation energy, since it was found that the rate constant of polymerization is directly proportional to the interface area of the two phase systems and the true activation energy of polymerization is only 26 kJ/mol. It is to be assumed that free radicals are involved in this rate determining step either by initiating the polymerization in the interface or by crossing the interface. The first order dependency of initiator concentration is difficult to explain if one does not postulate a monomolecular termination reaction which seems to be unlikely because of the observed gel effect. It could be explained with the assumption of homogeneous nucleation in the oil phase but, since the particle size of the disperse phase seems to be constant at least at conversions larger than 20 %, this assumption may also be doubtful. Another explanation of the first order dependency would be the assumption that radical pairs are entering the water phase by some reasons hard to understand. In any case for a better understanding of the start of polymerization further studies must be made. In the case of acrylamide polymerization with oil soluble initiators and sorbitan monooleate as emulsifier in toluene as organic phase, faster rates, smaller particles and higher molecular weights were observed at comparable reaction conditions as given in table 3. In toluene as well as benzene no aggregates of sorbitane monooleate could be detected. The kinetic expression measured in this system is probably due to the subdivision effect occuring in the very small water particles as observed in toluene as organic phase.

For further references see PhD thesis of W. Baade at the Technische Universitaet Berlin in 1986.

REPORT OF DISCUSSIONS AND RECOMMENDATIONS

1 - Polymerization in inclusion compounds
(M. Farina)

Polymerization in inclusion compounds is a valuable method for obtaining polymers with a well defined structure, sometimes different from that obtained by conventional processes. Such polymers can be produced at a laboratory scale and are useful as reference standards and for the evaluation of structure-property relationships, especially when the absence of metal impurities is strictly required (electrical properties, thermal and oxidative degradation, etc...).
A potential application of this method concerns the synthesis of products having peculiar properties, such as electroconductive polymers.

Future research should be directed to the extension of the scope of the process (search of new hosts, new monomers, new initiators and new types of reaction) to the study of the stability of the solid phase and to a better knowledge of the polymerization mechanism.

2 - Radical polymerizations along macromolecular templates
(Y.Y. Tan)

On polymerizations along macromolecular templates one obtains a polymer-polymer complex (polycomplex) which is in fact a new material, its properties being different from those of the individual components (the template macromolecule and the daughter polymer). Though various systems have been studied by different groups over the world, there are still many problems to solve with respect to their kinetics and mechanism and the structural features (molecular weight and microstructure) of the daughter polymers formed.

Since a proper understanding of a template process depends greatly on the degree of monomer preadaptation on the template, its adsorption constant K_M should be determined, or at least qualitatively the degree of

M. Fontanille and A. Guyot (eds.), Recent Advances in Mechanistic and Synthetic Aspects of Polymerization, 335–339.

preferential monomer adsorption. When this is known, rate enhancement due to monomer ordering along the template (notably in so-called type I systems where K_M is very large) could be estimated after correction for the local monomer concentration using the K_M value. Determination of rate constants, k_p and k_t, is generally hampered by the formation of precipitates, due to insoluble polycomplexes. By determination of overall ativation entropies (besides activation enthalpies), S , one can probably also gain some impression of monomer ordering because S should increase. Monomer ordering is very important with respect to its effect on the microstructure of the daughter polymer.

The determination of structural features (molecular weight, microstructure) of the daughter polymer is highly necessary (also for type II systems where K_M is small) and better means should be sought to characterize the daughter polymer directly or indirectly. Since there is a trend for the daughter polymers to increase in syndiotacticity (type II systems), this should be optimalized, especially with such monomers from which no syndiotactic configuration is known as yet e.g. poly-N-vinylpyrrolidone, poly-N-vinylimidazole and poly-2-vinylpyridine.

To deepen insight in the mechanism of template polymerization recourse may be taken to a) the determination of radical concentration by means of ESR (from which k_p could be estimated) and b) the copolymerization either with a non-interacting or interacting comonomer (with the template). In this connection it should be remarked that the usual assumption that initiator decomposition is unaffected by the template may not always be valid (adsorption by template !). It is advisable to employ templates of narrow distribution when the template effect is dependent on template chain length.

From a practical point of view, it is of interest to study template polymerization by applying photo-initiation of undiluted template systems. Finally, instead of homopolymers as templates, the use of copolymers or crosslinked polymers as templates should be examined.

3 - Formation of polymers and copolymers of -hydroxy butyric acid in bacterial cells

(D.G.H. Ballard)

Work at ICI has demonstrated that bacteria can be identified which, under the correct conditions, can be induced to produce novel polymers or polymer intermediates. The following examples demonstrate the potential for producing novel organic materials using this approach.

* The formation of polymers and copolymers of β-hydroxybutyric acid in bacterial cells

Many bacterial cells need to store carbon as a potential source of energy. In some bacteria the specific mechanism for this storage process is to form a precursor of β-hydroxybutyric acid and then polymerize this to give poly(β-hydroxybutyric acid) (PHB). It is possible to manipulate the system so that up to 80 % of the weight of the cell is polymer. It is then relatively straightforward to isolate the granules of PHB. An additional range of products can be obtained by introducing propionic acid to the carbon feed to the fermentor. The biosynthesis of copolymers has been achieved by feeding bacterial monocultures with carefully specified carbon substrates. For examples, if alcaligenes eutrophus is feed with a mixture of acetic and propionic acids under conditions of nitrogen or phosphorus depletion, then a copolymer of (R)-3-hydroxybutyrate and (R)-3-hydroxypentanoate is produced. This gives control over such properties as Tg, Tm and the level of crystallinity.

These polymers are produced in the cell as granules which consist of microfibrils with diameters of less than a micron. The surface area of the granulus is thus very large. The polymer can be isolated in this microfibrillar form and has a range of applications in this form.

The properties of PHB and its copolymers resemble, in some respects, those of polypropylene with the added novelty that they are biodegradable and biocompatible in specific circumstances. ICI is presently developing these materials and they are now readily available.

* Significance of this work for the nature of Polymer Science

The main impact of these studies is to make synthetic polymer chemists aware of the potential of microbiological routes for the syntheses of intermediates for polymer synthesis. Moreover, the routes can also be used for the syntheses of specialty macromolecules where chirality and a high degree of functionality are required. The technical possibilities are limitless but commercial opportunities can only be identified by very careful analysis of the market place. It is evident, however, that these materials will mainly be used in the bio-medical area.

4 - Polymerized vesicles
(S.L. Regen)

Phospholipid and surfactant vesicles serve as important models for biological membranes as potential carriers of drugs, and as devices for photochemical energy conversion. Over the past seven years, considerable progress has been made in the synthesis of a wide variety of polymerizable lipids and surfactants and their use in the construction of polymerized vesicles. Standard polymerizable moieties that have been employed thus far include conjugated diacetylene, methacryloyl, conjugate diene, and methacrylamide groups. In order to maximize their utility for biomechanistic studies (membrane modeling) and for biomedical applications (drug delivery) polymerized vesicles should be prepared under the mildest conditions possible, so that sensitive comembrane and entrapped components can be incorporated without degradation. Both UV irradiation and thermal procedures that are now commonly used are not mild, and there is need for further synthetic development. Further synthetic efforts should be aimed at preparing polymerized vesicles under mild conditions yielding biodegradable backbones. More detailed insight needs to be obtained into the influence of polymerization on lateral diffusion stability and permeability of lipid bilayer vesicles.

5 -Inverse emulsion polymerization
(K.H. Reichert)

Free radical polymerization in inverse emulsion is a suitable technique for production of water soluble polymers with high molecular weights at high rates of polymerization. Polymerization in inverse emulsion can also be applied for synthesis of hydrogels with special properties.

The understanding of polymerization in inverse emulsion is still rather poor although much work has been done in this field since the pioneering work of Vanderhoff in 1962.

Kinetic and analytic data published so far indicate that the polymerization process is very complex and strongly depends on the chemical system used. Since the system is of heterogeneous nature, mass transfer processes can also play an important role.

For further work in this field, it would be advisable to study first the basic feature of the nonpolymerizing disperse system. Attention should be paid to the properties of surfactant in solution, to the partition equilibria of the different materials and to kinetics of mass transfer process.

V - CARBENES AND POLYMERIZATION

Chairman : K.J. IVIN

METAL CARBENE COMPLEXES IN POLYMER SYNTHESIS

Robert H. Grubbs and Laura Gilliom
Arnold and Mabel Beckman Laboratories of Chemical Synthesis
California Institute of Technology, Pasadena, California 91125 USA

ABSTRACT. Cyclic olefin can be polymerized with stable metallacycle and carbene complexes. It has been shown that

$$(C_5H_5)_2\overline{Ti-CH_2-CHR-C}HR$$

complexes can initiate polymerization of norbornene at 65°C to yield a living polymer that can be stored at room temperature. The resulting polymer is monodispersed, can be end capped and blocked with other cyclic olefins. This and related tungsten carbene complexes can be used to polymerize a variety of monomers to give polymers which interacts with physical, electrical, and complexing properties.

NEW CATALYSTS

Metal carbene complexes can be used to ring open polymerize cyclic olefins. It has been found that titanium metallacycles and tungsten carbene complexes can be used to ring open polymerize cyclic olefins in a living process.[1] Since these catalysts are acid free they can be used to polymerize a variety of new monomers that deactivated the more classical systems.

Kinetics and Mechanism of Polymerization of Norbornene

Bis(cyclopentadienyl)titanacyclobutanes prepared from the precursors of "$Cp_2Ti=CH_2$" (1) and "$Cp_2Ti=CHC(CH_3)_2CHCH_2$" (2) are catalysts for the polymerization of norbornene and norbornene derivatives.

M. Fontanille and A. Guyot (eds.), Recent Advances in Mechanistic and Synthetic Aspects of Polymerization, 343–352.

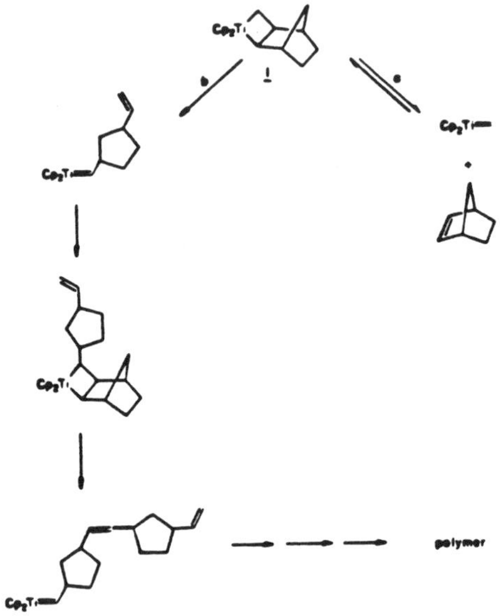

These metallacycles open on heating to 65° to yield a new carbene which is the propagating species. At room temperature, the trisubstituted metallacycle is stable and can be isolated. The kinetics of polymerization with both catalysts is zero order in monomer with $\Delta G^{\ddagger}_{.338} = 24$ kcal/mol, $\Delta H^{\ddagger} = 27$ kcal/mol and $\Delta S^{\ddagger} = 9$ eu. Since the initiation step using catalysts derived from (1) is the cleavage of a disubstituted metallacycle rather than a trisubstituted metallacycle as is the initiation starting with (2), the reaction starting with (1) shows an induction period. Initiating with (2) yields clean kinetics since both initiation and propagation both involve the decomposition of trisubstituted metallacycles.

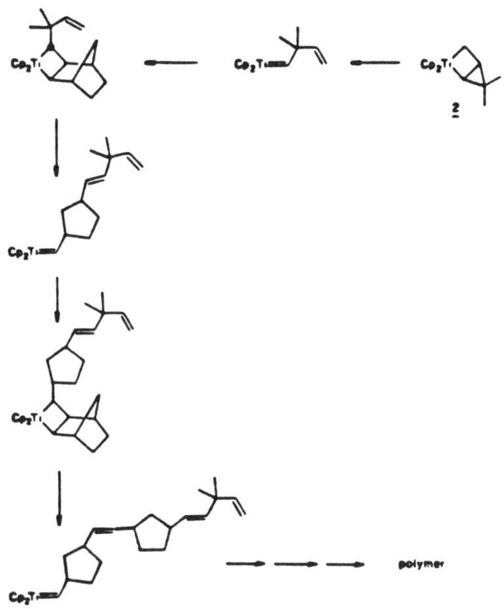

Molecular Weight Control

The molecular weight of the polymer increases with increasing conversion of monomer. A plot of Mn (polystyrene standard) vs conversion of the monomer is linear with a zero intercept. This observation conclusively demonstrates that this is a living system. When the polymerization is initiated with (2), the polydispersity at n = 50 is as low as 1.08. This system can also be used to prepare block copolymers. For example the following triblock has been prepared.

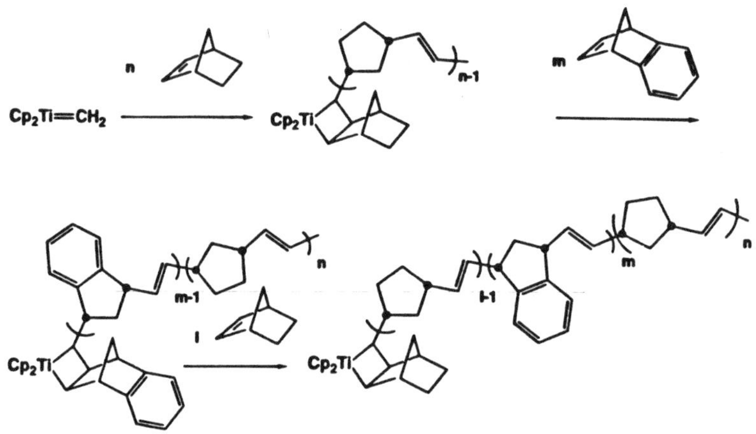

Functionalized Monomers

Other monomers can be polymerized using this catalyst. Polymerization of 3,4-diisopropylidenecyclobutene yields a transparent, highly soluble material that becomes conductive (10^{-3}(ohm cm)$^{-1}$) when highly doped with I_2.[3]

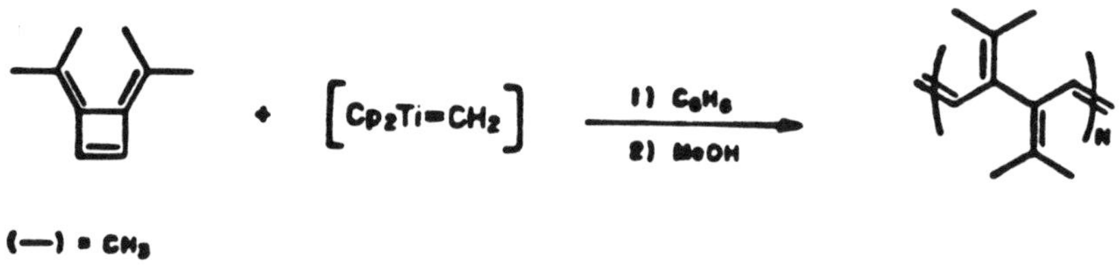

Basic monomers such as 7-oxa and aza benzonorbornadienes[4] can be polymerized to interesting new materials.

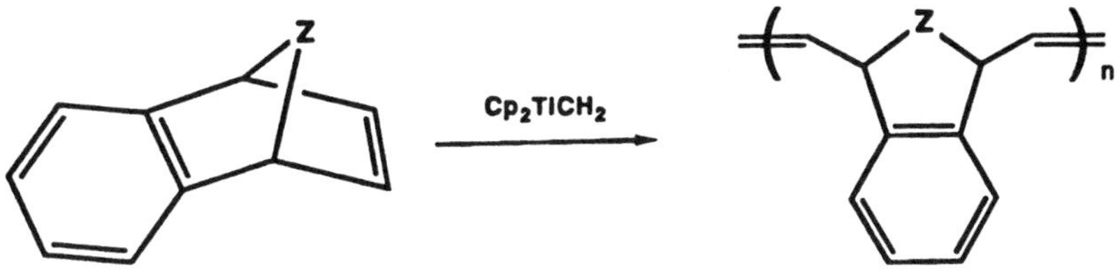

Cp$_2$TiCH$_2$

Z= O or NMe

In some cases the titanium system yields metallacycles that are too stable to initiate polymerization at temperatures where side reactions do not predominate. This is the case with the 7-oxa system shown below.[5] Tungsten alkylidene[6,7] complexes can be used to polymerize these systems. The polyfuran resulting from this reaction shows the ability to complex large organic and inorganic cations.[5]

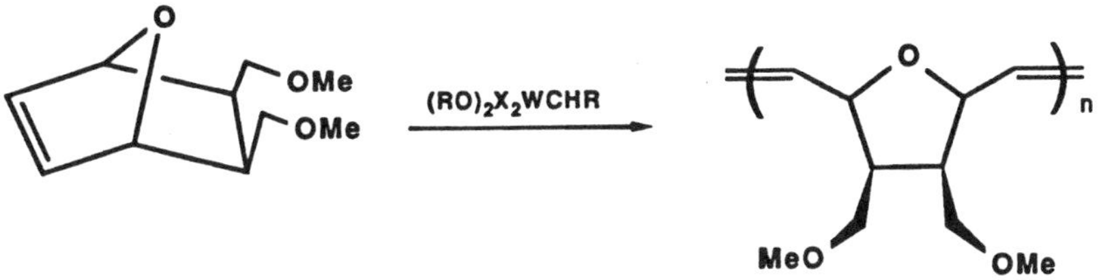

(RO)$_2$X$_2$WCHR

POLYMERIZATION OF CYCLOPENTENE

Catalyst Development

The metallacycle derived from addition of cyclopentene to"$Cp_2Ti=CH_2$" does not polymerize cyclopentene. The fact that quantitative loss of the olefin occurs on reduction of the resulting metallacycle with benzophenone shows that its cleavage does not yield any α-substituted carbene compound. No route to the chain carrying trisubstituted metallacyles exists from 1. The question of whether chain propagation requires high ring strain in the monomer was not addressed in studies with 1.

Initiation with **2**, however, would permit direct access to an α-substituted carbene and should circumvent the probelms of initiation with 1. The reaction of **2** with cyclopentene at 23° C was monitored by ^{1}H NMR spectroscopy. Exactly two equivalents of the olefin were consumed. This and the peaks characteristic of titanacyclobutanes observed in the cyclopentadienyl region (δ 5.36 and 5.30), the α-region (δ 0.44 and -0.21) suggest formation of metallacycle **3** as shown below. Detailed NMR assignment and stereochemical analysis were not performed.

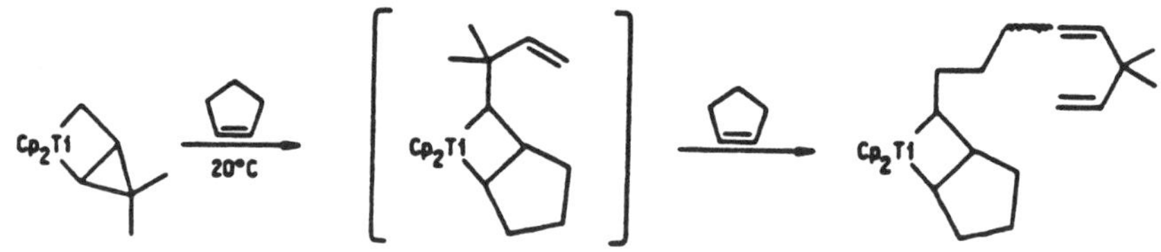

Trapping of Intermediates

Benzophenone reacted with **3** producing two major organic products. Although these products were not isolated, analogy with trapping experimetns on the norbornene metallacyle **1** and high resolution mass spectra support their formulation as olefins **4** and **5**. The formation of both

metallacycle cleavage products in a ratio of 3:1 indicates that, although the non-productive cleavage pathway predominates, reation in the direction of polymerization does occur.

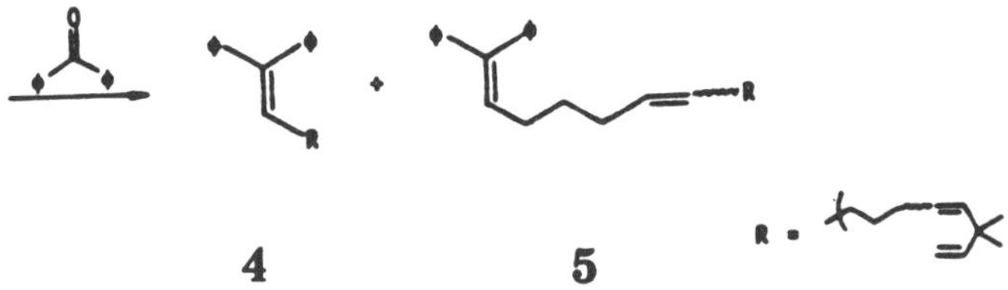

Nevertheless, when **3** was heated to 45° C in the presence of three equivalents of cyclopentene, the metallacycle decomposed without consuming any of the olefin. Subsequent polmerization of norbornene in the presence of the cyclopentene showned that **3** is active toward polymerization in the presence of strained olefins. No consumption of the cyclopentene occurred. These results are consistent with observations made in attempted cyclopentene polymerizations with a variety of other metathesis catalysts.[8] The large ΔH_0 for polymerization of strained olefin drives the reaction. Cyclopentene has a relatively low strain energy. At low monomer concentrations, the unfavorable ΔS_0 term overwhelms gain in enthalpy and polymerization is impossible.

Kinetics

Preliminary results indicate cyclopentene can be polymerized by titanocyclobutane **3** at high concentrations of monomer. Reaction of **3** with a 3.2 M solution of the olefin in benzene-d_6 was monitored by ^1H NMR spectroscopy of 40° C. Peaks attributable to ring-opened polymer[9] were observed The *cis:trans* ratio of 86:14 found for the isolated polymer is equal, within epxerimental error, to the expected thermodynamic *cis:trans* ratio of poly(1-pentylene). GPC analysis of the product polymer shows that its molecular weight is less than 2000 g/mol.

Although the conversion of monomer was low, the fact that this titanium system can polymerize cyclopentene is a significant and somewhat surprising result. The most common metathesis systems have low activation barriers. The activation barrier of 24 kcal/mol for the polymerization of norbornene with titanacycles **1** is high in comparison despite the relief of ring strain in the rate-limiting step. One factor which may contribute to the polymerizability of cyclopentene is the lowered energy of the intermediate α-substituted carbenes resulting from monosubstitution on the β-carbon rather than the disubstitution required in the polymerization of norbornene.

In contrast to the results of polymerizations with norbornene,[1] the thermodynamic ratio of double bond isomers was obtained rather than kinetic ratio. The reversibility of metallacycle formation as observed in the reaction with benzophenone and the probable reversibility of ring-cleavage may account for the observed *cis:trans* ratio. Formation of macrocyclic rings by reaction with double bonds in the polymer chain is also possible.[10]

Preliminary kinetic studies show that the polymerization of cyclopentene is 2nd order whereas the norbornene polymerization is 1st order in catalyst and 0 order in monomer. The polymerization of cyclopoentene, effected by catalyst2, appears to be thermodynamically controlled. More detailed studies of the effects of monomer concentration on the polymerization of cyclopentene are needed. In view of these preliminary results, studies of other less strained cyclic monomers such as cycloheptene and cyclooctene may prove useful.

These examples demonstrate the applications of the new catalysts that have resulted from organometallic synthesis in polymer synthesis. These results suggest that this area will provide a new level of control and flexibility in polymer synthesis.

REFERENCES

1. (a) L.R. Gilliom and R.H. Grubbs. *J. Am. Chem. Soc.* (1986) **108**, 733; (b) R.R. Schrock, J. Feldman, L.F. Cannizzo and R.H. Grubbs. *Macromolecules* submitted for publication.
2. L.R. Gilliom and R.H. Grubbs. *Organometallics* (1986) **5**, 721.
3. T.M. Swager and R.H. Grubbs. *J. Am. Chem. Soc.* (1987) **109**, 0000.
4. L. Cannizzo and R.H. Grubbs, unpublished results.
5. B. Novak and R.H. Grubbs, unpublished results.
6. J. Kress, J. Osborn, R. Green, K. Ivin and J. Rooney. *J. Chem. Soc., Chem. Commun.* (1985) 874.
7. C.J. Schaverier, J.C. Dewar and R.R. Schrock. *J. Am. Chem. Soc.* (1986) **108**, 2771.
8. E.A. Ofstead and N. Calderon. *MaKromol. Chem.* (1972) **154**, 21.
9. see K.J. Ivin, D.T. Laverty and J.J. Rooney. *MaKromol. Chem.* (1977) **178**, 1545.
10. J. Witte and M. Hoffmann. *MaKromol. Chem.* (1978) **179**, 641.

* Discussion of paper

(1) Polyfuran-type polymers can be obtained with a range of tacticities and cis/trans ratios by using different catalysts. The ability of the polyfuran to complex cations appears to depend more on the tacticity than on the cis/trans ratio. With analogous polyfuran-type polymers prepared some years ago (at 3M by Smid and Schwarz ?) from cis and trans polybutadiene, only the cis form was reported to have complexing ability.

(2) The polyacetylene films described in the abstract are all-trans, as shown by solid state NMR, in contrast to that made by Ziegler-Natta polymerization of acetylene which is cis and must be heated to convert it to trans. In the present case, the reaction is very exothermic so that some cis trans isomerization may occur during the course of the reaction. Alternatively there may be some sort of double bond shift along the chain, starting from the tungsten center, leading to all-trans product.

(3) The proposal that the Ti complex reacts via [Ti] = CH_2 is based mainly on mechanistic studies. The dimer and the PMe_3 adduct have also been made.

$$Cp_2Ti \longrightarrow Cp_2Ti \quad TiCp_2 \quad + 2$$

$$\xrightarrow{PMe_3} Cp_2Ti \begin{matrix} CH_2 \\ PMe_3 \end{matrix} \quad +$$

The question is whether the reactions of the Ti complex

with olefin involves prior dissociation of isobutene or whether the isobutene is displaced by the incoming olefin.

(4) Phosphines retard the polymerization of norbornene by the Ti catalyst, but do not stop it.

(5) It is possible to polymerize certain cyclic olefins containing ether, lactone or methyl amino groups in the ring, and even with two such groups in the one molecule. See abstract for examples of such bicyclic olefins.

(6) The α-oxa-norbornene system is very sensitive to Lewis acids. The propagating carbene complex whichy would be derived from this monomer would have a -oxygen which kills the Ti catalyst so that no polymer is obtained. Low molecular weight polymer was obtained with the Osborn catalyst in the absence of $GaBr_3$, but the catalyst appears to be killed by the monomer after a period of time. The Schrock catalyst is more stable. The monomer is sensitive to Lewis acid in the sense that it tends to open up to give allyl cations.

(7) Catalysis by $OsCl_3$, $RuCl_3$ and $IrCl_3$ is presumed to proceed via metal-carbene complexes but how these are formed is not known.

(8) Molecular weights in these metathesis polymerizations can be controlled by inclusion of 1-hexene which acts as a chain transfer agent.

(9) It should be possible to make difunctional initiators, thus providing a means of making telechelic polymers. If the monomer is bifunctional (i.e. contains two double bonds) then sometimes only one double bond will open, giving linear chains. In other cases, both double bonds will open and crosslinked polymers are formed.

(10) 2,3-disbustituted norbornenes have not been polymerized, but 2-methylnorobornene has been (Katz).

(11) Cyclic oligomers of norbornene and its derivatives have not normally been detected as by-products of ring-opening polymerization, although they have been observed by Reif and Hocker in the case of norbornene itself. It seems likely that the back-biting reaction to yield cyclic oligomers is knetically unfavorable relative to propagation, due to steric protection by the enchained cyclopentane rings. However, when the monomer concentration is kept very low (M/W = 3), it is sometimes possible to observe regeneration of the Osborn initiator after some time at room temperature ; this indicates a

back-biting reaction at the end of the chain to generate cyclic oligomer though the presence of the latter has not yet been proved directly (Ivin).

With W catalysts, Mw/Mn of the polymer recorded immediately after preparation is close to 1, but if the system is allowed to stand for some time before killing the value of Mw/Mn increases and the cis content changes indicating the occurence of slow secondary metathesis reactions, which may indicate the formation of cyclic oligomers. This does not, however, happen with the Ti catalyst.

These secondary reactions, including back-biting occur much more readily with monocyclic olefins. With cyclooctatetraene in concentrated solution, one makes polyacetylene but in dilute solution the product is benzene, formed by back biting.

It is also well known that when cis,trans-1,5-cyclodecadiene is polymerized, it tends to split out cyclohexene from the growing chain (Teyssié). Cyclohexene itself will only give cyclic oligomers at low temperature ; the critical concentration for the formation of long chain polymer is never attained (Patton 1986).

(12) Fluorocarbon ligands, as used by Schrock, change the basicity of the alkoxide and thus affect the reactivity at the W centre. Substituted phenoxide ligands can also be used (Basset). In this case, the polymers of norbornene derivatives are always tactic M(c/r, t/m) indicating a propagating species having a chiral metal centre. The Schrock catalyst, however, gives an atactic polymer indicating an achiral metal centre.

THE ROLE OF CARBENE COMPLEXES IN THE PHILLIPS OLEFIN POLYMERIZATION (SURFACE COMPOUNDS OF TRANSITION METALS, PART XXXI (1))

Hans L.Krauss, Erich Hums and Karin Weiss
Laboratorium fuer Anorganische Chemie der
Universitaet Bayreuth
Postfach 101251, 8580 Bayreuth, F.R.G.

ABSTRACT. Coordinatively highly unsaturated chromium(II) on silicagel surface - the typical active site for the catalytic olefine polymerization in the Phillips process - reacts at higher temperatures with olefins under <CH_2> transfer. This reaction can be rationalized by a sequence olefin-complex → carbene complex → metallacyclobutane; these intermediates allow to explain as well the observed isomerization reactions as start- and propagation steps of the polymerization. An irreversible transition from the metallacyclobutane to π-allyl-Cr(IV)-complexes terminates the activity of the sites, yielding alkanes and H_2 as reduction products.- Experiments at lower temperatures (< 120°C) show that this model is not appropriate to explain the "usual" polymerization, since oligomers with ±<CH_2> are not found and carbene complexes - made by reaction of the surface Cr(II) with diazocompounds - react with olefins to cyclopropane derivatives. Therefore we have to conclude that the carbene complex model may play a realistic role only in the higher temperature range, causing a redox termination of the catalytic reaction.

INTRODUCTION

Since the invention of the Phillips process - i.e. the polymerization reaction of ethylene with a supported chromium catalyst - by Hogan et al. (2), the mechanism of this reaction is a matter of controversy. The discussion includes the question of the nature of the catalytic centre, especially of the oxidation state of the chromium in the active sites: numbers from six down to two have been proposed, not to forget models in which the metal acts in groups of ions in different oxidation states (3). The "proof" of such a statement usually was the linear dependence of the reaction rate on the proposed parameter. Commonly accepted seems today that an original surface Cr(VI), bonded to the silica surface in monochromate units, is reduced in a first step of the reaction - be it by the olefin itself, by the solvent or by additional means like CO or aluminium alkyles (4). It is reasonable

M. Fontanille and A. Guyot (eds.), Recent Advances in Mechanistic and Synthetic Aspects of Polymerization, 353–362.

to assume that the reduced species is a "coordinatively unsaturated site" which offers a chance for the coordination of olefins. Cossee (5) postulated the propagation reaction then to proceed similar to the Ziegler-Natta process by insertion of a coordinated olefin into a somehow existing metal σ-alkyl bound; the terminating step would create another active center with a metal-hydrogen bond:

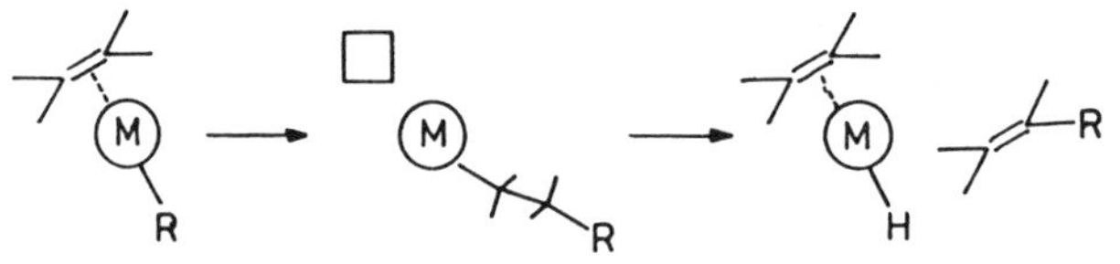

This model is widely accepted - but from where comes this one hydrogen atom to form R? Of course there are SiOH groups nearby which could act by oxidative addition, but a solid proof for this step was not delivered. In contrast, catalysts with a large number of SiOH groups (e.g. contacts on silica pretreated at T ≤ 500 °C) are less reactive than corresponding samples with low OH content (pretreatment temperatures of 800 or 850 °C (6)). M-H bonds were not proven experimentally. Obviously here lies one of the basic questions of the whole reaction. So we tried to elucidate this point in more detail, particularly under the aspect not to incorporate the formation of M-H or M-alkyl in the starting step.

MATERIALS, METHODS

To simplify the experimental basis, we worked with Cr(VI) impregnated silicagel without other metals as cocatalyst. For activation (dehydration of silica plus anchoring the chromate by two oxygens to the surface) the CrO_3/H_2O impregnated product (≤ 1 % Cr) was heated in a stream of oxygen to 800 or 500 °C. The mean oxidation number of the metal is then ≥ 5,5 and equals 6,0 at low Cr concentrations; some Cr(III) may be formed by thermal decomposition prior to anchoring the metal to the surface. (This Cr(III) shows to be inert to all further proceedings). - The reduction was carried out under controled conditions by means of either

- CO at higher temperatures (350 °C for products activated at 800 °C, 500 °C for products activated at 500 °C)

- olefins (preferentially ethylene and 1-octene) at different temperatures

- aluminum alkyles (AlR_3) or alumoxanes ($R_2AlOAlR_2$) with R = isobutyl.

The mean oxidation number of the metal is now between 2.2 and 2.0. The following polymerization was carried out with different monoolefins under a broad variation of the reaction conditions. If not mentioned in the following, see literature for experimental details.

RESULTS, DISCUSSION

The Catalyst

The characterization of the catalyst and its systematic variation was a first necessary step of the work. As described elsewhere (7) the relevant parameters for preparing different catalysts are Cr concentration, activation and reduction temperature. (Surprisingly, the special type of silicagel - "precipitated" or aerogel - did not work out as significant for the following experiments.)

While the Cr concentration was varied according to the later use, the activation was carried out at two standard temperatures as mentioned above. Except with CO, in the following reduction Cr(IV) was

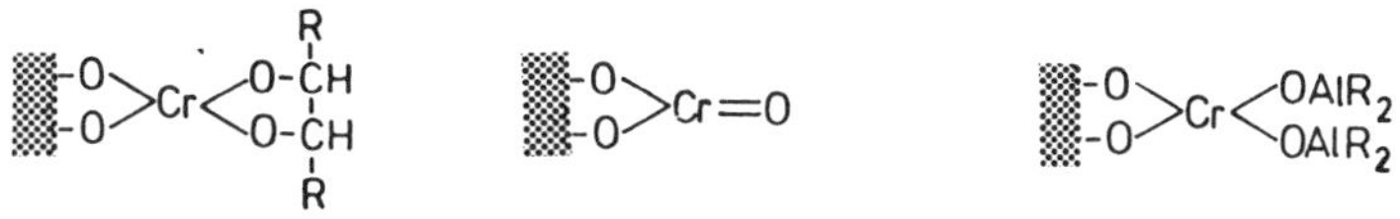

$(\equiv\text{-O})_2\text{Cr}(\text{O-CHR})_2$ $\qquad$ $(\equiv\text{-O})_2\text{Cr=O}$ $\qquad$ $(\equiv\text{-O})_2\text{Cr}(\text{OAlR}_2)_2$

found as a first intermediate, identified as by chemical means, magnetic measurements and reflection spectroscopy (4). The next step leads by a second 2e reduction to the surface Cr(II) *). Indeed, coordinatively unsaturated surface Cr(II) is formed as the predominant active species for the olefin polymerization (8). The property "coordinatively unsaturated" depends on the surrounding surface groups. Neighbouring SiOH especially reduces the activity considerably by occupying a third coordination site. Accordingly the active species form a "population profile" with more or less unsaturated (= more or less active) centres (7); the real catalysts can be represented by a mixture of the two species

- Cr(II)A: twofold coordinated to the surface
  pale green, highly reactive,
  predominating with 80 % in 800/350**) type catalysts

- Cr(II)B: threefold coordinated to the surface
  blue, less reactive
  predominating with 75 % in 500/500**) type catalysts

---

*) This oxidation state is reached directly with CO as reducing agent. A mean O.N. as low as 2.06 can be achieved.

**) Short for activation at T = 800, reduction in CO at T = 350 or activation at T = 500, reduction in CO at T = 500 ^{o}C respectively.

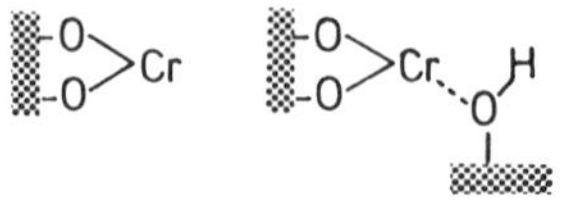

Cr(II) A   Cr(II)B

(A fourfold coordinated Cr(II) species was identified and characterized by Zecchina's group (9). Since this species does not react with olefins it shall not be included in the following considerations.)

Active surface chromium(III) can be prepared from the Cr(II) by reaction with liquid H_2O (complexation to surface $Cr(H_2O)_2$) followed by oxidation with O_2 and heating in inert gas to the former activation temperature (10). Other special redox procedures seem to give similar products (11). To follow the main stream in the following, we will discuss the Cr(II) type catalyst only. As well we shall omit a discussion of the support's influence in detail, but it should be mentioned that IR data show a considerable interaction of SiOH groups with the different olefins (12).

The Polymerization

The difficulty to characterize the single steps of the following polymerization reaction results partially from the fact that in the case of ethylene - the industrial case - a huge bulk of polymer, mostly crystalline already after very short time of propagation, is formed - obscuring e.g. the kinetic details (13). Higher alkenes were found to do no better if incorporated to a modest amount up to 5 % in copolymers with ethylene (14).

We tried two ways to overcome this problem. The first: while increasing the polymerization temperature we looked for low molecular weight products which might allow to trace back the way of their formation. (Of course there is a chance to collect and to identify just the products of those centers which are untypical for the real polymerization reaction.) Second, higher 1-olefins showed to be good candidates for a smooth polymerization with a well soluble polymer at all states of the proceeding reaction.

Following the first idea our experiments were carried out in a gas/solid system, using pulse technique at temperatures from 100 to 400 ^{o}C. Linear 1-olefins C_2 to C_{12} and the cyclic monoolefins C_5 - C_8 were reacted with the two standard catalysts. In all cases a puzzling multitude of low molecular weight products was formed (besides propagation to polymer with linear olefins), which can be related to two basic reactions (15) *):

---

*) For special cases, as with cyclohexene, see ref. (15).

- transfer of $\langle CH_2 \rangle$:
  methylation, demethylation, ring contraction;
  formation of low alkanes (see below)
- isomerisation:
  shift of double bond, branching.

The relative extent of the two reaction types depends primarily on the ratio Cr(II)A/Cr(II)B, the $\langle CH_2 \rangle$ transfer being preferentially connected with Cr(II)A, i.e. with high reactive catalysts. Furthermore, the first group of reactions decreases in first order, while the second does not seem to be weakened as a function of turnover.

It was tempting to construct a model attributing all these reactions to one single intermediate. And here the carbenes come into play. If we admit the formation of a carbene complex *) in the following manner and allow a further reaction with a second olefin molecule, we end up with a metallacyclobutane **). We should keep in mind

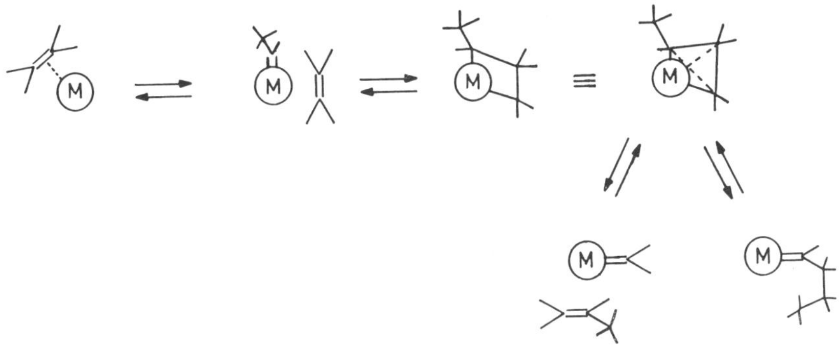

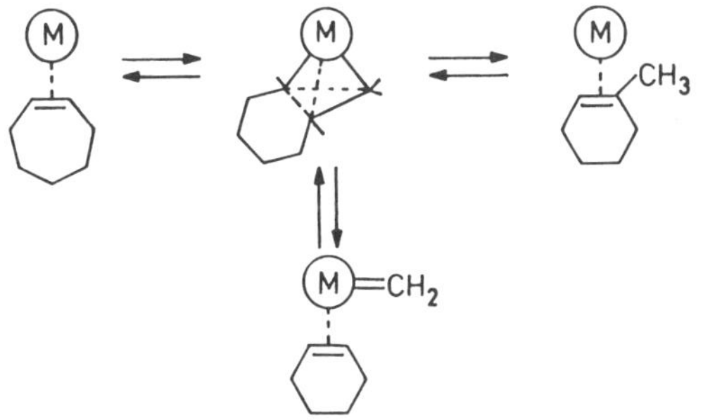

---

*) IR observations support the proposal of a carbene (alkylidene) complex as an intermediate of the reaction (16).

**) This pathway is analogous to the "non-pairwise" metathesis mechanism first proposed by Herisson and Chauvin (17).

that these metallacyclobutanes probably have a non-planer structure, as - for a tungsten compound - found by X-ray analysis by R.R.Schrock (18). Therefore an interaction of the β-carbon with the metal is not unlikely. By different ways of splitting the MC_3-unit we come to all the products found in the experiments (15). Here a ring contraction is given as example. 1-Olefins can act as a $\langle CH_2\rangle$ source toward other olefins in the presence of the catalyst: cyclopentene (which shows no ring contraction) is methylated by 1-pentene.

Although the model gives a plausible explanation for all the products observed, it is necessary to mention that only Cr(II)A is responsible for the first group of reactions (transfer of $\langle CH_2\rangle$) and that it is also this group which requires a carbene complex intermediate.

Redox-Reactions

As mentioned above the first group of reactions includes the formation of the lower alkanes C_1 to $^nC_5$ from all those olefine reactions which include the carbene complex step. Obviously the alkanes are produced from $\langle CH_2\rangle$ units; the support can act at least partially as source of the hydrogen since catalysts on deuterated silica give a certain amount of mono- or bis-deuterated alkanes. But where are the equivalent oxidation products? The step $\langle CH_2\rangle \rightarrow CH_4$ would correspond to an oxidation of Cr(II) → Cr(IV). Indeed Cr(IV) is formed and found: at a reaction temperature of 300 °C, ethylene produces mainly short chains (19) with π-allylic bond to the chromium. The transition from MC_3 units to π-allyl complexes is well known (20). The hydrogen, as far as not used up to form the alkanes from $\langle CH_2\rangle$, is evolved as H_2 (identified by mass spectroscopy). Of course the oxidized Cr is inactive for all reactions which require the carbene complex intermediate, including propagation.

It would have been difficult to prove the formation of the allylic unit, were not a special property of the complex: it can be split off as a whole from the surface by action of protons, e.g. with HCl gas, and dissolved in methanol, acetone, THF a.s.o., forming deep blue solutions. Since the chains can be kept rather short, it is easy to apply the usual methods of characterization to these solutions. With ethylene/HCl/methanol the analysis gave the formula

$CrL_2S_xCl_2$ with $L = C_3H_4\text{-}C_nH_{2n+1}$ and

S = solvent ($x \rightarrow 0$ if kept in high vacuum at 100 °C).

The complex is monomeric in benzene and dissociates in polar solvents giving a cationic chromium unit. The allylic nature of L was established by IR and H-NMR (19). The oxidation number 4,0 of the metal (titration) was confirmed by magnetic measurements (SQUID; high spin d^2) (21).

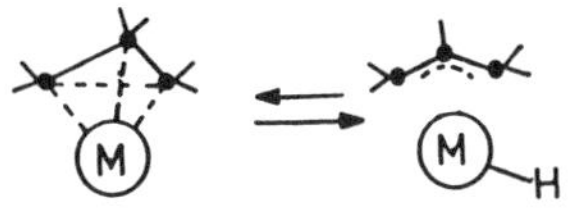

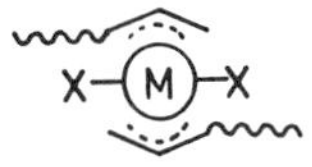

X= OSi< , Cl

It seems important to mention, that it is only the Cr(II)A species which is oxidized to Cr(IV) by the olefins; on the other hand by this way the whole Cr(II)A content of a given catalyst can be transferred to the π-allylic Cr(IV) complex and removed from the surface in form of the "blue solution".

Let us summarize the results so far. We can explain the reaction behaviour of surface Cr(II)/olefin by the assumption of carbene complexes/metallacyclobutanes as common intermediates for all products. The further proceeding of the reaction is due to different parameters, including the Cr(II)A/Cr(II)B character of the centres, the reaction temperature, the turnover a.s.o. As a consequence, with a monomer C_nH_{2n}

polymers (oligomers) with $(C_nH_{2n})_x(CH_2)$ beside $(C_nH_{2n})_x$

should be found, if the model is valid for the "normal" propagation. The second approach mentioned above should give an answer.

Polymerization of Higher Olefins

It was found that higher 1-olefins react at room temperature with surface Cr(II) smoothly to form comb polymers; the yield is almost 100 % for the monomers C_3 to C_{12} (22). This reaction opens an excellent way to a closer look on the polymerization, since - besides polymer with a typical WM of 10^4 to 10^5 - some oligomers are produced which allow a detailed (chromatographic) analysis.

All the polymers are monoolefins (sterically irregular). The molecular weight and its (very broad) distribution is found to be constant over the whole period of reaction - already in the very first minutes of turnover. Up to 1 % Cr content the specific activity of the centres is constant. The reaction rate does not depend on the chain length of the monomer; since there is no discrimination of the different incoming 1-olefins at the active centre, any co-polymerization works. Reaction rate and yield are dependend on parameters like catalyst quality (800/350 better then 500/500) and reaction temperature. CO and other good complex ligands for surface Cr(II) are competitive inhibitors; H_2 does not influence the molecular weight but works by hydrogenation of the monomer - at room temperature and higher pressure with yields $>$ 95 % (23).

The polymerization reaction is first order in monomer concentration (steady state polymerization experiments). There shows to be an induction period if the reactions is carried out as a batch run - a phenomenon which is repetitive with new addition of monomer to the same catalyst after the completion of the preceding run.

A similar induction period is observed with ethylene polymerization under pressure (batch type experiment) whereas here the apparent maxima and the declining part in the rate vs. time plot should be artefacts due to the formation of bulk polyethylene (13). It seems likely that the polymerizations of ethylene and of higher olefins, e.g. 1-octene proceed basically the same way. Corroborating this assumption, both olefins give "blue solutions" under appropriate reaction conditions. If we assume now the carbene complex/metallacyclobutane mechanism as a basis for the 1-octene polymerization (which is performed at room temperature as mentioned!), we should be able to observe oligomers with uneven C numbers. This was not the case. So we have to conclude that our model does not apply in the low temperature range.

The temperature dependence of the $\langle CH_2 \rangle$ transfer reactions (24) showed indeed, that with standard catalysts at $T < 120$ °C this pathway may be neglected, however gaining in weight linearly with increasing temperature. Since the irreversible redox reaction (formation of Cr(IV)) is linked with the assumed MC_3 intermediate, we may expect a more or less pronounced decrease of the conversion at Cr(II)A centres, leaving in so far finally an inactive catalyst at $T > 300$ °C already after a moderate turnover.

This leads to the question if at low temperatures olefins react with the catalyst via carbene complexes at all. Therefore it was interesting to look for the behaviour of complexes where a surface Cr(II) is bond to a carbene ligand independently from olefin reactions.

The Reactions of Carbene Complexes of Surface Chromium(II)

We started with a highly active catalyst (800/350) and produced the carbene complex (25) at 0 °C by reaction with diazoacetic ethylester, which is stable up to 150 °C in abscene of the catalyst. The behaviour of the carbene complex was studied via two reactions, which are characteristic for carbene ligands: the formation of cyclopropanes by 1+2 cycloaddition of olefinic double bonds and the Wittig-analogue reaction with ketones (exchange of oxo vs. alkylidene groups).

For the first type of experiments (25), the carbene complex was reacted in pentane suspension at temperatures from 0-20°C with methylstyrene or internal olefins; the corresponding cyclopropanes were isolated in good yield. If linear 1-alkenes are used, oligomeric and polymeric products are formed beside the simple cyclopropanes; by GC-MS analysis the oligomeric products were identified as poly-1-alkanes and as cyclopropanes formed by reaction of the carbene ligand with the poly-1-alkenes. Without addidion of alkenes the reaction of surface Cr(II) catalyst with diazoacetic ethylester at 0-20 °C gave only trimerisation products of the carbene ligand at 0-20 °C.

For the second type of experiments (25), diazoacetic ethylester and benzophenone were reacted with the catalyst in pentane suspension at temperatures from 0-20 °C. By a Wittig-analogue reaction 1-carbethoxy-2,2-diphenylene was formed which was identified by GC and GC/MS methods.

Both experiments are solid proof for the formation of a carbene complex intermediate from the diazo compound and the catalyst and its reactivity at low temperatures. Since with olefins alone no trace of a cyclopropane was observed, we have to conclude, that at moderate temperatures carbene complexes are not formed from Cr(II)A plus olefin.

Summarizing these results, we have to realize that carbene complexes will probably not play the role of a precursor in the catalytic reaction at lower to moderate temperatures, i.e. in a temperature range where the commercial polymerization process usually is carried out. On the other hand, at higher temperatures the sequence carbene complex → metallacyclobutane may become important - not only by yielding <CH_2> transfer products, but especially by desactivating the catalyst via the irreversible step to Cr(IV) and alkane/H_2.

ACKNOWLEDGEMENT

This work was supported by the Deutsche Forschungsgemeinschaft, Sonderforschungsbereich 213, and by the "Fonds der Chemischen Industrie".

REFERENCES

(1) Part XXX: E.Hums and H.L.Krauss, Z.anorg.allg.Chem. **537**, 154 (1986)

(2) A.Clark, J.P.Hogan, R.Banks and W.Lanning, Ind.Eng.Chem. **48**, 1152 (1956)

(3) P.Hogan, J.Polym.Sci. **8**, 2637 (1970)
K.G.Miesserov, J.Cat. **22**, 340 (1971)
D.D.Eley, C.H.Rochester and M.S.Scurrell, Proc.R.Soc.London, Ser. A, 329, 361 (**1972**)
P.Spitz, A.Revillon and A.Guyot, J.Cat. **35**, 335 (1974)
Yu.Yermakov and V.Zaccharov, Adv.Catal. **24**, 173 (1975)

(4) H.L.Krauss and H.Stach, Inorg.Nucl.Chem.Letters **4**, 393 (1968); Z.anorg.allg.Chem. **366**, 280 (1969)
L.M.Baker and W.L.Carrick, J.Org.Chem. **33**, 616 (1968)
M.P.McDaniel, J.Cat. **67**, 71, (1981)
U.S.Pat. 662044 (**1968**), 3959178 (**1974**)
Dissertation A.Schimmel, Universitaet Bayreuth, F.R.G., **1985**
H.L.Krauss and B.Hanke, Z.anorg.allg.Chem. **521**, 111 (1985)

(5) L.L. van Reijen and P.Cossee, Discussions Faraday Soc. **41**, 277 (1966)

(6) H.L.Krauss, B.Rebenstorf and U.Westphal, Z.anorg.allg.Chem. **414**, 97 (1975)

(7) H.L.Krauss and U.Westphal, Z.anorg.allg.Chem. **430**, 218 (1977)

(8) M.P.McDaniel, in ACS Symp. "Transition Metal Catalysed Polymerizations", Midland **1981**

(9) E.Garrone, G.Ghiotti, S.Coluccia and A.Zecchina, J.Phys.Chem. **79**, 984 (1975)
B.Fubini, G.Ghiotti, L.Stradella, E.Garrone and C.Morterra, J.Cat. **66**, 200 (1980)

(10) Dissertation D.Naumann, Freie Universitaet Berlin, **1979**

(11) M.P.Welch and M.P.McDaniel, J.Cat. 82, 110 (1983)

(12) K.Weiss and H.L.Krauss, 7th Intern.Symp.Olef.Metathesis and Polymerisation, Hull (England), **1987**, to be published

13) Dissertation B.Hanke, Universitaet Bayreuth, F.R.G., **1983**

(14) Belg.Pat. 609750 (**1962**)

(15) H.L.Krauss and E.Hums, Z.Naturforsch. **34b**, 1628 (1979); ibid. **35b**, 848 (1980); ibid. **38b** 1412 (1983)

(16) G.Ghiotti, E.Garrone, S.Coluccia, C.Morterra and A.Zecchina, J.Chem.Soc., Chem.Comm. 1032 (**1979**)

(17) J.L.Herisson and Y.Chauvin, Makromol.Chem. **141**, 161 (1970)

(18) R.R.Schrock, J.Organometall.Chem. **300**, 249 (1986)

(19) Dissertation K.Hagen, Universitaet Bayreuth, F.R.G., **1982**; H.L.Krauss, K.Hagen and E.Hums, J.Mol.Cat. **28**, 233 (1985)

(20) M.Ephritikhine, M.L.H.Green and R.E.Mackenzie, J.Chem.Soc., Chem.Commun. **1976**, 619 and 926

(21) G.Bayreuther, R.Hoepfl and H.L.Krauss, to be published

(22) K.Weiss and H.L.Krauss, J.Cat.**88**, 424 (1984)
Dissertation G.Langstein, Universitaet Bayreuth, F.R.G., **1986**

(23) H.L.Krauss and B.Janocha, to be published

(24) H.L.Krauss and G.Zeitler-Zuerner, to be published

(25) K.Weiss and K.Hoffmann, J.Mol.Cat. **28**, 99 (1985)

STEREOCHEMISTRY AND INTERMEDIATES IN THE RING-OPENING POLYMERIZATION OF NORBORNENE AND ITS DERIVATIVES

J. Kress[a], J.A. Osborn[a], K.J. Ivin[b] and J.J. Rooney[b]

a) Institut Le Bel
Université Louis Pasteur
4 rue Blaise Pascal
67000 Strasbourg, France

b) Department of Chemistry
Queen's University
Belfast BT9 5AG, U.K.

ABSTRACT. In the ring-opening polymerization of norbornene (NBE) and its derivatives, catalysed by $W(=CHCMe_3)(OR)_2Br_2/GaBr_3$ or $W[=\overline{C(CH_2)_3CH_2}](OR)_2Br_2/GaBr_3$ ($R = CH_2CMe_3$ or $CHMe_2$) in CD_2Cl_2 at 220-250K, ^1H and ^{13}C n.m.r. spectra reveal the presence of both tungstenacyclobutane complexes X and tungsten-carbene complexes P as chain carriers. In some systems one sees distinct spectra for X_1 and X_n ($n \geqslant 2$) or for P_1 and P_n ($n \geqslant 2$), where n denotes the number of monomer units added to the initiating carbene complex. When the monomer is unsymmetrically substituted, as in endo-5-methylnorbornene (endo-5-MNBE), both head and tail species are seen for X_1 (X_{1H} and X_{1T}) and P_1 (P_{1H} and P_{1T}), and also for X_n (X_{nH} and X_{nT}) and P_n (P_{nH} and P_{nT}), where H and T refer to the species in which the methyl substituent is closer to, and further away from, the tungsten centre respectively. X_{1T} is considerably more stable than X_{1H}, having a relatively long life (several hours) at 235K, but converting to P_{1T} at 250K. With cyclopentylidene as carbene ligand, P_1 is seen first, followed by X_n. The non-appearance of X_1 is attributed to a destabilising effect of the spiro structure at the α-carbon of the tungstenacyclobutane. With $R = CHMe_2$, X_1 but not X_n is seen, so that the reaction appears to be propagated by a very small number of centres. The tungsten-carbene complexes also initiate polymerization in the absence of $GaBr_3$ but the temperature must be raised by about 30-40° to obtain activity which is comparable with systems in which $GaBr_3$ is present. In this case, neither X_1 nor X_n are detected, but P_1 and P_n are again observed.

1. INTRODUCTION

It was first proposed in 1970 that olefin metathesis reactions proceeded by a chain mechanism involving metal-carbene and metallacyclobutane complexes as chain carriers [1]. One complete catalytic cycle for the metathesis polymerisation of bicyclo[2.2.1]hept-2-ene (norbornene, NBE), according to this mechanism, is represented by eqn. (1), where P_{n+1} and X_{n+1} denote metal-carbene and metallacyclobutane complexes, respectively, containing n+1 monomer units.

M. Fontanille and A. Guyot (eds.), Recent Advances in Mechanistic and Synthetic Aspects of Polymerization, 363–373.

(P_n) + (M) → (X_{n+1}) → (P_{n+1}) (1)

In the ensuing years much indirect evidence has been adduced in favour of this mechanism, to the extent that it has become generally accepted even when the precise nature of the initiating complex and the mechanism by which it is generated from the catalyst/cocatalyst/monomer system has remained obscure [2]. In the past few years the evidence for this mechanism has been strengthened still further by the discovery of well-defined metal-carbene [3-7] and metallacyclobutane [7-8] complexes which will catalyse these reactions. In several cases one or other of these two types of chain carrier was detected and characterised, but their simultaneous presence and their interconversion was never observed. However, we have recently reported that the use of the catalyst system $W(=CHCMe_3)(OCH_2CMe_3)_2Br_2/GaBr_3$ in CD_2Cl_2 at 220-250K, with the monomer norbornene, allows one to observe successively by ^1H and ^{13}C n.m.r. (1) the conversion of the initial metal-carbene complex (I) into an initial metallacyclobutane complex (X_1) by addition of one equivalent of monomer, (2) the orthogonal cleavage of X_1 to yield P_1, and (3) the formation and interconversion of the subsequent carriers X_n and P_n by reaction of further monomer [9].

We have now extended this work to include systems with and without $GaBr_3$, and in which the carbene ligand of the initiator was changed from $(=CHCMe_3)$ to $[=\overline{C(CH_2)_3CH_2}]$, the alkoxy ligands from (OCH_2CMe_3) to $(OCHMe_2)$, and the monomer from norbornene (NBE) to endo-5-methylnorbornene (endo-5-MNBE).

2. RESULTS AND DISCUSSION

2.1. The system $W[=\overline{C(CH_2)_3CH_2}](OCH_2CMe_3)_2Br_2/GaBr_3/NBE$ (1/1/3)

At 210K the initiator concentration fell to half its initial value ($[I]_o = 3.8x10^{-2}$ M) in 21 min while the monomer concentration was falling to 80% of its initial value ($[M]_o = 11.4x10^{-2}$M). The first product to be detected was P_1, characterised most readily by the doublet from the carbene proton at 12.47 ppm. Other readily identified peaks are listed in Table I.

After 70 min at 210K the concentration of P_1 species reached a shallow maximum ($[P]_{max} = 2.2x10^{-2}$ M) and then began to fall very slowly. At the same time the species X_n ($n \geqslant 2$) began to appear, characterised in particular by two overlapping doublets at 6.92 ppm and 6.87 ppm, J = 10 Hz. These doublets (Table II) were assigned to trans and cis isomers respectively on the basis of the trans/cis ratio (60/40) determined from the polymer spectrum at a later stage of the reaction.

TABLE I. ^1H n.m.r. assignments for P_1, δ ppm (numbering as for X_n, Table II).

Proton	δ ppm
H_2	12.47 (d,1H)*; $J_{1,2}$ = 10 Hz
H_1	5.11 (m,1H)
H_3	5.04 (d,1H)*
$H_{9,9'}$	4.44, 4.42 (s,2H)
H_{11}	2.14 (m,4H)
H_{12}	1.55 (m,4H)
H_{10}	0.99 (s,18H)

* These peaks were absent when NBE-2,3-d_2 was used as monomer.

On warming the reaction mixture to 225K the initiator disappeared completely after 90 min, $[X_n]$ built up to 3.2×10^{-2} M, [P] declined to 0.6×10^{-2} M while [M] fell to 3×10^{-2} M. The sum total of $[I] + [P] + [X_n]$ remained constant at 3.8×10^{-2} M throughout the reaction. Peaks identified as belonging to X_n are listed in Table II.

TABLE II. ^1H n.m.r. assignments for X_n, δ ppm

Proton	δ ppm	
H_2	t 6.91 (d) / c 6.85 (d)	1H*
$H_{9,9'}$	c 4.98, 4.52 / t 4.98, 4.54 / c 4.84, 4.30 / t 4.81, 4.30	4H** (AB quartets, J_{AB}=12 Hz)
H_8	4.94 (t,1H)*	
H_4	t 3.53 / c 3.50	(m,1H)
H_1	3.31 (m,1H)	
H_{11}	2.14 (m,4H)	
$H_{10,10'}$	0.92 (s,9H) / 0.85 (s,9H)	
H_3	0.73 (t,1H)*	

trans-X_n (n=2) : 60%

cis isomer (at bond a) : 40%

* These peaks were absent when NBE-2,3-d_2 was used as monomer.

** These peaks were absent when OCD_2CMe_3 was used as ligand, enabling clear observation of the H_8 signal.

Note that the precise ligand environment described in Tables I and II for gallium bromide adduct P_1 and cationic X_2 is also in agreement with the n.m.r. spectra of these species and has been discussed previously for the corresponding complexes obtained using $W(CHCMe_3)(OCH_2CMe_3)_2Br_2/GaBr_3$ as catalyst [9].

On raising the temperature still further to 235K the residual monomer was completely polymerized in 20 min, while X_n was gradually replaced by P over a period of a few hours.

The reaction was also followed by ^{13}C n.m.r., with similar results. The principal downfield assignments to P and X_n are shown in Table III. These are consistent with those obtained for the $W(=CHCMe_3)(OCH_2CMe_3)_2Br_2/GaBr_3$-initiated system and were confirmed by gated decoupling experiments.

TABLE III. Downfield ^{13}C n.m.r. shifts for P and trans-X_n (M = NBE), δ ppm (numbering as for ^1H n.m.r., tables I and II)

P			trans-X_n ($n \geqslant 2$)		
316.2	(J_{CH}=140 Hz)	C_2	154.2	(J_{CH}=155 Hz)	C_2
141.3		=<u>C</u>⬡	143.0		=<u>C</u>⬡
			134.5		C_8
134.4-131.0		-<u>C</u>H=<u>C</u>H-	133.0-131.0		-CH=CH-
124.6		-<u>C</u>H=C⬡	122.9	(J_{CH}=145 Hz)	-<u>C</u>H=C⬡
90.5, 90.2		$C_{9,9'}$	91.1, 90.8		$C_{9,9'}$

The observations on the rise and fall of the intermediate species are readily interpreted in terms of Scheme 1, in which the $GaBr_4^-$ ion is envisaged to move out of the coordination sphere for the conversion of P into X and back in again for the cleavage of X into P, as described before [9].

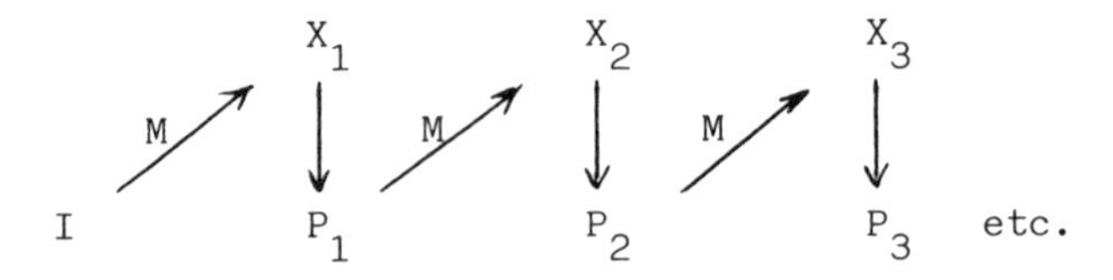

Scheme 1. Mechanism of addition polymerization by metathesis.

The fact that X_1 is much less stable than X_n ($n \geqslant 2$), and is not observed in this system, may be attributed to the spiro structure at the C_8 carbon (see Table II) for X_1. This will undoubtedly introduce some additional strain into the metallacycle, compared with that when the substituent is CMe_3 [9]. The relative rate constants for P_n + M $\rightarrow$

X_{n+1} and $X_{n+1} \longrightarrow P_{n+1}$ are evidently such that $[X_n]$ is much greater than $[P_n]$ at 225K while the reaction is progressing.

2.2. **The system** $W[=\overline{C(CH_2)_3CH_2}](OCH_2CMe_3)_2Br_2$**/NBE** (1/5)

Polymerization with this system required a somewhat higher temperature than that in the presence of $GaBr_3$ (section 2.1). However, it proceeded at a reasonable speed at 240K, where, with $[I]_o = 6.0 \times 10^{-2}$ M and an initial ratio $[M]_o/[I]_o$ of about 5, half the initiatior reacted in about 46 min. The first product species to appear was P_1, characterised in particular by the carbene proton doublet at 11.58 ppm (J = 10 Hz), and reached a maximum concentration of about 3.4×10^{-2} M after 130 min. About 10 min after the beginning of reaction, upfield shoulders developed on the side of the original P_1 doublet and gradually replaced it, more quickly when the temperature was raised to 250K. This upfield doublet (11.52 ppm, J = 10 Hz) is assigned to P_n ($n \geqslant 2$). On raising the temperature still further to 260K for 60 min consumption of initiator was complete, the monomer concentration had fallen to 30% of its initial value and P_n had become the dominant product species : 4.9×10^{-2} M compared with $[P_1] = 1.1 \times 10^{-2}$ M. The sum total of carbene species, $[I] + [P_1] + [P_n]$ remained constant at 6.0×10^{-2} M throughout the reaction. The trans/cis ratio in the product was about 60/40.

No X species was detected at any stage of the reaction. The hexacoordinate neutral tungstenacyclobutane intermediates expected in this case thus seem less stable than the pentacoordinate cationic species formed in the presence of $GaBr_3$. The higher temperature required to bring about initiation and propagation may however also be responsible for this lack of observation of X.

2.3. **The system** $W(=CHCMe_3)(OCHMe_2)_2Br_2/GaBr_3$**/NBE** (1/1/2)

In this system the monomer disappeared so fast that it was already 70% consumed by the time the first spectrum could be taken at 220K. The remainder of the monomer disappeared according to a first order law, $t_{1/2}$ = 3.9 min, reaction being complete in another 20 min. [I] fell from 6.0×10^{-2} M to a limiting value of 2.95×10^{-2} M during this period. At the same time, the metallacyclobutane X_1, characterised in particular by a doublet at 6.70 ppm, built up to a maximum concentration of 2.85×10^{-2} M. Only a trace of P (doublet at 12.54 ppm) had formed at this stage (0.2×10^{-2} M). The sum total of [I] + [X] + [P] remained at 6.0×10^{-2} M throughout the reaction. On raising the temperature X_1 was converted slowly into P_1 at 235K ($t_{1/2} \simeq 50$ min) and more rapidly at 250K. X_n was not observed. Note that the observed X species was readily identified as X_1 rather than X_n from the fact that its H_8 signal was a doublet, not a triplet [9]. This signal was here well separated from those due to the alkoxy protons $H_{9,9'}$, making the assignment (Table IV) easier than in the case where the ligands are OCH_2CMe_3 [9].

The fact that X_1 is observed but not X_n is the complete opposite of that described for the system in section 2.1. It means that most of the P species seen at 250K are P_1 and that the polymerization reaction has been carried by only a very small percentage of the original tungsten-carbene complex. Experiments are in hand to check that the molecular weight of the polymer is as high as would be predicted by this observation.

TABLE IV. ^1H n.m.r. assignment of X_1*, δ ppm

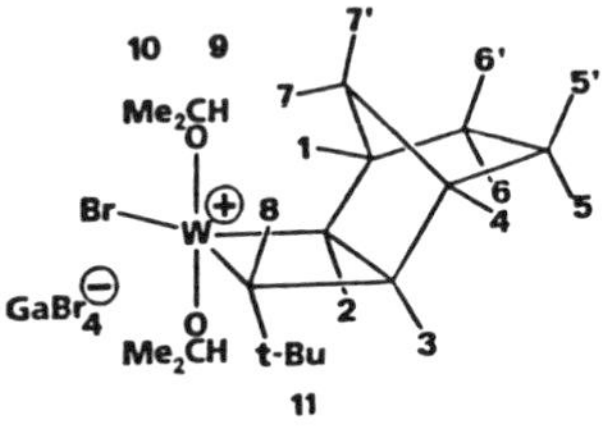

H_2	6.70 (d,1H)	$H_{5,6}$	1.98-1.81 (m,4H)
$H_{9,9'}$	5.70 (sept,1H) 5.40 (sept,1H)	$H_{10,10'}$	1.42 (d,6H) 1.39 (d,6H)
H_8	4.94(d,1H)	H_{11}	1.25 (s,9H)
H_4	3.46 (s,1H)	H_3	0.57 (t,1H)
H_1	3.19 (s,1H)		

* Assumed trans by analogy with the OCH_2CMe_3-containing system [9]

2.4. The system $W(=CHCMe_3)(OCH_2CMe_3)_2Br_2/GaBr_3$/endo-5-MNBE (1/1/3)

At 220K half the initiator ($[I]_o = 5.5x10^{-2}$ M) had reacted in 100 min, being converted initially into equal proportions of head and tail metallacyclobutane species X_{1H} and X_{1T}, characterised by H_2 doublets at 7.13 and 6.74 ppm respectively.

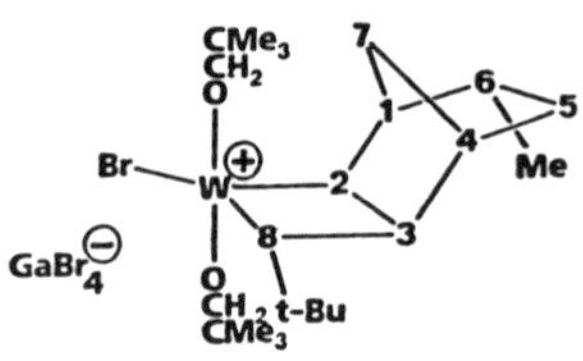

X_{1H} 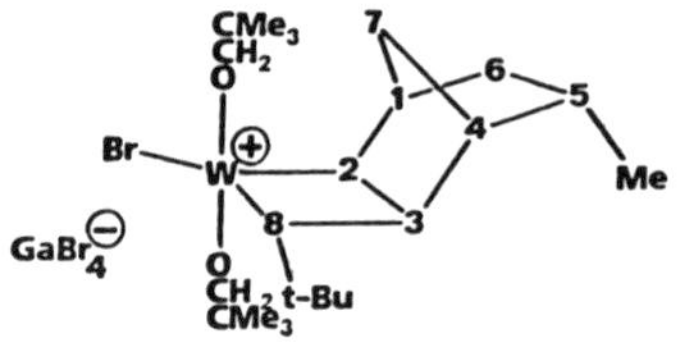 X_{1T}

After 35 min the head metal-carbene species P_H began to appear (doublet at 12.57 ppm) at the expense of X_{1H}. Even after 100 min there was no sign of the corresponding tail species P_T.

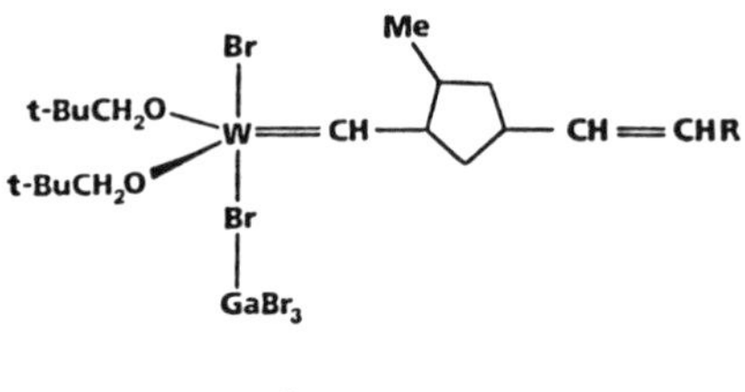

P_H 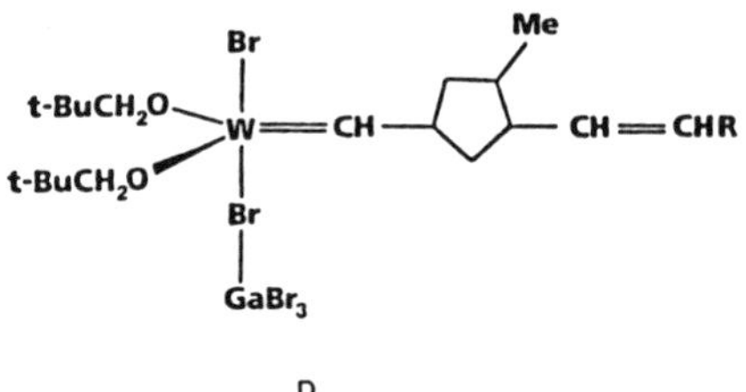 P_T

At this stage the concentrations were $[I]_2 = 2.7\times10^{-2}$ M, $[X_{1T}] = 1.4\times10^{-2}$ M, $[X_{1H}] = 0.9\times10^{-2}$ M, $[P_H] = 0.5\times10^{-2}$ M, the sum total being equal to $[I]_o$. The monomer concentration had fallen from $[M]_o = 16.2\times10^{-2}$ M to $[M] = 9.4\times10^{-2}$ M.

On raising the temperature to 235K the remainder of the monomer disappeared at a steady rate (zero order in monomer) in about 30 min, and [I] fell to 2.0×10^{-2} M. At the same time X_{1H} disappeared, X_{1T} passed through a shallow maximum while a very small amount of P_T (doublet at 12.41 ppm) was formed. At this stage there were roughly equal concentrations of X_{1T} and P_H (1.6×10^{-2} M) and a little P_T (0.2×10^{-2} M). On raising the temperature to 260K, X_{1T} was converted into P_{1T}. On close inspection of the later spectra at 235K it was noticed that there were several very weak doublets in the 6-7.5 ppm region. These were identified from the next system (section 2.5) as belonging to various isomeric forms of X_n. Their total concentration was low but not negligible in the present system.

The fraction of cis double bonds in the polymer was 0.3 as determined from the upfield region of the ^{1}H n.m.r. spectrum. Polymers made under similar conditions at 230K, but with an M/W ratio of 65 (M = monomer), had the same cis content and a random orientation of the methyl substituents with respect to both cis and trans double bonds as determined by ^{13}C n.m.r. [10]. This shows that the two propagation reactions $P_T + M \rightarrow X_T$ and $P_T + M \rightarrow X_H$ have similar rate constants but which are considerably higher than those for the other two propagation reactions $P_H + M \rightarrow X_H$ and $P_H + M \rightarrow X_T$. The zero order with respect to M at 235K can be explained in terms of such a mechanism and stems from the fact that $[P_H]$ and $[P_T]$ are both rising during this stage of the reaction, with $[X_{1T}]$ remaining almost steady.

It is evident that X_{1T} is considerably more stable than X_{1H}, both at 220K and 235K. This relationship was also found with the X_1 species derived from other monomers, such as endo, exo-5,6-dimethylnorbornene, where an endo-methyl group may occur either in the 5-position (X_{1T}) or 6-position (X_{1H}). Some degree of stability thus appears to be conferred on the metallacycle by the presence of an endo-methyl group on the C_5 carbon.

2.5. The system $W[=\overline{C(CH_2)_3CH_2}](OCH_2CMe_3)_2Br_2/GaBr_3$/endo-5-MNBE (1/1/4)

Preliminary experiments with this system under similar conditions to those used in the previous section show the following :

(1) At 210K, neither X_{1H} nor X_{1T} are detected, but high conversion of I into P_1 species is observed. This is to be expected from the behaviour of the corresponding NBE system (section 2.1). The P_1 species consist of 90% P_{1H} and 10% P_{1T}; the cyclopentylidene ligand thus induces a strong preference for the formation of the head carbene via the undetected X_{1H}.

(2) At 225K (monomer still present), P_{1H} and P_{1T} are slowly converted into at least six X_n species (doublets, 6.5-7.5 ppm) and the initiator is all consumed.

(3) At 235K (monomer still present) P_{1T} disappears completely and P_{1H} is gradually replaced by P_{nH} (carbene proton doublet slightly upfield

from P_{1H}), a behaviour which is similar to that described in section 2.2.
(4) Also at 235K, three of the X_n doublets begin to decline, being converted to P_{nH}. These X_n species are therefore X_{nH}.
(5) At 250K the residual monomer is consumed, the remaining X_n species disappear, and the P_{nT} signal strengthens (doublet upfield from that for P_{nH}). The final ratio $[P_{nH}]/[P_{nT}]$ is about 80/20.

It is not difficult to understand why so many as six X_n species are observed. Four possible structures for X_n are indicated below, taking account only of head and tail relationships. The bracketted part

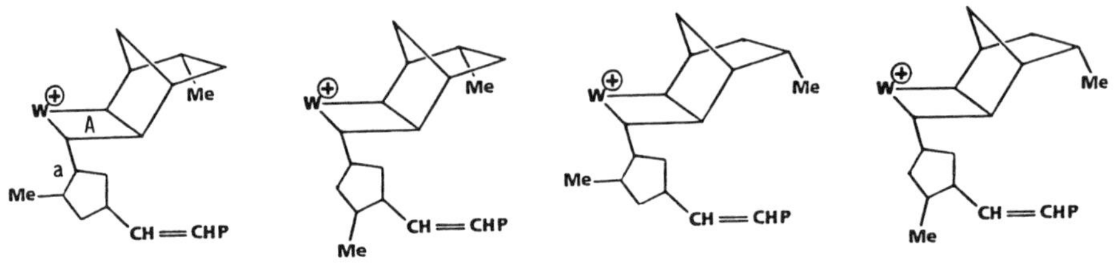

$X_{H(TH)}$ $X_{H(TT)}$ $X_{T(HH)}$ $X_{T(HT)}$

of the subscript indicates the head-tail, head-head, or tail-tail structure about the double bond that will be formed when X_n breaks down into P_n. Each of these is subject to possible cis/trans isomerism at bond A (as observed for NBE - see section 2.1) and configurational isomerism at C^a (see $X_{H(TH)}$), making a total of 16 structures which could, in principle, give rise to distinct n.m.r. spectra.

3. CONCLUSIONS

This work shows that a considerable variety of intermediate metal-carbene and metallacyclobutane complexes may be detected in these metathesis polymerizations which can show a wide range of kinetic behaviour. The X and P species observed are summarised in Table V.

The metal carbene complexes P are always observed as the final product. Sometimes both X_1 and X_n can be seen but sometimes only X_1 or only X_n depending on the relative stabilities of the metallacyclobutanes and the relative rates of the propagation reactions. Neither X_1 nor X_n has ever been seen when $GaBr_3$ is omitted from the catalyst system and X_1 has never been seen when the initiating carbene ligand is cyclopentylidene. The propagation reaction is exceptionally fast when isopropoxy ligands are used. Closer studies on selected systems should enable the determination of more precise information about the mechanism of reaction and the rate constants of some of the processes involved.

TABLE V. Types of species observed in metathesis polymerization (^1H n.m.r.)

Catalyst	Monomer	X_1 [a]	X_n [a,b]	P_1 [a,c]	P_n [a,c]
$W(=CHBu^t)(OCH_2Bu^t)_2Br_2/GaBr_3$	NBE [9]	+	+(c,t)	+	
" "	endo-5-MNBE	+(H,T)	trace	+(H,T)	
$W[=\overline{C(CH_2)_3CH_2}](OCH_2Bu^t)_2Br_2/GaBr_3$	NBE	–	+(c,t)	+	
" "	endo-5-MNBE	–	+(H,T)	+(H,T)	+(H,T)
$W[=\overline{C(CH_2)_3CH_2}](OCH_2Bu^t)_2Br_2$	NBE	–	–	+	+
$W(=CHBu^t)(OPr^i)_2Br_2/GaBr_3$	NBE	+	–	+	

[a] (H,T) means that both head and tail species are observed
[b] (c,t) means that both cis and trans species are observed
[c] A + between the columns means that P_1 and P_n were not resolved

4. EXPERIMENTAL

The tungsten-carbene complexes were prepared as described in previous papers [3,4] ; also endo-5-MNBE [10]. These complexes and $GaBr_3$ were weighed out in a dry box, a solution made in CD_2Cl_2 which was then transferred to an n.m.r. tube and closed with a serum cap. After the spectrum of the catalyst solution had been checked, the n.m.r. tube was immersed in dry ice (195K) and the liquid monomer (endo-5-MNBE) or monomer solution (NBE/CD_2Cl_2) injected into the top of the tube by means of a Hamilton microsyringe. The catalyst solution and monomer were then mixed by several rapid inversions of the n.m.r. tube while still contained within the dry ice. 200 MHz low temperature ^1H n.m.r. spectra (50 pulses) were then taken at regular intervals, the temperature being raised after a while according to the progress of the reaction. The instantaneous concentration in initiator, monomer and intermediates was determined from the integrated intensities of chosen peaks derived from these species, with reference to the known initial concentration in I. All chemical shifts were referred to the central solvent peak ($CHDCl_2$) taken as 5.33 ppm.

Acknowledgement. We thank the S.E.R.C. (U.K.) and C.N.R.S. (France) for supporting this collaborative work, and Dr. V. Amir-Ebrahimi for preparation of endo-5-MNBE.

REFERENCES

1. J.H. Hérisson and Y. Chauvin, Makromol. Chem., **141**, 161 (1970).
2. K.J. Ivin, *Olefin Metathesis*, Academic Press, London (1983), pp 399.

3. J. Kress, M. Wesolek and J.A. Osborn, J. Chem. Soc., Chem. Commun., 514 (1982).
4. J. Kress, A. Aguero and J.A. Osborn, J. Mol. Catal., 36, 1, 1986.
5. J. Kress, J.A. Osborn, R.M.E. Greene, K.J. Ivin and J.J. Rooney, J. Chem. Soc., Chem. Commun., 874 (1985).
6. F. Quignard, M. Leconte and J.M. Basset, J. Chem. Soc., Chem. Commun., 1816 (1985).
7. R.R. Schrock, J. Organometal. Chem., 300, 249 (1986) and references therein.
8. L.R. Gilliom and R.H. Grubbs, J. Am. Chem. Soc., 108, 733 (1986).
9. J. Kress, J.A. Osborn, R.M.E. Greene, K.J. Ivin and J.J. Rooney, J. Am. Chem. Soc., 109, 899 (1987).
10. K.J. Ivin, L.M. Lam and J.J. Rooney, Makromol. Chem., 182, 1847 (1981).

DISCUSSION

(13) The presence of $GaBr_3$ with the Osborn catalysts stabilises the metallacycle to the extent that it can be observed at 220-235 K where the propagation reaction proceeds at a measurable rate. Without $GaBr_3$ the catalyst is still active but requires a somewhat higher temperature to observe propagation (240 K). The failure to observe the metallacycle in the latter case suggests that the hexacoordinate neutral tungsten acyclobutane is less stable than the pentacoordinate cationic species formed in the presence of $GaBr_3$, but the higher temperature may also be a contributory factor.
It should also be borne in mind that $GaBr_3$ is involved in equilibria with I, X and P and that some redistribution occurs as I is consumed and when X is converted to P. This is indicated by the movements of the chemical shifts and is discussed in a recent paper (J. Kress et al, J. Am. Chem. Soc. 109 899 (1987)).

(14) In principle, the order with respect to monomer gives a direct indication of the nature of the slow step ; second order that $X_n \rightarrow P_n$ is the slow step. For the system described in section 2.4, there are four kinetically distinct X species : X_{IT}, X_{IH}, X_{nT} and X_{nH} ; likewise for P. Kinetic analsyis shows that the rate of removal of monomer in the steady state can be expressed as the sum of two terms : $k_I[X_{IT}]$ + $k_I[X_{nT}]$. Experimentally, $[X_{IT}]$ is observed to remain

approximately constant during the zero-order decay of the monomer, as expected. X_{nT} unfortunately cannot be seen, partly because of its low concentration in this system, and partly because the signal is spread over at least six isomers (see Section 2.5). However the zero-order decay of monomer means that its concentration must also be approximately constant. The corollary is that as [M] falls, both $[P_H]$ and $[P_T]$ must rise as indeed is observed. With the Ti catalyst, the order with respect to monomer is zero for norbornene, but one for cyclopentene (Grubbs).

(15) The Osborn catalyst without $GaBr_3$ can be used to polymerize derivatives of norbornene having polar substituents such as COOEt and CN. The catalyst system remains stable in the presence of these groups, although some complexing to the catalyst is observed.

(16) There is a formal analog between cis/trans blockiness in metathesis polymers of norbornene and cis/trans polymers of butadiene, formed with certain catalysts (Teyssié). Both have been attributed to the presence of two types of propagating species with different ligand geometry.

POLYMERIZATION OF ALKYNES INITIATED BY STABLE METAL CARBENE COMPLEXES

Alain SOUM, Michel FONTANILLE,

LABORATOIRE DE CHIMIE DES POLYMERES ORGANIQUES (associé au CNRS)
Institut du Pin - 351, cours de la Libération
33405 TALENCE Cédex - FRANCE

ABSTRACT :

The polymerization of alkynes (1-heptyne in particular) initiated by a transition metal carbene complex has been studied in detail from a kinetic and structural point of view.
The kinetic and thermodynamic parameters have been determined assuming a steady state process.
The existence of termination reactions (spontaneous, back-biting) has been demonstrated and a mechanism of polymerization is proposed.
The activation of the carbene complex by Lewis acids has also been investigated. The increase of both, the apparent activity and the apparent rates of polymerization, have been observed.
A mechanism of this activation, based on a coordination of the Lewis acid on one particular CO ligand of the carbene complex, is proposed.

. . .

The polymerization of alkynes initiated by catalytic systems used in metathesis reactions have been intensively studied over the last past few years (1) (2). The most active systems are transition metal halides (e.g. WCl_6 , $W(CO)_6$, $MoCl_5$), used alone or with a coinitiator (e.g. Lewis acids...). Considering the structural analogy between alkenes, alkynes and cycloolefins, authors proposed a mechanism of polymerization based on the existence of carbenic intermediates, as in the metathesis reactions of alkenes and in the ring-opening polymerization of cycloolefins (3) (4). Nevertheless, they could not demonstrate it directly.
KATZ and coll. have been the first authors who used, as initiators of the polymerization of alkynes, transition metal carbene complexes **(I)** **(II)**,thus corroborating the previous assumptions (5). Nevertheless, due to the low activity of complex **(I)** and to the low stability of complex **(II)**, these authors did not perform any detailed kinetic and mechanistic studies.

M. Fontanille and A. Guyot (eds.), Recent Advances in Mechanistic and Synthetic Aspects of Polymerization, 375–391.

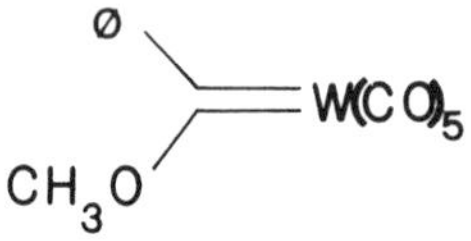

(I)  (II)

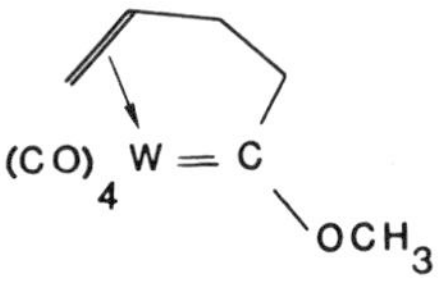

(III)  (IV)

In order to solve those problems, a new stable metal carbene complex **(III)** and analogous compounds **(IV)** synthesized by RUDLER and coll. (6) (7), were used to initiate the polymerization (8) (9). Although the activity of the complex **(III)** is as high as carbene complex **(II)**, it presents a higher thermal stability and thus easier handling.

Structure and features of the initiator (III)

The study of carbene complex **(III)** by ^{1}H, ^{13}C NMR and by X-Rays analysis (6) shows that :

. the carbenic bond is polarized positively on the carbenic carbon as in complexes **(I)** and **(II)** (Figure 8).

. the ethylenic double-bond is coordinated perpendicularly to carbenic bond and symmetrically to W atom **(V)**. The latter feature may explain the thermal stability of the complex.

Complementary studies were also carried out, in order to determine the effect of the intramolecular coordination, on the behaviour of **(III)** in presence of alkynes :

. Firstly, the existence of the coordination decrease the lability of CO ligands ; indeed, no exchange with ^{13}C-enriched carbon monoxide is observed, contrarily to what happened in uncoordinated carbene complex **(VI)**.

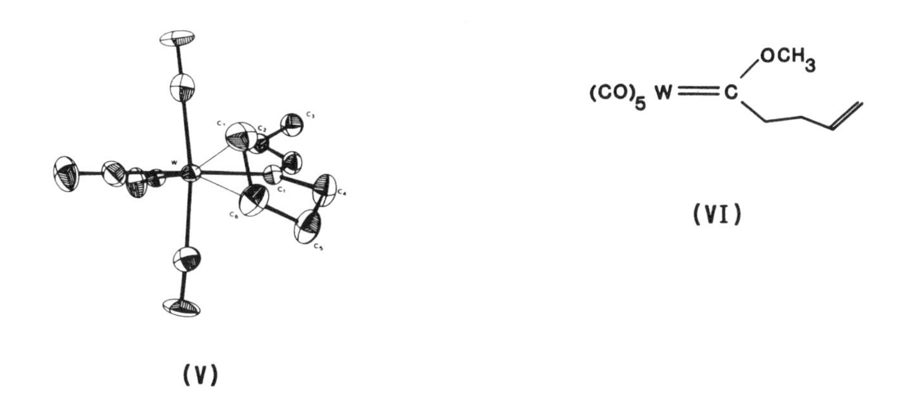

(V)

(VI)

. Secondly, uncoordinated complex **(VI)** does not initiate the polymerization

. Finally, the complexed ethylenic double-bond is easily replaced by strong ligands such as $P\emptyset_3$; the corresponding structure was determined by X-Rays analysis (10). With alkynes as ligands, no such complex has been isolated.

Thus, the intramolecular coordination existing in RUDLER carbene complex **(III)**, increases its thermal stability and, facilitating the complexation and the insertion of the first alkyne unit, maintains its potential activity.

Various alkynes have been polymerized by **(III)** : acetylene, 1-propyne, 1-pentyne, 3,3-dimethyl 1-butyne. Nevertheless, for experimental convenience, a comprehensive study was carried out with 1-heptyne as monomer.

RESULTS AND DISCUSSION

General behaviour of the polymerization reaction

Preliminary study of the reaction of 1-heptyne with initiator **(III)** was performed by GPC, HPLC and GC coupled with Mass Spectrometry (Figure 1).

GPC chromatograms showed that, at the beginning of the reaction, simultaneously with the polymer, some organic side products (in the low molar mass range) appear. Moreover, the low molar mass product concentration remains constant with conversion, whereas the polymer concentration increases.

During HPLC experiments, the slow disappearence of inital carbene **(III)** upon addition of 1-heptyne, and the appearence, after a short induction period, of the organic product **(VII)**, were observed.

GC and Mass Spectrometry experiments corroborated these observations. They also indicated the presence of a second side product which was isolated and characterized as a tri-substituted benzene **(VIII)**.

(VII) **(VIII)**

Thus, the results suggest that, besides the normal insertion of the monomer, there occurs, some side reactions yielding organic compounds.

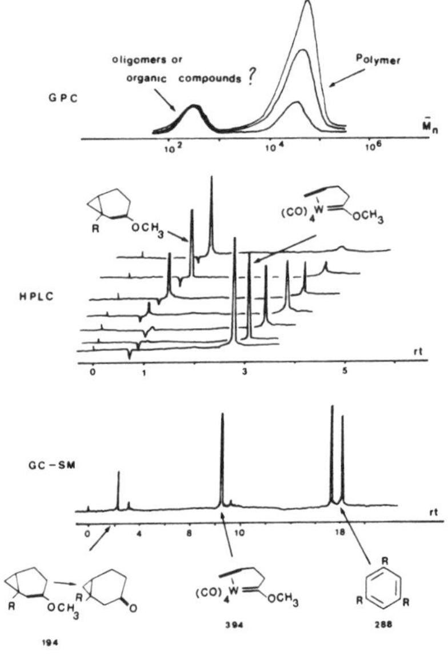

Figure 1. Chromatograms of the reaction of 1-heptyne with **(III)**, at 20°C, in hexane solution.

Kinetic study

The kinetic behaviour of the system was investigated in detail. The first order plots of monomer concentration versus time (Figure 2) indicated that :

. the initial rates of polymerization are maximal ones

. the curves are linear only over a few percent of conversion (10-20%) and then, the apparent rates of polymerization decrease.

. moreover, unexpected, the higher the initial monomer concentration, the higher the decrease of the apparent rates and the final yields.

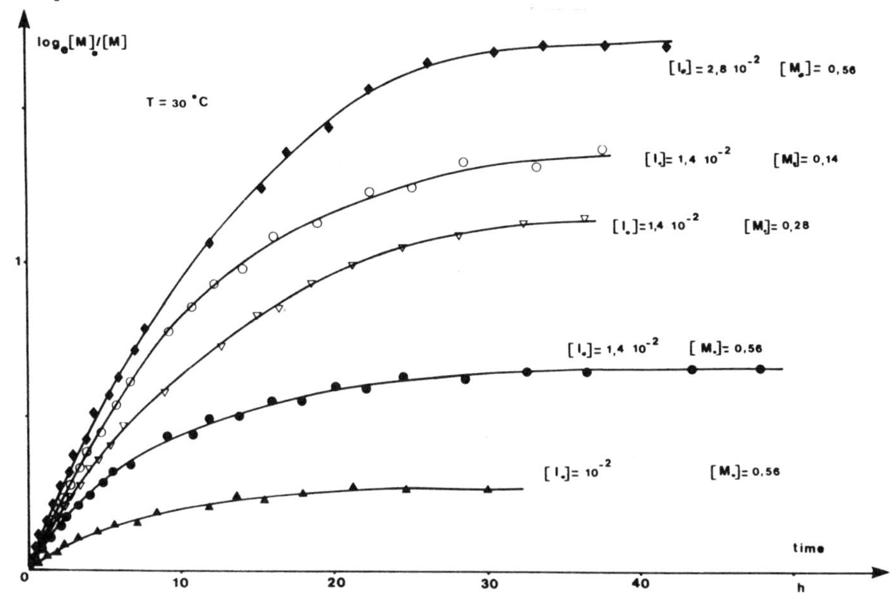

Figure 2. First order plots of monomer concentration versus time, in the polymerization of 1-heptyne initiated by **(III)**, at 20°C, in hexane solution.

From these observations, it can be inferred that active centers are unstable and that termination reactions including the monomer, occur. Therefore, the kinetic and thermodynamic parameters of the reaction, were determined at initial conditions. The initial kinetic orders relative to both monomer and initiator concentrations were found to be equal to unity. The initial apparent rate constant of propagation was equal to (Figure 3) :

$$(k_p)_o = 1.4\ \ 10^{-3}\ \ \text{l. mol}^{-1},\ \text{s}^{-1}$$

and the thermodynamic parameters to (Figure 3) :

$$\Delta H_p^{\neq} = 70\ \ \text{KJ.mol}^{-1} \qquad \Delta S_p^{\neq} = -67\ \ \text{J.mol}^{-1}.\text{K}^{-1}$$

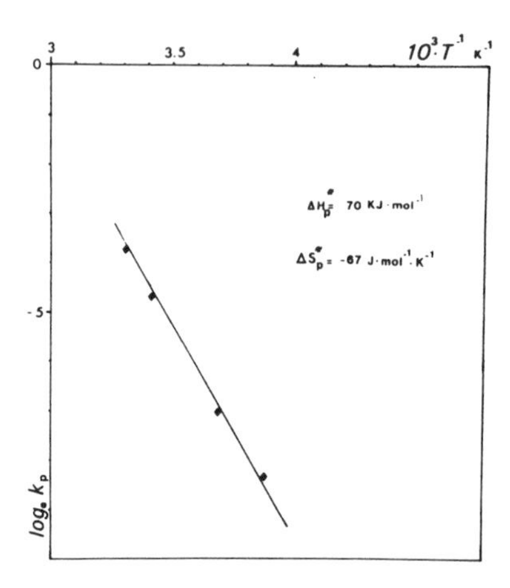

Figure 3. Variations of the apparent rate constant of propagation k_p, versus temperature, in the polymerization of 1-heptyne initiated by **(III)** in hexane.

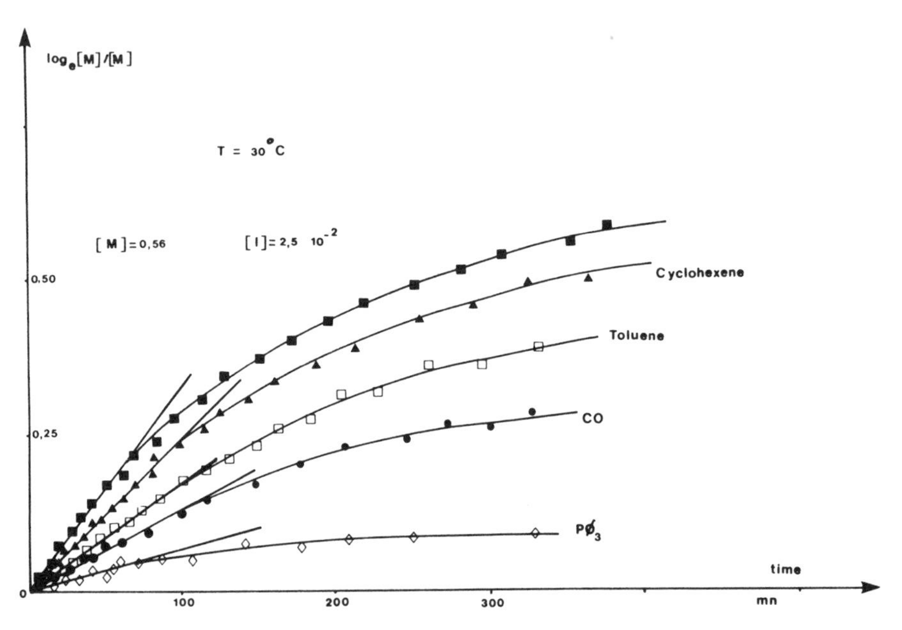

Figure 4. First order plots of monomer concentration versus time in the polymerization of 1-heptyne in presence of complexing additives. Initiator : **(III)** Solvent : Hexane Temperature : 30°C

Moreover, it is worthy to note that, in the presence of complexing additives ($P\phi_3$, CO ....) (Figure 4) the rate of reaction decreases strongly. Thus, the complexation of the monomer on the transition metal atom, appears to be a required step before the insertion

So, the following kinetic scheeme may be proposed :

Initiation $\quad A \xrightarrow{M} C^* \quad k_a$

$$R_a = k_a[A][M_o] = K_a\,[A] \qquad [A] = [A_o]\,\exp{-K_a t} \qquad (i)$$

Propagation

$$\sim\!\sim C_n^* + M \xrightarrow{k_p} \sim\!\sim C_{n+1}^*$$

$$R_p = -\frac{d[M]}{dt} = k_p\,[C]\,[M]$$

Termination :

$$\sim\!\sim C_n^* \xrightarrow{k_t} \sim\!\sim C_n$$

$$\sim\!\sim C_n^* + M \xrightarrow{k_{tm}} \sim\!\sim C_n$$

$$\sim\!\sim C_n^* + P \xrightarrow{k_{tp}} \sim\!\sim C_n$$

$$R_t = k_t[C] + k_{tm}[C][M] + k_{tp}([M_o]-[M])[C]$$

Taking into account the kinetic observations (initial rates maximal, slow initiation) a steady-state in active centers has to be assumed. Therefore :

$$R_a = R_t \text{ and } K_a[A] = k_t[C] + k_{tm}[C][M] + k_{tp}([M_o] - [M])[C]$$

Consequently,

$$[C] = \frac{K_a \quad [A]}{k_t + (k_{tm}-k_{tp})\ [M] + k_{tp}\ [M_o]}$$

Then,

$$- \frac{d\ [M]}{dt} = K_p\ [C][M] = \frac{K_p K_a \quad [A]\ [M]}{k_t + (k_{tm}-k_{tp})\ [M] + k_{tp}\ [M_o]} \qquad \text{(ii)}$$

with $K_p = k_p + k_{tm}$

Since initial kinetic orders are equal to one,

$$k_t << k_{tm} \quad [M_o]$$

Equation (i) and (ii) leads to :

$$(k_t + k_{tp}\ [M_o]\ )\ \ \mathrm{Log}\ \frac{[M_o]}{[M]} + (k_{tm} - k_{tp})\ (\ [M_o]-[M\ ])$$

$$= K_p\ [A_o]\ (\ 1\text{-}\exp\text{-}K_a t\ )$$

At infinite time :

$$\text{(iii)} \qquad \mathrm{Log}\ \frac{[M_o]}{[M_\infty]} = \frac{K_p\ [A_o]}{k_t + k_{tp}[M_o]} - \frac{k_{tm} - k_{tp}}{k_t + k_{tp}\ [M_o]}\ ([M_o] - [M_\infty]\ )$$

Experimentally, the relation (iii) is verified for different initiator concentrations (Figure 5).

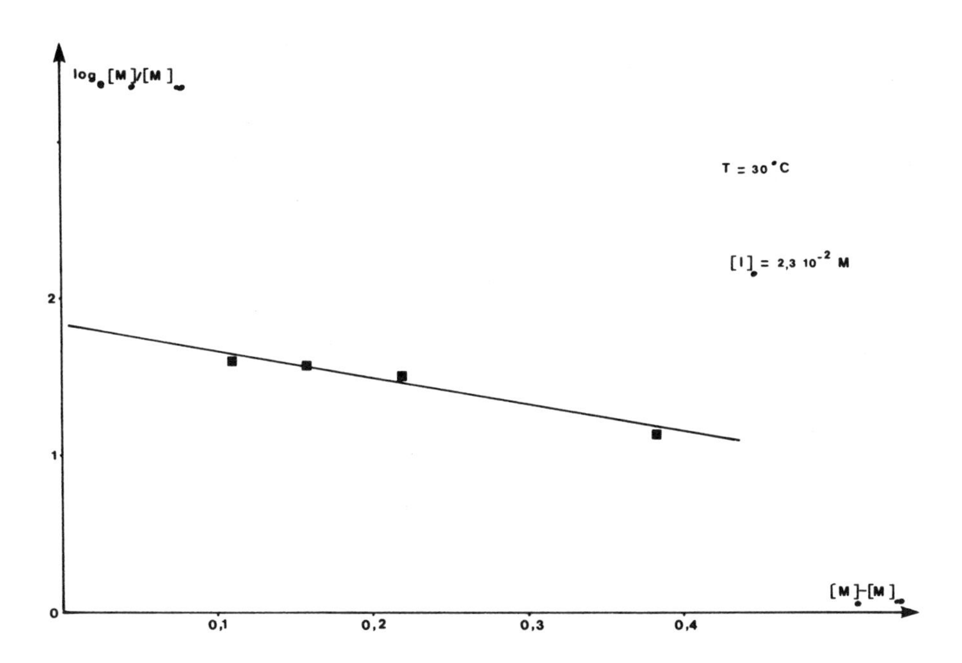

Figure 5. Experimental relation between initial ($[M_o]$) and final ($[M\infty]$) monomer concentrations.

Mechanism of polymerization

Considering all the experimental results and those previously obtained in ring-opening polymerization (4) (11) and in metathesis reactions (4) (3), a mechanism of the reaction may be proposed :

The first step of this mechanism (a) would be, first, the decoordination of the intramolecular coordinated double-bond in **(III)**; second, the complexation of the monomer and then, the insertion of this molecule through a metallacyclobutene intermediate, leading to the first carbenic active center **(IX)**.

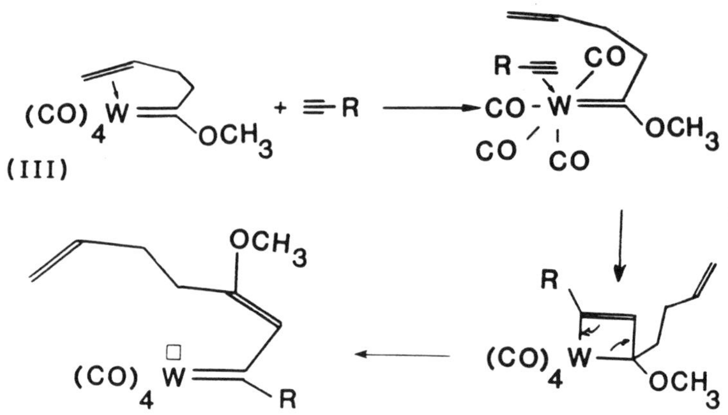

$(CO)_4W{=}C(OCH_3)$ (III) + $\equiv$-R → R-$\equiv$, CO-W(CO)$_3$=C(OCH$_3$)

(a)

R, $(CO)_4W$, OCH_3

(IX) $(CO)_4W{=}$, R, OCH_3

The new metal carbene complex **(IX)** is unstable because of the absence of heteroatom on the carbenic carbon; moreover, it has the adequat size to give cyclopronation reaction (b).

The cyclopropanation reaction (b) would lead to the enol ether **(VII)** and after hydrolysis to the corresponding ketone **(X)**.

The carbene complex **(IX)**, resulting from the insertion of the first monomer unit through a steady-state process (c), would give active polymer chains **(XI)**, which would be subject to termination reactions.

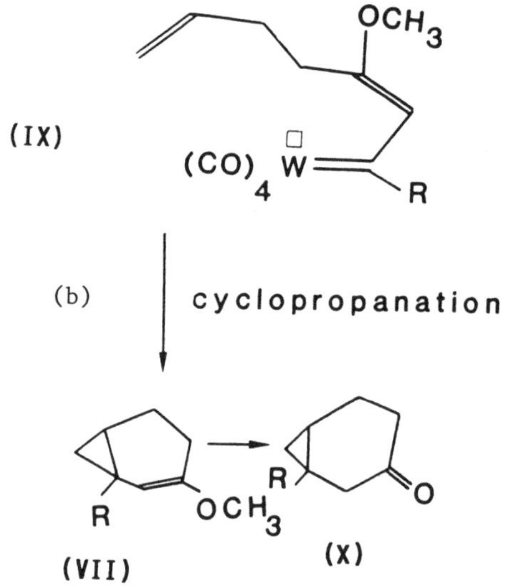

(IX) $(CO)_4W{=}$, R, OCH_3

(b) cyclopropanation

R, OCH_3 (VII) → R, O (X)

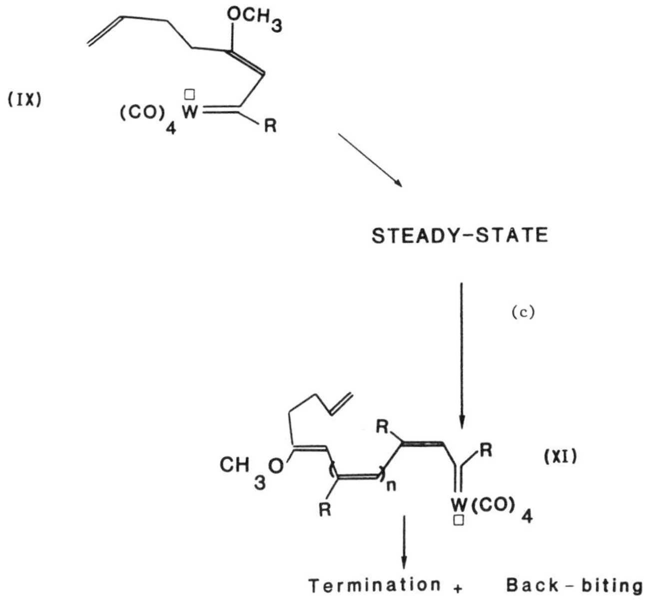

(IX)

STEADY-STATE

(c)

(XI)

Termination + Back-biting

The termination reactions (d), (e), (f), (g), would lead to inactive polymer chain **(XII)**, inactive tungsten derivatives and, through back-biting reactions, to tri-substituted benzene **(VIII)**.

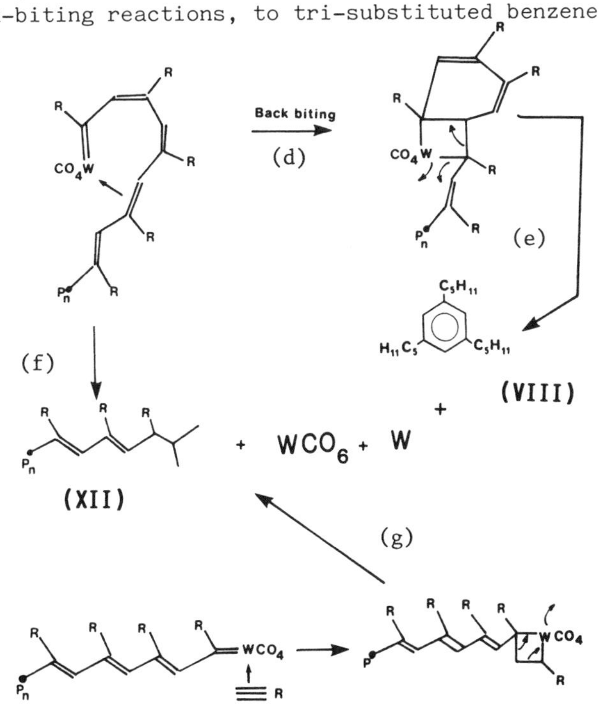

Back biting (d)

(e)

(f)

(VIII)

+ WCO_6 + W

(XII)

(g)

Activation of the reaction by Lewis acids

The chemical activation of carbenic complexes by Lewis acids, has already been reported by several authors (12) (13) (14). It appeared important to verify if the RUDLER complex **(III)** exhibits the same behaviour and to study the corresponding mechanism.

The apparent activity of complex **(III)** to polymerize alkynes, increases markedly in the presence of Lewis acids. This increase is a function of both, the nature of the acid and the ratio [Acid] / [W].
At this time, $AlEt_3$ and EtO_3B appears the most efficient acids and, in that case, the apparent activity reaches an asymptotic value for a ratio [Acid] / [W] equal to approximately 20 (Figure 6). The same behaviour is observed on the conversion curves (Figure 7); depending on the ratio, both the initial rates of polymerization and the final conversions increase.

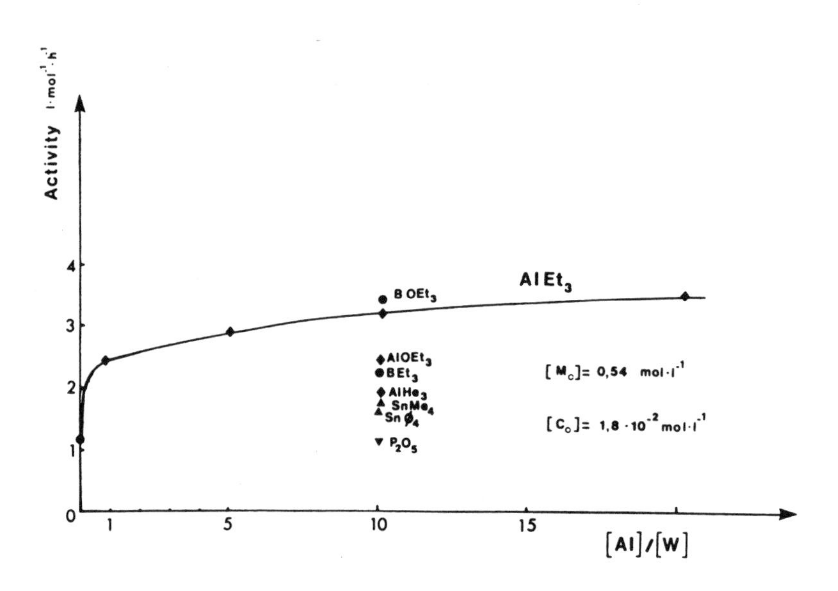

Figure 6. Variation of apparent activity of **(III)** along with [Acid] / [W] ratio in the polymerization of 1-heptyne, at 30°C, in hexane solution.

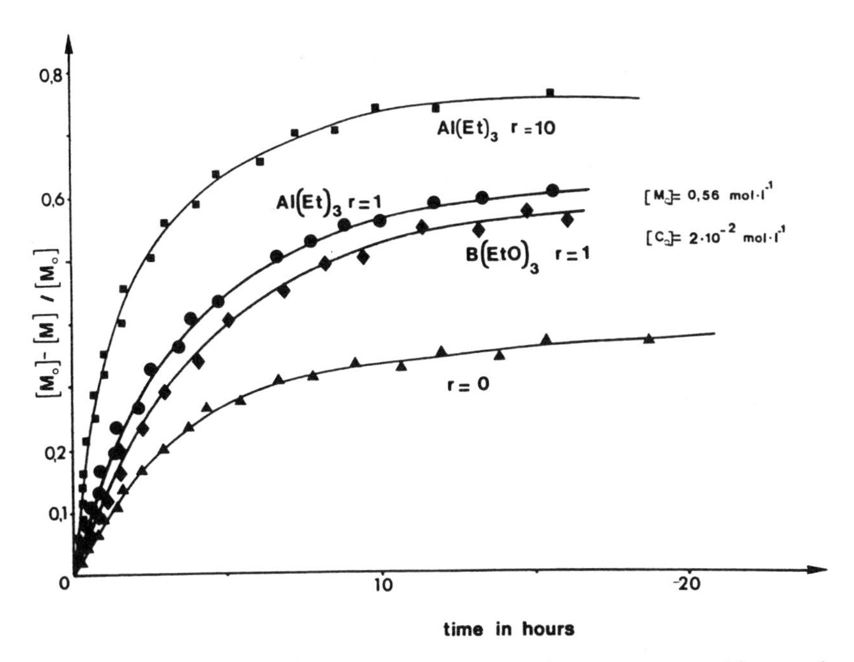

Figure 7. Variation of monomer conversion versus time, in the polymerization of 1-heptyne initiated by **(III)** and activated by Lewis acids, in hexane solution, at 30°C.

In order to explain the reaction of carbene complexes with Lewis acids, KATZ and coll. (13) proposed the formation of unstable and reactive carbyne . On the contrary, in the case of the reaction of AlX_3 on a carbene complex , OSBORN and coll. (14) observed a complexation of one ligand by AlX_3 and an increase of the electrophilic polarization of the [W = C] bond.

In the case of RUDLER complex **(III)** activated by $AlEt_3$, ^{1}H, ^{13}C NMR and UV studies indicated that the phenomenon is quite different :

. upon addition of $AlEt_3$, no change appears in the initial UV spectrum of the complex **(III)**, indicating the absence of any chemical modification in the complex.

.the ^{13}C NMR spectrum shows slight but significant differences (Figure 8); the signals of the carbons C_5 , C_6 of the double-bond and that of the ligand CO_9 , are shifted downfield, whereas the signal of carbenic carbon C_1 , remains unchanged. Moreover, this shift increases along with the ratio $[AlEt_3]/ [W]$ up to an asymptotic value as for the apparent activity of the system .(Figure 9).

Also, a broadening of C_5 , C_6 , C_9 NMR signals is observed at low temperature.

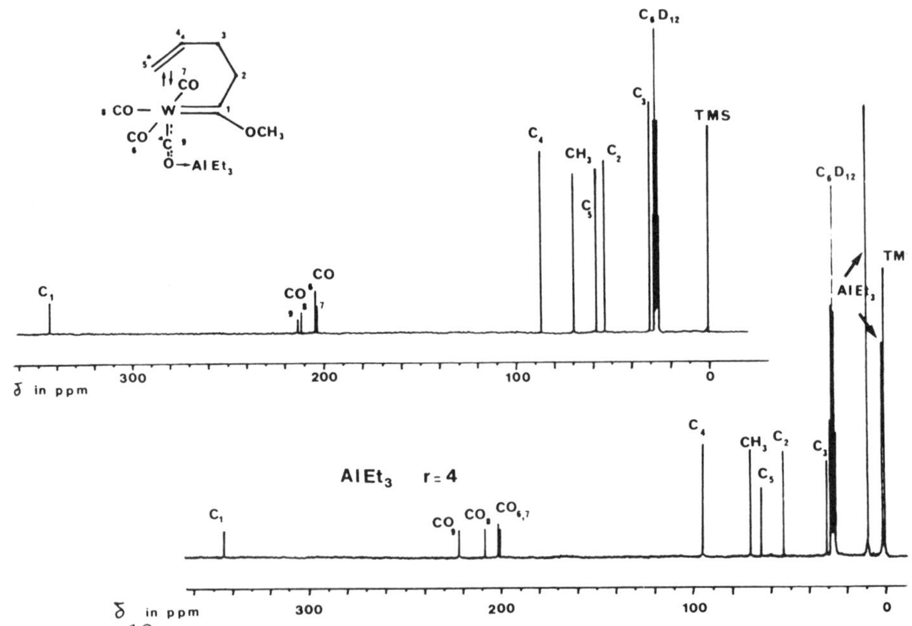

Figure 8. ^{13}C NMR spectrum (50MHz) of **(III)**, pure and in presence of $AlEt_3$.

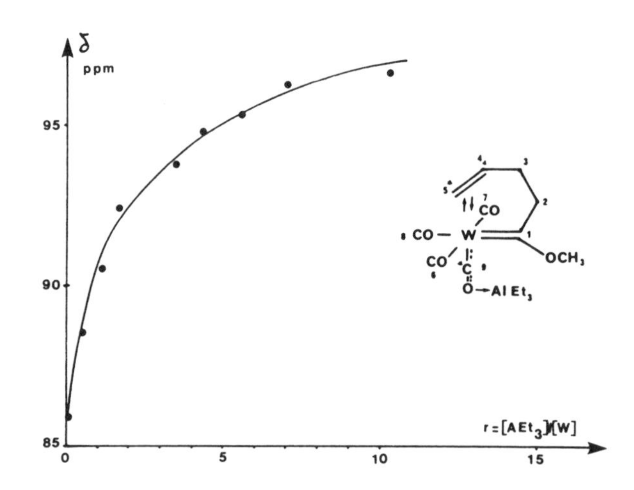

Figure 9. Variation of the chemical shift of the ethylenic carbon C_4 along with $[AlEt_3]/[W]$ ratio.

These observations show that the Lewis acid complexes preferably on the carbonyl ligand in 9 position (trans) and that, this complexation is an equilibrium reaction. Such complexation induces a decrease of the intramolecular coordination of the double-bond without perceptible modification of the electronic structure of the carbene complex.

Therefore, since the first step of the polymerization is the decoordination of the ethylenic bond, the acid increases both, the apparent activity and the apparent initial rate of polymerization.

Thus, although a comprehensive kinetic and structural study on the polymerization of alkynes initiated by transition metal carbene complexes was performed, several points are still unclear. In particular, the parameters which govern the efficiency and the absolute reactivity of carbenic active centers have to be determined; moreover, in such systems, the existence of transfer reactions may be assumed. Further experiments on these problems are in progress.

REFERENCES

(1). T. MASUDA, Y. OKANO, K. KUWANE, Polym.J., **12,** 907 (1980)

(2). T. MASUDA, T. HIGASHIMURA, Macromol. **8**, 6 (1975)
T. MASUDA, T. HIGASHIMURA, Macromol. **9**, 661 (1976)

(3). J.L. HERISSON, Y. CHAUVIN, Makromol. Chem. **141,** 161 (1970)

(4). V. DRAGUTAN, A.T. BALABAN, M. DIMONIE
'Olefin metathesis and ring-opening polymerization of cyclo-olefins' , Wiley, **49** (1985)

(5). T.J. KATZ, S.J. LEE, J.Am. Chem. Soc., **102,** 422 (1980)

(6). C. TOLEDANO, J. LEVISALLES, M. RUDLER, H. RUDLER,
J. Organomet. Chem., **228**, C 7 (1982)

(7). J. LEVISALLES, H. RUDLER, Y. JEANNIN, F. DARAN
J. Organomet. Chem. **178**, C8 (1979)

(8). D. MEZIANE, A. SOUM, M. FONTANILLE, H. RUDLER
Makromol. Chem., **186**, 367 (1985)

(9). D. LIAW, A. SOUM, M. FONTANILLE, A. PARLIER, H. RUDLER
Makromol. Chem. Rapid Commun., **6,** 309 (1985)

(10). H. RUDLER, A. PARLIER, C. ALVAREZ, J.C. DARAN,
Organometallics (in press)

(11). L.R. GILLIOM, R.H. GRUBBS,
J.Am. Chem. Soc., **108,** 733 (1986)

(12). E.O. FISCHER, Adv. in Organomet. Chem., **14,** 94 (1076)

(13). T.J. KATZ, T.H. HO, N.Y. SHIH, Y.C. YING, V.W. STUART,
J. Am. Chem. Soc., **106,** 2659 (1984)

(14). J KRESS, M. WESOLEK, J.A. OSBORN
J. Chem. Soc. Chem. Commun., 514 (1982)

DISCUSSION

(17) The chromium-carbene complexes show the same general behaviour as the corresponding tungsten complexes, but they are less stable and therefore more difficult to use.

(18) At [M] [W] = 1-3 mostly cyclopropanation products are obtained. At higher ratios mainly polymer is formed. There is thus a competition between cyclopropanation and polymerization immediately after the first addition of monomer.

(19) The intermediate metallacyclobutene has not been seen by NMR. It would probably be difficult to find conditions where its formation would be sufficiently fast, but its decomposition sufficiently slow, for it to be seen.

(20) The low molecular weight GPC peak corresponds to a number of compounds which have not been fully identified. Trisubstituted benzenes and ether enols are present and possibly some oligomers.

(21) The formation of trisubstituted benzenes can occur by a back-biting (transfer) reaction in which the activity of the tungsten centre is preserved.

REPORT OF DISCUSSIONS AND RECOMMENDATIONS

1 - INTRODUCTION

Metal-carbene complexes (titanium to iridium) are now well recognized as chain carriers in the metathesis polymerization of cycloalkenes and alkynes. The reaction proceeds via the formation and cleavage of a metallacyclobutane complex in the case of ring-opening polymerization, and a metallacyclobutene complex in the case of alkyne polymerization.

Until a few years ago, the vast majority of catalyst systems consisted of one (e.g. $RuCl_3$), two(e.g. $MOCl_5/Et_3Al$) or three components (e.g. $WCl_6/EtAlCl_2/EtOH$), not themselves being metal-carbene or metallacyclobutane complexes. The mode of generation of the active chain carriers was thus generally obscure, with no control over the number, type and disposition of the permanent ligands. Even metal-carbene complexes, when they were used, e.g. $Ph_2C{=}(CO)_5$ required activation by heat, light or use of a cocatalyst.

In the last few years, the situation has changed dramatically and several groups of workers have been able to prepare well defined metal-carbene and metallacyclobutane complexes capable of inducing metathesis polymerization either by themselves, or in the presence of a Lewis acid. These may be termed "second-generation" metathesis catalysts and rapid advances are now being made through the use and application of this group of catalysts, which give "living" systems. Examples are given in the three papers presented in this section of the Advanced Research Workshop.

2 - SECOND GENERATION CATALYSTS

These fall into two distinct groups :

M. Fontanille and A. Guyot (eds.), Recent Advances in Mechanistic and Synthetic Aspects of Polymerization, 393–395.

a) metal carbene complexes such as $W(=CHCMe_3)(OCH_2CMe_3)_2Br_2$

$$CH_2{=}CHCH_2CH_2 \searrow$$
and $(CO)_4\,W{=}C(OMe)$

b) metallacyclobutane complexes such as Cp_2Ti  also various W and Ta complexes

Synthetic methods are now available to allow the preparation of a wide variety of such complexes. Those which can act in the absence of a cocatalyst are to be preferred, since they generally allow a wide range of monomers to be used, especially those containing functional groups such as CO_2Et, CN, NH_2, CO. This is important in developing synthetic applications.

3 - INTERMEDIATES, KINETICS and MICROSTRUCTURE

In some systems one can now see, by NMR spectroscopy, the intermediate chain carriers at various stages of the reaction, for example the first metallacycle X, the first derived metal-carbene P, and the corresponding species X_n and P_n ($n > 2$) formed by subsequent addition of $(n-1)$ monomer molecules. When the monomer is unsymmetrical these are further resolved into head and tail and various sub-species. We now have the possibility of studying the various steps of the polymerization reaction both singly, e.g. X_1 P_1, and in combination. This will lead to a much better understanding and control of the mechanism of reaction. Conditions leading to fast initiation, relative to propagation, can readily be established. This is important in the preparation of polymers of narrow molecular weight distribution and block copolymers.

The cocatalyst, when used, modifies the structure and reactivity of the intermediates by complexation or ionization. The cocatalyst may interact not only with the initiatior I, but also with X_1, P_1, X_n and P_n, thus redistributing itself between these species as the reaction proceeds. A better understanding of such interactions may be obtained by the study of the small changes in chemical steps during the reaction, and other physical measurements.

Microstructure in these polymers takes three forms : cis/trans isomerism, m/r isomerism (ring tacticity) and head-head, head-tail, tail-tail isomerism. The use of particular monomers such as 7-methylnorbornene and 1-methylnorbornene allows a close study of microstructure by ^{13}C NMR spectroscopy. It will be important to determine the structure of polymers prepared under the same conditions as used to study the intermediates. An important question is whether the reaction X_n+M $\longrightarrow$ X_{n+1} can occur by a concerted displacement reaction rather than via $X_n \rightarrow P_n$, $P_n + M \rightarrow X_{n+1}$.

This can be determined from kinetic studies and may help to explain the observation of cis/trans blockiness in polymers of high cis content.

4 - BLOCK COPOLYMERS

These reactions provide yet another example of a living polymerization system. For the production of narrow MD polymers, or good block copolymers, it is essential that (a) the initiation is fast relative to propagation and (b) that there is no scrambling by secondary metathesis. This has already been achieved in at least one case and there is scope for rapid development in this area. As in other types of living system, block copolymers can be made by(a) sequential addition of monomers, (b) one-pot reactions of two or more monomers to give "tapered" block copolymers and (c) use of macromonomer initiator e.g. a cyclopropene unit with a polymeric tail. The variety of functional groups which can be attached to norbornene and other cyclic monomers makes possible the production of several interesting types of block copolymers.

5 - OTHER SYNTHETIC POSSIBILITIES

It has already been demonstrated that the new catalyst systems can be used to generate polymers containing regularly spaced ether functions which can act in concert, so as to complex certain cations.

All polymers made by metathesis polymerization are unsaturated, offering considerable scope for synthetic applications. Polyacetylenes can be made either directly for alkynes or undirectly by the ring opening reaction. End-capping, e.g. by the Wittig reaction of P_n with ketones such as $Ph_2C=O$, also offers considerable scope.

Statistical copolymers of alkynes and functionalized cycloalkenes can also be prepared.

6 - CONCLUSIONS AND RECOMMENDATIONS

1- New living catalysts (metal-carbenes and metallacyclobutanes) are required that :

a) allow a systematic variation of the ligands
b) act without the need for a cocatalyst
c) tolerate many functional groups
d) are easy to prepare

2- More detailed studies can now be made of the kinetics of the reaction, using these new catalysts. There is now the possibility of determining some absolute rate constants, as well as the relative rate constants which govern the microstructure of the polymers. This will be facilitated by collaboration of groups having (a) expertise in the preparation and handling of catalysts, (b) expertise in the preparation of monomers and determination of microstructure and (c) expertise in polymer characterization.

3- The further role of metal-carbenes in alkyne polymerization and -olefin polymerization should be explored.

4- Related complexes such as (M_t) = NR and (M_t) = PR also deserve attention.

VI - CHIRAL POLYMERS

Chairman : F. CIARDELLI

OPTICALLY ACTIVE VINYL POLYMERS WITH BACKBONE CHIRALITY

Günter Wulff
Institute of Organic Chemistry II
University of Düsseldorf
Universitätsstr. 1
D-4000 Düsseldorf 1
Federal Republic of Germany

ABSTRACT: By symmetry considerations it was found that certain stereochemical arrangements of vinyl monomers in homo- and copolymers would exhibit main chain chirality and hence should show optical activity. Several possible theoretical cases are discussed.

By an asymmetric cyclocopolymerization copolymers showing optical activity due to chirality of the backbone could be realized. The mechanism of this reaction was investigated.

Another possibility for obtaining optically active vinyl or vinylidene polymers is due to a chiral helical conformation stabilized by bulky substituents.

1. INTRODUCTION

Optical activity in polymers due to backbone chirality is known for quite some time. Beredjick and Schuerch (1) were able to synthesize an alternating copolymer from maleic acid and α-methylbenzyl methacrylate which after removal of the chiral side chain remained optically active due to main chain chirality. In this case the 1,2-disubstituted unit of maleic acid was considered to be the chiral unit. The stereochemistry of the comonomer is of less importance, its role is to separate the maleic acid units from each other, since a homopolymer of maleic acid could not be obtained in optically active form. The second role of the comonomer carrying the chiral side chain is to induce optical activity during asymmetric polymerization. Later on, the group of Natta had successfully achieved an optically active homopolymer based on a 1,2-disubstituted olefin carrying a ring with two different substituents using a chiral initiator (2,3). On the basis of this principle, many optically active polymers have already thus been obtained during last two decades by asymmetric induction during polymerization (4). The same concept holds true for polymers derived from substituted 1,3-dienes (4,5) where optical activity due to chirality of the main chain has also been realized.

Polymers from 1-substituted olefins CH_2=CHR termed as polyolefins, polyvinyl compounds or polyacrylates (depending on the substituent) or

M. Fontanille and A. Guyot (eds.), Recent Advances in Mechanistic and Synthetic Aspects of Polymerization, 399–408.

1,1-disubstituted olefins $CH_2=CRR'$ (accordingly termed as polyolefins, polyvinylidene compounds or polymethacrylates) play a very important role in today's modern science and technology. Polymers and copolymers of such type are produced of the order of millions of tons every year. Owing to their obvious importance, such type of polymers have been thoroughly and vividly investigated in large number of research laboratories all over the world. One of the main aspects of such studies has been the elucidation of stereochemical features of these compounds, followed by the discovery of stereoregular polymerization using Ziegler-Natta catalysts. As the stereochemistry for both classes of polymers is virtually same, they have been discussed together (6).

As has already been pointed out by Staudinger (7), during the polymerization of a 1-substituted olefin possessing a prochiral center, a chiral center is generated. Therefore with longer polymer chains a large number of diastereoisomers will be formed. This led to the idea of preparing optically active polymers of 1-substituted olefins, but the realization of such substances has met with unexpected difficulties. Though it has been easier to prepare polymers with optically active pendant groups, in spite of many attempts no polymers could be obtained which are optically active due to the chirality of the main chain. Polymers with chiral side groups have been systematically investigated using various chiroptical techniques and many valuable informations regarding the configuration and conformation of the polymers have been obtained (8).

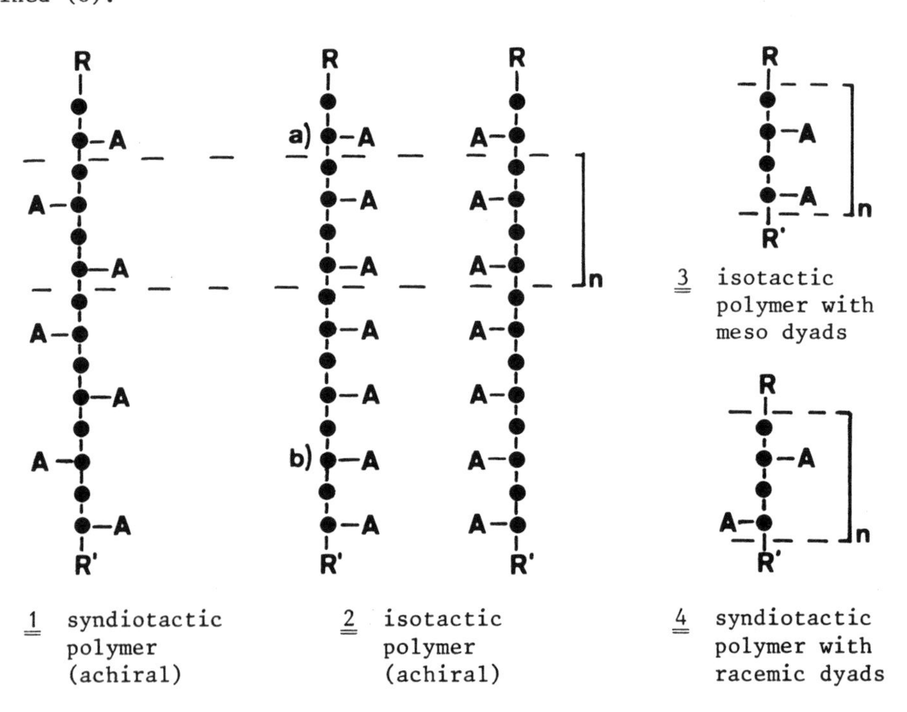

1 syndiotactic polymer (achiral)

2 isotactic polymer (achiral)

3 isotactic polymer with meso dyads

4 syndiotactic polymer with racemic dyads

Possible arrangements in regular homopolymers

The cause for the failure to achieve optically active polymers due to main chain chirality has been attributed to the peculiar situation of the stereochemistry associated with long polymer chains (4,6). A look at the three kinds of typical arrangements of polymer chain viz: atactic, isotactic and syndiotactic suggests that for syndiotactic species, the polymer chain 1 contains chiral centers with regularly alternating (R) and (S) configuration thereby resulting in a long polymer chain having no optical activity. Isotactic polymer chains 2 beginning with an (R) or an (S) configuration, would result in the formation of two enantiomorphic chains. If, however, the difference of the end groups can be neglected ($RCH_2 \simeq R'$), as always permissible for longer chain length, the molecule possesses a reflection plane (i.e., it is a mesoform) and hence cannot be chiral. This statement can also be expressed in the following different manner. The carbon atom a) of the chain 2 will be assigned an (R)-configuration. In this case the four substituents include A and H as well as the two chain ends of different length. If we compare the configuration of 'a' with that of carbon atom b) at the other side of the chain, now the previous shorter chain end is the longer one and vice versa. As a result this carbon atom possesses (S)-configuration. This implies in an isotactic chain one half has (R)- and the other has (S)-configuration.

Irregular (atactic) polymers contain randomly distributed alternating (R) and (S) configuration along the chain, an arrangement which usually would be chiral but not optically active. It has recently been pointed out by Green and Garetz (9) that atactic polymers with sufficiently high molecular weight (P_n>70) will usually be present as the molecules of single enantiomers since the number of possible diastereoisomers is so high that the probability of finding its optical antipode in the same sample is extremely low. Nevertheless, one can not expect optical activity, even with a single atactic chain since the configuration changes irregularly along the chain.

Furthermore, for copolymers too, as described by Arcus (10), the usual regular arrangements cannot be chiral. On the basis of these observations it was subsequently concluded that in general, all the polymers obtained from 1-substituted or 1,1-disubstituted olefins cannot in principle be optically active. This is the state of the art found frequently in today's standard textbooks of macromolecular chemistry and in relevant review articles.

2. SYMMETRY CONSIDERATIONS ON THE CHIRALITY OF POLYMERS

During our investigations for preparing cavities in crosslinked polymers by imprinting with templates (11,12) the problem of the existence of main chain chirality was encountered with us. The question arose whether despite of previous established findings there might be some regular arrangements in vinyl polymer chains which are chiral, thereby enabling the polymers to show optical activity. The problem was first approached by symmetry considerations of ideal, regular structures (13,14,15). Regular polymers can be described by the smallest repeating unit (called n-ads)(6). In polymer chemistry these n-ads are also used to describe irregular polymers. In this case the relative proportion of certain triads or pentads (determined mostly by NMR investigations) is used for characterizing a polymer. Earlier described syndiotactic and isotactic polymers can be described by the dyads 3 and 4, which are called "racemic" and "meso" in polymer chemistry.

In contrast to low molecular weight substances the identification of chirality in polymers is more difficult since the substituents along the entire chain have to be taken into consideration (otherwise 3 would be a chiral polymer). Furthermore, the differences of the end groups can usually be neglected. Polymer 2, in a strictest sense, is of course chiral but it is a good approximation for very long chains to neglect this difference.

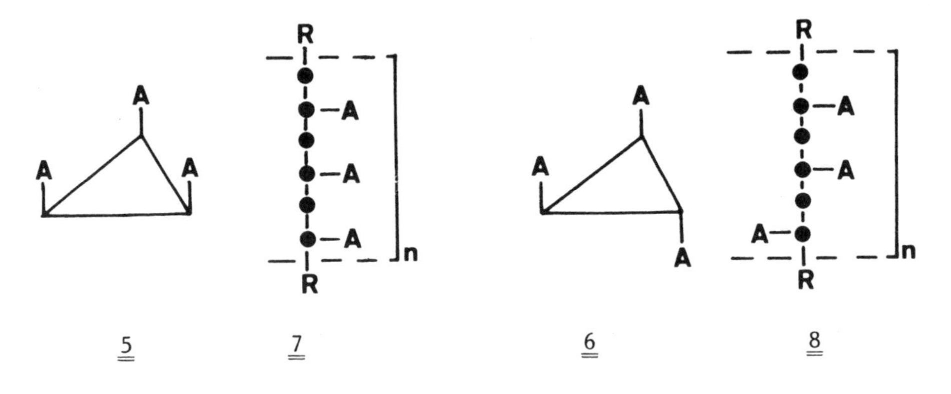

Possible triads in homo polymers shown in ring form and in the usual open chain form

Symmetry properties of high molecular weight, regular chain polymers can best be represented by placing the repeating unit in the form of a ring structure (3). The ring possesses comparable symmetry properties as a hypothetical infinitely long chain. It is also a good approximation for the real polymer chains. As examples, the two possible triads 5 and 6 represent the chain structures for the two polymers 7 and 8. The substituents are placed in a three membered ring, the CH_2-groups can be neglected since they are insignificant from a stereochemical point of view. Triad 5 describes an isotactic polymer which has already sufficiently been described by dyad 4. Triad 6 representing a new type of arrangement possesses a plane of symmetry and is therefore achiral.

If this principle is extended to regular polymers built up of tetrad and pentad repeating units all possible arrangements are found to be achiral, until a regular repeating unit of six monomers wherein out of eight possible hexads one is chiral with C_2-symmetry. Polymer 9 built up of such hexads should be able to show optical activity. Inspection of longer regular repeating sequences reveals the possibility of more number of chiral arrangements.

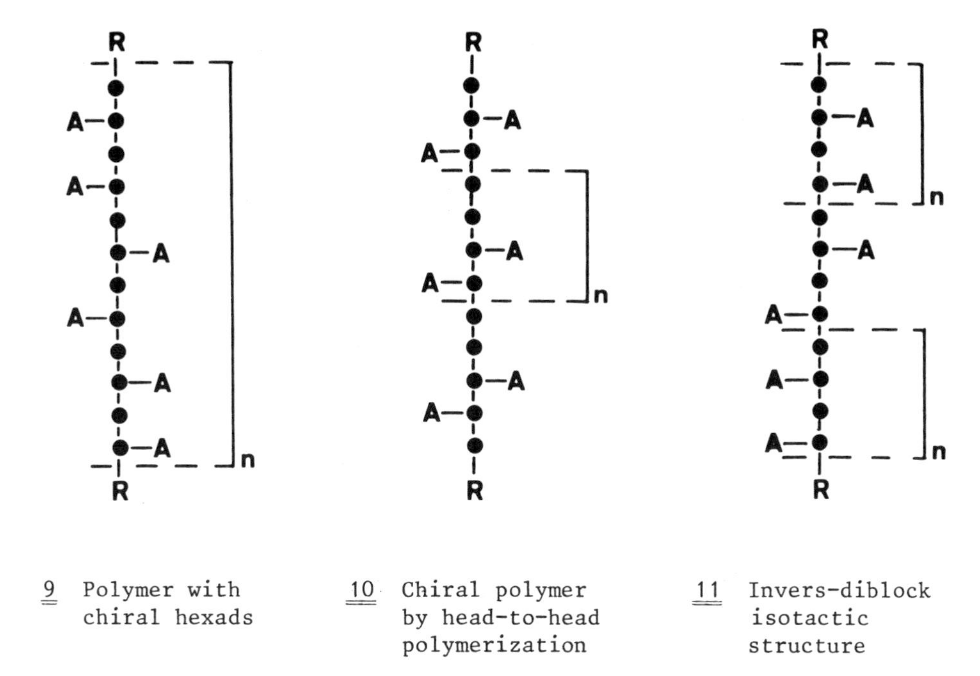

9 Polymer with chiral hexads

10 Chiral polymer by head-to-head polymerization

11 Invers-diblock isotactic structure

Chiral arrangements in homopolymers

In addition to the above examples, there are also other structural arrangements in homopolymers which would result in backbone chirality. One possibility would be the head-to-head polymerization of the monomers, wherein 1,2-disubstituted chiral segments can be formed with a structure 10 similar to that of the copolymer of the 1,2-disubstituted monomers. Chirality in the main chain would also be possible, if from the center of an isotactic polymer chain, the substituents change to the other side of the main chain with respect to the Fischer projection (see 11). Since each part of the chain has in itself an exclusive isotactic structure, we propose to designate such an arrangement as "inverse-diblock isotactic" structure.

Structural arrangements are of considerable interest in case of copolymers. In an alternating copolymer both possible arrangements for the dyads 12 and 13 are achiral. With regular copolymers three triad arrangements viz: 14, 15 and 16 are possible, of which 16 is chiral and asymmetric. In this case the dyad carrying the two A substituents constitutes the chiral part and is separated from other dyads by the B-unit for which the configuration is stereogenic meaningless.

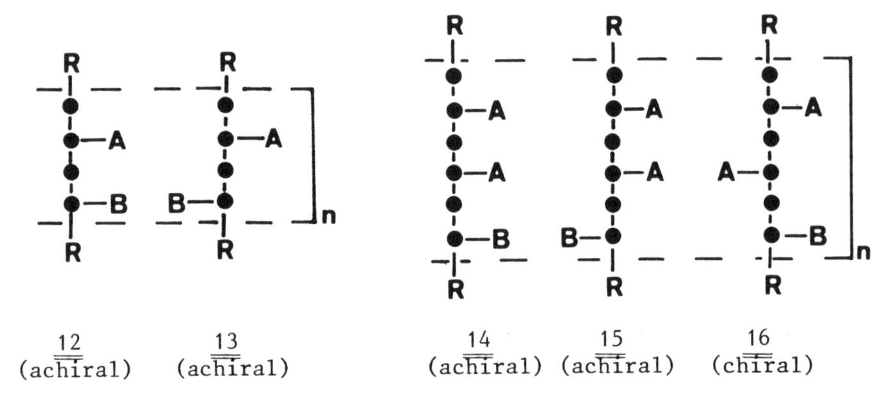

Possible dyad and triad arrangements in regular copolymers

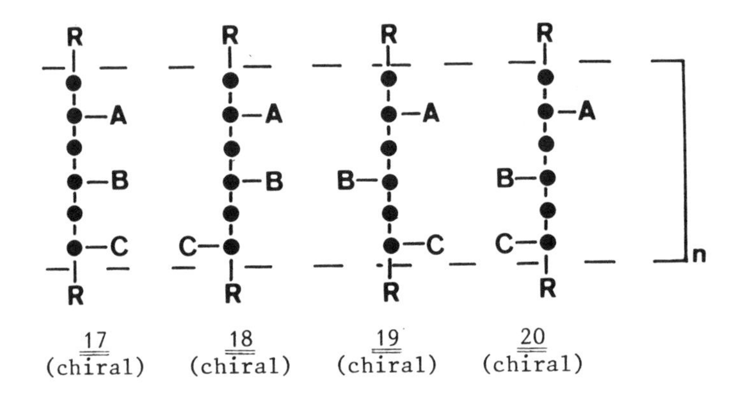

Possible triad arrangements in terpolymers

All four possible triads 17, 18, 19 and 20 of a terpolymer are chiral and asymmetric. If we look at longer sequences in regular copolymers the percentage of possible chiral structural units is increasing.For tetrads of the type AABB there are two achiral and three chiral arrangements.

From these theoretical considerations it appears, polymers of 1-substituted olefins, particularly copolymers, possessing regular structures can exhibit backbone chirality. If these structures can be obtained as one optical antipode through well designed synthetic scheme, they should be optically active. Hence, the prevailing view regarding the impossibility of realizing optically active vinyl polymers due to backbone chirality needs to be changed.

3. CHIRAL CONFORMATIONS

Chirality in vinyl or vinylidene polymers on the basis of an entirely different principle has been realized recently. Okamoto et al. have described the preparation of optically active, isotactic poly (trityl methacrylate) possessing a single-handed helical conformation by anionic polymerization in the presence of (-)-sparteine. In this case the optical activity is not brought about by optically active side groups or backbone chirality rather due to a chiral helical conformation stabilized by bulky trityl groups (16,17). It remained unclear whether during the formation of optically active poly(trityl methacrylate) all the chiral centers generated along the polymer chain possess the same absolute configuration. This ambiguity has been clarified by isolation of uniform, optically active oligomers of trityl methacrylate those have been transformed into optically active oligomers of methyl methacrylate. This investigation showed that growth of the helix is likely to proceed only with identical chain configuration (18).

Optically active, helical conformations in polymers of other methacrylates (19) and of 1-substituted olefins (20) have been observed more recently. For other classes of polymers like (trichloroacetaldehyde) (21) and poly(isocyanides) (22) this type of structure has been reported earlier.

4. AN ASYMMETRIC CYCLOCOPOLYMERZATION

Since it has been deduced theoretically that optical activity due to backbone chirality should be possible, it was of considerable interest to verify these hypotheses by synthesizing the polymers bearing some of the structures shown in section 2. To begin with, we have attempted to synthesize regular copolymers consisting of triads of the type 16. To achieve such triads, we fixed two monomeric units in a definite geometry on a chiral template molecule and tried to copolymerize this template monomer with various other vinyl monomers. After polymerization, the template molecule was intended to be splitt off, from the polymer, resulting in the desired chiral polymers.

For this purpose 3,4-substituted D-mannitol-1,2; 5,6-bis-O-(4-vinylphenylboronate)e.g. 21 was copolymerized with simple comonomers like methyl methacrylate, methacrylonitrile or styrene. After complete removal of the chiral template, the 3,4-substituted D-mannitol, from the polymer, a copolymer of 4-vinylphenylboronic acid with e.g. methyl methacrylate having molecular weight of around 100 000 was obtained. It showed a strong negative optical rotation compared to the starting monomer 21 which exhibits positive rotation (13,14,23). These substances are first examples of polymers derived from 1-substituted olefins or 1,1-disubstituted olefins showing optical activity due to main chain chirality.

By polymer analogous reaction they can be transformed to copolymers of styrene with e.g. methyl methacrylate or methacrylonitrile which are also optically active. The detail stereochemical feature of these polymers can be investigated by ^{13}C-NMR spectroscopy (24).

Systematic investigations of the mechanism of this asymmetric

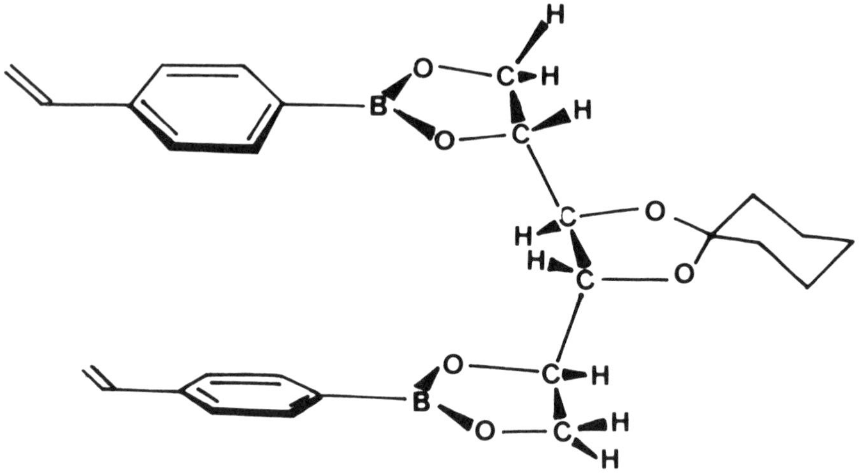

21 Template monomer for asymmetric copolymerization

copolymerization suggests that this polymerization proceeds through a cyclopolymerization involving the monomer 21 (25). In this case both the double bonds of 21 are incorporated in the polymer chain one after another forming a 19-membered ring. This is the first example of a cyclopolymerization (26) involving a large ring, although it is well documented for five or six membered rings. The asymmetric induction occurs during this cyclopolymerization step and chiral dyads of a styrene derivative are formed which are separated from each other by the comonomer. The absolute configuration of the chiral dyads has been determined by a conformational analysis and by the synthesis of appropriate low molecular weight model compounds (25,27).

5. CONCLUSION AND OUTLOOK

Optically active polymers with main chain chirality seem to gain increasing attention in future. By designing suitable synthetic strategies, it will certainly be possible to realize the above described regular polymers possessing chiral structures. Besides being an interesting class of materials from the view point of fundamental investigations, these polymers could be utilized as novel chiral auxiliaries for various asymmetric chemical processes such as for resolution of racemates or asymmetric catalysis. For this purpose, it was of considerable interest to transform these polymers to different chiral functional copolymers through chemical modifications (28). These polymers bearing functional groups arranged in a stereoregular and chiral environment can be used for the investigation of conformational properties by chiroptical methods. Further applications of these polymers may include as materials showing non-linear optical properties used for different purposes.

ACKNOWLEDGEMENT

These investigations were supported by the "Deutsche Forschungsgemeinschaft" and "Fonds der Chemischen Industrie".

REFERENCES

1. N. Beredjick, C. Schuerch, *J. Am. Chem. Soc.* 80 (1958) 1933.
2. G. Natta, M. Farina, M. Peraldo, G. Bressan, *Makromol. Chem.* 43 (1961) 68.
3. M. Farina, M. Peraldo, G. Natta, *Angew. Chem.* 77 (1965) 149.
4. For reviews on optically active polymers see: C.L. Arcus, *Prog. Stereochem.* 3 (1962) 264; R.C. Schulz, E. Kaiser, *Adv. Polym. Sci.* 4 (1965) 236; P. Pino, Ibid. 4 (1965) 393; M. Farina, G. Bressan, *Stereochem. Macromol.* 3 (1967) 181; E. Selegny, Ed. *Optically Active Polymers*; D. Reidel Publishing Comp.: Dordrecht, Holland 1979; P. Pino, G.P. Lorenzi, in:*Preparation and Properties of Stereoregular Polymers*, R.W. Lenz, F. Ciardelli, Eds.; D. Reidel Publishing Comp.: Dordrecht, Holland, 1980.Chapter 1.
5. G. Natta, M. Farina, M. Donati, M. Peraldo, *Chim. Ind. (Milan)*42 (1960) 1363.
6. For reviews on the stereochemistry of vinyl polymers see: A.D. Ketley, Ed. *The Stereochemistry of Macromolecules.* 3 Volumes, M. Dekker, New York, 1967-1968; F.A. Bovey, *Polymer Conformation and Configuration,* Academic Press, New York, 1969; R.W. Lenz and F. Ciardelli, Eds., *Preparation and Properties of Stereoregular Polymers,* D. Reidel Publishing Comp, Dordrecht, Holland, 1980.
7. H. Staudinger, *Die hochmolekularen, organischen Verbindungen,* J. Springer, Berlin, 1932, pp. 113-114.
8. F. Ciardelli, P. Salvadori, *Pure Appl. Chem.* 57, (1985) 931.
9. M.M. Green, B.A. Garetz, *Tetrahedron Lett.* 25, (1984) 2831.
10. C.L. Arcus, *J. Chem. Soc.* 1957, 1189.
11. G. Wulff, A. Sarhan, K. Zabrocki, *Tetrahedron Lett.* 1973, 4329; G. Wulff, W. Vesper, R. Grobe-Einsler, A. Sarhan, *Makromol. Chem.* 178, (1977) 2799.
12. For a review see: G. Wulff, in: W.T. Ford, Ed., *Polymeric Reagents and Catalysts,* ACS-Symposium Series, 308, 1986, p. 186.
13. G. Wulff, K. Zabrocki, J. Hohn, *Angew. Chem.* 90, (1978) 567; *Angew. Chem. Int. Ed. Engl.* 17, (1978) 535.
14. G. Wulff, J. Hohn, *Macromolecules,* 5 (1982) 1255.
15. For a short review see: G. Wulff, *Nachr. Chem. Techn. Lab.* 33 (1985) 956.
16. Y. Okamoto, K. Suzuki, K. Ohta, K. Hatada, H. Yuki, *J. Am. Chem.*

Soc. 101 (1979) 4763.
17. Y. Okamoto, K. Ohta, K. Hatada, H. Yuki, in: *Anionic Polymerization: Kinetics, Mechanism and Synthesis,* J.E. McGrath, Ed., ACS-Symposium Series 166, 1981, p. 353.
18. G. Wulff, R. Sczepan, A. Steigel, *Tetrahedron Lett.* 27 (1986) 1991.
19. D.J. Cram, D.Y. Sogah, *J. Am. Chem. Soc.* 107 (1985) 8301.
20. W. Kaminsky, S. Niedoba, presented at the *Makromolekulares Kolloquium,* Freiburg, March 1986 in Freiburg, F.R.G.; Abstract page 33.
21. W.J. Harris, O. Vogl, J.R. Havens, J.L. Koenig, *Makromol. Chem.* 184 (1983) 1243 and earlier papers.
22. J.M. Van der Eijk, V.E.M. Richters, R.J.M. Nolte, W. Drenth, *Recl. Trav. Chim. Pays-Bas* 103 (1984) 46 and earlier papers.
23. G. Wulff, R. Kemmerer, J. Vietmeier, H.-G. Poll, *Nouv. J. Chim.* 6 (1982) 681.
24. G. Wulff, R. Kemmerer, P.K. Dhal, A. Steigel, Manuscript under preparation.
25. G. Wulff, R. Kemmerer, B. Vogt, Manuscript under preparation.
26. see e.g.: G.B. Butler, *J. Polym. Sci. Polym. Symp.* 64 (1978) 71.
27. G. Wulff, R. Kemmerer, B. Vogt, presented at the *Makromolekulares Kolloquium,* March 1986 in Freiburg, F.R.G., Abstract page 31.
28. P.K. Dhal, G. Wulff, Manuscript under preparation.

CONFORMATIONAL OPTICAL ACTIVITY IN POLYMERS

F. Ciardelli, M. Aglietto, and G. Ruggeri
Dipartimento di Chimica e Chimica Industriale, Università di Pisa
Centro CNR Macromolecole Stereoordinate ed Otticamente Attive
Via Risorgimento 35
56100 Pisa
Italy

ABSTRACT. Structural order along a polymer chain involves in general monomolecular chirality, but even in highly isotactic polymers intramolecular compensation occurs due to equimolar amounts of sections with opposite handness. Predominance of one screw sense conformations can be induced by polymerizing optically active monomers, which incorporate in the chain an excess of asymmetric carbon atoms with a single absolute configuration. Whereas in low molecular weight compounds such an excess is proportional to the enantiomeric purity - and then to the magnitude of optical activity -, in polymers the asymmetric effect can be cooperatively transmitted along the chain. This effect in isotactic macromolecules can be achieved by chiral initiators or by copolymerizing chiral monomers with achiral comonomers. As the consequence, the dissymmetric conformational ordering of units from achiral monomers can provide a surplus of optical rotation (conformational optical activity). Synthetic approaches to polymers showing these particular properties and requisites for conformational optical activity are presented as well as experimental evidences deriving from measured and calculated chiroptical properties of typical polymers.

1. INTRODUCTION

In this paper the term conformational optical activity referred to macromolecular compounds is used to indicate the presence of secondary structures, involving the macromolecule as whole or substantial fraction of it, with a predominance handness. This last is not to be due simply to stereogenic centers (asymmetric centers) with a single absolute configuration in each repeating unit.

Such conformational optical activity must be observable at molecular level in the isolated macromolecule, that is in the diluted solution, and should not be due to intermolecular interactions, as in the aggregates or crystallites.

Detection of the phenomenon can be performed by determination of chiroptical properties (measurement of optical rotation and circular dichroism). However direct evidence is not simply coming out from these

M. Fontanille and A. Guyot (eds.), Recent Advances in Mechanistic and Synthetic Aspects of Polymerization, 409–423.

measurements but conformational analysis and comparison with suitable low molecular weight analogs are needed (1).

Indeed existence of purely conformational optical activity is not an unique macromolecular requisite, being well known in low molecular weight atropisomerism (2). However in polymers it assumes a very specific characteristic connected with the occurence of cooperative effects which allow transmittance of molecular asymmetry along the chain up to very long distances. Moreover these concepts can be used in order to obtain new polymeric materials with high oriented order and useful chemical or physical properties.

In this paper molecular requisites for conformational optical activity in polymers are presented as well as the available synthetic methods. Finally the particular chiroptical properties of these systems are discussed with reference to relationship between optical activity and conformation in synthetic macromolecules.

2. REQUISITES FOR CONFORMATIONAL OPTICAL ACTIVITY IN POLYMERS

The most straightforward possibility to obtain conformational optical activity in macromolecules is offered by polymers whose macromolecules can assume conformations with opposite chirality which cannot rapidly interconvert. Let us consider for sake of simplicity chains, which due to primary structure, preferentially assume helical conformations. This is the typical case of isotactic vinyl polymers for which the minimum energy conformation is a helix (3). Isotactic macromolecules derived from achiral monomers have no preference for right-or left-handed screw senses and the two are perfectly balanced at inter- and intramolecular level. However distribution of left- and right-handed helical secondary structures affect markedly the free energy of the system, alternation of the two senses in the same chain being favoured for entropic reasons (4,5).

If this last situation takes place, conformational optical activity cannot be obtained due to intramolecular conformational compensation which hinders any isolation of chains with a predominant handness.

In order to make one screw sense largely predominating in a single macromolecule the intramolecular equilibration must be hindered by building very rigid chains. In the limiting case the chains will results as rods with either left- or right-handed helicity. Even if hindering of equilibration can be considered as a kinetic effect, it cannot be excluded that thermodynamic contributions are involved particularly when rigidity is due to bulky side chains and conformational reversals have a very high internal energy (5,6). In other words bulky sides chains in a vinyl type structure forinstance can favour the formation of longer helical chain sections as the lower entropy is balanced by the gain in internal energy due to the minimization of conformational reversal number. Indeed these last have in case of macromolecules with bulky and branched side chains, larger internal energy per structural unit than the same unit in the helical conformation (7). Which is the actually operating mechanism cannot be established as a general rule depending on primary structure, an increase in temperature favouring in both cases the equilibration of the two screw senses within each macromolecular chain.

As mentioned in the introduction the absence of stereogenic centers (8) in the macromolecule or macromolecular section displaying conformational optical activity simplify the stereochemical analysis and the understanding of the origin of the chiral arrangement of the backbone.
In most cases stereogenic centers are indeed present, but again it will be possible to call for conformational optical activity if these centers do not affect the position of the conformational equilibrium between right- and left-handed conformers. Typically isotactic vinyl polymers with high enough molecular weight can be considered as meso structure and the tertiary carbon atoms in the backbone cannot therefore induce any optical activity unless a dissymmetric conformation (one screw sense helix) does not predominate within a single macromolecule (9).

In such a case according to the previously mentioned definition, the polymer could show conformational optical activity after isolation of a polymer consisting of chains with a predominant handness.
As discussed by Wulff et al. (10) a suitable distribution of asymmetric carbon atoms in copolymers can provide optical activity in the polymer, which however should not be mainly of conformational origin.

3. SYNTHETIC APPROACHES

According to general principles discussed in the previous paragraph the synthetic route must provide a polymeric structure consisting of macromolecules disposed up to a certain extent in a chiral conformation with a predominant handness.

Clearly if the catalyst or initiator are achiral or racemic and the same holds for the monomer, the obtainement of an optically active polymer is realized by separation of macromolecules with one or the other handness. On the other side if optically active initiator or catalyst are used, the polymer with conformational optical activity can be directly obtained when starting with a structurally suitable monomer. Schematically the three following routes have been up to now followed succesfully.

3.1. Polymerization of achiral monomers

Polymers from achiral isonitriles obtained in the presence of Ni-catalysts

have been separated into fractions with opposite optical rotation due to the one sense helical conformation of the backbone (11).

The direct production of an optically active polymer by optically active catalyst has been only partially successfull(12), while if R contain a stereogenic center of a single configuration the large predominance of one screw sense can be achieved (13) by homopolymerization of a single enantiomer.

3.2. Polymerization of prochiral monomers with bulky side groups

In the case of tritylmethacrylate (TrMA) the polymerization with an optically active catalyst gives polymers with chiroptical properties related to the one screw sense helical conformation (14,15).

$$CH_2{=}C(CH_3)-C(=O)-OTr \xrightarrow{\quad} -CH_2-C(CH_3)(C{=}O\,OTr)- \qquad Tr = C(Ph)_3$$

No attempt has been made in this case to separate into fractions with opposite rotation the optically inactive polymer obtained with a nonoptically active or better racemic catalyst. In the last case one could expect that left- and right-handed helices can be obtained having conformational optical activity of opposite sign.

3.3. Copolymerization of an achiral monomer with a chiral comonomer

In this case the inclusion of the two monomers in the same chain induces the structural units of the achiral monomer to assume a chiral conformation. Thus if only one enantiomer of the chiral monomer is used, the units of the achiral comonomer can display induced optical activity mainly of conformational origin.

Examples of this method are offered by the co isotactic copolymers of optically active α-olefins with achiral comonomers as 4-methyl-1-pentene (16), styrene (17) and vinylnaphthalenes (18) obtained by Ziegler-

$$CH_2{=}CH-\overset{*}{C}H(CH_3)R + CH_2{=}CH-C_6H_5 \longrightarrow -CH_2-CH(\overset{*}{C}H(CH_3)R)-CH_2-CH(C_6H_5)-$$

Natta catalyst and anionic copolymers of TrMA with (S)-α-methylbenzylmethacrylate (19,20).

A similar approach was been more recently estended to free radical copolymerization of several achiral chromophoric monomers with (-)-menthylacrylate or methacrylate (21).

Moreover cationic and free radical initiators have been used for copolymers of 9-vinyl carbazole with several optically active monomers (22).

4. CHIROPTICAL EVIDENCES OF CONFORMATIONAL OPTICAL ACTIVITY

4.1. Isotactic polymers

Chirooptical properties (molar rotatory power $|\Phi|$ and molar ellipticity $|\Theta|$) result from a weighed average of the contributions from different conformers as shown in equations 1 and 2.

$$|\Phi| = \Sigma_i N_i \; |\Phi|_i \qquad |1|$$

$$|\Theta| = \Sigma_i N_i \; |\Theta|_i \qquad |2|$$

where N_i indicates the molar fraction and $|\Phi|_i$ (or $|\Theta|_i$) the molar rotatory power (or molar ellipticity) of the ith conformer.
In macromolecules the molar entities are referred to one residue, thus the chiroptical properties are indipendent of molecular weight.

It has been demonstrated that in isotactic polymers of optically active α-olefins the molar optical rotation per monomeric residue can be interpreted in terms of the prevalence of few conformations with very high optical rotation of the same sign corresponding to those allowed to the structural unit inserted in an one screw sense helix (23).
Thus with respect to the low molecular weight analog the stereospecific polymerization to isotactic chain implies the selection of few highly dissymmetric conformers which are stabilized by cooperative interactions among chiral side chains (24).

Such cooperative effect can be transmitted also along sequences of units with non chiral side chain in a coisotactic copolymer.
Indeed in copolymers of (S)-4-methyl-1-hexene (4MH) with 4-methyl-1-pentene (4MP)

```
            H             H
            |             |
   —CH2 -C ——CH2 -C —
            |             |
          CH2           CH2
            |*            |
       H -C-CH3      H -C-CH3
            |             |
          C2H5          CH3
```

it has been shown that 4MP units contribute to the copolymer optical rotation to an extent increasing in absolute value with increasing isotacticity degree and content of optically active co-units. However even in the copolymer with only 11 % mol of 4MH, the 4MP blocks have still a predominance of 22 % of the favoured screw sense (16) (Table I).

An even more straightforward evidence is offered by the investigation of circular dichroism of coisotactic copolymers of optically active α-olefins with vinylaromatic monomers (TableII). It is known that the oblique orientation of transition dipoles of at least two suitable chromophores can give rise to band splitting as esemplified in Fig. 1.

TABLE I
Conformational optical activity of 4-methyl-1-pentene (4MP) units in coisotactic copolymers with (S)-4-methyl-1-hexene (S4MH) (16)

Stereoregularity	Molar fraction of 4MP in the copolymer	Average molar rotatory power of 4MP units in the copolymer [a]	Molar fraction of 4MP units inserted in left handed helical sections [b]
Partially isotactic	0.29	+137	0.78
	0.50	+136	0.78
	0.74	+76	0.68
	0.89	+55	0.61
Highly isotactic	0.30	+160	0.83
	0.56	+155	0.82
	0.77	+98	0.70

a) Calculated by comparison with homopolymers mixtures having the same composition.
b) Evaluated on the basis of the molar rotatory power of +240 calculated for purely isotactic poly(4MP)in the left-handed helix.

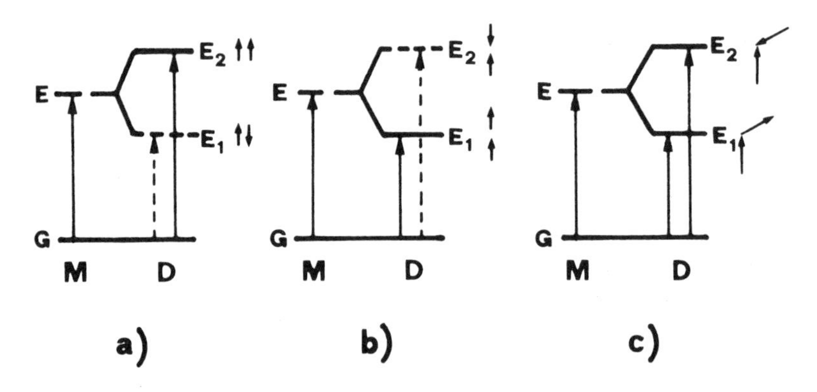

Fig. 1 Energy levels for dimers D with different relative geometrical disposition with respect to monomer M.

a) parallel (hypsochromic effect)
b) head-to-head (bathochromic effect)
c) skewed (band splitting ≡ exciton)

TABLE II
Conformational optical activity of the aromatic chromophore in coisotactic random copolymers of (-)(R)-3,7-dimethyl-1-octene with vinylaromatic comonomers

Vinyl aromatic monomer		Circular dichroism						
Type	molar fraction in the copolymer	exp. $\lvert\Theta\rvert \cdot 10^{-3}$ (λ)		calc.a) $\lvert\Theta\rvert \cdot 10^{-3}$ (λ)		right-handed helix (calc.)b) $\lvert\Theta\rvert \cdot 10^{-3}$ (λ)		Ref.
Styrene	0.20	-63 (195)	+36 (187)	-54.5 (193)	+33 (186)	-310 (194)	+165 (183)	(17,25)
o-methyl-styrene	0.20	-97	+74	-47	+36.3	-264 (196)	+221 (189)	(17,25)
	0.46	-85 (200)	+71 (192)	-45.7 (195)	+34.3 (187)			
1-vinyl-naphtalene	0.03	-68	+35.3	-44.3	+13.3	-5740	+1380	(18,26)
	0.10	-145	+103	-165	+50			
	0.36	-175 (232)	+130 (223)	-363 (231)	+132 (220)			

a) For a statistical coisotactic copolymer having the same composition (see refs. 18 and 25)
b) Assuming for the right-handed helical homopolymer the same CD as the tetramer (see refs. 18 and 25)

This behaviour can be expected at least for the strong allowed highest energy $\pi \to \pi^*$ electronic transition of the aromatic chromophores in the subject copolymers. Moreover, if the oblique orientation has predominantly the left- or right-handed skewness (Fig. 2) a couplet, characterized by two dichroic bands of opposite sign, is observed in the same absorption region, when recording the CD-spectrum.

This expected couplet has been actually observed for several coisotactic copolymers of (R)-3,7-dimethyl-1-octene. The comparison between observed and calculated ellipticity values and sign are in agreement with a large predominance of right-handed helical conformations which are favoured by the (R)-monomer (Table II). Indeed for the calculation of copolymer CD it has been assumed that non isolated aromatic unit contribute to the CD-couplet as in a homopolymer right-handed helix (25).

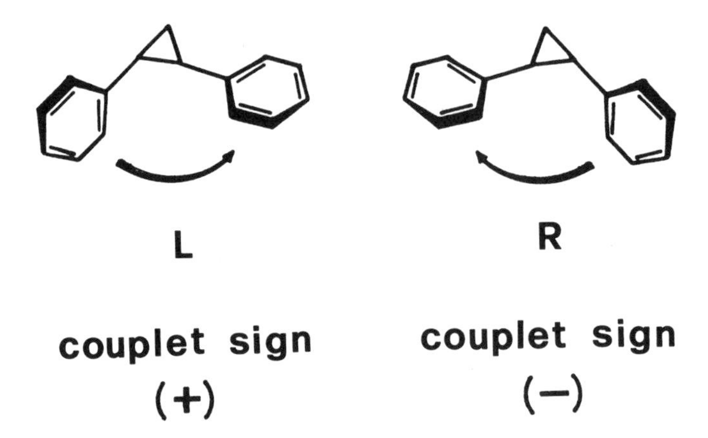

Fig. 2 Relative chiral disposition of aromatic chromophores giving rise to exciton CD couplet in left- (L) and right-handed (R) 3_1 helices (25)

A substantially similar situation seems to hold for copolymers between tritylmethacrylate (TrMA) and optically active α-methylbenzylmethacrylate (19,20). Both optical rotation and CD show remarkable dependence on composition and change of sign when TrMA content increases. The maximum absolute values of chiroptical properties are reached with small amounts of the optically active comonomer (0.05 mol) when correspondingly the isotacticity degree is the highest (84%) (Table III). Moreover the fully isotactic TrMA homopolymer obtained in the presence of the anionic initiator (−)-sparteine / n.BuLi complex (14) has much higher optical rotation in spite of the complete absence of any residue from optically active starting material (Table III).
These results are interpreted on the basis of the single screw sense helical conformation of the isotactic TrMA chains.

The substantially conformational origin of the high optical activity observed in isotactic poly(TrMA) has been nicely confirmed very recently (27). Anionic polymerization of TrMA using 1,1-diphenylhexyl lithium / (−)-sparteine gave oligomers with predominantly negative optical rotation whereas the polymer (DP≃30) showed strong positive rotatory power 10 to 30 times larger in absolute value. This supports the Okamoto et al. (14) original idea that, apart asymmetric induction at level of monomer insertion, the repetition of this act with identical stereochemistry gives rise to one screw sense helical chain when DP > 10÷30.

Also consistent with this general picture are the chiroptical properties of polymers obtained starting with diphenyl-2-pyridylmethylmethacrylate (D2PyMA) and several organolithium complexes of chiral ligands (28). In this case optical rotation of the polymer depended greatly on polymerization conditions and in some cases extraction with THF or gel permeation chromatography allowed to fractionate the polymers in frac-

TABLE III
Conformational optical activity in polymers from triphenylmethylmethacrylate (TrMA)

Polymer	Molar fraction of TrMA units	Isotacticity %	Molar rotatory power for TrMA units	Ref.
Copolymer of TrMA with (S)-α-methylbenzylmethacrylate	0.22	9	-74	(19,20)
	0.42	13	-64	
	0.68	36	-61	
	0.81	57	+187	
	0.95	84	+435	
Polymer from TrMA initiated by (-)-sparteine / n.BuLi complex	1.00	100	+1239	(14,15)

tions with opposite rotatory powers. The order of magnitude of these last was very close to that of 100% isotactic poly(TrMA). Considering that the asymmetric induction with a determined optically active initiator should be of constant handness, the origin of these polymer fractions with opposite optical rotation has been attributed to the different propeller structures which may be originated by the restricted rotation of the pyridyl group. Indeed the only possibility existing in this system for obtaining polymers with optical rotation of opposite sign is the formation of distinct structures with enantiomeric chirality, such as helices with right and left screw senses, thus providing a very characteristic case of "conformational optical activity".

4.2. Non stereoregular copolymers

In the previous section it has been clearly shown that the isotactic structure isan important prerequisite to grant one screw sense helical sections, and then conformational optical activity, both in copolymers of an optically active monomer with a nonchiral one and in homopolymers of suitable hindered achiral monomers obtained in the presence of optically active initiators. However in the last ten years it has been shown (21) that in free radical atactic copolymers of the type mentioned above optical activity can be evidenced in the electronic transition of several chromophoric achiral comonomers inserted in the chains with (-)-menthylacrylate (MtA) or other optically active comonomers (21,22). A typical example is offered (29) by copolymers of styrene with MtA showing optical activity in the lowest energy $\pi\rightarrow\pi^*$ electronic transition of the aromatic chromophore. The molar ellipticity even in the copolymer containing only 28% mol of MtA is larger in absolute value than in the low molecular weight model where two menthyl groups per phenyl group are present. Moreover the ellipticity increases with increasing the average length of MtA sequences and is about 50% of that of the corresponding coisotactic copolymer (Table IV). These results are consistent with the conformational rather than configurational origin of the observed CD as the ellipticity is not directly related to the average distance of the phenyl group from the asymmetric carbon atoms present in the menthyl groups. Unfortunately, the presence of the ester groups absorbing around 210-220 nm hinders the investigation of CD properties of phenyl chromophore in the spectral region where the allowed $\pi \rightarrow \pi^*$ electronic transition is located and a possible CD-couplet deriving from exciton splitting could be detected.

The investigation was then extended to MtA copolymers with comonomers having more polarizable chromophores and having strong absorption bands in a well distinct region with respect to the ester absorption. The more informative systems investigated up to now are certainly MtA copolymers with p-vinyltrifluoroacetophenone (VTFA) (30) and trans-vinylstylbene (VS) (31).

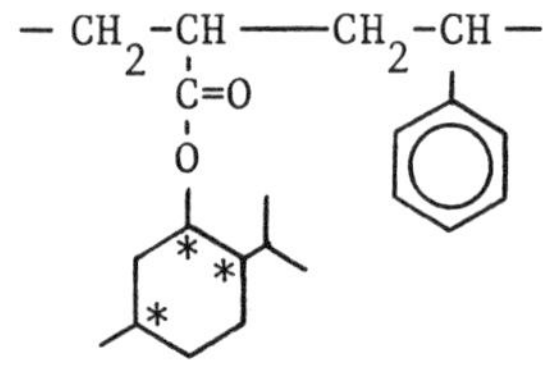

— CH_2-CH ——CH_2-CH —   — CH_2-CH ——CH_2-CH —

C=O   C=O

O   O

H— C = C —H

MtA/VTFA 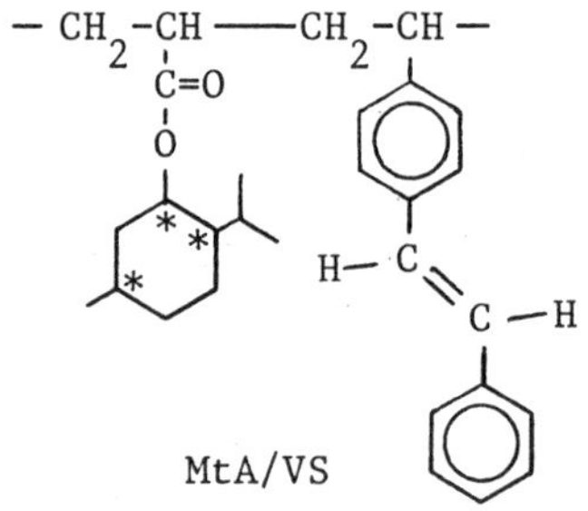   MtA/VS

In both cases the units derived from the achiral comonomer, VTFA and VS respectively, contribute to the polymer optical rotation, the sign of this contribution at 589 nm being positive and then of opposite sign with respect to that of MtA units. However whereas in case of the VTFA/MtA copolymers the values of the rotatory power of VTFA units decreases monotonically with decreasing the content of the optically active comonomer (Table V), a substantially asymptotic value for a content of MtA larger than 60% mol is observed for the molar rotation coming from VS units (Table VI).

These indications in favour of conformationally induced optical rotation are strongly substantiated by CD measurements in the absorption region of the respective trifluoroacetylphenyl and 4-stilbenyl groups. In particular a positive couplet is detected in both cases the amplitude of which reaches a maximum at about 50/50 composition for both copolymers (Tables V and VI). Unfortunately lacks of analogous stereoregular polymers and of chiroptical properties calculations of definite conformation do not allow at present to evaluate the extent of dissymmetric

TABLE IV
Conformational optical activity of styrene units in atactic copolymers with (-)-menthylacrylate (29).

Structure	Sequence distribution	Molar fraction of styrene units	Average sequence length of styrene units	Molar ellipticity at 262 nm	
				these compounds	isotactic left-handed elix a)
$-CH_2-CH(C_6H_5)-CH_2-CH(CO-O-menthyl)-$	Random	0.06	1.01	+333	+660
		0.14	1.03	+294	+460
		0.35	1.15	+228	+380
		0.54	1.59	+109	+280
		0.72	2.96	+ 76	+165
	Alternating	0.49	1.00	+191	-
Model: $CH_2(CO-O-menthyl)-CH_2-CH(C_6H_5)-CH_2-CH_2(CO-O-menthyl)$		0.33	1.00	- 59	-

a) See refs. 17 and 25

TABLE V
Conformational optical activity of p-vinyltrifluoroacetophenone units (VTFA) in random copolymers with (-)-menthylacrylate (MtA) (30)

Molar fraction of VTFA units in the copolymer	Average sequence length ($\bar{\ell}$)		Molar rotatory power of VTFA units [a]	Molar ellipticity · 10^{-3} of VTFA units in the couplet region	
	$\bar{\ell}_{VTFA}$	$\bar{\ell}_{MtA}$		276 nm	255 nm
0.10	1.1	9.4	+110	+1.48	-0.56
0.25	1.3	4.4	+ 12	+3.30	-1.68
0.41	1.7	2.6	+ 22	+2.90	-2.90
0.61	2.6	1.7	+ 26	+2.51	-1.88
0.81	5.3	1.3	+ 11	+1.15	-0.59

a) See footnote a in Table I .

TABLE VI
Conformational optical activity of 4-vinylstylbene units (VS) in random copolymers with (-)-menthylacrylate (MtA) (31)

Molar fraction of VS units in the copolymer	Average sequence length ($\bar{\ell}$)		$\bar{M}w$ 10^{-3}	Molar rotatory power of VS units a)	Molar ellipticity · 10^{-3} of VS units in the couplet region	
	$\bar{\ell}_{VS}$	$\bar{\ell}_{MtA}$			322 nm	285 nm
0.14	1.1	5.3	69.0	+107.0	+1.58	-1.15
0.26	1.2	3.4	32.0	+121.0	+3.73	-1.75
0.41	1.4	2.2	30.5	+114.0	+4.88	-3.17
0.57	1.9	1.6	31.0	+ 91.5	+5.02	-2.84
0.67	2.5	1.3	35.0	+ 71.0	+4.62	-2.01
0.78	4.0	1.2	50.0	+ 47.0	+4.02	-1.55
0.89	10.5	1.1	53.0	+ 20.5	+0.86	-0.53

a) See footnote a in Table I .

conformationsinvolving the achiral chromophoric units. However the values of chiroptical properties compared to low molecular weight compounds and the exciton splitting of the longest wavelength absorption bands are consistent with the presence, even in thesesubstantially atactic random copolymers, of ordered chiral structures with a single sense of skewness.

5. FINAL REMARKS

In a time during which one of the main goal of polymer science and technology is the production of new highly specialized materials, the present work is intended to suggest possible approachesto prepare optically active macromolecular compounds by using relatively small amounts of expensive optically active starting materials.

Indeed the proper choice of monomers for copolymerization with esigue fractions of optically active comonomers or by starting the polymerization with optically active initiators or catalysts can allow to end up with macromolecules where the units from the originally achiral monomer have assumed optical activity due to insertion in chiral conformations with a predominant handness (conformational optical activity).

In this paper the molecular requisites of the monomers to be used and the need of controlled polymerization stereochemistry when starting with a bulky monomer and optically active initiators are shown.
The copolymerization method on the other side is more flexible as conformational optical activity can be obtained also in free radical processes where the steric control of the polymerization is very low if any.

Finally experimental approaches based on measurements of chiroptical properties, both optical rotatory power and circular dichroism in solution, have been indicated. Evaluation of the extent of conformational optical activity achieved is not easy at the present but it is possible in some cases in which it seems to be rather large even if detected in solution where macromolecules display a large flexibility.
It is possible to expect that these effects are even larger in the solid state as reported for polymers from chloral (32). In this last aggregation state conformational optical activity can be extended even in flexible structures such as polypropylene (33) when these polymers are prepared starting with optically active initiators or catalysts.

6. REFERENCES

1. F. Ciardelli and P. Salvadori, Pure & Appl. Chem., **57**, 931 (1985)
2. see for instance : M. Oki, Topics in Stereochemistry, **14**, 1 (1983)
3. G. Natta, Makromol. Chem., **35**, 93 (1960) and references therein
4. G. Allegra, P. Corradini and P. Ganis, Makromol. Chem., **90**, 60 (1966)
5. T. M. Birshstein and P. L. Luisi, Vysocomol. Soed., **6**, 1238 (1964)
6. P. L. Luisi and F. Ciardelli in Reactivity,Mechanism and Structure in Polymer Chemistry, A. D. Jenkins and A. Ledwith Eds., J. Wiley & Sons, N. Y. (1974), p. 471

7. P. L. Luisi and P. Pino, J. Phys. Chem., 72, 2400 (1968)
8. K. Mislow and J. Siegel, J. Am. Chem. Soc., 106, 3319 (1984)
9. M. Farina, Chimica & Industria, 68, 62 (1986)
10. G. Wulff, R. Kemmerer, J. Vietmeier and H.-G. Poll, Nouv. J. Chim., 6, 681 (1982)
11. R. J. M. Nolte, A. J. M. van Beijnen and W. Drenth, J. Am. Chem. Soc., 96, 5932 (1974) ; W. Drenth and R. J. M. Nolte, Acc. Chem. Res., 12, 30 (1979)
12. P. C. J. Kamer, R. J. M. Nolte and W. Drenth, J. Chem. Soc., Chem. Commun., 1798 (1986)
13. A. J. M. van Beijnen, R. J. M. Nolte, A. J. Naaktgeboren, J. W. Zwikker and W. Drenth, Macromolecules, 16, 1679 (1983)
14. Y. Okamoto, K. Suzuki, K. Hota, K. Hatada and H. Yuki, J. Am. Chem. Soc., 101, 4763 (1979)
15. Y. Okamoto, H. Shohi and H. Yuki, J. Polym. Sci. Polym. Lett. Ed., 21, 601 (1983)
16. C. Carlini, F. Ciardelli and P. Pino, Makromol. Chem., 119, 244 (1968)
17. F. Ciardelli, P. Salvadori, C.Carlini and E. Chiellini, J. Am. Chem. Soc., 94, 6536 (1972)
18. F. Ciardelli, C. Righini, M. Zandomeneghi and W. Hug, J. Phys. Chem., 81, 1948 (1977)
19. H. Yuki, K. Ohta, Y. Okamoto and K. Hatada, J. Polym. Sci. Polym. Lett. Ed., 15, 589 (1977)
20. Y. Okamoto, K. Suzuki and H. Yuki, J. Polym. Sci. Polym. Chem. Ed., 18, 3043 (1980)
21. F. Ciardelli, M. Aglietto, C.Carlini, E. Chiellini and R. Solaro, Pure & Appl. Chem., 54, 521 (1982) and references therein
22. E. Chiellini, R. Solaro, G. Galli and A. Ledwith, Adv. Polym. Sci., 62, 143 (1984)
23. P. Pino, F. Ciardelli, G. P. Lorenzi and G. Montagnoli, Makromol. Chem., 61, 207 (1963)
24. P. Pino, F. Ciardelli and M. Zandomeneghi, Ann. Rev. Phys. Chem., 21, 561 (1970) and references therein
25. W. Hug, F. Ciardelli and I. Tinoco, Jr., J. Am. Chem. Soc., 96, 3407 (1974)
26. F. Ciardelli, P. Salvadori, C. Carlini, R. Menicagli and L. Lardicci, Tetrahedron Letters, 16, 1779 (1975)
27. G. Wulff, R. Sczepan and A. Steigel, Tetrahedron Letters, 27, 1991 (1986)
28. Y. Okamoto, H. Mohri, M. Ishikura, K. Hatada and M. Yuri, J. Polym. Sci. Polym. Symp., 74, 125 (1986)
29. R. N Majumdar and C. Carlini, Makromol. Chem., 181, 201 (1980)
30. A. Altomare, C. Carlini, F. Ciardelli and E. M. Pearce, J. Polym. Sci. Polym. Chem. Ed., 21, 1693 (1983)
31. A. Altomare, C. Carlini, M. Panattoni and R. Solaro, Macromolecules, 17, 2207 (1984)
32. O. Vogl and G. D. Jaycox, Chem. Tech., 698 (1986)
33. W. Kaminsky in History of Polyolefins, R. B. Seymour and T. Cheng Eds., D. Reidel Publishing Company (1986), p. 257

CHIRAL THERMOTROPIC LIQUID CRYSTALLINE POLYMERS: AN OVERVIEW

Emo Chiellini and Giancarlo Galli
Dipartimento di Chimica e Chimica Industriale
Università di Pisa
56100 Pisa, Italy

ABSTRACT. The present contribution is aimed at providing an up-to-date overview on the ever growing area of polymer chemistry dealing with mesomorphic macromolecular compounds. Attention is focused on thermotropic liquid crystalline polymers characterized by the presence of chirality in the repeat units and with mesogenic groups in either the side chain or the main chain. The strategies relevant to the synthetic routes applied for the preparation of chiral thermotropic liquid crystalline polymers are discussed and information is provided on their structural and thermal-optical features. As a conclusion one may find indications to the production, tuning of the bulk properties and possible applications of a novel class of polymeric materials.

1. INTRODUCTION

In recent years a great deal of attention has been devoted by several polymer reasearch groups to the synthesis and characterization of new macromolecular compounds displaying mesomorphic properties either in the melt (thermotropic) or in solution (lyotropic).

The extension to those materials of the concepts and fundamental knowledge well established in the field of low molar mass 'liquid crystalline' (LC) compounds has led to the coniation of the term 'liquid crystalline polymers' as suited to classify macromolecular compounds containing mesogenic groups in either the side chain or the main chain. Accordingly, they are able to show, under conditions controlled by external parameters, phase behaviour intermediate between isotropic liquids and ordered solids (1).

Depending upon specific structural features (2), essentially two types of ordering can be adopted by the mesogenic units in the mesomorphic state, namely orientational order (nematic) or orientational and positional order (smectic) with several modifications (A, B, ..., K). When the mesogenic units are asymmetrically perturbed, either directly by the

M. Fontanille and A. Guyot (eds.), Recent Advances in Mechanistic and Synthetic Aspects of Polymerization, 425–450.

presence of a chiral center on them or indirectly by simple mixing with a chiral compound, the nematic phase tends to twist and assemble in a helical array characterized by a prevalent screw sense (cholesteric) of the orientation of the director of each nematic plane (3). Similarly, the tilted smectic phases, when asymmetrically perturbed, give rise to a stacking of layers with a preferential twist of the tilt, and the bulk material turns to be ferroelectric in character (4).

The helical twisting power (Φ) of the cholesteric phase |eq.1|, as generated by dipole-quadrupole interactions of chiral molecules (5), may vary in thermotropic binary systems with the content of one component in either a linear (6,7) or non-linear (8,9) fashion.

$$\Phi = 2\pi d/p \qquad |1|$$

d: distance between two successive nematic planes
p: length of the helical pitch

As a consequence of their uncompensated helical structure, cholesterics exhibit quite unusual optical rotation and circular dichroism properties (10,11). Moreover, homogeneously oriented planar cholesteric textures, when radiated with light parallel to the optical axis, i.e., normal to the nematic planes, give rise to a Bragg-type scattering accompanied by selective reflection (12), the wavelength (λ_R) of the reflected light being:

$$\lambda_R = \bar{n}p \qquad |2|$$

$\bar{n}$: average refractive index of the cholesteric
p: length of the helical pitch

In particular, depending on the handedness of the cholesteric helix, a right-handed or left-handed polarized light can be reflected, that is the component of the ordinary light mismatching the helical handedness, whereas the matching component passes through unaltered in intensity and direction of propagation. From this behaviour one may envisage for cholesterics and chiral smectics with short relaxation times applications in optoelectronics, as thermal indicators and radiation sensors, while for systems with long relaxation times the field of passive devices, such as polarizers, electromagnetic filters, reflective displays and storage displays appears rather promising.

The first kind of applications has mainly found room in the field of low molar mass LC compounds (13-16) and there are also attempts with side chain LC polymers (17,18). For the second set of applications, particularly attractive appear polymeric LC that can retain for fairly long times or indefinitely their long range conformational order after removal of the perturbing factors.

The interest and properties of lyotropic LC polymers, whose discovery dates back to the 50's (19,20), have been recently reviewed (21,22). The importance of the LC order in living systems and in model substances is also well documented (23-29). However, any comment on this special area is beyond the scope of this paper, that is intended to focus on thermotropic LC polymers characterized by the presence of intrinsically chiral elements.

We are not going to consider here the principles that are to be taken into account for the assessment of molecular architectures best suited to provide mesomorphic responses, as authoritative contributions on the subject are already available (1,30-35). We rather pay attention to the synthetic strategies adopted for the preparation of new polymeric materials whose properties in the bulk and in solution can be tuned on a structural basis. After giving in the first part an up-to-date outline of the work that has been rather extensively reviewed in previous surveys (36,37), we wish to focus in the second part on the results obtained in our laboratories on optically active thermotropic polyesters containing mesogenic aromatic diads or triads based on p-oxybenzoic acid. As an ultimate goal, information should be gained on the general issue of structure-property relationships for synthetic polymers designed to fulfill the 'specialty' attribute.

2. OUTLINE OF CHIRAL THERMOTROPIC LC POLYMERS FROM OTHER LABORATORIES

According to the commonly accepted subdivision of LC polymers based on the placement of the mesogenic groups in either the side chain or the main chain, we shall present the two classes in separate sections, including in each one both homopolymers and copolymers. There are substantial differences between the two classes of polymers in the mutual role exerted by the macromolecular chain and topology of the mesogenic group. In the side chain systems the tendency is to decouple the segmental motion of the mesogenic group from the polymer backbone, which should act as a mere carrier with specific influences only on the thermal transitions (Tg or Tm) to the onset of the LC state and on the viscoelastic properties of the polymer system. These last, in fact, cannot be strictly equated to those of bulks or solutions having comparable constitution and content of low molar mass mesogens.

In the main chain systems, the backbone behaviour is intimately related to the nature of the rigid mesogenic cores that are usually interconnected by relatively soft segments and give rise to semiflexible LC polymers. The specific polymer behaviour of main chain polymers is, therefore, heavily affected by the nature and relative extension of the mesogenic group.

2.1. Side chain thermotropic LC polymers

Polymers containing chiral groups with a prevalent chirality in the side chain have been studied extensively by several research groups (38-59). In Schemes 1 and 2 there are represented the most common structures in terms of backbone repeat unit, spacer, chiral and achiral group for the homopolymers and copolymers investigated.

It is possible to stress that acrylates and methacrylates are the most used monomeric precursors, while siloxane backbone polymers, that are obtained by polymer-analog reactions, constitute the only alternative. The chiral component are essentially confined to cholesterol derivatives, with the only exception of a few (S)-2-methylbutyl containing residues.

Cholesterol containing monomers are usually mesogenic and incorporation into a polymer backbone does not necessarily disrupt their mesogenic character. By starting from either a cholesteric or smectic monomer, generally a smectic polymer is obtained. This indicates that the polymerization process of an anisotropic monomer tends to stabilize the LC order (42,50). As a consequence, if twisted nematic polymers have to be prepared, isotropic optically active monomers with mesogenic propensity should be used (51).

Scheme 1. Chemical structures of most common chiral LC side chain homopolymers.

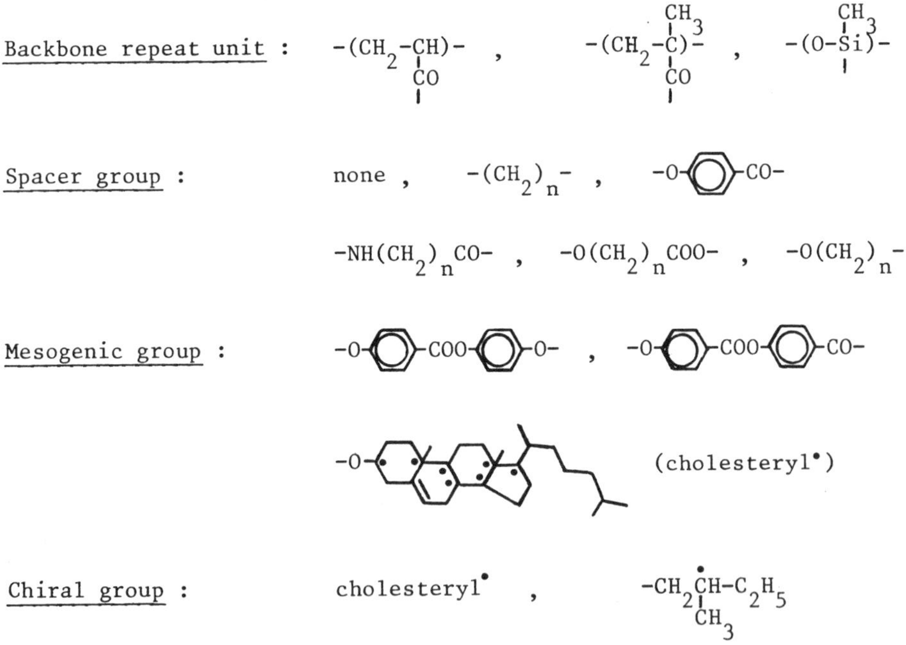

Backbone repeat unit : $-(CH_2-CH)-$ with CO below the CH, $-(CH_2-C(CH_3))-$ with CO below the C, $-(O-Si(CH_3))-$

Spacer group : none , $-(CH_2)_n-$, -O-⟨phenylene⟩-CO-

$-NH(CH_2)_nCO-$, $-O(CH_2)_nCOO-$, $-O(CH_2)_n-$

Mesogenic group : -O-⟨phenylene⟩-COO-⟨phenylene⟩-O- , -O-⟨phenylene⟩-COO-⟨phenylene⟩-CO-

-O-[cholesteryl ring structure] (cholesteryl•)

Chiral group : cholesteryl• , $-CH_2\overset{\bullet}{C}H(CH_3)-C_2H_5$

Scheme 2. Chemical structures of most common chiral LC side chain copolymers.

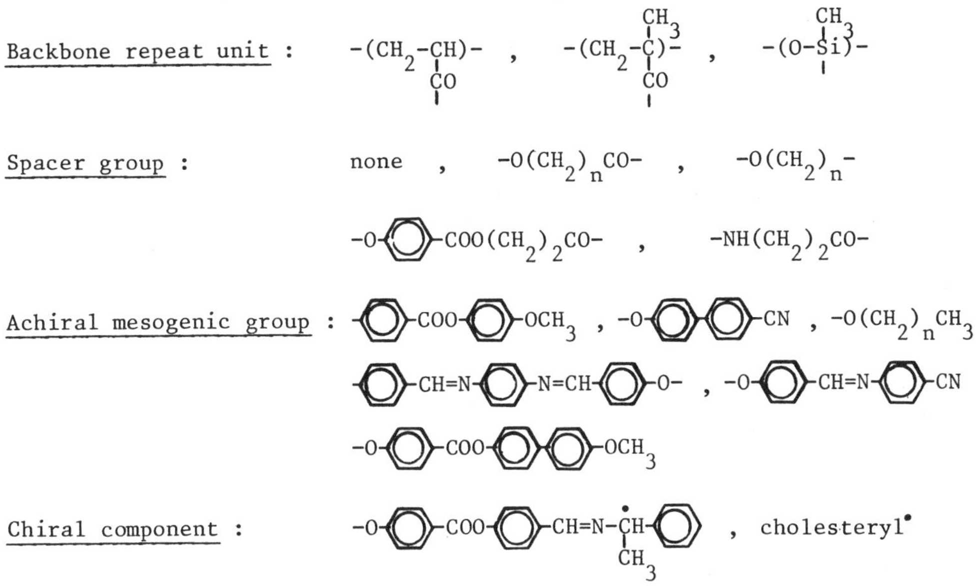

Backbone repeat unit : $-(CH_2-CH)-$ (with pendant CO), $-(CH_2-C(CH_3))-$ (with pendant CO), $-(O-Si(CH_3))-$

Spacer group : none , $-O(CH_2)_nCO-$, $-O(CH_2)_n-$

$-O-C_6H_4-COO(CH_2)_2CO-$, $-NH(CH_2)_2CO-$

Achiral mesogenic group : $-C_6H_4-COO-C_6H_4-OCH_3$, $-O-C_6H_4-C_6H_4-CN$, $-O(CH_2)_nCH_3$

$-C_6H_4-CH{=}N-C_6H_4-N{=}CH-C_6H_4-O-$, $-O-C_6H_4-CH{=}N-C_6H_4-CN$

$-O-C_6H_4-COO-C_6H_4-C_6H_4-OCH_3$

Chiral component : $-O-C_6H_4-COO-C_6H_4-CH{=}N-C^{\bullet}H(CH_3)-C_6H_5$, cholesteryl•

The stereoirregular polymers obtained possess relatively high glass transition temperatures and normally, above Tg, a marked anisotropic character associated with the onset of smectic orders. Wider temperature ranges of mesophase existence are recorded for polymers with higher overall flexibility connected with either less main chain rigidity or higher conformational freedom of the side chains.

The values of the isotropization temperatures in corresponding samples of the two series of homopolymers based on acrylate or methacrylate monomers are comparable, provided that the length of the spacer between the mesogenic group and the backbone is sufficiently high. There is clear indication, therefore, that the nature of the mesogenic moiety is solely responsible for the mesomorphic behaviour. Samples with the mesogenic unit relatively close to the polymer backbone do not show clear evidence of LC order in the melt phase (41-43,45).

The synthetic route based on post-reactions carried out on poly(siloxane hydride)s, that has been widely applied to the preparation of achiral LC side chain polymers, led to the first example of cholesteric homopolymers (50). However, reactions performed even on structurally simple polymers never reach total conversion of functional groups, and one must take into account that the presence of any structural irregularity along the backbone might play a role in affecting the ultimate properties of the polymer bulk (60).

The reported polysiloxanes are characterized by typical reflection of incident light in accordance with the establishment of a right-handed helical structure (50). The pitch of the helix decreases with increasing temperature with a concurrent shift of the maximum of the reflected wavelength from near IR towards the visible range.

Very recently, the preparation of a chiral smectic C polymer has been obtained by polymerization of a shortly spaced acrylate monomer based on hydroquinone oxybenzoate etherified with a (S)-2-methylbutyl group (57).

A more valuable route to polymers displaying cholesteric phases appears to be that based on the formation of copolymers (38,43-46,52-55) by copolymerization of: i) two optically active, mesogenic precursors; ii) an optically active, not mesogenic monomer with an achiral, mesogenic comonomer; iii) an optically active, mesogenic monomer with an achiral comonomer, mesogenic or not.

The first enantiotropic cholesteric copolymer was prepared by copolymerization of appropriate mixtures of mesogenic cholesterol derivatives with the chiral unit spaced from the methacrylate backbone through spacer segments of different length and rigidity (53). The copolymer effect tends to suppress the smectogenic tendency of corresponding homopolymers and either intrinsic or induced twisted nematic phases can be originated in rather wide ranges of temperature. The onset and stability of the cholesteric phases are markedly affected by the chemical structure of the two comonomers (nature and length of the spacer and character of the mesogenic core) and chemical composition in copolymers. It appears, therefore, evident the great value of the copolymerization methods to realize the synthesis of polymeric materials characterized by structurally modulated mesomorphic properties.

Based on polymer-analog reaction procedures, a series of linear (49) and crosslinked (elastomeric) (55) copolymers with highly flexible siloxane backbone has been prepared and investigated in great detail. In all cases the copolymers adopt cholesteric structures which reflect the visible light. It is important to note, from an applicative standpoint, that amorphous materials not possessing a smectic phase at low temperatures can preserve their cholesteric structure in the glassy state (44,49). This may allow the manufacture of films and coatings provided with intrinsic reflecting properties and bright colours not imparted by any dye or chromophore in the polymer matrix (17).

2.2. Main chain thermotropic LC polymers

In this section chiral thermotropic polymers of semisynthetic (cellulose derivatives) and synthetic (polypeptides and polyesters) origin are included.

Cellulose derivatives (61) and polypeptides (27) are well known to exhibit lyotropic LC properties in a large variety of solvents, and only recently have been found to possess additionally thermotropic characteristics.

The systems based on cellulose (62-67) can be classified as cellulose ethers, namely hydroxypropyl cellulose, with some adjunctive ester functions normally localized on the ether branches. The idealized cellobiose repeating unit is represented as follows:

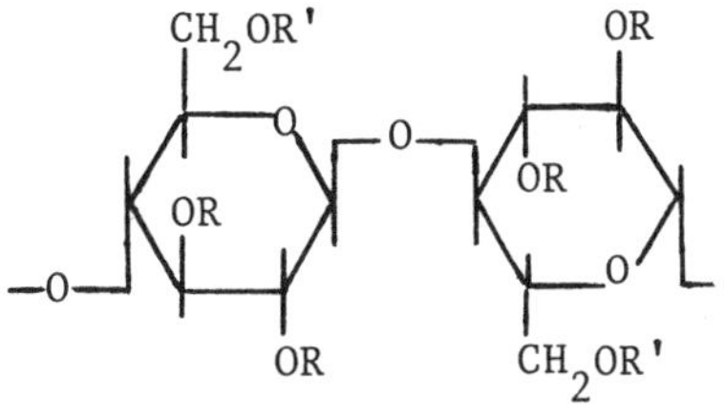

$R = -CH_2CH(OX)CH_3$ $\quad$ $R' = -CH_2CH(CH_3)OCH_2CH(OX)CH_3$

$X = H,\ CH_3CO-,\ CF_3CO-,\ CH_3CH_2CO-,\ C_6H_5CO-$

The molar average degrees of etherification (allowing also for the ether functions in R) and esterification are 6-8 and 5-6, respectively. The introduction of ester groups on the side branches inhibits the formation of intermolecular hydrogen bonding, with consequent rather marked decrease of the melting temperature of the ester derivative relative to the unmodified hydroxypropyl cellulose. Materials exhibiting thermotropic mesomorphism even at room temperature are thus obtained (63,65-67).

The cellulose backbone appears to be sufficiently stiff to guarantee the formation of ordered mesophases with nearly parallel orientation of single stranded macromolecules. Preferential chirality of the repeating unit imparts a twist to the parallel arrangement of the macromolecules, thus inducing the helical structure typical of cholesteric materials. Nature and extent of side substituents on the repeating unit control the average spacing among the rigid segments with a consequent influence on the pitch of the cholesteric array and, hence, on the optical properties of anisotropic melt. Reflection of visible light is observed with samples containing less bulky acyl residues (acetate and propionate esters) (63,65), while longer pitches exist in the benzoate derivative (66,67).

In synthetic polypeptides the thermotropic behaviour has been recognized for the first time in copoly(α-aminoacid)s consisting of γ-n-alkyl-L-glutamates (68), as shown in Scheme 3.

Scheme 3. Chemical structure of thermotropic LC polypeptides.

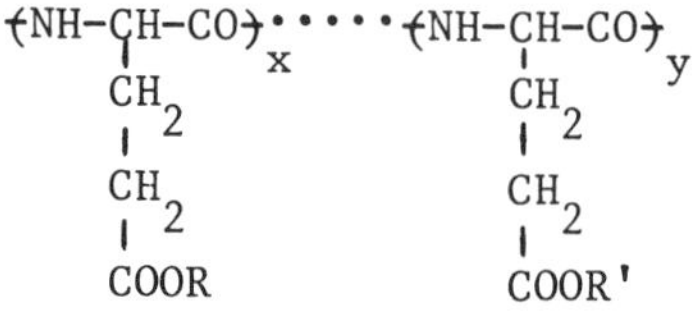

$$\text{-(NH-CH-CO)}_x\cdots\cdots\text{-(NH-CH-CO)}_y$$

$$\underset{\substack{| \\ CH_2 \\ | \\ CH_2 \\ | \\ COOR}}{} \qquad\qquad \underset{\substack{| \\ CH_2 \\ | \\ CH_2 \\ | \\ COOR'}}{}$$

R	R'
CH_3	$\underline{n}C_6H_{13}$
CH_3	$\underline{n}C_8H_{17}$
$\underline{n}C_3H_7$	$\underline{n}C_8H_{17}$

The prepared copolymers give rise (68) to anisotropic melts characterized by selective reflection of visible light in substantial ranges of composition (50-80% of shorter side chain substituent), consistent with the existence of a right-handed cholesteric structure. The temperature intervals of mesophase persistence are relatively narrow (30°C at most) and depend on the chemical structure of the alkyl substituent, minimum values being reached with longer lateral chains. Within these ranges, the wavelength of the reflected light increases, at any given copolymer composition, with increasing temperature, suggesting a significant unwinding of the helical structure. Incidentally, it may be noted that this trend coincides with that shown by lyotropic poly(γ-benzyl-L-glutamate) (69,70) which adopts a right-handed helical conformation in chloroform and dioxane solutions (69,71). In that respect, the side chain aliphatic component in copolypeptides and the ether-ester component in modified celluloses can exert a 'solvent' role in a virtually highly concentrated solution of single stranded macromolecules, that might be, therefore, regarded as lyotropic systems.

Chiral LC polyesters constitute so far the most investigated class of synthetic polymers capable of assuming cholesteric-type structures. Independent of their homopolymer or copolymer nature, they can be classified as semiflexible LC polymers whose prevalent chirality resides in any case in the soft segment. For convenience they can be grouped in two homogeneous classes based on the nature of the diacid component:
i) hard aromatic dihydroxy compound/soft aliphatic diacid compound; ii) hard aromatic dicarboxylic compound/soft aliphatic diol.

For the sake of completeness, one should also consider the chiral composite materials as obtained by addition of a compatible low or high molecular weight optically active compound to LC polyesters to produce cholesteric or chiral smectic phases (72-77).

To the first group belong LC polyesters derived from commercially available (R)-3-methyladipic acid and a variety of mesogenic diols or diphenols based on diads or triads of aromatic nuclei bridged with each other by different functional groups or a terephthalate residue (78-87) (Scheme 4).

Scheme 4. Chemical structures of chiral LC polyesters.

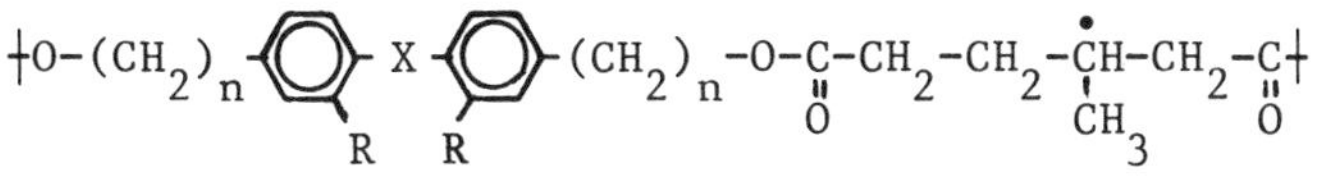

$$\left[O-(CH_2)_n-\underset{R}{C_6H_3}-X-\underset{R}{C_6H_3}-(CH_2)_n-O-\underset{\underset{O}{\|}}{C}-CH_2-CH_2-\underset{CH_3}{\overset{\bullet}{C}H}-CH_2-\underset{\underset{O}{\|}}{C} \right]$$

n = 0, R = H, X = none, -CH=CH-COO-, $-CH=C(CH_3)-$, -N=N-,

-N=N(O)-, -COO-, $-OOC-C_6H_4-COO-$.

n = 0, R = CH_3, X = -N=N(O)-.

n = 2, R = H, X = -N=N(O)-.

Copolyesters containing the same mesogenic cores and variable mixtures of (R)-3-methyladipic acid and C_6-C_{12} unbranched aliphatic diacids have also been investigated (78-87).

The second group of LC polyesters comprises a few homopolymer and copolymer samples derived from (R)-3-methyl-1,6-hexanediol and linear diols and 4,4'-(terephthaloyldioxy)dibenzoic acid (HTH) (Series HTH-$C_6^{\bullet}$) (86):

$$\left(HTH-OCH_2CH_2\underset{CH_3}{\overset{\bullet}{C}H}CH_2CH_2CH_2O \right)_x \left(HTH-O(CH_2)_nO \right)_{1-x}$$

n = 6,10 $\qquad 0 \leqslant x \leqslant 1$

$$HTH : \; -\underset{\underset{O}{\|}}{C}-C_6H_4-O\underset{\underset{O}{\|}}{C}-C_6H_4-\underset{\underset{O}{\|}}{C}O-C_6H_4-\underset{\underset{O}{\|}}{C}-$$

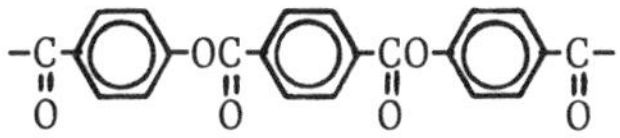

A series of polyesters constituted by oligosiloxane blocks in the main chain has also been prepared (88) by starting from a diallyl dibenzenoid mesogen and estradiol, as the chiral component:

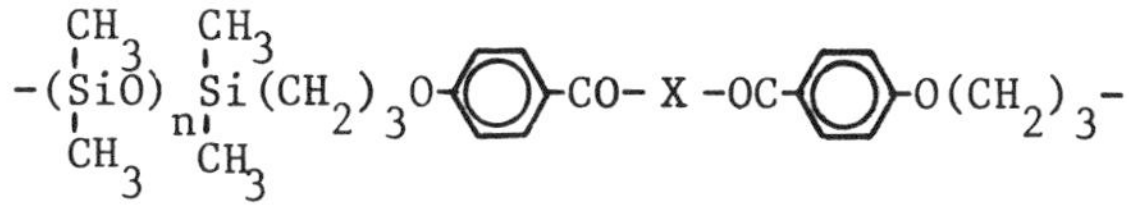

$$-(\underset{CH_3}{\overset{CH_3}{Si}}O)_n\underset{CH_3}{\overset{CH_3}{Si}}(CH_2)_3O-C_6H_4-CO-X-OC-C_6H_4-O(CH_2)_3-$$

X : -O-(estradiol residue)-O- $\qquad$ n = 2,3,4,5

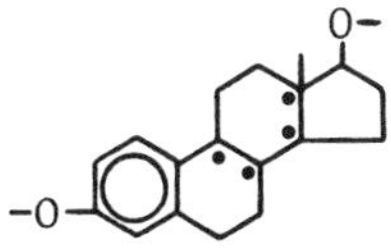

In neither class of macromolecular systems is information available about the relative position and orientation of the non-symmetric residues (89). However, owing to the expected identical reactivity, e.g. of the two carboxy groups in the chiral component and of the hydroxy groups in the mesogen, it is reasonable to assume a random sequencing of the orientations of the repeating units in polyesters, and even random distribution of the different diacid residues in copolyesters.

Depending on the nature of the mesogenic core, the clearing temperatures of the investigated homopolyesters vary from 150 to 300°C. Samples incorporating either substituted or less rigid mesogens are chracterized by lower values of the isotropization temperature. The stability ranges are rather broad (50-140°C), the widest breadth being reached for samples containing less flexible mesogens. Typically, the copolyesters show improved solubility properties and expanded ranges of mesomorphic behaviour. Probably the randomness of different residues in the polymer backbone helps to disrupt the three-dimensional order of polymer segments in the semicrystalline phase, which results in a marked depression of the melting temperature relative to corresponding homopolyesters. Consistently, the breadth of the mesophase is enlarged, the widest range usually occurring at the equimolar composition of chiral and achiral diacid residues in the copolymer, i.e. corresponding to a maximum of structural randomness.

In several homopolymer and copolymer samples hot stage polarizing microscopy allowed to evidence Grandjean textures with oily streaks (90), thus providing strong indication of the existence of a cholesteric mesophase. Sometimes the planar reflecting textures could be frozen in the solid state by quenching (80). In a few cases (79,81,86,87,91) there are semiquantitative UV or CD absorption measurements of the helical pitch and twisting power of the cholesteric structure, and of their variation with chemical nature and composition (copolymers). Depending upon the structure of the mesogenic group, an increase of the chiral component proportion or of the temperature results in a compression of the helical pitch, analogous to low molar mass cholesterics, (80,83,87), whereas in other cases the opposite trend occurs (85,86). As a consequence, appropriate combinations of the effects caused by an external parameter (temperature) and by an intrinsic factor (chemical composition) can offer a simple route to achieve a wide variety of polymeric materials with suitably differentiated physical properties in the thermotropic melt.

No indication is given of the chiroptical properties in dilute solutions of these polyesters, nor are data available on the effect of the enantiomeric purity of the chiral component on the bulk properties. Despite the rather few data for the reported classes of polyesters, we may stress at least qualitatively the structural effects of the various mesogens in terms of the following: i)Extension of the mesogen. Mesogens containing three aromatic rings exhibit comparatively high melting points and clearing temperatures, while the interval of mesomorphic behaviour is restricted relative to the highest values in polymers comprising binuclear aromatic mesogens (92). ii) Conformational rigidity of binuclear aromatic mesogens. In systems based on mesogens with a bridging group, the increase in flexibility or steric crowding produces a narrowing of the mesophase. iii) Bridging group in binuclear aromatic mesogens. In mesogens with structurally identical aromatic rings interconnected by different polarizable substituents, the effectiveness in extending the mesomorphic behaviour follows the order $-CH=C(CH_3)- > -N=N(O)- > -N=N- > -CH=CH-COO-$, while as far as the stability of the mesophase is concerned the order is $-N=N(O)- > -CH=C(CH_3)- > -N=N- > -CH=CH-COO-$. These trends are basically in agreement with those reported for low molar mass mesogens (1) and for achiral polymers based on the same mesogenic cores (77). iv) Chemical composition of copolyesters. Mesophase persistence is increased and the maximum extension is reached at equimolar amounts of chiral and achiral diacids, even though this particular feature remains to be better established.

3. OUTLINE OF CHIRAL THERMOTROPIC LC POLYMERS FROM OUR LABORATORIES

In the present section we report on the results stemming from a part of our own contributions in the field of LC polyesters in view of preparing segmented polycondensation polymers consisting of flexible segments of varying hydrophilic character and rigid segments of either linear or non-linear structure (93-95).

From the standpoint of the efforts provided in the assessment of structure-property relationships in a novel class of polymeric materials, information was somewhat lacking on the effects connected with the structural features of the chiral spacer, enantiomeric excess of the chiral spacer, and orientational sequencing of non-symmetric flexible or rigid segments, in respect also to the study of the behaviour in dilute solution, including chiroptical properties, of typically thermotropic polymers (96,97).

The synthetic strategy selected by us to that purpose was based on testing several routes for the production of materials with modified solubility and thermal properties. Accordingly, while keeping the mesogen structure constant, variations were allowed within the chiral segment. Three main series of chiral polyesters derived from 4,4'-(terephthaloyldioxy)dibenzoic acid (HTH) and propylene glycol (PG) or its head-to-tail oligomers $(PG)_n$, propylene glycol homologs (PGH), and glycerol ethers (GE) (Table I) were prepared and studied.

TABLE I. General structures of LC polyesters derived from 4,4'-(terephthaloyldioxy)dibenzoic acid (HTH) and chiral diols.

$$— \text{HTH}\text{-O-}[(CH_2)_m\text{-}\overset{\bullet}{C}H\text{-O}]_n— \quad (\text{R on } \overset{\bullet}{C}H)$$

R	m	n	series
CH_3	1	1-20	HTH-$(PG)_n$
CH_3	2-4	1	HTH-PGH
CH_2OR'	1	1	HTH-GE

Two series of chiral copolyesters were obtained from various mixtures of isomeric triads HTH/HPH (P = phthaloyl) or HTH/HIH (I = isophthaloyl) and the same optically active (S)-1,2-propanediol (PG) (Scheme 5).

Scheme 5. General structures of LC copolyesters based on HTH units.

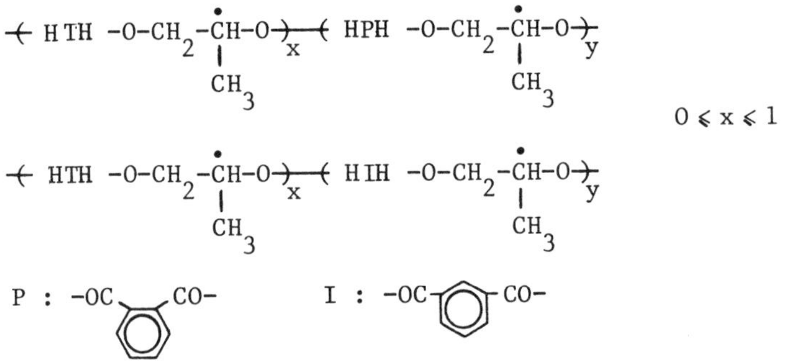

$$\text{-(HTH -O-}CH_2\text{-}\overset{\bullet}{C}H(CH_3)\text{-O-})_x\text{-(HPH -O-}CH_2\text{-}\overset{\bullet}{C}H(CH_3)\text{-O-})_y$$

$$0 \leqslant x \leqslant 1$$

$$\text{-(HTH -O-}CH_2\text{-}\overset{\bullet}{C}H(CH_3)\text{-O-})_x\text{-(HIH -O-}CH_2\text{-}\overset{\bullet}{C}H(CH_3)\text{-O-})_y$$

P : -OC-(o-phenylene)-CO-   I : -OC-(m-phenylene)-CO-

All the polymer samples were synthesized by polycondensation in solution starting from stoichiometric amounts of the diol of choice and aromatic diacid chloride, or mixtures of isomeric diacid chlorides. Details relevant to typical polymerization runs, purification and characterization of the polymeric products have been already described (see for instance 98).

By using a very versatile synthetic procedure based on the Michael-type addition of bis-active-hydrogen compounds to mesogenic diacrylates applied for the first time by us to the preparation of functional LC polyesters (99,100), two series of chiral β-sulfide containing polyesters have been prepared (101,102), as represented in Scheme 6.

Scheme 6. Synthesis and general structure of chiral LC poly(β-thioester)s.

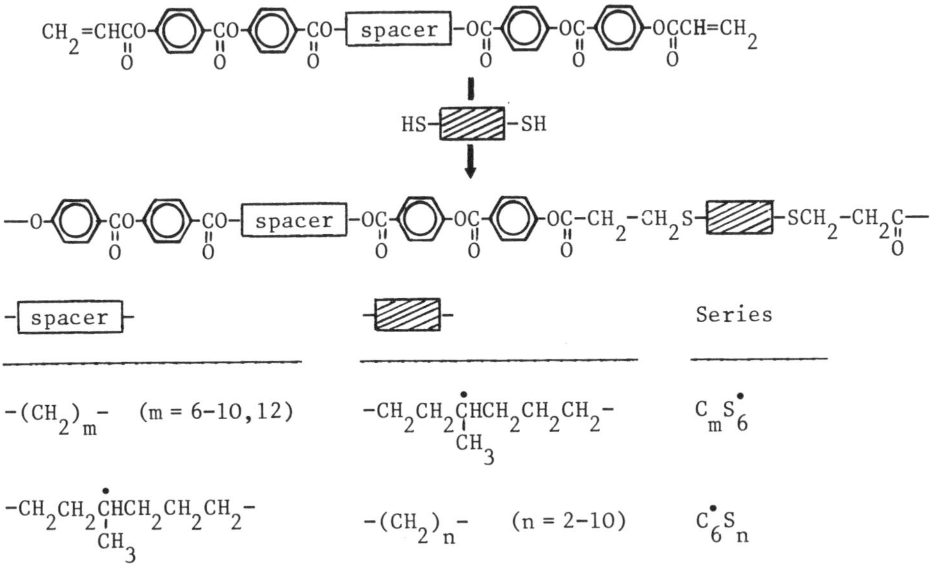

$CH_2=CHCO$-⟨Ph⟩-$\underset{O}{\overset{\|}{C}}O$-⟨Ph⟩-CO-[spacer]-OC-⟨Ph⟩-OC-⟨Ph⟩-OCCH=CH_2 (each CO / OC group with =O)

↓ HS-[▨]-SH ↓

—O-⟨Ph⟩-CO-⟨Ph⟩-CO-[spacer]-OC-⟨Ph⟩-OC-⟨Ph⟩-OC-CH_2-CH_2S-[▨]-SCH_2-CH_2C— (each CO / OC group with =O)

-[spacer]-	-[▨]-	Series
$-(CH_2)_m-$ (m = 6-10,12)	$-CH_2CH_2\overset{\bullet}{C}H(CH_3)CH_2CH_2CH_2-$	$C_mS^{\bullet}_6$
$-CH_2CH_2\overset{\bullet}{C}H(CH_3)CH_2CH_2CH_2-$	$-(CH_2)_n-$ (n = 2-10)	$C^{\bullet}_6S_n$

The analysis of the results obtained allows a rationalization of the polymer behaviour in terms of the length of the spacer, nature of the substituents, structural isomerism, enantiomeric excess, and chemical composition and distribution of different repeating units in copolymers.

Spacer length. The effect of the spacer length has been investigated in three series of polyesters based on the HTH mesogenic triad and diols originated by either backbone homologation (103) or oligomerization of propylene glycol (104,105) (main chain effect) (Series I and II) or extension of side substituents in glycerol ethers (106) (side chain effect) (Series III):

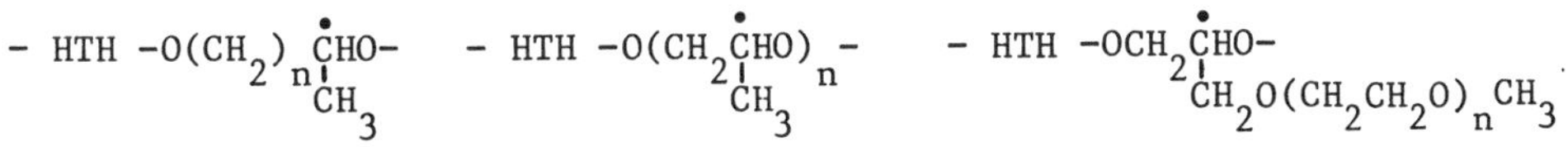

$- HTH -O(CH_2)_n\overset{\bullet}{C}H(CH_3)O-$ (I, n = 1-4)

$- HTH -O(CH_2\overset{\bullet}{C}H(CH_3)O)_n-$ (II, n = 1-20)

$- HTH -OCH_2\overset{\bullet}{C}H(CH_2O(CH_2CH_2O)_nCH_3)O-$ (III, n = 0-3)

An analogous evaluation has also been performed on two series of poly(β-thioester)s derived from twin bisacrylates and chiral or achiral bisthiols (Series $C^{\bullet}_6S_n$ and $C_mS^{\bullet}_6$).

In Figures 1 and 2 are reported the phase diagrams for Series I and III, respectively, while some significant data for the samples of Series II are collected in Table II.

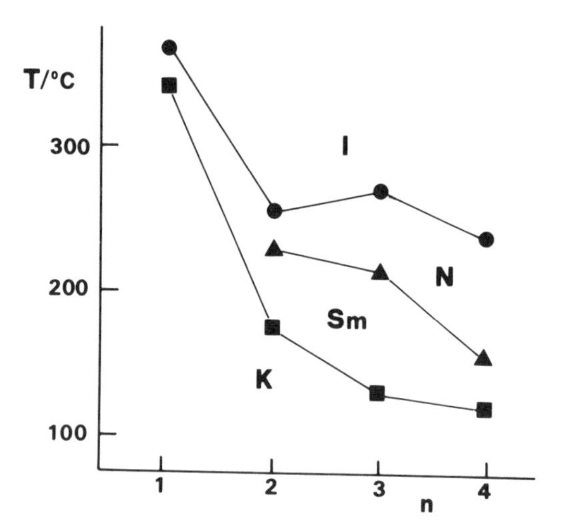

Figure 1. Trends of phase transition temperatures vs. length n of the chiral spacer for thermotropic polyesters of Series I: -■- melting, -▲- smectic (Sm) to nematic (N) or cholesteric (Ch), -●- isotropization.

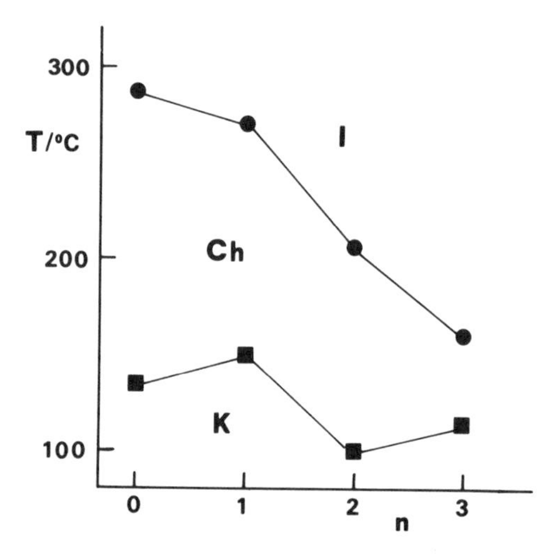

Figure 2. Trends of phase transition temperatures vs. length n of the side substituent for thermotropic polyesters of Series III (for symbols see Figure 1).

As a general consideration, a common descending trend in the phase transition temperatures is observable for all the investigated series

TABLE II. LC properties of chiral polyesters based on HTH diacid and propylene glycol oligomers $(PG)_n$ (Series II).

run	propylene glycol		polyester	
	n	configuration	T_m/°C	T_i/°C
PG-1	1	(S)	334	362[a]
PG-2	2	(S,S)	105	290[b]
PG-3	3	(S,S,S)	275	321[c]
PG-7	7	racemic	181	> 300
PG-20	20	racemic	188	not LC

[a] Partial decomposition. [b] Smectic to cholesteric at 236°C.
[c] Smectic to cholesteric at 305°C.

with more or less evident and regular odd-even effects (107,108). The isotropization temperature T_i is comparably or less markedly influenced by the spacer length gradient, thus helping mesomorphism to establish at lower temperatures and to maintain a considerable width of the mesophase. It is also worth noting how the length of the spacer may influence the incidence of LC polymorphism (Figure 1), and that the existence of a mesophase is still possible at spacer lengths comprising sequences of as much as 20 atoms (Table II).

By comparing the two analogous series of sulfide-containing polyesters (Figures 3 and 4), one may stress that in the field of LC polymers even subtle changes in the topology of structural isomers may cause significant variations in the incidence and nature of the mesomorphic behaviour.

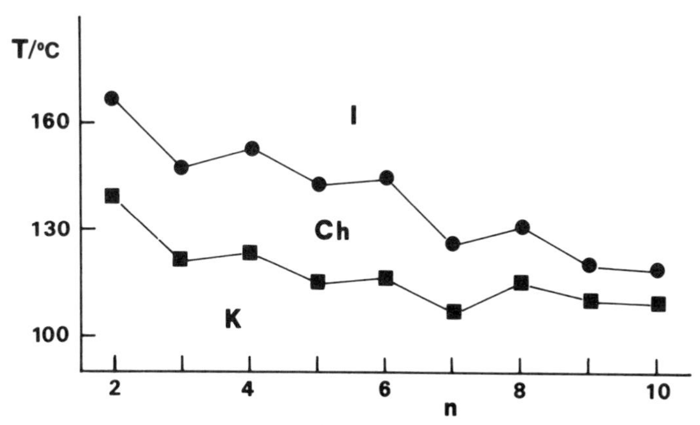

Figure 3. Trends of phase transition temperatures vs. length n of the sulfide spacer for thermotropic poly(β-thioester)s $C_6^{\bullet}S_n$ (for symbols see Figure 1).

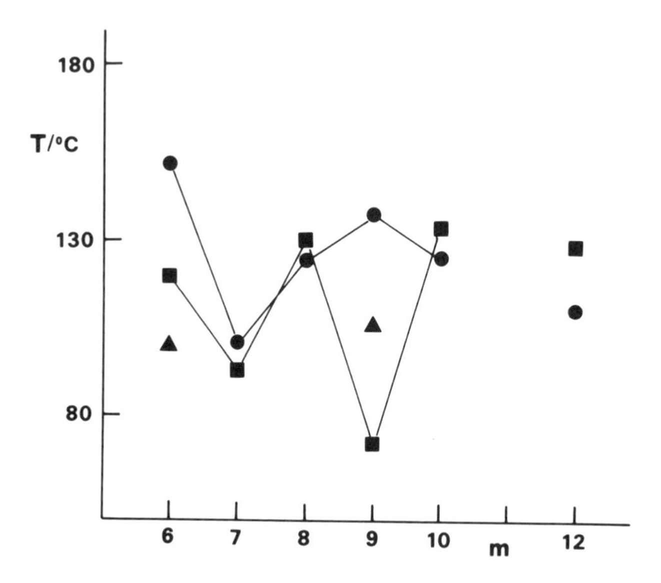

Figure 4. Trends of phase transition temperatures vs. length m of the alkylene segment for thermotropic poly(β-thioester)s $C_mS_6^{\bullet}$: -■- melting, -▲- cholesteric to smectic, -●- isotropization (monotropic cholesteric phase for m = 8,10,12).

Structure and nature of the substituent. This effect, as referred to monosubstituted glycols, has not been much investigated until now. Nevertheless, one may consider the trend already reported for Series III based on the methyl derivatives of variously ethoxylated glycerols. Within the limitations of the small number of runs, a slightly pronounced odd-even alternation of the isotropization temperature with increasing length of the side chain substituent is established (Figure 2). The adoption of local zig-zag planar conformations with different symmetry parameters can affect the bulk properties of the polymer (109), apparently analogous to the effect exerted by flexible spacers inserted in the main chain.

In Table III are collected the melting and isotropization temperatures of HTH polyesters (Series IV) ordered according to the increasing bulkiness of the substituent on the ethylene glycol residue (103):

$$-\ \mathrm{HTH}\ -\mathrm{OCH_2}\overset{\bullet}{\underset{\mathrm{R}}{\mathrm{C}}}\mathrm{HO}-$$

$$(\mathrm{IV},\quad \mathrm{R} = \mathrm{H}, -\mathrm{CH_3}, -\mathrm{CH_2CH_3}, -\mathrm{CH_2OCH_3}, -\mathrm{CH_2OCH_2C_6H_5})$$

TABLE III. LC properties of chiral polyesters based on HTH diacid and R substituted ethylene glycols.

R	T_m/°C	T_i/°C
H	342	365[a]
CH_3	334	362[a]
CH_2CH_3	162	265
CH_2OCH_3	135	287
$CH_2OCH_2C_6H_5$	145	205

[a]Partial decomposition.

A dramatic drop in the melting temperature is observed in going from the unsubstituted or methyl substituted term to samples with slightly bulkier groups such as ethyl or methoxymethyl, while fairly high isotropization temperatures T_i are retained. However, the mesophase polymorphism seems to be favoured by the presence of short or less bulky side chains, while the increase of length and steric hindrance of the pendant group allows the onset of one mesophase (nematic or cholesteric).

Structural isomerism. Within the polymer samples belonging to Series II, a detailed study of the thermal-optical characteristics in the bulk and optical rotation in solution against structural isomerism of dipropylene glycol has been reported (37,110). In Table IV are summarized some physicochemical characteristics for the three optically active polymers and for the not optically active polymer obtained from a racemic mixture of the three isomers of commercial source.

TABLE IV. Physicochemical characteristics of chiral polyesters based on HTH diacid and dipropylene glycol (DPG) isomers.

run	dipropylene glycol		polyester		
	isomer	$\lvert\Phi\rvert^{25}_D$	$\lvert\Phi\rvert^{25}_D$	T_m/°C	T_i/°C
DPGht	head-to-tail[a]	+84	+143	105	290
DPGhh	head-to-head[b]	+49	+360	108	213
DPGtt	tail-to-tail[c]	+ 2	+ 7	203	280
DPGrm	racemic mixture[d]	-	-	112	260

[a](2S,5S)-1,5-dihydroxy-2-methyl-3-oxy-hexane (ht, e.e. ⩾95%). [b](2S,6S)-2,6-dihydroxy-4-oxy-pentane (hh, e.e. > 95%). [c](2S,4S)-1,5-dihydroxy-2,4-dimethyl-3-oxy-pentane (tt, e.e. ≃ 10%). [d]Mixture of the

(TABLE IV contd) three isomers: ht 71% + hh 21% + tt 8%.

The molar optical rotation in solution of the starting dimers is greatly increased in corresponding HTH polyesters, and these data combined with circular dichroism absorption measuraments (97) strongly suggest the establishment, even in dilute solution, of preferential conformations for the mesogenic units as forced by the chiral flexible segments.

The melting temperatures are in the proximity of 100°C, with the sole exception of the sample based on the tail-to-tail dimer for which a value of over 200°C is found, in accordance with conformational constraints imposed to the repeating unit by the presence of two methyl groups in 1,3 positions. Clearing temperatures are in all cases higher than 200°C and are consistently maintained over wide temperature ranges (80-190°C). In this respect, the symmetrical placement of two methyl groups close to the aromatic core in the head-to-head dimer is particularly effective in destabilizing the mesophase. The mesomorphic behaviour was found to be complex and characterized by extended polymorphism with incidence of tilted and orthogonal smectic phases with increasing temperature (110). For the two polyesters incorporating the head-to-tail and head-to-head dipropylene glycol residues, a smectic H phase (centred rectangular lattice) appears to exist in the approximate range 110-190°C. Such a phase presents the elements of tilt required to produce a chiral phase, the repeating unit possessing an inherently asymmetric structure. It is envisaged, however, that a chiral smectic H phase with a loose helical arrangement of tilt directions might only be formed, in that the interlayer correlations are strong and the twisting power of the molecules does not seem sufficiently high. In all cases below isotropization, a cholesteric phase (optically active samples) or a nematic phase (racemic sample) was observed.

Enantiomeric excess. The effect of the enantiomeric excess of the chiral diol on the properties of the corresponding polymer in the bulk and in solution has been studied for two series of polyesters based on the HTH mesogen and (R)-1,3-butanediol (111) or (2S,3S)-2,3-butanediol (103). In both cases the molar optical rotation in dilute solution follows a linear trend as against the enantiomeric excess (see for example Table V), thus indicating that no specific cooperative effect dependent on the optical purity is operating.

In Figure 5 are represented the trends of the transition temperatures vs. the enantiomeric excess of the chiral spacer for the series based on 1,3-butanediol. No marked influence can be pointed out, a slight tendency to increase with increasing optical purity (0-80%) being observed in the smectic-cholesteric and cholesteric-isotropic transitions. That behaviour is even more evident in the comparison among the three samples

TABLE V . Physicochemical properties of chiral LC polyesters based on HTH diacid and 2,3-butanediol.

run	diol		polyester		
	e.e.	$\|\Phi\|_D^{25}$	$\|\Phi\|_D^{25}$	$T_m/°C$	$T_i/°C$
BD99	99	-15	-230	217	>310
BD50	50	- 7	-119	222	>310
BDrac	0	-	-	220	≃300

characterized by extremely different values of the enantiomeric excess of the 2,3-butanediol precursor (Table V). Apparently steric effects connected with the presence of two methyl groups on adjacent carbon atoms overwhelm any subtle influence related to specific copolymer effects as established by different distributions of enantiomers along the polymer backbone. On the other hand, fundamental variations depending on the optical purity should arise in the determination of the twisting power in cholesterics. In this connection it is worth mentioning that the cholesteric phase of the polyester from 1,3-butanediol with 40% enantiomeric excess reflects the blue light at 230°C, while the polyester from 1,3-butanediol with 20% enantiomeric excess shows blue and violet reflections at 230 and 245°C, respectively. Assuming the refractive index to be independent of temperature within the narrow ranges of temperatures examined, it is concluded that the length of the cholesteric pitch in this class of polyesters can vary with the temperature and extent of chirality of the spacer segment. Moreover, the X-ray structural data (111) are consistent with the occurrence in this series of smectic A phases with a monolayer or bilayer structure (optical purity 0 and 80%). It is interesting to speculate whether even fairly short sequences of repeating units with one single absolute configuration or placement of orientationally asymmetric diol residues can couple adjacently to sequences of units of opposite configuration building up a bilayer structure. Uncoupled units, on the contrary, can be packed in a different monolayer structure with a high degree of translational disorder. Therefore, while the thermodynamic characteristics (transition temperatures and enthalpies) of this class of thermotropic polyesters do not vary significantly with the enantiomeric excess in the range 0-80%, specific details of the smectic and cholesteric phases may be affected in an uneven way by the intrinsic chirality of the polymer.

The tuning of the cholesteric pitch on an enantiomeric excess base is still a pending problem whose assessment may offer good opportunities to

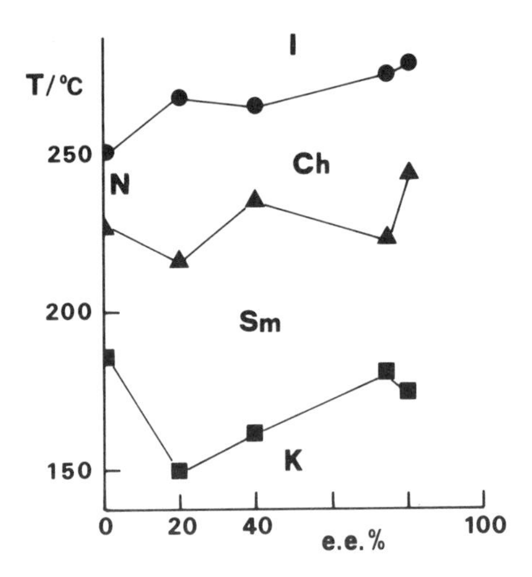

Figure 5. Trends of phase transition temperatures vs. enantiomeric excess (e.e.%) of 1,3-butanediol spacer in HTH polyesters: -■- melting, -▲- smectic (Sm) to cholesteric (Ch) or nematic (N), -●- isotropization.

synthetic polymers as useful candidates for any given application connected with the peculiar alignment of the macromolecules in mesomorphic textures.

Copolymer effect. The random introduction of kinks or different structural units in a polymer backbone results in a lowered tendency of macromolecular chains to close packing with consequent lowering of the melting temperature and increased solubility. These general concepts have also been applied in the field of LC polymers to improve their tractability and processability in the bulk or in solution. We have already mentioned that the thermal properties of chiral LC polyesters are not much affected by the copolymer effect due to random distribution of the enantiomeric diol residues along the polymer backbone.

More significant effects have been obtained for random copolymers in which non-linear HPH or HIH isomers of the HTH residues (Scheme 5) have been introduced in the main chain containing the (S)-1,2-propanediol (PG) residue, as the chiral component (98) (Table VI).

Substantially, on increasing the content of non-linear units, T_m is continuously depressed much more significantly than T_i. A wide range of existence of mesomorphic properties occurs at intermediate compositions and extends down to about 75-80% of distorted units. This demonstrates the possibility of producing stable and persistent mesophases even in copolymers derived from rigid non-mesogenic precursors. Samples comprising 10% content of HPH or HIH units develop planar textures, that

TABLE VI. Physicochemical properties of chiral LC copolyesters based on HTH units and non-linear HPH (runs TP) or HIH (runs TI) units.

run	% HTH	$\|\Phi\|_D^{25}$	T_m/°C	T_i/°C
T	100	+ 42[a]	334	362[b]
TP1	90	+337	208	290
TP2	70	+283	182	271
TP3	50	+259	130	277
P	0	+123	115	not LC
TI1	90	+409	222	325
TI2	70	+364	145	285
TI3	50	+367	139	220
TI4	30	+297	112	115
I	0	+297	130	not LC

[a]In fuming sulfuric acid, with possible degradation.
[b]Partial decomposition.

selectively reflect the visible light in appropriate ranges of temperatures (112,113), accompanied by an expansion of the cholesteric helical pitch with increasing temperature, analogous to the trend observed in the already mentioned copolymers based on the same HTH mesogen and different amounts of chiral and achiral diols (86). Such anomalous behaviour with respect to low molecular weight systems (3) can be imputed to the population of conformers characterized by reduced anisotropy in the intermolecular potential (114). In the present case the peculiar feature observed may be caused by the introduction of small amounts of distorted HPH or HIH units. Hysteresis phenomena occur on cooling, and quenching of the anisotropic melt give coloured solid films that retain their own reflecting characteristics with ageing.

4. CONCLUDING REMARKS

Among the several liquid crystalline polymers of synthetic and semisynthetic origin that have been studied in recent years, those containing chiral elements within the repeating unit constitute a class of new polymeric materials in which the chirality can play a key role in defining specific molecular properties. Chirality in its extent and nature appears in fact an important tool in tuning the ultimate responses of the material in the bulk and in solution.

Within the class of thermotropic liquid crystalline polymers, major attention has been devoted to main chain systems, because not only of our particular interest in the field, but also for the need of a better insight into their structure-property relationships. The assessment of this has revealed to be cumbersome and less predictable than that achievable for side chain polymers. Side chain liquid crystalline polymers deserve attention for their potential of fast responses to thermal and radiative inputs and for an apparently easier tendency to display ferroelectric properties (chiral smectics). Synthetic chiral main chain polymers are essentially polyesters based on commercially available (R)-3-methyladipic acid or the diol or dithiol derived therefrom.

The results obtained with these systems coupled with those derived from our specifically synthesized chiral homo- and copolyesters have been analyzed in terms of some particular features. The effects due to the nature of the mesogen, structure and nature of the chiral precursor, spacer length, structural isomerism, enantiomeric excess, and copolymer effect allow to stress the following conclusions. i) Depending upon suitable combinations of the above factors, a large variety of new liquid crystalline polymers can be obtained. ii) Stability and range of existence of the mesophase, as well as tendency to give twisted nematic structures able to reflect selectively the light can be tuned on structural bases. iii) A more or less direct control of the helical superstructure in cholesterics can be exerted by the nature and enantiomeric excess of the chiral component, while complementary behaviours may be induced by external parameters, such as temperature. These conclusions are schematically summarized in Figure 6 for five series of chiral liquid crystalline HTH polyesters. In consideration of the influence of temperature and long relaxation times of main chain cholesterics, chiral polyesters will possibly find applications in thermochromic or electrooptic devices with memory properties.

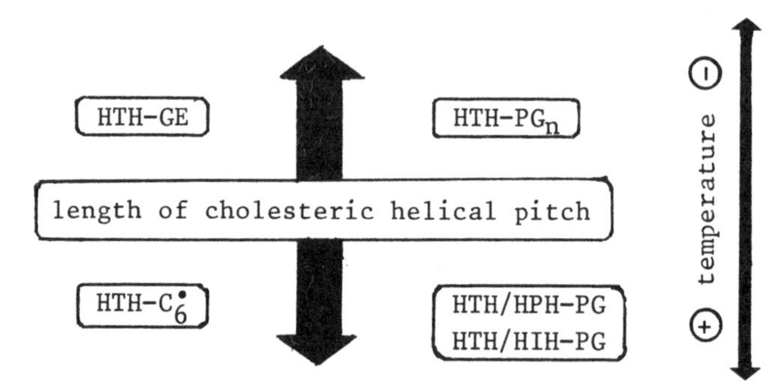

Figure 6. Schematic trend of length of helical pitch vs. temperature in chiral LC polyesters based on 4,4'-(terephthaloyldioxy)dibenzoic acid (HTH) and glycerol ethers (GE), propylene glycol (PG) or its oligomers (PG_n), 3-methyl-hexanediol ($C_6^{\bullet}$) (see text).

5. REFERENCES

1). M.Gordon, N.A.Platé eds., Adv.Polym.Sci., 59-61 (1984)
2) G.W.Gray, in 'The Molecular Physics of Liquid Crystals', G.R. Luckhurst, G.W.Gray eds., Academic Press, New York, 1979, Chaps. 1 and 12
3) S.Chandrasekhar, 'Liquid Crystals', Cambridge University Press, Cambridge, 1977
4) R.B.Meyer, L.Liebert, L.Strzelecki, P.Keller, J.Phys., 36, L69 (1975)
5) W.J.A.Goossens, Mol.Cryst.Liq.Cryst., 89, 23 (1982)
6) H.Baessler, M.M.Labes, J.Chem.Phys., 52, 631 (1970)
7) D.Dolphin, Z.Muljani, J.Cheng, R.B.Meyer, J.Chem.Phys., 58, 413 (1973)
8) H.Stegemeyer, H.Finkelmann, Chem.Phys.Lett., 23, 227 (1971)
9) H.Huason, A.J.Dekker, F.Van der Wonde, Mol.Cryst.Liq.Cryst., 42, 15 (1977)
10) F.D.Saeva, Pure Appl.Chem., 38, 25 (1974)
11) G.Solladié, R.Zimmermann, Angew.Chem.Int.Ed.Engl., 23, 348 (1984)
12) H.L.deVries, Acta Cryst., 4, 219 (1951)
13) N.A.Clark, S.T.Lagerwall, Appl.Phys.Lett., 36, 899 (1980)
14) F.Kahn, J.Appl.Phys.Lett., 18, 231 (1971)
15) J.E.Adams, W.E.Haas, J.Daily, J.Appl.Phys., 42, 4096 (1971)
16) T.J.Scheffer, J.Phys.D, 8, 1441 (1975)
17) H.Finkelmann, G.Rehage, Adv.Polym.Sci., 60/61, 99 (1984)
18) G.Decobert, J.C.Dubois, S.Esselin, C.Noel, Liquid Crystals, 1, 307 (1986)
19) A.Elliot, E.J.Ambrose, Discuss.Faraday Soc., 9, 246 (1950)
20) C.Robinson, Trans.Faraday Soc., 53, 571 (1956)
21) I.Uematsu, Y.Uematsu, Adv.Polym.Sci., 59, 37 (1984)
22) S.P.Papkov, Adv.Polym.Sci., 59, 75 (1984)
23) C.Robinson, Mol.Cryst., 1, 467 (1966)
24) E.Iizuka, J.T.Yang, in 'Liquid Crystals and Ordered Fluids', vol.3, J.F.Johnson, R.S.Porter eds., Plenum Press, New York, 1978
25) B.Gallot, in 'Liquid Crystalline Order in Polymers', A.Blumstein ed., Academic Press, New York, 1978, p.191
26) G.H.Brown, J.J.Wolken eds., 'Liquid Crystals and Biological Structures', Academic Press, New York, 1979
27) D.B.DuPré, E.T.Samulski, in 'Liquid Crystals', F.D.Saeva ed., Marcel Dekker, New York, 1979, p.203
28) L.Gros, H.Ringsdorf, J.Skura, Angew.Chem.Int.Ed.Engl., 20, 305 (1981)
29) A.C.Neville, Mol.Cryst.Liq.Cryst., 76, 279 (1981)
30) A.Blumstein ed., 'Liquid Crystalline Order in Polymers', Academic Press, New York, 1978
31) A.Ciferri, W.R.Krigbaum, R.B.Meyer eds., 'Polymer Liquid Crystals', Academic Press, New York, 1982

32) A.C.Griffin, J.F.Johnson eds., 'Liquid Crystals and Ordered Fluids', vol.4, Plenum Press, New York, 1984
33) A.Blumstein ed., 'Polymeric Liquid Crystals', Plenum Press, New York, 1985
34) L.L.Chapoy ed., 'Recent Advances in Liquid Crystalline Polymers', Elsevier Applied Science, London, 1985
35) Faraday Discuss.Chem.Soc., 79 (1985)
36) E.Chiellini, G.Galli, in Ref.34, p.15
37) E.Chiellini, G.Galli, in Ref.35, p.241
38) L.Strzelecki, L.Liebert, Bull.Soc.Chim.France, 597 (1973)
39) E.C.Hsu, R.B.Clough, A.Blumstein, J.Polym.Sci., Polym.Lett.Ed., 15, 545 (1977)
40) A.Blumstein, Macromolecules, 10, 872 (1977)
41) V.P.Shibaev, A.V.Kharitonov, Ya.S.Freidzon, N.A.Platé, Vysokomol. Soed.Ser.A, 21, 1849 (1979)
42) V.P.Shibaev, in 'Advances in Liquid Crystal Research and Application', L.Bata ed., Pergamon Press, Oxford, 1980, p.869
43) V.P.Shibaev, N.A.Platé, Ya.S.Freidzon, J.Polym.Sci., Polym.Chem.Ed., 17, 1655 (1979)
44) Ya.S.Freidzon, S.G.Kostromin, N.I.Boiko, V.P.Shibaev, N.A.Platé, ACS Polym.Prep., 24(2), 279 (1983)
45) N.A.Platé, V.P.Shibaev, J.Polym.Sci., Polym.Symp., 67, 1 (1980)
46) H.Finkelmann, H.Ringsdorf, W.Siol, J.H.Wendorff, Makromol.Chem., 179, 829 (1978)
47) V.P.Shibaev, V.M.Moisenko, N.Yu.Lukin, N.A.Platé, Dokl.Akad.Nauk. SSSR, 237, 401 (1977)
48) H.Finkelmann, G.Rehage, Makromol.Chem., Rapid Commun., 1 31 (1980)
49) H.Finkelmann, G.Rehage, Makromol.Chem., Rapid Commun., 1, 733 (1980)
50) H.Finkelmann, G.Rehage, Makromol.Chem., Rapid Commun., 3, 859 (1982)
51) H.Finkelmann, G.Rehage, ACS Polym.Prep., 24(2), 277 (1983)
52) A.M.Mousa, Ya.S.Freidzon, V.P.Shibaev, N.A.Platé, Polym.Bull., 6, 485 (1982)
53) H.Finkelmann, J.Koldehoff, H.Ringsdorf, Angew.Chem.Int.Ed.Engl., 17, 935 (1978)
54) V.P.Shibaev, H.Finkelmann, A.V.Kharitonov, M.Portugall, H.Ringsdorf, N.A.Platé, Vysokomol.Soed.Ser.A., 23, 919 (1981)
55) H.Finkelmann, H.J.Kock, G.Rehage, Makromol.Chem., Rapid Commun., 2, 317 (1981)
56) P.A.Gemmell, G.W.Gray, D.Lacey, ACS Polym.Prep., 24(2), 253 (1983)
57) G.Decobert, F.Soyer, J.C.Dubois, Polym.Bull., 14, 179 (1985)
58) Ya.S.Fridzon, N.I.Boiko, V.P.Shibaev, N.A.Platé, Eur.Polym.J., 22, 13 (1986)
59) P.J.Shannon, Macromolecules, 17, 1873 (1984)

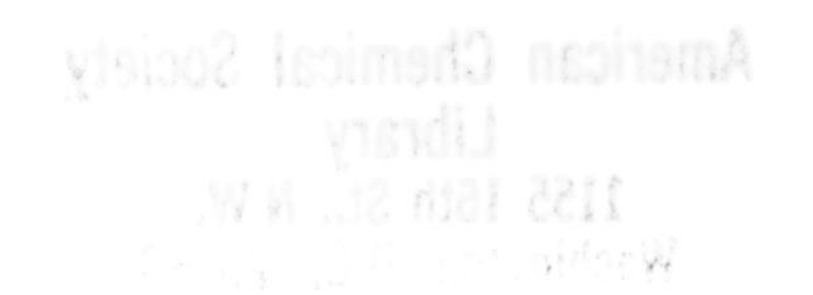

60) G.W.Gray, D.Lacey, G.Nestor, M.S.White, Makromol.Chem., 187, 71 (1986)
61) D.G.Gray, Appl.Polym.Symp., 37, 179 (1983)
62) K.Shimamura, J.L.Fellers, J.Appl.Polym.Sci., 26, 2165 (1981)
63) S.L.Tseng, A.Valente, D.G.Gray, Macromolecules, 14, 715 (1981)
64) S.M.Aharoni, J.Polym.Sci., Polym.Lett.Ed., 19, 495 (1981)
65) S.L.Tseng, G.V.Laivins, D.G.Gray, Macromolecules, 15, 1262 (1982)
66) S.N.Bhadani, D.G.Gray, Makromol.Chem., Rapid Commun., 3, 449 (1982)
67) S.N.Bhadani, S.L.Tseng, D.G.Gray, Makromol.Chem., 184, 1727 (1983)
68) S.Kasuya, S.Sasaki, J.Watanabe, Y.Fukuda, I.Uematsu, Polym.Bull., 7, 241 (1982)
69) D.B.DuPré, R.W.Duke, J.Chem.Phys., 63, 143 (1975)
70) H.Toriumi, Y.Kusumi, I.Uematsu, Y.Uematsu, Polym.J., 11, 863 (1979)
71) C.Robinson, Tetrahedron, 13, 219 (1961)
72) B.Fayolle, C.Noel, J.Billard, J.Phys.C3, 40, 485 (1979)
73) C.Noel, J.Billard, L.Bosio, C.Friedrich, F.Laupetre, C.Strazielle, Polymer, 25, 263 (1984)
74) C.Noel, F.Laupetre, C.Friedrich, B.Fayolle, L.Bosio, Polymer, 25, 808 (1984)
75) J.Billard, A.Blumstein, S.Vilasagar, Mol.Cryst.Liq.Cryst., 72, 163 (1982)
76) J.I.Jin, E.J.Choi, K.Y.Lee, Polym.J., 18, 99 (1986)
77) A.S.Angeloni, M.Laus, C.Castellari, G.Galli, P.Ferruti, E.Chiellini, Makromol.Chem., 186, 977 (1985)
78) D.VanLuyen, L.Liebert, L.Strzelecki, Eur.Polym.J., 16, 307 (1980)
79) S.Vilasagar, A.Blumstein, Mol.Cryst.Liq.Cryst.Lett., 56, 203 (1980)
80) W.R.Krigbaum, A.Ciferri, J.Asrar, H.Toriumi, J.Preston, Mol.Cryst.Liq.Cryst., 76, 79 (1981)
81) A.Blumstein, S.Vilasagar, S.Ponrathnam, S.B.Clough, R.B.Blumstein, G.Maret, J.Polym.Sci., Polym.Phys.Ed., 20, 877 (1982)
82) K.Iimura, N.Koide, Y.Tsutsumi, M.Nakatami, Rep.Progr.Polym.Phys.Jpn., 25, 297 (1982)
83) W.R.Krigbaum, T.Ishikawa, J.Watanabe, H.Toriumi, K.Kubota, J.Polym.Sci., Polym.Phys.Ed., 21, 1851 (1983)
84) J.Asrar, H.Toriumi, J.Watanabe, W.R.Krigbaum, A.Ciferri, J.Preston, J.Polym.Sci., Polym.Phys.Ed., 21, 1119 (1983)
85) C.K.Ober, J.I.Jin, R.W.Lenz, Adv.Polym.Sci., 59, 103 (1984)
86) H.J.Park, J.I.Jin, R.W.Lenz, Polymer, 26, 1301 (1985)
87) J.Watanabe, W.R.Krigbaum, Mol.Cryst.Liq.Cryst., 135, 1 (1986)
88) C.Aguilera, Mol.Cryst.Liq.Cryst.Lett., 2(6), 185 (1985)
89) U.W.Suter, P.Pino, Macromolecules, 17, 2248 (1984)
90) D.Demus, L.Richter, 'Textures of Liquid Crystals', Verlag Chemie, Weinheim, 1978
91) J.Watanabe, W.R.Krigbaum, J.Polym.Sci., Polym.Phys.Ed., 23, 565 (1985)

92) C.K.Ober, R.W.Lenz, G.Galli, E.Chiellini, Macromolecules, 16, 1034 (1983)
93) E.Chiellini, G.Galli, F.Ciardelli, R.Palla, F.Carmassi, Inf.Chim., 176, 221 (1978)
94) E.Chiellini, G.Galli, R.W.Lenz, C.K.Ober, Preprints XXVIII Macromolecular Symposium, Amherst, 1982, p.365
95) G.Galli, P.Nieri, C.K.Ober, E.Chiellini, Makromol.Chem., Rapid Commun., 3, 543 (1982)
96) A.Blumstein, G.Maret, S.Vilasagar, Macromolecules, 14, 1543 (1981)
97) E.Chiellini, G.Galli, Makromol.Chem., Rapid Commun., 4, 285 (1983)
98) E.Chiellini, G.Galli, Macromolecules, 18, 1652 (1985)
99) G.Galli, M.Laus, A.S.Angeloni, P.Ferruti, E.Chiellini, Makromol. Chem., Rapid Commun., 4, 681 (1983)
100) M.Laus, A.S.Angeloni, P.Ferruti, G.Galli, E.Chiellini, J.Polym.Sci., Polym.Lett.Ed., 22, 587 (1984)
101) E.Chiellini, G.Galli, A.S.Angeloni, M.Laus, R.Pellegrini, Liquid Crystals, in press (1987)
102) M.Laus, A.S.Angeloni, E.Chiellini, G.Galli, Preprints IUPAC Symposium on Non Crystalline Order in Polymers, Naples, 1985, p.53
103) S.Carrozzino, Thesis, University of Pisa, 1985
104) G.Galli, E.Chiellini, C.K.Ober, R.W.Lenz, Makromol.Chem., 183, 2693 (1982)
105) E.Chiellini, G.Galli, C.Malanga, N.Spassky, Polym.Bull., 9, 336 (1983)
106) E.Chiellini, P.Nieri, G.Galli, Mol.Cryst.Liq.Cryst., 113, 213 (1984)
107) A.Roviello, A.Sirigu, Makromol.Chem., 183, 895 (1982)
108) A.Blumstein, O.Thomas, Macromolecules, 15, 1264 (1982)
109) Q.F.Zhou, R.W.Lenz, J.Polym.Sci., Polym.Chem.Ed., 21, 3313 (1983)
110) B.Gallot, G.Galli, E.Chiellini, Makromol.Chem., Rapid Commun., in press (1987)
111) E.Chiellini, G.Galli, S.Carrozzino, S.Melone, G.Torquati, Mol.Cryst.Liq.Cryst., in press (1987)
112) R.Caciuffo, E.Chiellini, G.Galli, F.Rustichelli, G.Torquati, Mol.Cryst.Liq.Cryst., 127, 129 (1985)
113) G.Torquati, R.Caciuffo, S.Melone, E.Chiellini, G.Galli, Mol.Cryst.Liq.Cryst., in press (1987)
114) T.V.Samulski, E.T.Samulski, J.Chem.Phys., 67, 824 (1977)

Acknowledgement. The authors wish to thank the Ministero Pubblica Istruzione of Italy (Fondi 40%) for financial support of this work.

ATROPISOMERIC POLYMERS

Roeland J.M. Nolte and Wiendelt Drenth
Department of Organic Chemistry,
University at Utrecht,
Padualaan 8, 3584 CH Utrecht,
The Netherlands

ABSTRACT. Restricted rotation around single bonds in a polymer main chain can generate a stable helix. In order for this to occur, the main chain must consist of bulky repeating units which are all similarly rotated with respect to each other. Polymers which display these features have recently been discovered. They are called atropisomeric polymers. Examples are poly(isocyanides), poly(chloral), and poly(methacrylates) carrying bulky ester substituents. Atropisomeric polymers with an excess of one screw sense can be prepared from optically active monomers. They can also be obtained from achiral monomers by one of the following methods: (i) screw sense selective polymerization using optically active initiators, (ii) screw sense selective copolymerization with small amounts of optically active comonomers, (iii) chromatographic resolution of the racemic polymer of an achiral monomer.

1. INTRODUCTION

Stereoisomerism resulting from a restricted rotation around a single bond is a well-known phenomenon in organic chemistry [1]. This type of isomerism was discovered in the early 1920's in biphenyl derivatives [2]. The general term atropisomerism (Greek ατροπος = not to be turned) is used to denote any kind of stereoisomerism that is the result of restricted rotation around single bonds where the isomers can actually be isolated [1].

Atropisomerism in polymers was reported for the first time in 1974 in polymers of isocyanides [3]. It was demonstrated that poly(tert-butyl isocyanide), $[>C=N-C(CH_3)_3]_n$, could be resolved into fractions that show positive and negative optical rotations. Subsequent work showed that the observed optical rotation is due to a helical configuration of the polymer backbone [4]. A mechanism of the polymerization was proposed and methods were worked out to achieve screw sense selective polymerization [5].

To date, two other examples of atropisomerism in polymers have been reported. There is evidence that polychloral, prepared from chloral with a chiral initiator, forms a stable helix with a preference of one helical screw sense over the other [6]. Yuki et al. [7] and later Cram and Sogah [8], showed that methacrylate esters polymerize in the presence of chiral anionic catalysts to give optically active helical polymers. Depending on the bulkiness of the ester side chains, the one-handed helical structures of the polymer main chains were either retained or randomized after some time.

Here we wish to highlight the most interesting features of atropisomerism in polymers. As this field is still undeveloped, we hope that the present review will stimulate scientists to do further research in this fascinating area.

M. Fontanille and A. Guyot (eds.), Recent Advances in Mechanistic and Synthetic Aspects of Polymerization, 451–464.

2. POLY(ISOCYANIDES)

2.1. Synthesis and structure

In the presence of nickel(II)salts, isocyanides readily polymerize to give polymers called poly(isocyanides), poly(iminomethylenes) or poly(carbonimidoyls) (Chart I) [9]. A different method to prepare these polymers, which makes use of an acid-coated glass system, was developed by Millich [10]. Depending on the reaction conditions the molecular weights can vary between M_v 5,000 and 500,000. The heat of polymerization (Chart I) is considerable, and reflects the conversion of a formal bivalent carbon atom in the monomer to a tetravalent carbon in the polymer. The high value of the exponent in the Mark-Houwink equation (Chart I) indicates that the polymers have a very rigid structure [11].

Polymers of isocyanides are unusual polymers as they carry a side chain on each atom of their main chain. As can clearly be seen from space-filling molecular models, this feature causes a restricted rotation around the single bonds that connect the main chain carbon atoms. Just as in the case of hindered biphenyls, two configurations are possible around each of the single bonds viz. R or S [1]. If the poly(isocyanide) chain is highly isotactic (meaning that the configuration around all the single bonds is the same) a stable helix can be formed. This helix is right-handed (P) if the aforementioned configurations are all S and left-handed (M) if they are all R, see Figure 1.

CHART I

Isocyanide: $[\text{R-}\overset{+}{\text{N}}\equiv\overset{-}{\text{C}} \leftrightarrow \text{R-N=C}]$

Polymerization: $n\ \text{R-}\overset{+}{\text{N}}\equiv\overset{-}{\text{C}} \rightarrow [\text{R-N=C<}]_n$

Range of molecular weights: $\overline{M}_v = 5{,}000 - 500{,}000$

Mark-Houwink equation: $[\eta] = 1.4 \times 10^{-9}\, \overline{M}_w^{\ 1.75}$

Heat of polymerization: 81.4 $kJ.mol^{-1}$

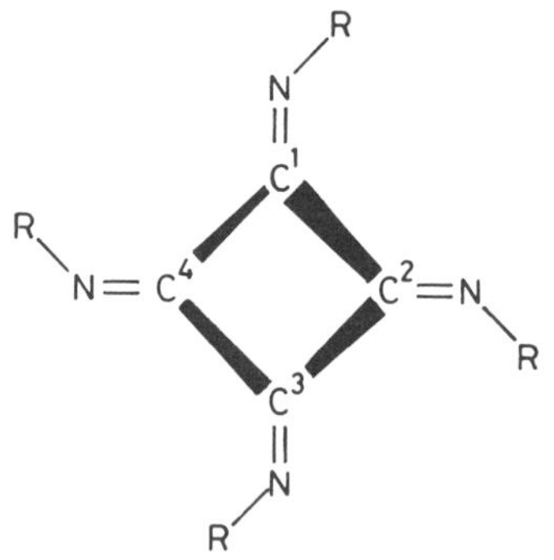

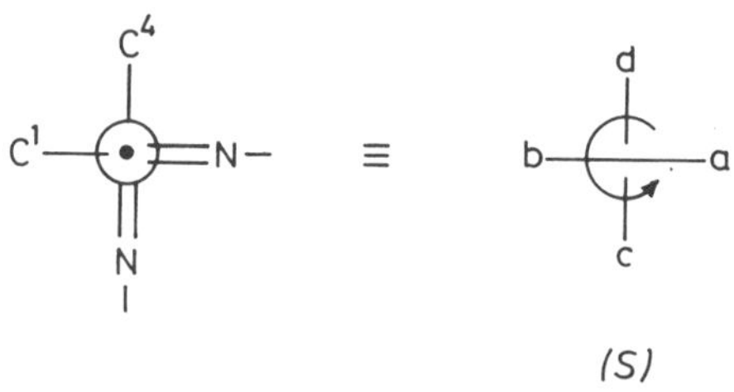

Figure 1. View along the helical axis of a poly(isocyanide) molecule with a right-handed screw sense; unit 5 is behind unit 1, etc. (A). According to the Cahn-Ingold-Prelog nomenclature rules, the configuration around each of the single bonds connecting the main chain carbon atoms is S, see Newman projection along C^2-C^3 (B); a-d denotes the priority sequence.

The polymer chain would have adopted a zig-zag structure if the single bonds had had alternating R and S absolute configurations. Such a polymer would have been named syndiotactic. Molecular models, however, show that such an arrangement is not possible for steric reasons.

Evidence that polymers of isocyanides have a helical structure and consist of equal mixtures of left-handed and right-handed screws came from chromatographic experiments. Poly(tert-butylisocyanide) was resolved by running its solution in chloroform over a chiral column [3,12]. The support of the column consisted of glass beads coated with an insoluble, high-molecular weight polymer of optically active (S)-sec-butyl isocyanide. The completely resolved polymer of tert-butyl isocyanide showed a specific rotation that increased with increasing molecular weight up to a value of $[\alpha]_{578} = 56^{\circ}$ when a chain length of 20 repeating units was reached. This observation of increasing $[\alpha]$-values with increasing molecular weight is in line with calculations which predict that the optical rotation per repeating unit in any helical polymer increases with increasing molecular weight until a constant value is reached at 16-20 units [13].

The question, which screw sense belongs to the (+)-rotating enantiomer of poly(tert-butylisocyanide) and which to the (-)-rotating one, was solved by circular dichroism (CD). The CD spectra of the two enantiomers have, besides sign inversion, a similar shape. In the region 240-400 nm a so-called exciton couplet is visible, which is positive (S-shape) for the (+)-rotating enantiomer and negative (Z-shape) for the (-)-rotating enantiomer (Figure 2).

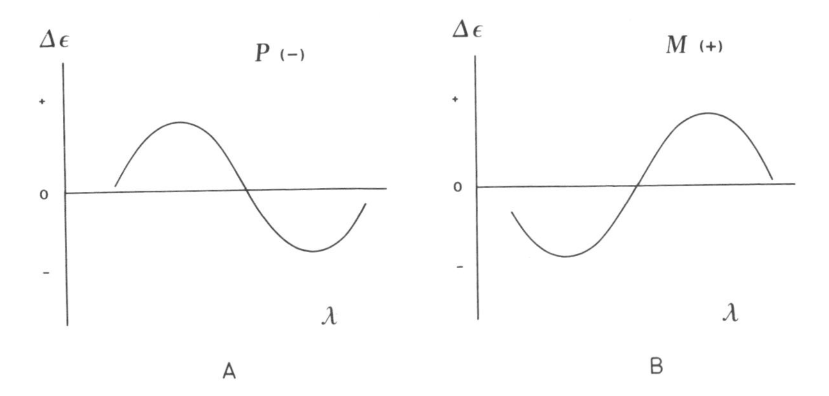

Figure 2. CD-spectra of (-)-rotating (P-screw,A) and (+)-rotating (M-screw, B) enantiomers of poly(tert-butyl isocyanide).

This couplet is due to the $n \rightarrow \pi^*$ transition of the imino functions in the polymer main chain [4]. According to Tinoco's exciton theory, the CD-spectrum of a helical polymer can be described by a two-term equation [14]. One term represents the so-called conservative and the other the non-conservative contribution. The two contributions have been calculated as a function of the number of repeating units per turn for a right-handed and for a left-handed helix. The experimental CD curves were also analyzed in terms of conservative and non-conservative contributions. On comparison of the calculated and experimental spectra it was concluded that the (+)-rotating enantiomer has a left-handed screw and the (-)-rotating one a right-handed screw. The number of repeating units per turn of the helix was confined to the region between 3.6 and 4.6 [4]. Recently, more extensive CD calculations have been performed using the polarizability theory of DeVoe [15]. These calculations reveal that the number of repeating units per turn in poly(tert-butyl isocyanide) is close to four, i.e. 3.81 [16]. A similar value (3.75) is obtained from consistent force field (molecular mechanics) calculations on oligomers of tert-butyl isocyanide [16]. In Table I some configurational parameters of poly(tert-butyl isocyanide) are listed, which resulted from the molecular mechanics and CD-calculations. So far it has not yet been possible to get these parameters from an X-ray structure determination.

TABLE I. Configurational parameters of poly(tert-butyl isocyanide)[a]

Number of repeating units per turn	3.6-4.6[b], 3.81[c] 3.75[d]
Dihedral angle between two repeating units	76-77° [c]
Radius of helix from center to chain carbon atoms	0.0305 nm
Symmetry operation connecting two neighboring repeating units: [e]	
i translation along helix axis	0.10 nm
ii rotation around helix axis	94.5°

[a] Data have been taken from Refs. 4 and 16.
[b] From CD-calculations using Tinoco's exciton theory.
[c] From CD-calculations using DeVoe's polarizability theory.
[d] From consistent force field calculations on hexamer of tert-butyl isocyanide.
[e] Based on 3.81 repeating units per helical turn

2.2. Polymerization mechanism

A mechanism was proposed for the polymerization of isocyanides by nickel(II) salts. It involves a sequence of insertion reactions around the nickel center (Scheme I) [17]. The reaction starts from the square-planar nickel complex, $Ni(CNR)_4^{2+}$ (**1**), which is formed when an excess of isocyanide is added to nickel chloride or nickel perchlorate. A common reaction of a metal-coordinated isocyanide is nucleophilic attack on its terminal carbon atom [20]. Therefore, a conceivable initiation step of the polymerization is attack by a nucleophile X^- on one of the four isocyanide ligands. This nucleophile can be a solvent molecule (e.g. water or methanol), an anion (e.g. Cl^-) or an initiator which is added purposely, e.g. an amine (vide infra). In the resulting complex **2**, which has been isolated and characterized for X = 1-phenylethylamine [21], the plane of the ligand C(X)=NR is approximately perpendicular to the plane of the isocyanide carbons and nickel. For steric reasons there is no rotation around the C^1-Ni bond. Carbon atom C^1 has gained in nucleophilicity, allowing it to attack a neighbouring isocyanide ligand. Such an attack is facilitated when a new isocyanide ligand C^5=NR is substituted for C^1(X)=NR. Attack can occur on either C^2 or C^4. In **3** it has occurred on C^2. When the sequence of attack continues in the direction $C^1 \rightarrow C^2 \rightarrow C^3 \rightarrow C^4$ a left handed helix is formed (**4**). In a similar way the sequence $C^1 \rightarrow C^4 \rightarrow C^3 \rightarrow C^2$ will result in a right handed helix. The foregoing mechanism has been named "merry-go-round" mechanism [5]. It accounts for the enantioselective polymerization of chiral and achiral isocyanides, which will be discussed in the next sections. It also explains why the tightly coiled helix of a poly(isocyande) is so easily formed; the monomers are preorganized in the nickel complex and with a minimal amount of movement are knitted together. Since one turn of the helix consists of approximately 4 repeating units, each rotation around nickel contributes one turn to this helix. The entropy of activation amounts to $\Delta S^{\neq}$ - 54 J.mol^{-1}.K^{-1} [17]. This value is in the range found for ligand substitutions in square planar d-8 metal complexes [22].

SCHEME I

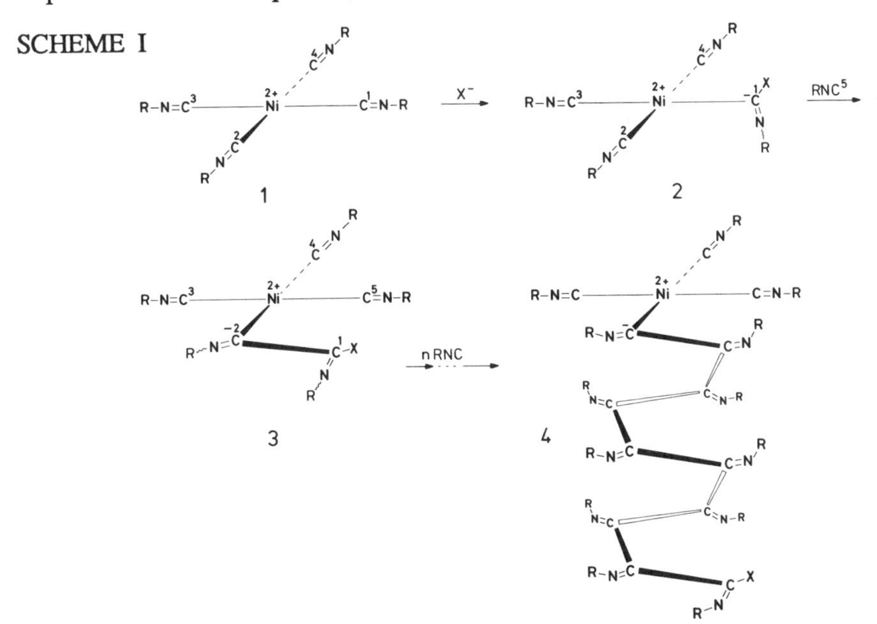

2.3 Screw sense selective polymerization of chiral isocyanides

The resolution of poly(tert-butyl isocyanide) into enantiomers indicates that polymerization of isocyanides proceeds stereoselectively with respect to the screw sense. When the monomer is achiral, a racemic mixture of P and M-screws is formed. However, when the monomer is one enantiomer of a chiral isocyanide, its polymer can be expected to be a mixture of diastereoisomers, and P and M screws will not be obtained in equal amounts. This hypothesis was tested on approximately 20 different optically active isocyanides and found to be correct [18,19,23]. As an example we present data on the polymerization of optically active alkyl isocyanides **5**, which contain substituents R^1, varying in steric

R^1 | H-C-$\overset{+}{N}$≡$\overset{-}{C}$ | CH_3

5-(R)

R^1 | H-C-N=C(H)(t-C_4H_9) | CH_3

6-(R)

requirement.(Table II) [19]. Polymers of **5** displayed positive and negative optical rotations.

These rotations are the sum of two contributions: one from the main chain and a second one from the side chain. The optical rotation due to the side chain can be estimated from model imines **6**. In Table II the difference between the molar optical rotation value of the polymer

TABLE II. Screw sense selective polymerization of (R)-$R^1CH_3CH\overset{+}{N}{\equiv}\overset{-}{C}$ (**5**) [a]

R^1	$\Delta[M]_{578}$/deg [b]	$\Delta\varepsilon$/[L/(mol cm)] [c]	Screw sense, % e.e.	λ_{R^1}
C_2H_5	-	-	P(?)	1.05
n-C_6H_{13}	81.8	0.040	M, 20	1.1
i-C_4H_9	127.1	0.115	M, 56	1.20
i-C_3H_7	59.9	0.125	M, 62	1.27
(t-C_4H_9NC)$_n$	56	0.20	M, 100	

[a] Data have been taken from Ref. 19.
[b] Difference between the molar optical rotation values of polymers of **5** and model compounds **6**.
[c] $\Delta\varepsilon$ of positive band at 300-400 nm in CD-spectrum.

and that of the model compound, $\Delta[M]$, is presented. For various substituents R^1 this difference is positive, suggesting that the contribution by the main chain is dextrorotatory. Applying the relation found for poly(tert-butyl isocyanide) - viz an M-helix gives rise to a (+) sign of optical rotation and a P-helix gives rise to a (-) sign - it can be concluded that on polymerization, monomers **5** preferentially form left-handed helices. The same conclusion is drawn from the CD-spectra of polymers of **5**, which display positive exciton couplets in the wavelength region 240-400 nm, characteristic of left-handed helices. The intensity of the couplets varies with the substituent R^1 in **5**. They can be used to estimate the enantiomeric excess (e.e.) of left-handed screws. As can be seen in Table II the e.e.-value increases with increasing bulkiness of R^1, as expressed by the so-called λ-steric constant [24].

The polymerization mechanism discussed in the previous section, allows one to predict the screw sense that is preferentially formed when one enantiomer of a chiral isocyanide is polymerized. The procedure will be illustrated for compounds **5.** The steric requirements of the substituents at the chiral carbon center in **5** are denoted by small (S), medium (M), and large (L). The first intermediate in the polymerization process will have the configuration **7-Z** or **7-E** (Scheme II). Attack by C^1 on C^2 or C^4 in **7-Z** and **7-E** depends on the steric interactions in the two transition states. With regard to these interactions the relative sizes and the arrangement of the substituents at the chiral carbon atom of the C^1-ligand are of interest.

SCHEME II

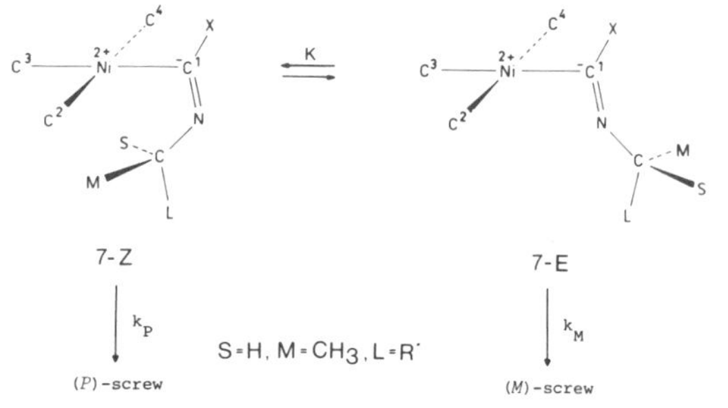

7-Z ⇌(K) 7-E

7-Z →(k_P) (P)-screw

7-E →(k_M) (M)-screw

S=H, M=CH_3, L=R'

In **7-Z** and **7-E** these substituents are positioned in such a way that the least steric hindrance occurs in the transition state, i.e. L pointing away from the nickel center. For **7-Z** attack will preferentially occur on C^4 as this is in the direction of the least-hindered side. A P-screw is formed. Similarly, in **7-E** the attack will be on C^2 and an M-screw is generated. With sterically more demanding substituents R^1 (in Table II reading from top to bottom) the equilibrium **7-Z** ⇄ **7-E** shifts to the right, since in **7-E** the steric interactions are less severe than in **7-Z**. Consequently, the preference for M-helices increases in the same direction. By similar reasoning the screw senses of a large number of optically active poly(isocyanides) have been predicted and found to coincide with the screw senses derived from optical rotation data and CD-spectra. From these findings it has been concluded that the screw sense induction in the polymerization of isocyanides, is a process controlled kinetically at the catalyst center [19].

2.4. Screw sense selective polymerization of achiral isocyanides

Stereo-selective polymerization of achiral isocyanides in order to form polymers with an excess of one screw sense has been tried using chiral catalysts, chiral ligands, and chiral solvents [25]. These attempts have not yet been successful: the chiral additive either expelled the isocyanide from the nickel center, blocking the polymerization, or was not able to do so, in which case no chiral induction occurred. Screw sense selective polymerization, however, could be achieved by using an optically active amine as the initiator (Table III). Samples of poly(*tert*-butyl isocyanide) showing an enantiomeric excess of 85% of right-handed screws, have been obtained by this method [21,26]. The screw sense that is induced by the chiral initiator can be predicted from the polymerization mechanism, by reasoning in a way analogous to the one described for the polymerization of chiral isocyanides [26].

TABLE III. Screw sense selective polymerization of *tert*-butyl isocyanide by nickel(II) and optically active initiators.[a]

Initiator[b]	Screw sense	e.e./%
S-(+)-*sec*-$C_4H_9NH_2$	P	7
L-prolinol	M	36
L-phenylalaninol	M	37
S-(-)-1-$CH_3CH(C_6H_5)NH_2$	P	61
S-(-)-1-$CH_3CH(C_6H_5)NH_2$[c]	P	85

[a] Data have been taken from Refs. 21 an 26.
[b] Catalyst: $Ni(t\text{-}C_4H_9NC)_4\ (ClO_4)_2$
[c] Catalyst: $Ni[2\text{-}(t\text{-}C_4H_9)C_6H_4NC]_4\ (ClO_4)_2$

In a different procedure, bulky, optically active co-monomers have been used to induce a screw sense selective polymerization of achiral isocyanides [25]. One of such bulky co-monomers is the *tert*-butyl ester of (S)-2-isocyanoisovaleric acid (**8**). In the presence of nickel(II) salts this isocyanide slowly polymerizes to give a homopolymer with

$$(CH_3)_2CHCH(CO_2\text{-}t\text{-}C_4H_9)\overset{+}{N}{\equiv}\overset{-}{C} \quad \mathbf{8\text{-}(S)}$$

$$Y\text{-}C_6H_4\text{-}\overset{+}{N}{\equiv}\overset{-}{C} \quad Y = H, CH_3, CH_3O, Cl \quad \mathbf{9}$$

predominantly left-handed screws (see CD-spectrum A in Figure 3). When isocyanide **8** is mixed with an achiral isocyanide, e.g. a substituted phenyl isocyanide **9**, and the mixture is polymerized, polymer samples are obtained which show high negative optical rotations, $[a]^{20}D$ -350 to -650°. These samples mainly contain the homopolymer of the achiral isocyanide. Remarkably, CD-spectra reveal that these homopolymers have right-handed screws (see Figure 3). This finding suggests that the polymerization is not a common

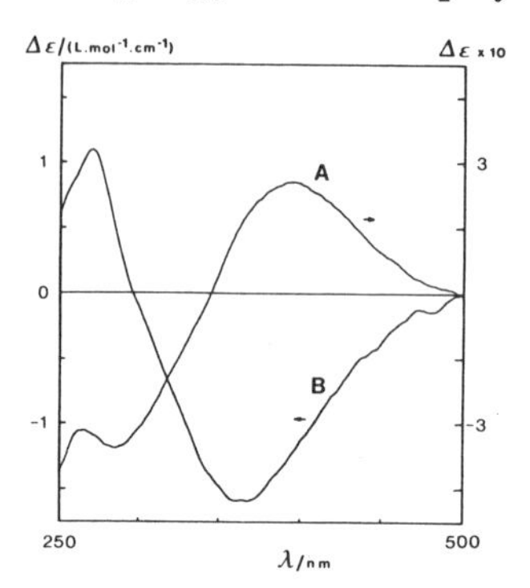

Figure 3. CD spectra of homopolymers of **8** (curve A) and **9** (Y = MeO, curve B). Curve A indicates a left-handed helix, curve B indicates a right-handed helix.

copolymerization reaction. If this were the case, a left-handed screw sense for polymers **9** would have been expected. The mechanism is probably as follows [27]. The bulky chiral monomer **8** preferentially forms an M-screw, its rate of polymerization being low, $k_p << 10^{-7}s^{-1}$, (Figure 4A). Monomers **9** are fast polymerizing isocyanides ($k_p \cong 10^{-4}s^{-1}$), and form racemic mixtures of P and M screws (Figure 4B). When **8** and **9** are copolymerized the former one has a preference for inclusion into the M-helices and retards the formation of these helices out of **9.** The P-helices continue to grow on and ultimately consume all the achiral monomers (Figure 4C). In line with this mechanism, gel-permeation chromatography experiments show that polymerization mixtures of **8** and **9** contain low molecular-weight

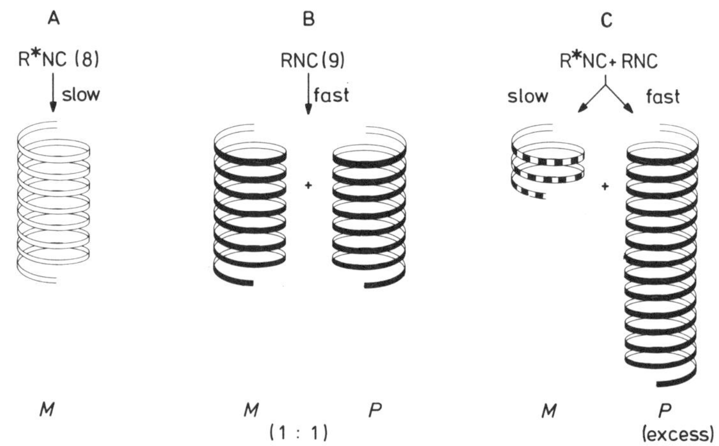

Figure 4. Screw sense selective polymerization of achiral isocyanide in the presence of optically active co-monomer.

fractions with a high content of **8** and the high-molecular weight fractions with a low content of this monomer [27].

3. POLY(CHLORAL)

Chloral can easily be polymerized by cationic and anionic initiators, the latter procedure being the method of choice [28,29].

$$n\ CCl_3CH{=}O \xrightarrow{LiO\text{-}\underline{t}\text{-}Bu} \left(CH(CCl_3)\text{-}O\right)_n$$

The polymer formed is insoluble and infusable, and depolymerizes around 220°C. According to X-ray studies the polymer backbone should have an isotactic 4/1-helical structure in the crystalline state [28]. This structure has recently been confirmed by molecular mechanics calculations [30]. The preferred helical arrangement of the polymer chain is caused by the bulky trihalomethyl side chains, which restrict the rotation around the carbon-oxygen single bonds.

Vogel and coworkers polymerized chloral with optically active initiators and observed that films of the obtained polymers show high optical rotations (table IV) [6,31]. Polymer films prepared with optically inactive initiators, such as benzyltriphenylphosphonium bromide, also displayed optical rotation [6]. This rotation, however was ascribed to

TABLE IV Optical rotation of poly(chloral) films prepared with various chiral initiators[a]

Initiator	$[\alpha]_D$
(-)-tetramethylammonium-o-acetylmandelate	+ 1180
(-)-tetramethylammonium-o-methylmandelate	+ 210
Li alkoxide of methyl mandelate	- 4030
Li 2-octanoxide	+ 4280
Li β-cholestanoxide	+ 3000

[a] Data have been taken from Ref. 31.

birefringence and had a much smaller value. Since there is no data available on optical rotation in solution, it was not possible to confirm that poly(chloral) is a true atropisomeric polymer, whose chain remains coiled in solution.

4. POLY(METHACRYLATES)

Atropisomerism in vinyl polymers was reported for the first time in 1979 by Yuki and coworkers [7]. They polymerized triphenylmethyl methacrylate **10a** (Chart II) using the initiators lithium (R)-N-(1-phenylethyl)anilide (**11**) and (-)-sparteine (**12**)-butyl lithium. The resulting polymers showed a high optical rotation and were also found to be highly isotactic.

CHART II

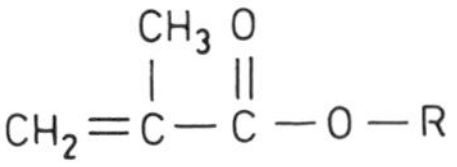

$CH_2{=}C(CH_3){-}C({=}O){-}O{-}R$

10

a, R = Ph_3C

b, R = Ph_2C–(2-pyridyl)

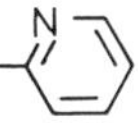

c, R = Ph_2C–(2-naphthyl)

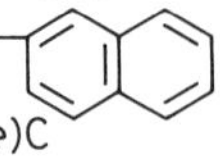

d, R = $Ph_2(Me)C$

e, R = *t*-Bu

f, R = $PhCH_2$

g, R = Me

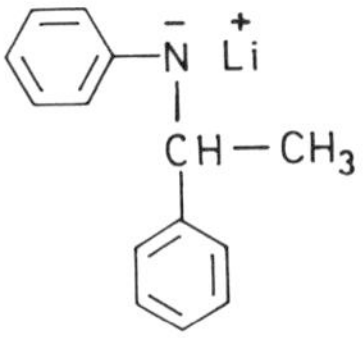

$PhN^{-}Li^{+}$–$CH(CH_3)Ph$

11

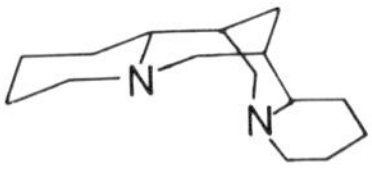

12

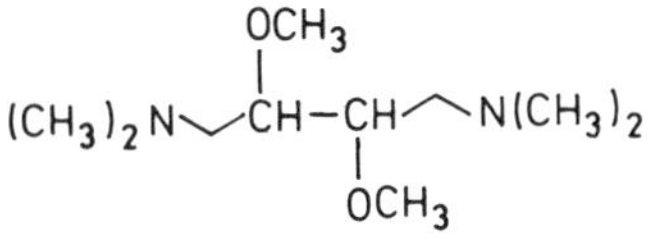

$(CH_3)_2N{-}CH_2{-}CH(OCH_3){-}CH(OCH_3){-}CH_2{-}N(CH_3)_2$

13

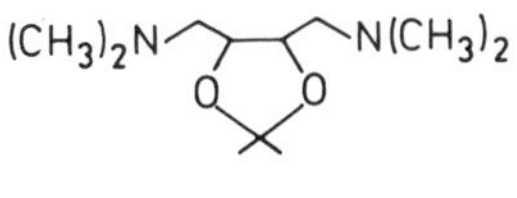

$(CH_3)_2N$ ... $N(CH_3)_2$

14

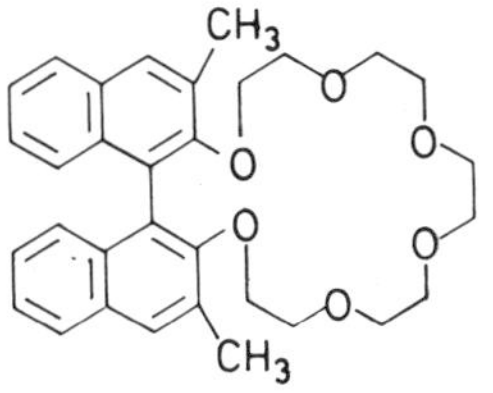

CH_3 CH_3

15

CH_3 CH_3

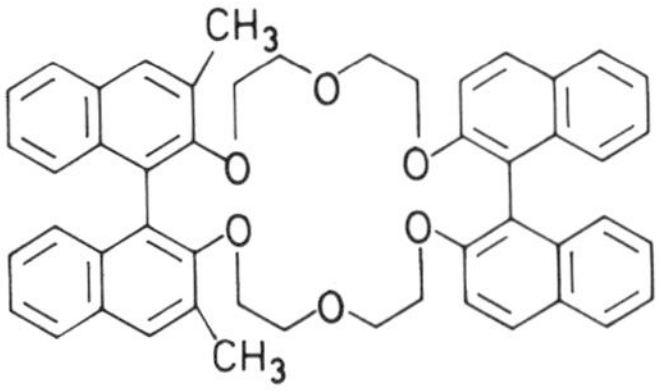

16

Various optically active isotactic methacrylate polymers have been prepared since, using different chiral anionic initiators [8,32-35].

TABLE V. Polymerization of methacrylate esters by chiral anionic initiators in toluene

Monomer	Initiator	isotacticity/%	[a]/°	Ref.
10a	(R)-**11**	100	-85[a]	32
10a	(-)-**12** .BuLi	100	+363[a]	32
10a	(R,R)-**13**.BuLi	-	-335[b]	33
10a	(S,S)-**14**/(R,S)-**11**	-	+259[b]	33
10b	(R,R)-**13**.DPEDA Li[c]	91	-307[d]	34
10c	(-)-**12**.BuLi	100	+299[b]	35
10d	(-)-**12**.BuLi	small	+ 0.7[b]	35
10e	(S)-**15**.KO-t-Bu	90	+117[e,f]	8
10f	(S)-**15**.KO-t-Bu	-	+350[e]	8
10g	(S)-**15**.KO-t-Bu	78	+248[e,f]	8
10g	(S,S)-**16**.KO-t-Bu	88	-180[e,f]	8

[a] $[a]^{20}_D$ toluene
[b] $[a]^{25}_D$ THF
[c] DPEDA = N,N'-diphenylethylenediamine
[d] $[a]^{25}_{365}$ $CHCl_3$-2,2,2-trifluoroethanol
[e] $[a]^{25}_{578}$ THF
[f] Optical rotation becomes zero after standing at room temperature for 5-40 hrs.

Representative examples are compiled in Table V.

The chiral carbon atoms formed on polymerization of a methacrylate monomer do not give rise to measurable optical activity [36,37]. The reason for this is that they become pseudo-chiral when the polymer chain has grown long. Therefore, the observed large optical rotations in the polymers, can only be due to a helical conformation of the polymer backbone. Such a helix can be relatively stable when the ester functions are sufficiently bulky to prevent uncoiling. This was found to be the case for the triphenylmethyl and diphenylnapthyl ester polymers but not for the tert-butyl and methyl ester ones. Cram and Sogah have shown that the helices of the last two polymers are thermodynamically unstable kinetic products of the polymerization reaction. These helices slowly unwind over a period of hours to give small residual optical rotations [8].

Poly(triphenylmethacrylate) prepared with BuLi as the initiator, can be separated into fractions of positive and negative optical rotation by column chromatography using insoluble (+)-poly(triphenylmethylmethacrylate) as the support [38]. This suggests that polymerization of methacrylate esters with bulky substituents proceeds stereoselectively and leads to racemic mixtures of left-handed and right-handed screws. It is not known yet which optical rotation belongs to which screw. From theoretical considerations using models, it has been predicted that the dextrorotatory polymer samples contain right-handed screws and the laevoratatory ones left-handed screws [8].

5. APPLICATION

The combination chirality and a well-defined structure, makes atropisomeric polymers attractive reagents to use in enantioselective experiments. Optically active polymers and copolymers of isocyanides have been synthesized carrying catalytically active imidazole functions [23b]. Polymer 17, derived from the dipeptide L-alanyl-L-histidine, was used as catalyst in the hydrolyses of L- and D-esters of amino acids e.g. **18**.

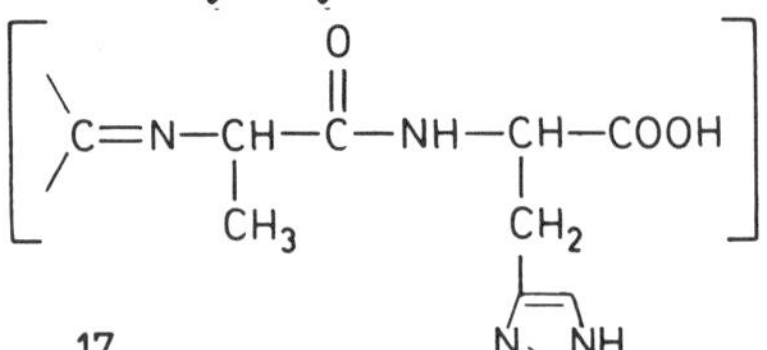

$$\left[\begin{array}{c} \\ \end{array} \text{>C=N-CH(CH}_3\text{)-C(=O)-NH-CH(CH}_2\text{-imidazolyl)-COOH} \right]_n \qquad CH_3C(=O)\text{-NH-CH(CH}_3\text{)-C(=O)-O-}C_6H_4\text{-}NO_2$$

17 (C=N–CH(CH₃)–C(O)–NH–CH(CH₂–imidazole, N/NH)–COOH, repeat unit n) 18

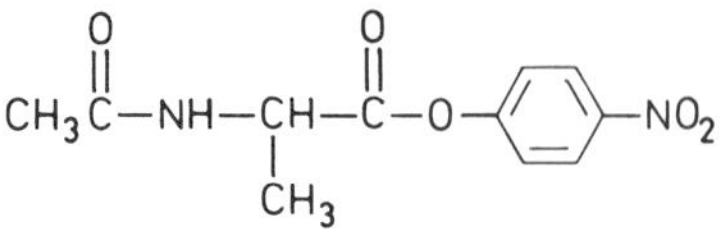

In the presence of surfactant molecules, e.g. N-cetylpyridinium chloride, polymer **17** catalyzes the hydrolysis of the L-enantiomer of **18** more efficiently than the hydrolysis of the D-enantiomer. The enantioselectivity ratio, k_L/k_D, amounts to 3 [39]. By varying the type of polymeric catalyst and surfactant an even higher enantioselectivity can be obtained with a maximum value of $k_L/k_D = 12$ [40]. The observed enantioselectivity depends on the screw sense of the polymeric catalyst as well as on the chirality of the side chain. Which of these factors dominates is not known yet [41].

Optically active polychloral [42] and in particular, optically active poly(triphenylmethyl-methacrylate) [43-49] have been applied as chiral supports for the chromatographic resolution of racemates. Table VI exemplifies the type of compounds that have been partly or completely resolved with dextrorotatory poly(triphenylmethyl methacrylate) on macroporous silica gel. This column material is now also available commercially [33].

TABLE VI. Resolution of racemic compounds by (+)-poly(triphenylmethyl methacrylate) coated on silica gel

Racemate	α [a]	R_s [a]	Ref.
2,2'-dihydroxy-1,1'-binaphtyl	2.37(+) [b]	3.83	45
trans 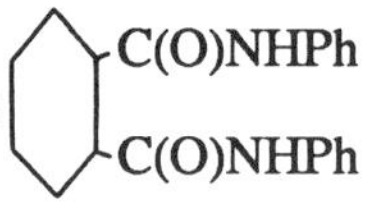 cyclohexane with C(O)NHPh, C(O)NHPh	6.31(-)	2.30	45
trans  cyclobutane with C(O)NHPh, C(O)NHPh	1.78(+)	3.03	45
styrene oxide	1.22(-)	0.87	45
trans-stilbene oxide	5.21(-)	5.30	45
$Co(III)(Acac)_3$ [c]	1.32		45
$Al(III)(Acac)_3$	- (-)		46
$(C_6Cl_5)_3N$	2.9 (+)		47
ArP=C=PAr ($Ar = 2,4,6\text{-}(t\text{-}C_4H_9)_3C_6H_2$)	- (-)		48

[a] α and R_s stand for separation and resolution factor, respectively; see Ref. 45.
[b] Sign of rotation of enantiomer eluting first. [c] Acac = acetylacetonate.

6. REFERENCES

1. Eliel, E.L., Stereochemistry of Carbon Compounds, McGraw-Hill: New York, 1962; p. 156.
2. Christie, G.H.; Kenner, J. J. Chem. Soc. 1922, **121**, 614.
3. Nolte, R.J.M.; Beijnen, A.J.M. van; Drenth, W. J. Am. Chem. Soc. 1974, **96**, 5932.
4. Beijnen, A.J.M.; Nolte, R.J.M.; Drenth, W.; Hezemans, A.M.F. Tetrahedron 1976, **32**, 2017.
5. Drenth, W.; Nolte, R.J.M. Acc. Chem. Res. 1979, **12**, 30.
6. Corley, L.S.; Vogl, O. Polym.Bull. 1980, **3**, 211.
7. Okamoto, Y.; Suzuki, K.; Ohta, K.; Hatada, K.; Yuki, H. J. Am. Chem. Soc., 1979, **101**, 4763.
8. Cram, D.J.; Sogah, D.Y. J. Am. Chem. Soc. 1985, **107**, 8301.
9. Nolte, R.J.M.; Stephany, R.W.; Drenth, W. Recl. Trav. Chim. Pays-Bas 1973, **92**, 83; Nolte, R.J.M.; Beijnen, A.J.M. van; Zwikker, J.W.; Drenth, W. Polym. Synth. 1985, **9**, 81.
10. Millich, F.; Sinclair, R.G. J. Polym. Sci. 1968, **C22**, 33.
11. Millich, F. Macromol. Rev. 1980, **15**, 207.
12. Beijnen, A.J.M. van; Nolte, R.J.M.; Drenth, W. Recl. Trav. Chim. Pays-Bas 1980, **99**, 121.
13. Aslanyan, V.M. J. Polym. Sci. 1968, **C16**, 3145. Brahms, J.; Michelson, A.M.; Van Holde, K.E. J. Mol. Biol. 1966, **15**, 467. Rich, A.; Tinoco, I. J. Am. Chem. SOc. 1960, **82**, 6409.
14. Tinoco, I. J. Chim. Phys., Phys. Chim. Biol. 1968, **65**, 91.
15. De Voe, H. J. Chem. Phys. 1964, **41**, 393; De Voe, H. Ibidem 1965, **43**, 3199.
16. Huige, C.J.M., Thesis, Utrecht (1985).
17. Nolte, R.J.M.; Drenth, W. Recl. Trav. Chim. Pays-Bas 1973, **92**, 788.
18. Nolte, R.J.M.; Zwikker, J.W.; Reedijk, J.; Drenth, W. J. Mol. Catal. 1978, **4**, 423; Beijnen, A.J.M. van; Nolte, R.J.M.; Zwikker, J.W.; Drenth, W. Ibidem 1978, **4**, 427.
19. Beijnen, A.J.M. van; Nolte, R.J.M.; Drenth, W.; Hezemans, A.M.F.; Coolwijk, P.J.F.M. van de Macromolecules 1980, **13**, 1386; Beijnen, A.J.M. van; Nolte, R.J.M.; Naaktgeboren, A.J.; Zwikker, J.W.; Drenth, W.; Hezemans, A.M.F. Ibidem 1983, **16**, 1679.
20. Treichel, P.M. Adv. Organomet. Chem. 1973, **11**, 21; Yamamoto, Y.; Yamazaki, H. Coord. Chem. Rev., 1977, **8**, 225.
21. Kamer, P.C.J.; Nolte, R.J.M.; Drenth, W., unpublished results.
22. Basolo, F.; Pearson, R.G. Mechanism of Inorganic Reactions; Wiley: New York, 1967; p. 404.
23. (a) Nolte, R.J.M; Zomeren, J.A.J. van; Zwikker, J.W. J. Org. Chem. 1978, **43**, 1972. (b) Visser, H.G.J.; Nolte, R.J.M.; Zwikker, J.W.; Drenth, W. Ibidem 1985, **50**, 3133 and 3138.
24. Kagan, H.B., Stereochemistry; Thieme; Stuttgart, 1977; Vol 3, p. 26.
25. Harada, T.; Cleij, M.C.; Nolte, R.J.M.; Hezemans, A.M.F.; Drenth, W. J. Chem. Soc. Chem. Commun. 1984, 726.
26. Kamer, P.C.J.; Nolte, R.J.M; Drenth, W. J. Chem. Soc. Chem. Commun. 1986,in press.
27. Kamer, P.C.J.; Cleij, M.C.; Harada, T.; Nolte, R.J.M.; Drenth, W., manuscript in preparation.
28. Vogl, O.; Miller, H.C.; Sharkey, W.H. Macromolecules 1972, **5**, 658.
29. Vogl, O.; Corley, L.S. Polym. Preprints 1978, **19**, 210.
30. Abe, A.; Tasaki, K.; Inomata, K.; Vogl, O. Macromolecules, 1986, **19**, 2707.

31. Vogl, O.; Jay Cox, G.D. Chem. Tech. 1986, 698.
32. Okamoto, Y.; Suzuki, K.; Yuki, H. J. Polym. Sci. Polym. Chem. Ed. 1980, **18**, 3043.
33. Okamoto, Y.; Shohi, H.; Yuki, H. J. Polym. Sci. Polym. Lett. 1983, **21**, 601.
34. Okamoto, Y.; Ishikura, M.; Hatada, K.; Yukij, H. Polym.J. 1983, **15**, 851.
35. Okamoto, Y.; Shohi, H.; Ishikura, M.; Yuki, H. Abstracts 28th IUPAC Macromolecular Symposium, Amherst, 1982, p. 233.
36. Frisch, H.L.; Schuerch, C.; Szwarc, M. J. Polym. Sci. 1953, **11**, 559.
37. Pino, P. Adv. Polym. Sci. 1965, **4**, 393.
38. Okamoto, Y.; Okamoto, I.; Yuki, H. J. Polym. Sci. Polym. Lett. 1981, **19**, 451.
39. Visser, H.G.J.; Nolte, R.J.M.; Drenth, W. Macromolecules 1985, **18**, 1818.
40. Cleij, M.C.; Nolte, R.J.M.; Drenth, W., unpublished results.
41. Visser, H.G.J., Thesis, Utrecht 1983.
42. Hatada, K.; Shimizu, S.; Yuki, H.; Harris, W.; Vogl, O. Polym. Bull. 1981, **4**, 179.
43. Yuki, H.; Okamoto, Y.; Okamoto, I. J. Am. Chem. Soc. 1980, **102**, 6356.
44. Okamoto, Y.; Okamoto, I.; Yuki, H. Chem. Lett. 1981, 835.
45. Okamoto, Y.; Honda, S.; Okamoto, I.; Yuki, H.; Murata, S.; Noyori, R.; Takaya, H. J. Am. Chem. Soc. 1981, **103**, 6971.
46. Okamoto, Y.; Yashima, E.; Hatada, K. J. Chem. SOc. Chem. Commun. 1984, 1051.
47. Okamoto, Y.; Yashima, E.; Hatada, K.; Mislow, K. J. Org. Chem. 1984, **49**, 557.
48. Yoshifuji, M.; Toyota, K.; Niitsu, T.; Inamoto, N.; Okamoto, Y.; Aburatani, R. J. Chem. Soc. Chem. Commun. 1986, 1550.
49. Kissener, W.; Vögtle, F. Angew. Chem. 1985, **97**, 227.

REPORT OF DISCUSSIONS AND RECOMMENDATIONS

General Aspects

Chirality can be introduced into a macromolecule according to several different ways ; however chiral polymers of interest are substantially those in which the chain backbone has assumed a substantial chiral structure and each macromolecule one hardness is prevailing. Accordingly, chiral polymers can be isolated in optically active form and the study of their chiroptical properties (rotatory power and circular dichroism) allows to detect the type of chiral structure present.

Early investigation in this area was mainly due to the interest for chiral polymers as simplified models for biopolymers -proteins, nucleic acid and polysaccharides are indeed optically active- and to fundamental stereochemical studies.

The present interest for obtaining specialized polymers and new organic materials however can provide a renewed interest for chiral polymers and suggest new applications. Thus in addition to the development of established applications as reagents and catalysts for reaction where the stereochemical control is necessary and as sationary phase for chromatographic resolution of racemates, additional ones can be predicted in biomedical and pharmaceutical areas, for the preparation of chiral liquid crystals and of organic materials with non linear optical properties.

Indeed, the more specific molecular recognition allowed by the proper chirality should allow more specific interactions of these man made polymers with biological molecules and biopolymers. Moreover, the possibility of obtaining high molecular asymmetry should be helpful in designing new materials where orientation can be fundamental for the ultimate properties.

M. Fontanille and A. Guyot (eds.), Recent Advances in Mechanistic and Synthetic Aspects of Polymerization, 465–470.

New Synthetic Methods and Chiral Structures

The conventional synthesis based on the direct polymerization of chiral monomers is in general limited by the limited availability of optically active monomers and by their relative high cost. Moreover, in several cases, the presence of chirality in the monomer is not sufficient to warrant a chiral polymer structure. The new method available and discussed during the Workshop offers on the other side better versatility and produce macromolecules where the backbone is in disymmetric arrangement starting in general with relatively low cost materials.

Synthesis + Structure

The new synthetic approaches for the preparation of optically active polymers encountering the above requisites are :

* Synthesis of stereoregular copolymers possessing asymmetric triads of the type AAB-(1,2)

Symmetry considerations on regular vinyl polymers has shown that certain arrangements in homo- and copolymers are chiral due to backbone chirality and hence should show optical activity as pure enantiomers. These copolymers have been prepared by copolymerization of 3.4-O-cyclohexylidene-D-mannitol 1.2:5.6-bis-O [(4-vinylphenyl)borate] with several comonomers followed by the quantitative removal of the chiral template D-mannitol. The resulting copolymers of 4-phenyl boric acid with e.g. styrene methyl methacrylate and methacrylonitrile show negative optical rotation, large in absolute value. Polymer analog reactions allowed to transform these polymers into different optically active functional copolymers with chiral centers in the backbone. The chiral template D-mannitol is not incorporated in the chiral copolymer and can be recycled. Thus the preparation of these chiral macromolecules is performed without consuming optically active materials.

* Atropisomeric polymers

The steric strain imposed by the side chains may force the polymer backbone to assume a rigid helical structure. Chiral polymers (showing this feature) can be obtained by polymerizing suitable isocyanides in the presence of nickel (II) catalysts. In particular, optically active polymers can be obtained starting from achiral isocyanides by one of

the following methods :
a) screw sense selective polymerization using nickel (II) catalysts and optically active intiatiors ;
b) screw sense selective copolymerization with small amounts of optically active comonomers ;
c) chromatographic resolution of the racemic polymer of an achiral isocyanide-(3).

This group also includes polymers derived from conventional unsaturated monomers bearing very bulky side chains. A typical example is offered by triphenylmethylmethacrylate which can be obtained in rigid helical conformation. Also in this case, the use of an optically active initiator allows the preparation of helices of one screw sense only (4).

* Copolymers of chiral monomers with achiral comonomers

This method is based on the fact that the units derived from the chiral comonomer can induce a chiral conformation into the units, either isolated or in sequences derived from the achiral comonomer. Even if the method is in principle very general, it has been applied up to now to a limited number of cases. Thus, isotactic structures have been obtained by copolymerization of optically active -olefins with achiral vinylaromatic monomers, such as styrene and vinylnaphtalenes, and it has been shown that these last have assumed in the copolymer a helical conformation (5). A similar situation has been obtained for stereoregular copolymers of triphenylmethacrylate with (S)- -methylbenzylmethacrylate(6).
More recently, the applicability of the method also to free radical copolymerization has been shown for several fucntional monomers (7).

<u>Supermolecular structures</u>

Chiral arrangements in polymeric materials can also be obtained due to the specific ordering of a large number of macromolecules in larger domains. One possibility is the crosslinking around chiral templates, another the existence of cholesteric phases in certain macromolecules.

* Chiral templates to which polymerizable binding groups were attached are polymerized in the presence of large amounts of crosslinking agent. After splitting off the template, chiral cavities remain, whose shape and arrangement of the functional groups

correspond to the original templates. These polymers can be used for racemic solution and as models for the active site of enzymes (8).

* Liquid crystal polymers with mesogenic groups in the main chain and side chain, either lyotropic or thermotropic, containing intrinsically chiral elements holds a particular position for their properties in bulk and in solution.

Polymers with mesogenic groups in the side chains and displaying cholesteric mesophases are in general based on copolymerization of chiral and mesogenic monomers (9). Chiral thermotropic polymers with mesogenic groups in the main chain on the other side can be obtained by polycondensation. Typically segmented polycondensation polymers consisting of flexible chiral segments (optically active diols) and rigid segments can be prepared in this way (10).

The presently available synthetic method allows to modulate in large range the chirality and relevant structural features of this class of polymeric materials.

RECOMMENDATIONS

The present status of the research in this area suggest that future activity should be focused on the improvement of synthetic approaches and to develop structures (particularly supermolecular ones) designed for potential specific applications. The following indications are given below :

* Synthetic problems

1-Copolymerization of chiral and chiral monomers

new optically active monomers derived from cheap readily available natural product (compounds should be synthetized and used for the preparation of polymers with high enantiomeric excess)

new functional and chromophoric achiral monomers including dyes

improvement of control of sequence distribution and stereoregularity

2-Preparation of synthetic chiral polymers where dissymmetric structures are stabilized by internal forces (hydrogen bonding, dipolar interactions) in addition to steric effects

3-New rigid achiral monomers which can produce chiral structures with chiral catalyst or initiators

4-Preparation of polymers with defined chiral stereochemistry in the backbone by template polymerization. High optical purity and stereoregularity should be obtained.

5-The mechanism of the formation of polymer which are helical due to atropisomerism should be investigated in more details. Then it might be possible to design new syntheses for similar substances.

6-Crosslinked polymers with chiral cavities should be further exploited as polymeric reagents and catalysts in asymmetric reactions.

* Supermolecular properties

1-In chiral liquid crystalline polymers more work is needed to correlate the extent of macromolecular chirality to the cholesteric and smectic structure (twist power and pitch length)

2-Study of spectroscopic and optical properties of extended chromophores (including dyes and photochromic groups) in a chiral polymeric matrix.

REFERENCES

1 G. WULFF, J. HOHN - Macromolecules 5 (1982) 1255

2 G. WULFF - Nachr. Chem. Techn. Lab. 33 (1985) 956

3 A.J.M. VAN BEYNEN, R.M.J. NOLTE, N.J. NAAKTGEBOREN, J.W. ZWIKKER, W. BRENTH, A.M.F. HESEMANS - Macromolecules 16 (1983) 1679

4 Y. OKAMOTO, K. SUZUKI, K. OHTA, K. HATADA, H. YUKI - J. Am. Chem. Soc. 101 (1979) 4763

5 F. CIARDELLI and P. SALVADORI - Pure and Appl. Chem. 57 (1985) 931

6 Y. OKAMOTO, K. SUZUKI and H. YUKI - J. Polym. Sci. Chem. Ed. 18 (1980) 3043

7 F. CIARDELLI. A. ALTOMARE, C. CARLINI, G. RUGGERI and E. TAMORI - Gasz. Chim. Ital. 116 (1986) 533

8 G. WULFF in .T. Ford Ed. "Polymeric Reagents and Catalysts" - ACS SYmp. Ser. 308 (1986) p. 186

9 H. FINKELMANN, H.J. KOCK and G. REHAGE - Makromol. Chem. Rapid Commun. - 2 317 (1981)

10 E. CHIELLINI and G. GALLI - Macromolecules 18 (1985) 1652

VII - POLYMERIZATIONS BY COORDINATION CATALYSTS

Chairman : A. GUYOT

SOME ASPECTS OF THE POLYMERIZATION OF OLEFINS WITH HIGH ACTIVITY HETEROGENEOUS CATALYSTS

U. Giannini, G. Giunchi and E. Albizzati
Istituto G. Donegani S.p.A.
Via G. Fauser 4
28100 Novara
Italy
P.C. Barbè
Himont Inc., G. Natta Research Center
Piazzale Donegani
44100 Ferrara
Italy

ABSTRACT. The presence of Lewis bases is essential to improve the stereospecificity of high activity catalyst systems for olefin polymerization based on AlR_3 and titanium halides supported on activated $MgCl_2$.

This paper reviews the main hypotheses formulated in order to elucidate the role of the Lewis bases. The structure of the support, the models of catalytic sites and the chemical interactions of the Lewis bases with the other components of the catalytic systems are discussed in relation to the proposed mechanisms.

INTRODUCTION

It is well recognized that the discovery of new catalytic systems having much greater activity than the traditional ones has led to important development in the industrial production of polypropylene (1, 2).

Since the first discovery a number of academic and industrial research laboratories have been actively engaged to try to explain the reasons which determine the high activity and the steric control of the polymerization.

These catalysts, irrespective of the methods of synthesis and the starting reagents, mainly consist of titanium halides and Lewis bases supported on activated $MgCl_2$; AlR_3 alone or combined with an additional Lewis base (LB) is used as cocatalyst (3, 4).

The presence of a Lewis base is essential to reach a high stereospecificity however as the ratio LB/AlR_3 increases, overall polymerization activity is reduced (5).

The choice of suitable Lewis bases and of appropriate methods to prepare the catalyst allows the achievement of exceptional high activity together with outstanding stereospecificity.

It is the purpose of the present paper to discuss the various hypotheses formulated in order to explain the stereoregulating effect of the Lewis bases

M. Fontanille and A. Guyot (eds.), Recent Advances in Mechanistic and Synthetic Aspects of Polymerization, 473–484.

taking into account the structure of activated $MgCl_2$, the models of catalytic sites present on the surface of the support and the chemical interactions of the Lewis bases with the other components of the catalytic systems.

STRUCTURE OF ACTIVATED $MgCl_2$

α-$MgCl_2$ has a crystal structure similar to that of γ-$TiCl_3$, consisting of layers piled one on top of the other according to a cubic close packing of the chlorine atoms. Each layer can be viewed as a sandwich in which two planes of chlorine atoms embed a plane of octaedrally coordinated magnesium atoms. In further analogy with $TiCl_3$, $MgCl_2$ may also exhibit, because of shear forces, e.g. by dry grinding, a disordered structure resulting from translation and rotation of the layers which destroy the cubic order in the stacking direction (3).

During the activation process $MgCl_2$ crystals are broken down and the resulting product consists of agglomerates of a large number of primary crystallites whose size are of only few layers in thickness and of 50-100 Å in the basal plane.

Experimental data indicate that preferential lateral cuts of $MgCl_2$ correspond to (110) and (100) planes. As a matter of fact, while only angles between lateral faces of 60° and 120° are observed in well formed crystals of $MgCl_2$, angles of 30°, 60°, 90°, 120° and 150° are present in fractured platelets of $MgCl_2$, as expected by assuming that the layers are cut both along planes (100) and along planes (110) (6).

The relative amounts of the two different fracture planes were not quantitatively evaluated; however theoretical calculations of the lattice electrostatic energies for ionic models laterally terminated by (100) or (110) planes give a lower fracture energy for the (110) cut. Therefore (110) planes could prevail in $MgCl_2$ highly activated by grinding (6).

These two kinds of lateral cuts lead to different local situations, electroneutrality conditions impose an average coordination number 5 for Mg atoms on (100) cuts and 4 for Mg atoms on (110) cuts.

Additional different situations occur at the corners and edges. Therefore the presence on the surface of activated $MgCl_2$, of Lewis acid sites having different acid strength and steric requirements was predictable.

The reaction of Lewis bases with activated $MgCl_2$ has allowed to verify this hypothesis.

Indeed the amount of Lewis base complexed with $MgCl_2$ increases by enhancing the basic strength of the base, but keeping constant his steric hindrance (Table I).

TABLE I. Complexation of different acyl chlorides R-C_6H_4-COCl with activated $MgCl_2$ (7)

R	r (1)
Cl	8.04
H	8.47
CH_3	9.43
OCH_3	13.32

(1) molar ratio $\frac{\text{complexed acyl chloride}}{MgCl_2} \cdot 100$

Moreover, following by IR the reaction of activated $MgCl_2$ with increasing amounts of ethylbenzoate we observe that the maximum of the $\nu_{C=O}$ absorption band shifts from 1620 cm^{-1} to 1685 cm^{-1}; at the same time the absorption band becomes broader as expected by assuming the presence of a variety of coordination complexes with different bond strength (Table II).

TABLE II. Carbonyl stretching frequencies in the IR absorption spectra of $MgCl_2$ complexed with different amounts of Ethyl benzoate (7).

$\tau^{(1)}$	$\nu_{C=O}$ cm^{-1}
0.33	1620
1.56	1660
15.3	1685
30.8	1685

(1) molar ratio $\frac{EB}{MgCl_2} \cdot 100$

MODELS OF CATALYTIC SITES

There is experimental evidence (8) that the polymerization of ethylene on well formed $MgCl_2$ crystals reacted with $TiCl_4$ occurs essentially on the corners, edges and surfaces corresponding to lateral cuts of the layers but not on the basal planes of the crystals. Therefore it can be supposed that, during the reaction with the titanium halides single $TiCl_4$ molecules or bridged dimers Ti_2Cl_8 may be epitactically placed on the corners, edges and on the lateral surfaces of $MgCl_2$ crystallites giving rise to reliefs crystallografically coherent with the matrix.

In Fig. (1) is only represented a hypothetical epitactic placement of $TiCl_4$ and Ti_2Cl_8 units on the lateral surfaces of $MgCl_2$ crystallites.

The reaction with aluminum alkyls should lead to the reduction of titanium at least to the trivalent state thus generating $TiCl_3$ and Ti_2Cl_6 reliefs on the support.

In accordance to this description it appears evident that in these catalytic systems various classes of catalytic centres could be present differing in steric structure, Lewis acidity, activity and stereospecificity.

Corradini and coworkers have carried out a number of theoretical studies on the stereospecificity of the different active sites (9), located on the lateral surfaces.

By accepting the monometallic model of the active centre, on the basis of the evaluation of the non-bonded interactions for the proposed catalytic sites, these authors believe that only the epitactic placement of dimeric Ti_2Cl_6 species on (100) lateral termination of the layer of $MgCl_2$ could lead to the formation of stereospecific active sites.

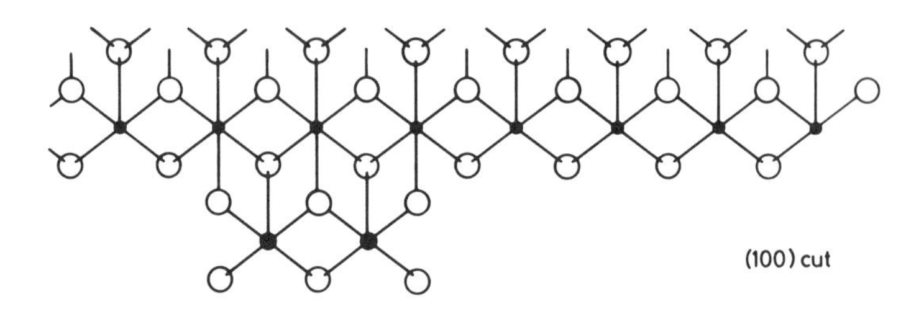

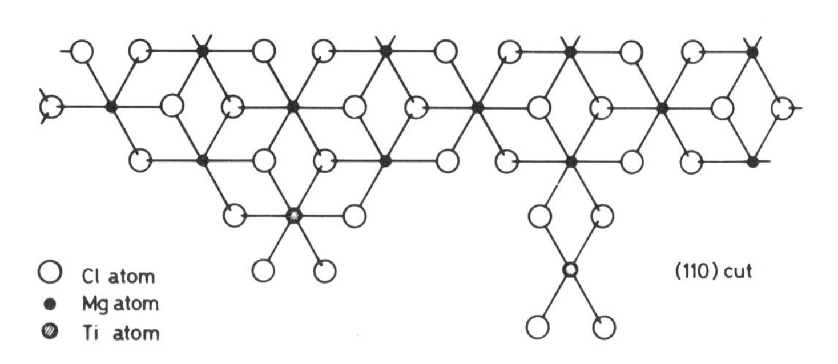

Figure 1. $TiCl_4$ and Ti_2Cl_8 epitactically placed on a (100) and on a (110) cut of $MgCl_2$.

Non stereospecific models result from single $TiCl_3$ groups coordinated with the same lateral termination and also from epitactic coordination of $TiCl_3$ and Ti_2Cl_6 groups on the (110) cut.

Moreover, considering the two kinds of catalytic sites present on the (100) face of $MgCl_2$, Corradini has evaluated the relative fractions of dimeric and monomeric titanium species by simple estimation of the elecrostatic interactions and has found that the dimeric stereospecific Titanium species prevail over the monomeric ones and this preference increases at high coordination energies of the Ti atoms (9). This result can be explained by considering that each dimeric species occupies only three coordination sites while two monomeric Titanium units occupy four sites.

Therefore on the basis of this model, in the absence of Lewis bases, the (100) faces should be mostly covered by bridged dinuclear stereospecific titanium species, whereas non stereospecific sites should be present on the (110) faces.

Some experimental evidences are in accord with this hypothesis.

Catalysts obtained by reaction of $TiCl_4$ with well formed $MgCl_2$ crystals terminated predominantly by (100) faces are much more stereospecific than catalysts prepared from dry-milled $MgCl_2$ crystals terminated by both (100) and (110) faces.

On the other hand catalysts obtained by supporting $TiCl_4$ on dry-milled $TiCl_3$ give polypropylene whose stereoregularity is almost the same as that of the polymer prepared with $TiCl_3$ (Table III).

TABLE III. Influence of morphology of support on the stereospecificity of catalysts in propylene polymerization.

Catalyst	Productivity (g PP/g cat.)	Isotactic Index (%)
$MgCl_2$ macrocrystals/$TiCl_4$	12	74
$MgCl_2$[a]/$TiCl_4$	220	59
$TiCl_3HR$[a]	61	90
$TiCl_3HR$[a]/$TiCl_4$	310	84

a = activated by grinding
Polymerization conditions: $[Al(C_2H_5)_3]$ = 5 Mol/l;
$P_{C_3H_6}$ = 5 atm; T = 70°C; time = 1 h.

It was shown that, whereas $MgCl_2$ crystals are broken both along (100) and (110) planes, $TiCl_3$ crystals are fractured only along (110) planes. Catalytic sites on (110) surfaces of $TiCl_3$ are believed to be stereospecific and similar, concerning the stereospecific behaviour, to sites resulting from epitactic placement of Ti_2Cl_6 units on (100) lateral terminations of $MgCl_2$ crystals (9).

CHEMICAL INTERACTIONS BETWEEN THE LEWIS BASES AND THE OTHER COMPONENTS OF THE CATALYTIC SYSTEM

The poor stereospecificity of the $MgCl_2/TiCl_4/AlR_3$ catalytic system can be increased by conditioning Lewis bases.

A number of Lewis bases have been investigated in order to improve the stereospecificity of the supported catalysts. They differ in basic strength, steric hindrance, electron donor character (σ or π donors), hardness, functionality, ect.

Most studies have been carried out with esters of aromatic acids like ethylbenzoate (EB) or hindered amines as 2,2,6.6-tetramethylpiperidine (TMPP).

Efficient solid catalysts can be prepared reacting with $TiCl_4$ anhydrous $MgCl_2$ previously ground with EB. During this second step a part of the electron donor previously complexed with $MgCl_2$ is displaced by $TiCl_4$ (10).

The more obvious reactions of the Lewis bases with the other components of the catalytic system are their coordination with the Lewis acid sites present on the surface of the catalyst and with the aluminium alkyl.

Furthermore the Al-C, Ti-C and Ti-H (arising from β -hydrogen elimination and hydrogenolysis of T-C bonds) present in the system before and during the polymerization may react with the functional groups present in the base.

The Lewis base may also compete with $TiCl_4$ toward some Lewis acid sites present on the surface of $MgCl_2$.

The interactions of aromatic esters with the solid catalyst have been generally studied in the absence of the cocatalyst.

Most of the authors, on the basis of I.R., far I.R., TGA and XPS analyses, did conclude that the ester is coordinated only with $MgCl_2$ through the oxygen of the carbonyl groups (11, 12, 13).

However according to Rytter and al. (14) carbonyl stretching frequency similar as for the 2 $TiCl_4 \cdot C_6H_5COOC_2H_5$ complex appears as a weak band in the I.R. spectrum of the catalyst; hence following these authors a part of the ester is coordinated with titanium on the surface of the $MgCl_2$ support.

The presence of Ti^{+3} ions having EB as ligand has been demonstrated by ESR spectroscopy in the catalytic system $MgCl_2/TiCl_4/EB$ + Al(i-Bu_3)/EB (15).

Several papers have already described complexes between AlR_3 and aromatic esters.

Contradictory results are reported in the literature about the stoichiometry of these complexes.

1/1 and 2/1 AlR_3/aromatic esters complexes have been detected by I.R. and their stoichiometry seems to depend on the nature of the ester and of the aluminum alkyl (16).

Some authors, on the contrary, exclude the presence of 2/1 AlR_3/ester complexes (17).

The discrepancy between the results may be ascribed to different reaction conditions.

The reduction of the ester group by the aluminium alkyls is a well known reaction.

A nucleofile attack of uncomplexed AlR_3 molecule to the C = O group of the 1/1 or 2/1 AlR_3/ester complexes has been postulated. The reaction may further proceed and 2 moles of dialkylaluminium alkoxide are formed from one mole of ester.

$$AlR_3 + C_6H_5\text{-}\underset{\underset{O \rightarrow AlR_3}{\|}}{C}\text{-}OC_2H_5 \longrightarrow C_6H_5\text{-}\underset{\underset{O\,AlR_2}{|}}{\overset{\overset{R}{|}}{C}}\text{-}OC_2H_5 \xrightarrow{AlR_3} C_6H_5\text{-}\underset{\underset{R}{|}}{\overset{\overset{R}{|}}{C}}\text{-}OAlR_2 + R_2AlOC_2H_5$$

The reaction rate is enhanced in concentrated solution and upon heating. The presence of electron releasing groups in para position to the ester group (18) and of sterically demanding alkyl groups in AlR_3 (19) decreases the reaction rate.

Aluminium dialkylmonohydrides reduce instantaneously the aromatic esters.

An analogous reactivity toward the ester group is to be expected also for the Ti-C (20) and Ti-H bonds.

The presence of Ti^{+3} ions containing a Ti-OR bond has been ascertained by ESR in the catalytic system $MgCl_2/TiCl_4/EB/AlR_3$, in analogy to similar species resulting from the interaction of EB with surface benzyl compounds of Ti^{+3} (15).

The more pronounced reduction with time of polymerization rate, observed in the presence of H_2 as molecular weight regulator, has been ascribed to the high reactivity, toward the ester groups, of the Ti-H bonds resulting from the hydrogenolysis of Ti-C bonds (21).

PROPOSED MECHANISMS OF STEREOREGULATION BY THE LEWIS BASES

On the basis of:

- the models of catalytic sites on the surface of $MgCl_2$;

- the over reported different reactions occurring between the Lewis bases and the components of the catalytic system;
- the catalytic behaviours of the different catalysts

three main mechanisms have been hypothesized in order to explain the stereoselective control of the polymerization by the electron donors.

a) Selective poisoning and/or modifying of aspecific catalytic sites through their complexation with Lewis bases

As previously reported the addition of a Lewis base to the catalytic system $MgCl_2/TiCl_4/AlR_3$ causes a decrease of the overall polymerization rate. By increasing the Lewis base/AlR_3 molar ratio, the activity of the atactic sites declines continously, while the activity of the isotactic sites, at least in the presence of some Lewis bases (22, 23, 24, 16), goes through a maximum and then decreases but more gradually than that of aspecific sites (Table IV).

TABLE IV. Polymerization of propylene with the catalytic system $MgCl_2/TiCl_4$-$AlEt_3$/ethyl benzoate (EB) at different EB/Ti molar ratios (28).

[EB]/[Ti] (mol/mol)	Productivity (g PP/mmoli Ti) overall	C_7-insol.	C_7-sol.	I.I. (wt %)
0	230	81	147	35
2.5	225	131	94	58
5.0	195	152	43	78
10.0	177	150	27	85
15.0	104	94	10	90

The reversibility of this effect has been proved (25).

This phenomenology has been interpreted by assuming the presence on the support of many families of centers, containing catalytically active Ti-C bonds, differing in Lewis acidity, steric hindrance, number of vacancies in the octahedral coordination shell of the metal, activity and stereospecificity.

Many equilibria can be established between the Lewis base (LB), the aluminium alkyl and the different active sites present on the support as follows (5):

$$LB + AlR_3 \rightleftharpoons AlR_3 \cdot LB$$
$$Ca + AlR_3 \cdot LB \rightleftharpoons Ca \cdot LB + AlR_3$$
$$Cb + AlR_3 \cdot LB \rightleftharpoons Cb \cdot LB + AlR_3$$
$$Cc + AlR_3 \cdot LB \rightleftharpoons Cc \cdot LB + AlR_3$$
$$Cc \cdot LB + AlR_3 \cdot LB \rightleftharpoons Cc \cdot 2LB + AlR_3$$

where Ca = stereospecific center,
Cb = aspecific center,
Cc = aspecific center with two coordination vacancies,
Cc•LB = stereospecific center,
Ca•LB, Cb•LB and Cc•2LB are inactive centers.

The occurring of exchange reactions between the LB present in the solid catalyst and the LB complexed in solution with the aluminium alkyl has been demonstrated (26).

The aspecific sites would be deactivated by the Lewis bases more quickly than isospecific ones because of their higher Lewis acidity. Alternatively the selectivity of some LB (e.g. TMPP) has been ascribed to their steric hindrance, which is sufficient to prevent complex formation at single vacancy isospecific sites, but allows to block both vacancies of a divacant nonspecific site (19).

The observed enhanced productivity of steroregular polymer at low LB/AlR_3 ratios can be explained by assuming the presence of aspecific catalyst sites (Cc) having two coordination vacancies one of which accessible to LB complexation; a modified isospecific active site (Cc·LB) should result. This hypothesis also accounts for some results obtained in stereoselective polymerization of 3,7-dimethyl-1-octene (5).

Other authors suggest that aspecific sites are reversibly transformed into isospecific ones upon association with the LB-complexed AlR_3 (16).

Following this hypothesis the Lewis bases function as simple electron donors and their reactivity toward Ti-C bonds is not taken into account.

b) Selective deactivation and reactivation of active Ti-C bonds.

The reactions between active Ti-C bonds present in the catalytic sites and reactive groups of the Lewis base, disregarded in the above hypothesis, become the key issue of the mechanism proposed by Zambelli et al. (27).

These authors, using labelled $C_6H_5-{}^{14}COOCH_3$ as external Lewis base in polymerization of propylene with AlR_3 and the solid catalyst $MgCl_2/TiCl_4/EB$, have shown the presence of radioactive fragments in the polymer.

It was excluded that incorporation of ^{14}C in the polymer occurs through reaction of the labelled molecule and Al-Polymer bonds resulting from chain transfer reactions. A reaction between methyl benzoate and Ti-Polymer (Ti-P) bonds is suggested according the equation I:

$$\text{Ti-P} + C_6H_5-{}^{14}\underset{\underset{O \rightarrow Al(C_2H_5)_3}{\|}}{C}\text{-}OCH_3 \xrightarrow{Al(C_2H_5)_3} \text{Ti-}OCH_3 + C_6H_5-{}^{14}\overset{\overset{P}{|}}{\underset{\underset{O\,Al(C_2H_5)_2}{|}}{C}}\text{-}C_2H_5 \qquad \text{I}$$

This reaction is a termination process. Since the polymerization is slowed but not completely inhibited by the ester it was suggested that the deactivated sites can become again active through the proposed mechanism shown in equation II:

$$\text{Ti-}OCH_3 + Al(C_2H_5)_3 \longrightarrow \text{Ti-}C_2H_5 + Al(C_2H_5)_2\,OCH_3 \qquad \text{II}$$

The stereospecificity of the polymerization should depend on the different rates of the reactions I and II for the isotactic specific and the non stereospecific sites.

The proposed mechanism does not account for the stereoregulating effect of Lewis bases not containing reactive groups as e.g. tertiary amines; however the above authors do not consider their hypothesis as a complete rationalization of the problem.

c) Competition of the Lewis bases with $TiCl_4$ for selective coordination of unsaturated Mg ions on the different faces of $MgCl_2$

A further hypothesis on the role of the Lewis bases which is not alternative but complementary to the other mechanisms previously discussed has been suggested by Corradini et al. (26).

This hypothesis is indeed related to the influence of the LB on the structural features of the catalytic sites during the synthesis of the catalyst and is based on the ability of $TiCl_4$ to compete with LB for selective coordination on the different lateral faces of $MgCl_2$ crystals.

Taking into account the method of catalysts, synthesis and the proposed models of catalytic sites, the role of the LB in increasing the stereospecificity of the catalyst has been related to the capability of $TiCl_4$ to remove, at least in part, the LB only from the (100) cuts of $MgCl_2$.

This suggestion was supported by energetic calculations of electrostatic interactions of ionic model species with the lateral termination of $MgCl_2$ crystals. The resulting data lead to the conclusion that (100) faces of $MgCl_2$ should behave as more basic than (110) faces as far as binding to $TiCl_4$ is concerned.

Therefore on the basis of this oversimplified model, in catalysts prepared in the presence of Lewis bases, dimeric stereospecific titanium species should be present on the (100) faces whereas the LB should saturate the vacancies of Mg atoms present on the (110) faces,thus avoiding the formation on these planes of non stereospecific sites (fig. 2).

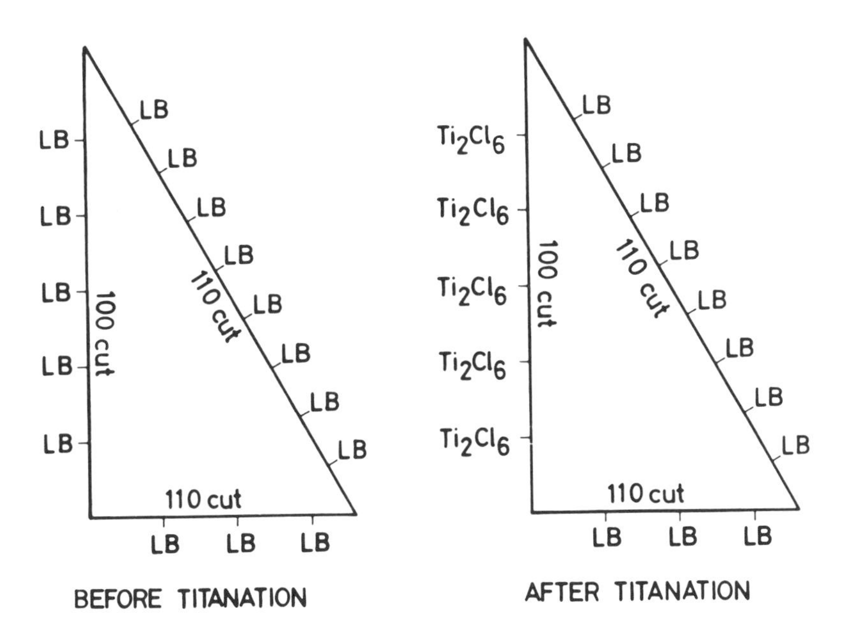

Figure 2. Schematic drawing of a hypothetical distribution of the Lewis Base and Ti halide on the (110) and (100) cuts of $MgCl_2$ before the reaction with $TiCl_4$ and after the reaction with $TiCl_4$ and AlR_3.

CONCLUSIONS

The manifold reactivity of the Lewis bases toward the various species present in the catalytic system and their conceivable influence on the morphology of the support and on the structure of the active centers during the synthesis of catalyst, lead to the assumption that the Lewis bases play a multiplicity of functions whose relative weight depends on their nature.

Therefore the individual hypotheses, up to now formulated and discussed in the previous section, account, in our opinion, only for specific aspects of the whole problem.

Nevertheless the proposed mechanisms are a useful base to investigate new more efficient electron donors and to design crucial experiments aimed at verifying the validity of the hypotheses previously mentioned.

REFERENCES

1) G. Di Drusco and R. Rinaldi, Hydrocarbon Process, Int. Ed. **60**, 153 (1981).
2) N.F. Brockmeier in Transition Metal Catalyzed Polymerizations: Alkenes and Dienes, R.P. Quirk et al. (Ed.), Harwood Academic Publishers, New York, 1983, p. 671.
3) U. Giannini, Makromol. Chem. Suppli. **5**, 216 (1981).
4) P. Galli, P.C. Barbè and L. Noristi, Angew. Makromol. Chem. **120**, 73 (1984).
5) P. Pino, G. Guastalla, B. Rotzinger and R. Mülhaupt in Transition Metal Catalyzed Polymerizations; Alkenes and Dienes, R.P. Quirk et al. (Ed.), Harwood Academic Publishers, New York, 1983, p. 435.
6) G. Giunchi, unpublished results.
7) G. Fochi, unpublished results.
8) P. Galli, Proc. IUPAC 28th Macromolecular Symposium, Amherst, July 1982, p. 248.
9) P. Corradini, U. Barone, R. Fusco and G. Guerra, Gazz. Chim. Ital. **113**, 601 (1983).
10) L. Luciani, N. Kashiwa, P.C. Barbè and A. Toyota, Ger. Offen. 2,643,143 to Montedison S.p.A. and Mitsui Petrochemical Industries.
11) Xiao Shijing, Cai Shimian, Chen Zeugbo and Liu Huanquin in Catalytic Polymerization of Olefins, T. Keii and K. Soga (Eds.), Elsevier, Amsterdam, 1986, p. 431.
12) M. Teramo, T. Kataoka and T. Keii in Catalytic Polymerization of Olefins, T. Keii and K. Soga (Eds.) Elsevier,Amsterdam, 1986, p. 407.
13) N. Kashiwa, Polymer J. **12**, 603 (1980).
14) E. Rytter, O. Nirissen and M. Ystenes, Int. Symp. 'Transition Metal Catalyzed Polymerization', Akron, June 16-20 (1986), p. 48.
15) S.E. Sergeev, U.A. Poluboyarov, V.A. Zakharov, V.F. Anufrienko and G.D. Bukatov, Makromol. Chem. **186**, 243 (1985).
16) A. Guyot, R. Spitz, L. Duranel and J.L. Lacombe in Catalytic Polymerization of Olefins, T. Keii and K. Soga (Eds.), Elsevier, Amsterdam, 1986, p. 147.
17) G. Fochi, J. Organomet. Chem. **307**, 145 (1986).
18) K.B. Starowieyski, S. Pasynkiewicz, A. Sporzynsky and K. Wisniewska, J. Organometal. Chem. **117**, C1 (1976).

19) A.W. Langer, T.J. Burkhardt and J. Z. Steger in Transition Metal Catalyzed Polymerizations; Alkenes and Dienes, R. P. QUirk et al. (Ed.) Harwood Academic Publisher, New York, 1983, p. 421.
20) B.L. Goodall in Transition Metal Catalyzed Polymerizations; Alkenes and Dienes, P.P. Quirk et al. (Ed.), Harwood Academic Publishers, New York, 1983, p. 355.
21) A.W. Langer, T.J. Burkhardt and J.J. Steger, Proc. IUPAC, 28th Macromolecular Symposium, Amherst, July 1982, p. 244.
22) N. Kashiwa and J. Yoshitake, Makromol. Chem. **185**, 1133 (1984).
23) P. Pino, B. Rotzinger and E. von Achenbach in Catalytic Polymerization of Olefins, T. Keii and K. Soga (Eds.), Elsevier,Amsterdam, 1986, p. 461.
24) I. Tritto, P. Locatelli and M. Sacchi, Int. Symp. Transition Metal Catalyzed Polymerization, Akron, June 16-20 (1986), p. 44.
25) P. Pino, B. Rotzinger and E. von Achenbach, Makromol. Chem. Suppl. **13**, 105 (1985).
26) V. Busico, P. Corradini, L. De Martino, A. Proto, V. Savino and E. Albizzati, Makromol. Chem. **186**, 1279 (1985).
27) A. Zambelli, L. Oliva and P. Ammendola, Gazz. Chim. Ital. **116**, 259 (1986)
28) N. Kashiwa, M. Kawasaki and J. Yoshitake in Catalytic Polymerization of Olefins T. Keii and K. Soga (Eds.), Elsevier, Amsterdam, 1986, p. 43.

DISCUSSION

The reactivation of Ti-OCH_3 bonds through an exchange reaction with AlR_3 was suggested by Zambelli et al (Gazz. Chim. Ital. 116 259 (1986))

The activation by H_2 is an enhancement of the initial olefin polymerization rate. The same effect is observed in the presence of Lewis bases.

The stereospecificity of the catalyst may be explained in terms of theoretical ground, using monometallic models of catalytic site ; however, owing to the variety of catalytic centers, the presence of aluminium alkyl ligands would be essential to induce stereospecificity in some cases. Stereospecificity and also activity are strongly governed by the nature and the composition of the external cocatalytic solution : alkyl aluminium and Lewis base ; in these cases, the bimetallic nature of the active sites seems to be established. On the other hand, the crucial effect of the internal Lewis base on the stereospecificity is to be pointed out.

Although all the oxidation states of the titanium (from 2 to 4) have been observed to be active, it seems that the oxidation degree of the most active sites is 3.

The increased activity of the supported catalyst may be due to both a large increase in the number of active centers and an increase in the rate constant of the active sites. However, only the product of kp x C can be measured. Another effect of the support may be to eliminate the mutual deactivation of vicinal sites, owing to their separation.

About the effect of Lewis base which increases the productivity in isotactic polymer, it is noted that, because the addition of Lewis bases does not change the molecular weight of the two fractions (isotactic and atactic), the values of the rate constants should not be changed ; then most probably aspecific centers are changed into isospecific centers.

The reason for the special behavior of silane derivatives as cocatalyst promoter were discussed ; the complex with alkyl aluminium seems to be more stable and involves less side-reactions. Although the same stereospecificity may be obtained with other compounds, a lower amount of silane is needed.

The enhanced activity in ethylene polymerization caused by the introduction of propylene might be explained as a mass transfer effect, the polymer being made less crystalline.

MOLECULAR WEIGHT DISTRIBUTION IN ZIEGLER-NATTA OLEFIN POLYMERIZATION.

R. SPITZ
CNRS - LABORATOIRE DES MATERIAUX ORGANIQUES
BP 24
69390 . VERNAISON (France)

ABSTRACT. The molecular weight distribution (MWD) of polyoelfins obtained with Ziegler-Natta catalysis can range from 2 to 20 or more depending on the monomers, the process and the catalytic systems. It is rather easy to show that in most cases the broad MWD are not due to diffusion effects but to a distribution of active centers. The chemical and process factors controlling the MWD are numerous. This results in a complex patent literature with more weight on the catalyst formula and on the process.

A discussion of these different parameters is presented in order to distinguish the true chemical effects corresponding to changes in the process or in the polymerization conditions. An example of chemical modification of MWD by a reduction treatment is detailed. A comparison of polymerization conditions between gas phase and slurry is also presented. An attempt is made to correlate the selectivity of the catalyst, as it is used for stereospecific polymerization for instance to the narrowing of MWD.

I. INTRODUCTION

Polyethylene and polypropylene are thermoplastic semi-crystalline polymers. The second aspect (crystallinity) is clearly related to the regularity of the chain structure. The first one is dependant on the molecular weight (MW) and MW distribution (MWD). All has been said about the unique character of activity and selectivity of the Ziegler-Natta catalysts (1,2). The corresponding research work resulted since 1953 in thousands of patents and a significant even if less developed number of publications. A detailed survey of all the theory and practice of Ziegler-Natta is outside the scope of this paper but MW and MWD are concerned with all aspects of domain. For a long time the problem of MW and MWD was restricted to its pratical aspects, chiefly as patent claims defining processability or properties improvement without relation to any theory. Introducing a detailed review on this subject, Zucchini and Cecchin (3) following Boor (1) observed that only the trend of homogeneous catalysts to give rise to narrower distribution than heterogeneous

M. Fontanille and A. Guyot (eds.), Recent Advances in Mechanistic and Synthetic Aspects of Polymerization, 485–502.

one do was commonly admitted. Nethertheless the conclusions of the same paper established how narrow or broad MWD are obtained and related it to the chemistry of the catalytic system and not to a diffusion control on the polymerization. But a general theory of MWD is not available at this time.

MW, MWD AND POLYMER PROPERTIES

The ability of a thermoplastic to be processed is a function of its molten viscosity, elasticity and strength. The viscoelastic properties of a molten polymer depends on the MW but not only on it, as it can be seen on fig. 1.

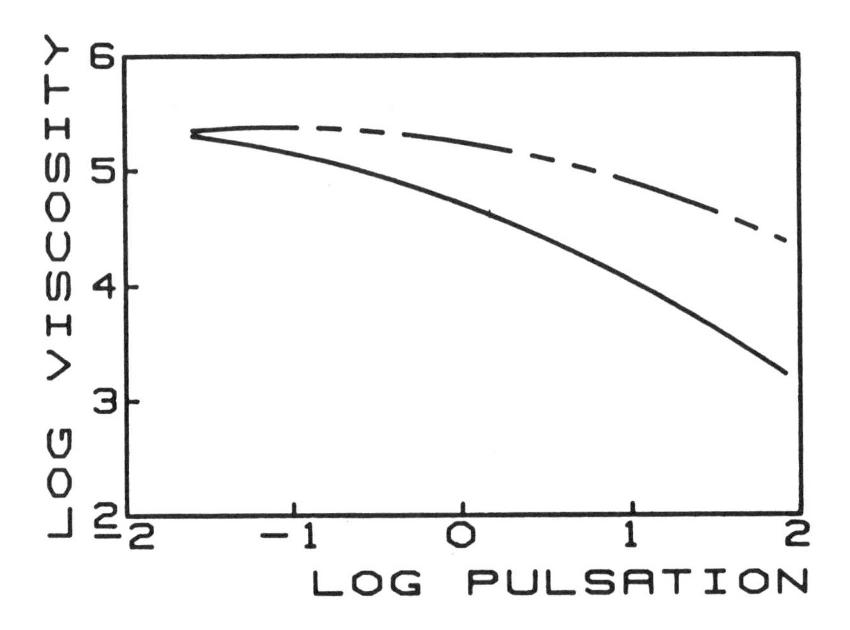

Figure 1. Complex viscosity (poises) at 443 K as a function of pulsation (rad/s) in log-log scale for two polyethylenes.
—— - - —— Q = 8.5 ——— Q = 19.9

The complex dynamic viscosity is a function of temperature, shear, MW and MWD. For the same MW, (expressed as Mw) at high shear the viscosities may differ by 10, 100 times or more between a narrow and a broad MWD polyethylene. The equivalent Mw ratios range from 2 to 5 or more : this allows the extrusion of longer chains with the same screw conditions and enhanced strength. The counterpart is that to the broader MWD corresponds a larger elastic modulus, the first consequence being dye swell. So, the broader are the MWD, the easier is the polymer processed and the worst are the dimensions adjusted. Narrow MWD are then required for precise molding.

The mechanical properties relation to MW and MWD are more difficult to describe (4,5). Polyethylene (PE) and polypropylene (PP) are generally used far over the glass transition temperature. The mechanical properties result then from two contribution : in the amorphous phase, the long chains give rise to chain entanglement and the overall structure of the solid is physically crosslinked by the crystal organisation. The real properties are then very sensitive to the thermal and mechanical conditions of the transformation, the extent of crystallization and the size of the crystallites.

Polypropylene is more sensitive to crystal structure variations. Till the last years, narrow MWD, when required, were obtained by thermomechanical and thermochemical post-treatment after the synthesis. The recent high-yield third generation catalysts allow better control and variation of MWD from narrow to wide (6). Strain hardening properties (6) can thus be obtained from broader MWD (7,8). Narrow low MW PP are suited for the production of spun-bonded fibers for non-woven application (6).

The case of PE is more complex. The crystallinity can be varied by incorporation of a low amount of α-olefin, giving rise to a wide range of copolymers comprising the linear low density polyethylene (LLDPE). All combinations of MW, MWD and crystallinity can thus be obtained. The balance of crystalline and plastic properties has favored the growing of LLDPE in the last ten years (9).

A careful control of the processing may develop extreme properties an exemple being the high modulus fibers (5,10). But generally, the ultimate properties are not reached and the better tensile properties expected for larger MW are difficult to observe : "this lies in the fact that crystallites help hold the material together much the same as chain entanglement. Molecular weight dependance can further be confused because the degree of crystallinity can decrease as MW increases" state Nunes, Martin and Johnson (5). All kinds of polymers cannot be used : for instance, very broad PE are only produced as copolymers, the environmental stress cracking resistance of the homopolymer being too low.

MW and MWD MEASUREMENT AND EVALUATION

Exact MWD are obtained by size exclusion chromatography (SEC), especially when associated to viscosity or light scattering. But in the case of polyolefins, the measurements are made at high temperature, the polymers are very difficult to dissolve and the SEC chromatograms are often very far away from the expected values, chiefly for high molecular weight polymers.

The non-linear sensitivity of flow rate to shear is used in the common evaluation of chain length through chain entanglement properties reflected by the Melt-Index measurement, according to different standarts for PE or PP. These fast measurements permit the on-line control of molecular-weight during polymer production.

As exemplified in fig. 2, Melt Indexes can be correlated to Mw values obtained from SEC in a large useful range even with varying MWD for polymers obtained with the same catalytic system.
Melt index ratios at different shear rates are dependent on the broadness of MWD. But the correlation holds only if the lowest Melt index remains in a narrow window of values as illustrated on fig. 3 for PE. The conclusion holds only if the long chain branching remains low.

MOLECULAR WEIGHT CONTROL

Compared to the Phillips catalysts appearing nearly at the same date (11) MW are easier to adjust to a convenient value with Ziegler-Natta catalysts : molecular hydrogen acts as a true transfer agent (12,13). Other component of the reaction medium like the cocatalyst may also act as chain grow limiter (14). Molecular weight where reported to vary as a function of the square root of alkylaluminium concentration. Zinc diethyl is by far more effective (15). But hydrogen is the cheapest agent with generally limited effect on catalytic activity.

In the absence of hydrogen, the polymers present insatured (vinylene or vinylidene) chain ends, especially at high temperatures (16), showing the spontaneous cleavage of the metal-carbon bond at which occurs monomer insertion. Hydrogen, when present, saturates the chain-ends reducing the insaturation content to almost zero. More details on different chain termination mechanism can be found in ref. 1.

The effect of hydrogen is discused in a recent paper by KEII (17). The observation of Natta (18) that hydrogen concentration appears as square root in rate or MW variations suggests that molecular hydrogen is implied in the transfer process (19) but hydrogen effect is in fact very complex and hydrogen interacts in different ways with the catalyst, producing activity enhancement or inhibition as observed by many authors (20,21).

The case of propylene is very complicated : the isotactic heptane insoluble and the atactic heptane soluble fraction have MW in a ratio up to 5 (22,23).

The time dependance of the molecular weight is also a controversial question. Keii et al. (22) found the MW of the two fractions of polypropylene to be constant in the range 5s-3 hours with supported catalysts. With slightly different catalyst Chien et al. reported recently some variation (24,25). Nethertheless, it seems clear that with high-activity catalyst, long chains are built up in a very short time.

THE ORIGIN OF MOLECULAR WEIGHT DISTRIBUTION

This point was extensively discussed in ref. 3. Several authors

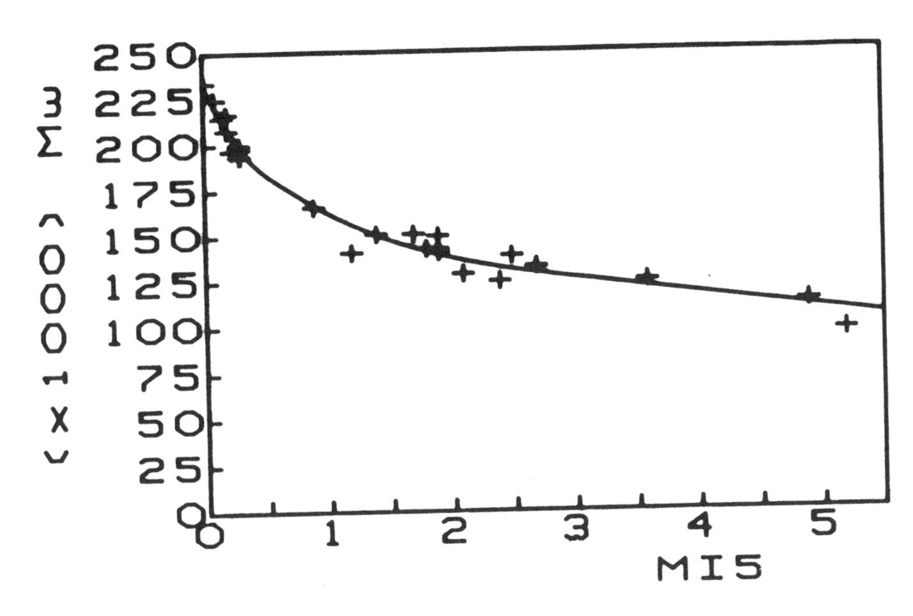

Figure 2. Correlation between Mw and Melt-Index (190° C - 5 kg) for a series of polyethylenes obtained with the same catalyst.

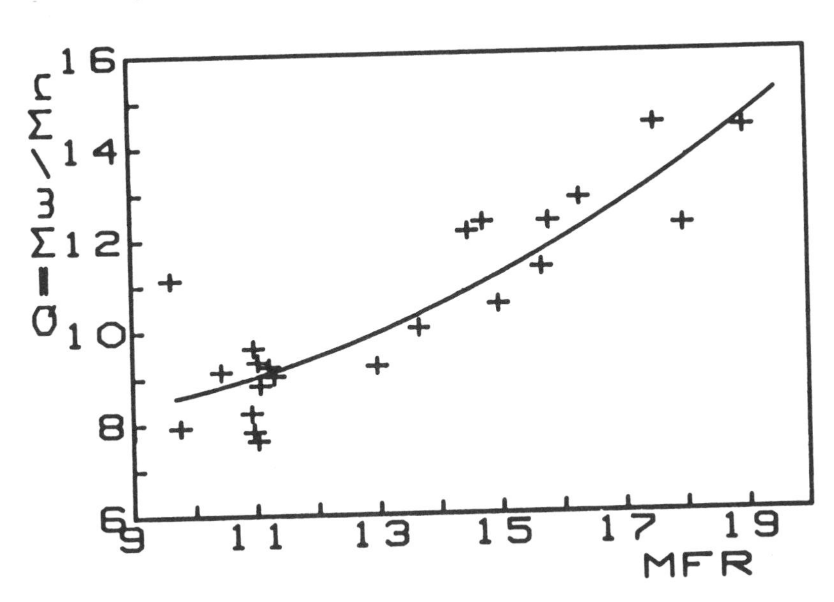

Figure 3. Correlation between the polydispersity and the melt flow ratio (MFR) $MI_{21.6}/MI_5$ (190°C ; 21.6 kg and 5 kg) for a series of polyethylenes obtained with the same catalyst.

have tried to describe all the kinetic process (including : initiation, propagation and the many possible termination reactions (26,28)). Even with different termination mechanisms the MWD is supposed to be of Shultz-Flory type with a Q factor (Q = Mw/Mn) close to 2. In a very particular situation (oligomerization of ethylene in homogeneous liquid phase) Fink (29) observes the theoretical distributions. But generally the MWD are broad to very broad when compared to the theoretical value of 2. Till the last years two theories were opposed in order explain the broad MWD : the diffusion theory and the chemical theory.

The diffusion theory :

To summarize the diffusion theory : in the case of an heterogeneous polymerization medium (solid/gas or solid/liquid system), inside the catalyst or polymer particles the ratio of the reagent is not constant, the monomer being consumed faster than the other species (hydrogen, aluminium alkyl). So, polymers with different MW may be produced at the same time in different layers of the solid and this brings out a first kind of heterogeneity. During the polymerization in the case of slurry or gas phase the polymer particle expands with production, the diffusion barrier changes modifying the MWD of the polymer with time. The theory was detailed by Ray et al. (30) on the basis of realistic diffusion parameter in the case of propene polymerization. Different models of the dispersion of the catalyst in the polymer particle have been discussed : solid core : catalyst surrounded by the solid polymer ; polymeric core (like the precedent with permeable fragmented polymer) ; multigrain model (dispersion of catalyst fragments in the polymer grain) and polymeric flow model (assembling catalyst fragments in a fragmented polymer). All these models lead to a polydispersity up to Q = 9, the last permitting the best fit of the expected variations of MWD.

But in fact a great number of simple experiences show that generally there is no significant diffusion resistance in most usual polymerization conditions and processes, even with the high yield catalysts. The idea of a diffusion resistance was suggested by the fact that the polymerization rate often decreases with time. A careful examination of the rate decrease in the case of propene polymerization with 3rd generation catalyst shows that rate decreases even if no polymer is produced (31-33). A new example is presented in fig. 4.

Another way to prove this is to show that an appropriate choice of the cocatalyst leads to constant high activities. In fig. 5 are compared suspension kinetics for propene with triethylaluminium cocatalyst (decreasing) and with isoprenylaluminium (constant). The same result is obtained for gas phase kinetic for ethylene with trihexylaluminium (decreasing) and isoprenylaluminium (constant).

Examination of the molecular weight gives another way to characterize diffusion effects. Constant molecular weight with expanding particle proves that diffusion is not an important

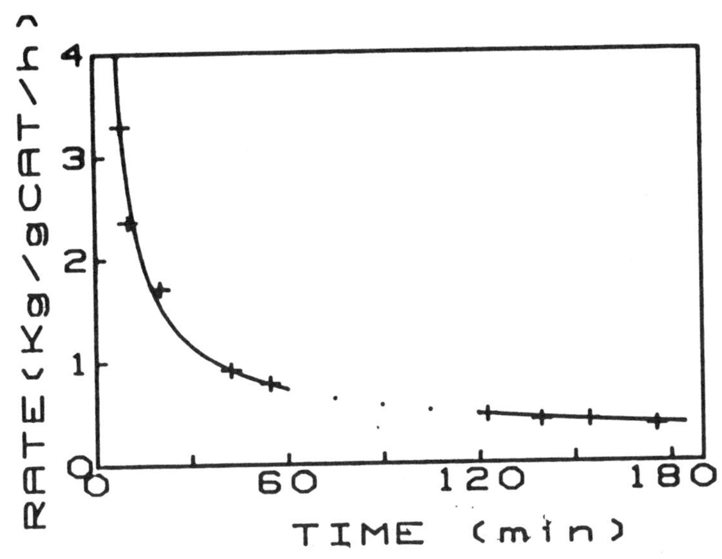

Figure 4. Kinetics for propene polymerization with a 3 rd generation catalyst ($MgCl_2$-ethylbenzoate-$TiCl_4$; cocatalyst triethylaluminium : ethylparatoluate ; ratio 4 : 1) at 63°C, 4 bars pressure. No monomer was added from 60 min to 120 min (...)

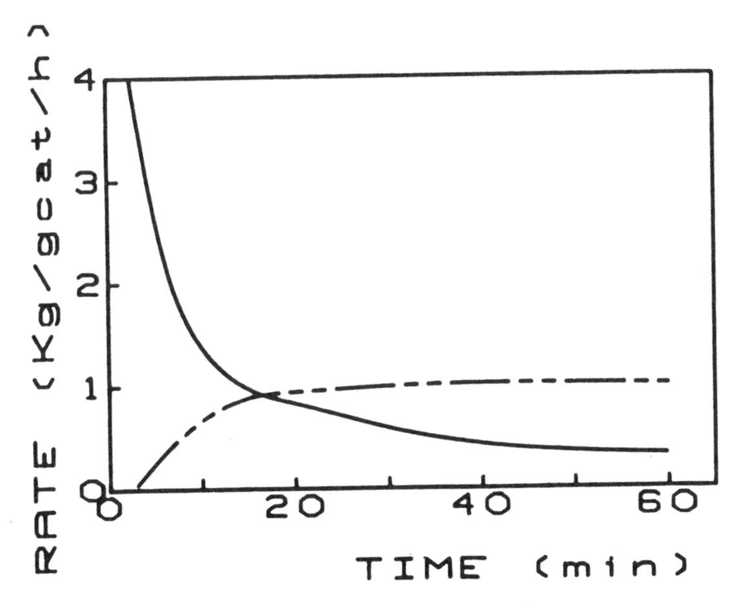

Figure 5. Kinetics for propene polymerization with 3 rd generation catalyst. Same conditions as for fig. 4
—— cocatalyst : triethylaluminium
– – – cocatalyst : isoprenylaluminium

feature of the reaction. This was reported in the case of propene (22) and can often be observed in the case of ethylene polymerization. Even if MW vary with time with unsteady polymerization rate, a molecular weight variation doesn't prove that a diffusion limit is reached.

A different method giving access to the overall diffusion is obtained by comparing the MW of the largest and smallest particles size fraction from the particle size distribution at the end of the polymerization. The MW are generally the same for diameter ratios up to 10 (corresponding to productivity ratios up to 1000 between the particles).

The so-called "replica" polymers reported for PP, PE and LLDPE with the same spherical shape than the catalyst and same particle size distribution, expanded in the ratio of polymer production (34) suppose that the same polymer is obtained at the same rate per volume unit, independantly of the size of the growing particle for all polymerization times.

A direct measurement of the kinetic constants (kp) for different monomers can also be used to refute diffusion effect. Comparing a short and a long chain monomer Chien, Ang and Kuo think to have found "definitive evidence against diffusion limitation" (35) : isotactic propene polymerization occurs at a eight times greater rate than isotactic 1-decene polymerization, but the rate decay are the same for the two reactions.

The chemical theory :

Very narrow MWD are almost only observed with homogeneous catalytic systems with one kind of active species (3). The industrial catalysts are heterogeneous and rather broad when compared to a limit Q value close to 2. Only particular claims of so low Q values can be found for heterogeneous systems (25, 36, 38). A distribution of active centers was proposed for a long time to explain broad MWD (39). The distribution can provide from different propagation or different termination constants or from the two effects together. A convenient method to test the hypothesis is to observe what happens to the MWD when the concentration of a transfer agent (H_2 for instance) is varied. This is discussed by Zucchini and Cecchin (3). For instance, a model from Gordon and Roe (40) supposed that the termination is due to chain desorption from the catalyst surface, dependent on the chain length. The MWD is then supposed to get narrow when a transfer agent is added-and this is not the case (41). In fact, several authors reviewed in ref. 3 have found that MWD are not dependent on hydrogen concentration "this supports the hypothesis of polymerization centres non uniformity with regard to the respective rate constants of propagation and termination ; each of these centres yields polymers having the "most probable" MWD but different average molecular weights ..." as developed by Clark and Bailey (42) for Phillips catalysis (3).

Zucchini and Cecchin have also discussed different arguments

in favor of a distribution of active centers :
- heterogeneity of active centers : stereospecific and aspecific centers giving both broad MWD in the case of propene (43)
- selective poisoning selects active centers with different kinetic constants (PP ; δ-$TiCl_3$ + DEAC + CS_2) (44)
- selection of active centers on the basis of their lewis acidity by poisoning with lewis bases (2)
- heterogeneity of the crystal structure : Corradini et al. have established different theoretical surface structures for defects on $TiCl_3$ or $TiCl_4$ supported on Mg Cl_2 (45)

A remark can be added to all these arguments : the heterogeneity of the active center distribution can also be revealed by copolymerization reactions (ethylene + propene, propene + ethylene, ethylene + butene). In most cases these copolymers can be fractionated in fractions with very disperse ratios of the two comonomers (46). For instance, LLDPE are easily fractionated by hexane extraction (47). A more sophisticated fractionation : T. R. E. F. (temperature rising elution fractionation) (48) gives access to the distribution of copolymerization reactivities. On the contrary copolymers obtained with vanadium catalysts have sometimes narrow MWD and homogeneous composition with for instance a melting point in the range of LLDPE composition more than 20 K lower than commercial copolymers (49, 50).

HOW CAN MWD BE CONTROLLED :

Among the numerous factors which can modify the MWD (51) many are of restricted use : the industrial requirements (activity, economicity, simplicity ...) forbid many receipts thinkable at laboratory scale.

We will first try to give examples of modification of MWD in relation with the idea that the MWD reflects the distribution of the active centers.

According to Pino (2, 21) for the $TiCl_4$ supported on $MgCl_2$ precatalyst, the lewis base selects the active centers on the basis of lewis acidity and steric hindrance. In table 1 are reported two sets of typical results obtained in propene and ethylene slurry polymerization. The heptane soluble atactic PP fraction diminishes as expected from the active centers selection when the precatalyst is modified (the corresponding MWD are not very different for the iso and atactic fraction respectively). The PE obtained with the same precatalysts show no difference (column 2) with triethylaluminium cocatalyst. With trihexylaluminium cocatalyst, the MWD is very broad with the $MgCl_2$-$TiCl_4$ precatalyst. On the contrary a "medium" distribution is obtained with a lewis base. Changing the cocatalyst to trioctyl or commercial triisoprenylaluminium (IPRA) gives the same kind of result : the uncomplexed precatalyst has a broad distribution of active centers which is revealed by the use of an aluminium alkyl with a not too strong reducing power. With a strong one a fraction

TABLE 1

Precatalyst	hot heptane[a)] soluble PP %	$Q^{b)}$ (PE)	$Q^{c)}$ (PE)
$MgCl_2$-$TiCl_4$	11	9.5	16
Mg Cl_2-ethylbenzoate-$TiCl_4$	5	8.5	

a) cocatalyst : triethylaluminium/ethylenzoate 3 : 1 ; 63°C

b) cocatalyst : triethylaluminium-80° C

c) cocatalyst : trihexylaluminium-80° C

EFFECT OF ACTIVE CENTERS SELECTION ON THE PRECATALYST FOR ETHYLENE AND PROPENE SLURRY POLYMERIZATION.

TABLE 2

PRECATALYST	process	Polymer	Q	MW
SiO_2-$MgCl_2$-$TiCl_4$	slurry	PE	15.0	190.000
SiO_2-$MgCl_2$-$TiCl_4$	gas phase	LLDPE	7.5	158.000
SiO_2-$MgCl_2$-ethylbenzoate-$TiCl_4$	gas phase	LLDPE	5.3	148.000

cocatalyst : IPRA

EFFECT OF PROCESS AND ACTIVE CENTER SELECTION ON MWD

TABLE 3

Precatalyst	preparation	% Ti	$MI_{21.6}/MI_5$
$MgCl_2$-$TiCl_4$	impregnation	0.5	14.9
"	cogrinding	1.43	14.3
"	improved cogrinding	1.8	19.7
"	cogrinding	2	10.2

Polymerization conditions : heptane slurry ; 80° C ; cocatalyst : trihexylaluminium. Total pressure 8 bars

CONTROL OF MWD FROM CATALYST PREPARATION

of the population of active centers is destroyed and all the precatalysts appear as to have almost the same MWD.

In table 2 is shown a comparison between gas phase and slurry polymerization using a supported catalytic system supported on silica (bisupported catalyst). In gas phase, the conditions of temperature and aluminiumalkyl at the beginning of the polymerization favor the selection of active centers : the distribution is narrow in gaz phase and can be narrowed by complexation of the catalyst.

The broadness of the active center distribution can be increased by a convenient choice of the preparation conditions : $TiCl_4$ impregnation on ground $MgCl_2$ gives a less heterogeneous catalyst than $TiCl_4$ - $MgCl_2$ cogrinding.

An improved cogrinding process can extend the MWD to very broad values (table 3).

An example of MWD tayloring somewhat different from the

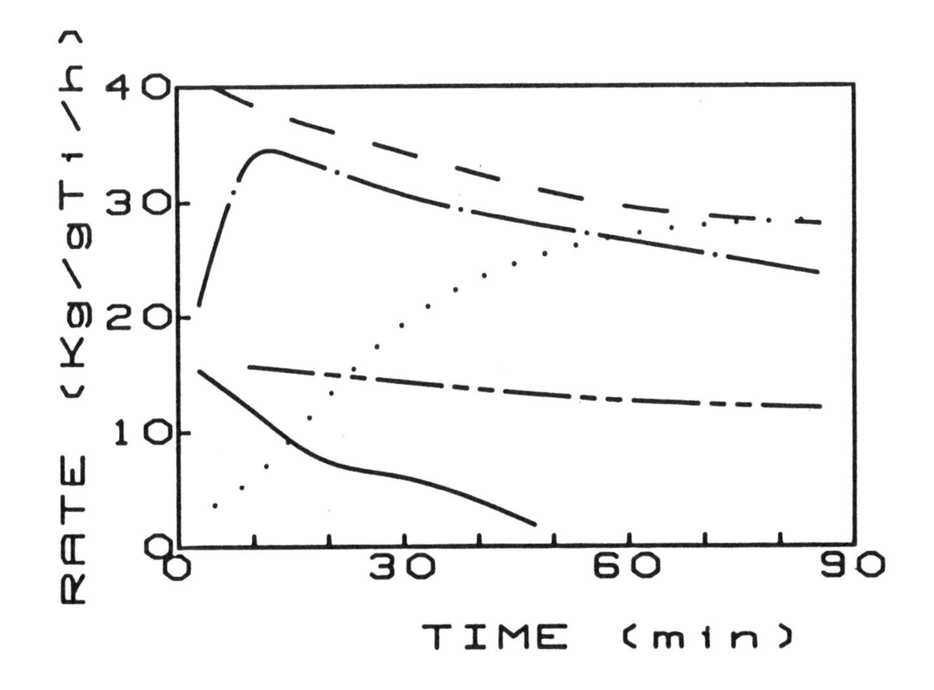

Figure 6. Kinetics of ethylene polymerization (heptane slurry ; 80°C ; 11 bars total pressure) for different cocatalysts
- - - - trioctylaluminium —.— tetrabutyldialuminoxane
. . - - — triisobutylaluminium — triethylaluminium
. . . isoprenylaluminium

preceding will be detailed in order to show how difficult it is to bring out universal rules. The starting catalyst is a classical β-$TiCl_3$ obtained by careful reduction of $TiCl_4$ by commercial isoprenylaluminium (IPRA). The productivity of the catalyst is extremely high (typical productivity 20000 to 30000 g/g Ti/hour at 80° C and 7 bars monomer pressure).

As it can be seen on a typical kinetic curve (fig. 6) a long time is necessary before the full activity is reached with IPRA. The Q factor is then 9. Similar results are observed with trioctyaluminium or tetrabutyldialuminoxane (table 4) but the kinetics are rather constant. With triisobutylaluminium the Q factor reaches 10.8 and 13.1 with triethylaluminium with a change in kinetic shape (fig. 6) and productivity.

In fact, the Mw, the melt-indexes, the Q factors and productivies vary with the expected reducing power of the corresponding alkylaluminium, the Mn remaining rather constant.

We have developed the idea that the potential of broad MWD is not developed with the weakest alkylaluminium but the activity is decreased to lower values with the strongest alkylaluminium.

In a second set of experiments (table 5) : the MW increase when the [Al] concentration increases. An opposite result is expected from transfer if H_2 acts only as transfer agent.

In the table 6 are described the effects of a reducing treatment before the polymerization : the catalyst is heated in the presence of IPRA and H_2 : the productivities are increased (the system reaches faster a maximum activity) the MW increase with more effect on Mw than on Mn : so the Q factors increase also. A maximum value of more than 13 can be obtained. A too strong treatment destroys the catalytic activity. The solid turns to a black color typical of Ti (II). When IPRA or H_2 are used alone, no effect is observed.

So, depending on the catalyst, the same kind of reducing treatment leads to MWD broadening or narrowing : the active center distribution may change with the polymerization conditions or with chemical treatments.

USE OF TWO POPULATIONS OF ACTIVE CENTERS

The convenient choice of the catalytic system is then the first and determing step of MWD tayloring. The patent review contained in ref. 3 clearly supports that conclusion. Recently, Böhm et al (52) show how MWD are broadned when "the Mg - Ti catalyst is coated with a further transition metal component " in order to increase the heterogeneity. A slightly different reagents mixing sequence (the catalyst is aged several minutes in the presence of the cocatalyst or not) gives rise to two different behaviors : in one case MWD constant with time, in the second case MWD varying from broad bimodal to narrow unimodal with time. A similar kind of result was

TABLE 4

Cocatalyst	a	b	c	d	e
[Al] mmol/l	2.34	2.33	2.35	2.35	1.174
H_2 (bars)	3.6	4	3.6	3.6	3.6
Productivity g/g Ti/h	19600	17100	29700	6600	26700
Mw	128000	168000	121000	199000	116000
Mn	14300	15600	14700	15200	13100
Q	9	10.8	8.2	13.1	8.8

a : isoprenylaluminium b : triisobutylaluminium c : trioctylaluminium
d : triethylaluminium e : tétrabutyldialuminoxane
Polymerization conditions : heptane slurry ; 80°C ; 90 min.

EFFECT OF THE CHOICE OF THE COCATALYST ON PRODUCTIVITY AND MW

TABLE 5

[Al]/[Ti]	Al mmol/l	M_w	Melt flow index (10kg/190°C)	Q	Productivity g/g Ti/h
15	2.3	165 000	2.9	9.3	26 700
50	7.7	172 000	2.6	10	27 300
100	15.2	169 000	2.3	10.2	28 300
200	29.4	198 000	0.8	10.1	29 720

Polymerization conditions are found on table 4. Cocatalyst isoprenylaluminium

EFFECT OF COCATALYST CONCENTRATION ON MOLECULAR WEIGHT

TABLE 6

Ageing conditions					
Duration (h)	0.5	0.5	1	1	1
[Al] mmol/l	2.31	13.3	2.3	1.2	0.6
Productivity (g/gTi/h)	25500	18100	25500	26900	26400
Mw ($x10^3$)	197	214	224	228	185
Mn	17700	21400	22700	18500	15200
Q	11.1	10	9.9	13.3	12.2

Polymerization conditions are found on table 4.
Cocatalyst : isoprenylaluminium
Ageing conditions : hydrogen (3.6 bars), 80°C, no monomer.

EFFECT OF AGEING ON MWD

TABLE 7

Temperature (Celcius)	duration (min.)	Productivity g/gTi/h	$MI_{21.6}/MI_5$
80	90	10200	11.2
80	240	15400	17.5
90[a)]	90	11100	15.6

a) the catalytic systems aged 1h at 80° C in the presence of hydrogen (3.6 bars)

The catalyst is obtained by reacting $TiCl_4$ with a mixture of isoprenylaluminium and alumoxane. The other polymerization conditions are found on table 4. Cocatalyst : isoprenylaluminium

EFFECT OF POLYMERIZATION OR AGEING CONDITIONS ON THE MWD REFLECTED BY THE MELT FLOW RATIO $MI_{21.6}/MI_5$

observed in our laboratory : if the catalyst in table 4 is prepared with a mixture of IPRA and alumoxane, the MWD continuously broadens with time (table 7). A reducing treatment combined with higher polymerization temperature gives rise to a broad MWD in a shorter time.

The precedent values are measured on batch designs. With non stationnary MWD, the use of a continuous polymerization process is expected to change the MWD, depending on the residence time distribution. On the contrary, the same MWD obtained with the two kind of designs proves that the MW are not time dependent.

CONCLUSION

The problem of the fine control of MWD cannot be separated from any aspect of the polymer synthesis with Ziegler-Natta catalysts. Depending on the system, the same causes can lead to opposite effects. The signification of the example given here remains simple : to a broader active center distribution corresponds a broader MWD. But how the result is got depend on details of the chemistry of the active centers which are generally unforseable because at the present time, the exact chemical nature of the active centers remains unknown.

Diffusion is generally not the factor determining alone the broad MWD. Nethertheless the possibility of diffusion effects must be kept in mind and in a recent communication Ray (53) suggests that chemical and diffusion effects have to be associated in modelization computations.

The easiest way to obtain broad MWD is to increase the chemical heterogeneity of the catalytic system. It may seem easy to obtain such a result with only the MWD request, leaving free the process and uncontrolled the other polymer properties and in fact Q factors from about 2 to more than 20 are really obtained in the case of PE. But with a second requirement for instance : stereospecific polymerization for PP or statistical copolymerization (LLDPE) or with process- dependant conditions of temperature or concentration (high temperature, gas phase) it becomes very difficult to adjust the MWD. For instance, no convincing examples of broad MWD in one step reaction are reported for gaz phase ethylene polymerization with titanium catalysts. Even the "multi-step" processes are not always easy to work out and lead to broad distribution only after a convenient choice of the catalytic system (3).

Acknowledgement

The author is indebted to Michel PLISSON, Patrick MASSON and Véronique PASQUET for a part of presented results.

This work was supported by ATOCHEM - Groupe ELF AQUITAINE.

REFERENCES :

1/ J. BOOR, Ziegler Natta Catalysis and Polymerizations Academic Press 1979, Chap. 1

2/ P. PINO and R. MULHAUPT Transition metal catalyzed polymerizations R.P. QUICK Ed. MMI press 1983 vol.A p.1

3/ V. ZUCCHINI and G. CECCHIN Adv. polym. Sci. 51, 101 (1983)

4/ J.R. MARTIN, J.F. JOHNSON and A.R. COOPER J. Macromol Sci. Revs. Macromol. chem. C8 57 (1972)

5/ R.W. NUNES, J.R. MARTIN and J.F. JOHNSON Polym. Eng. Sci. 22, 205 (1982)

6/ S. DANESI, T. SIMONAZZI and F. ZEGERA 7e convegno Italiano di scienza delle macromolecole Invited lectures proceedings p.181 (1985)

7/ European Patent 0098077A1 to Chisso corp. (1981)

8/ W. MINOSHIMA, J.L. WHITE and J.E. SPRULL Polym. Eng. Sci. 24, 544 (1984)

9/ D. LASSALLE, J.L. VIDAL, J.C. ROUSTANT and P. MANGIN 5th European Plastics and Rubber conference Paris (1978)

10/ I.M. WORD Adv. Polym. Sci. 70, 3 (1985)

11/ J.P. HOGAN and R.L. BANKS Belgian Patent 530617 (1955) (to Phillips Petroleum Company)

12/ E.J. VANDENBERG US Patent 3051690 (1962) (to Hercules Powder Company)

13/ B. ETTORE and L. LUCIANO Italian Patent 554013 (1957) (to Montecatini)

14/ G. NATTA and I. PASQUON Adv. Catal. 11, 1 (1959)

15/ G. NATTA, I. PASQUON and L. GIUFFRE Chim. Ind. (Milan) 43, 871 (1961)

16/ P. LONGI, G. MAZZANTI, A. ROGGERO and A.M. LACHI Makromol. Chem. 61 63 (1963)

17/ T. KEII in Catalytic polymerization of olefins T. KEII and K. SOGA ed. Kodansha - Elsevier (1986) p.1

18/ G. NATTA Chim. ind. (Milan) 41, 519 (1959)

19/ T. KEII Kinetic of Ziegler-Natta polymerization, Kodansha Tokyo (1972) p. 121

20/ U. GIANNINI, G. GUASTALLA Makromol chem.Rapid. Comm. 4, 429 (1983)

21/ P. PINO and B. ROTZINGER, Makromol. Chem. suppl. 7, 41 (1984)

22/ E. SUZUKI, M. TAMURA, Y. DOI and T. KEII Makromol. chem. 180, 2235 (1979)

23/ K. SOJA and T. SIONO Polym. Bull 8, 261 (1982)

24/ J.C.W CHIEN and C. I. KUO J.Polymer. Sci. Chem. 24, 1779 (1986)

25/ J.C.W. CHIEN and P.L. BRES J. Polym. Sci. Chem. 24, 2483 (1986)

26/ L.L. BOHM Polymer 19, 545 (1978)

27/ L.L. BOHM Makromol. Chem. 182, 3291 (1981)

28/ ZAKHAROV, G.D. BUKATOV, N.B. CHUMEVSKII and Y.I. YERMAKOV Makromol. Chem. 178, 967 (1977)

29/ G. FINK, D. SCHELL Makromol. Chem. 105, 15 and 31 and 39 (1982)

30/ T.W. TAYLOR, K.Y. CHOI, H. YUAN and W.H. RAY in Transition metal catalyzed polymerization ibid. ref.2 vol. A p.191

31/ R. SPITZ, J.L. LACOMBE, M. PRIMET and A. GUYOT ibid. vol.A p.389

32/ R. SPITZ, J.L. LACOMBE and A. GUYOT J. Polym. Sci. Chem. 23 2625 (1984)

33/ Y. DOI, M. MURATA, K. YANO Ind. Eng. Chem. Prod. res. dev. 21, 580 (1982)

34/ P. GALLI, P.C. BARBE and L. NORISTI Angew Makromol. Chem. 120, 73 (1984)

35/ J.C.W. CHIEN, T. ANG and C.I. KUO J. Polym. Sci. 23 723 (1985)

36/ J.C.W. CHIEN and C.I. KUO J. Polym. Sci. Chem. 24, 1779 (1986)

37/ H. SINN and W. KAMINSKY Adv. Organomet. Chem. 18, 99 (1980)

38/ W. KAMINSKY ibid. ref. 17 p.293

39/ G. NATTA J. Polym. Sci. 34, 21 (1959)

40/ M. GORDON, R.J. ROE Polymer 2, 41 (1961)

41/ R.J. ROE Polymer 2, 60 (1961)

42/ A. CLARK and G.C. BAILEY J. Catal. 2, 230 (1963) and 2, 241 (1963)

43/ Y. DOI, T. KEII Makromol. Chem. 179, 2117 (1978)

44/ L.A. RISHINA and E.I. VIZEN Eur. Polym. J. 16, 965 (1980)

45/ a) P. CORRADINI, G. GUERRA, F. FUSCO and V. BARONE Eur. Polym. J. 16 835 (1980)
b) P. CORRADINI, G. GUERRA, V. BARONE Preprint IUPAC 28th Makromol. Sympos. Amherst p.235 (1982)

46/ E. SPITZ, L. DURANEL, P. MASSON, M.F. LLAURO-DARRICADES and A. GUYOT Paper presented at the international Symposium Transition metal catalyzed polymerization Akron june 1986

47/ V.GASCHARD-PASQUET thesis Lyon 1985

48/ L. WILD, T.R. RYLE, D.C. KNOBELOCH and I.R. REAT J. Polym. Sci. 20 441 (1982)

49/ B.K. HUNTER, K.E. RUSSEL, M.V. SCAMMELL and S.L. THOMSON J. Polym. Sci. Chem. 22 1383 (1984)

50/ C. COZEWITH and G. VER STRATE Macromolecules 4, 482 (1971)

51/ ref. 3 p.137

52/ L.L. BOHM, J. BERTHOLD, R. FRANK, W. STROBEL and U. WOLFMEIER ibid. ref.17 p. 29

53/ W.H. RAY ibid. ref 46

STEREOSPECIFIC POLYMERIZATION OF OLEFINS WITH HOMOGENEOUS CATALYSTS

W. Kaminsky , M. Buschermöhle
Institute for Technical and Macromolecular Chemistry
University of Hamburg
Bundesstrasse 45
D-2000 HAMBURG 13
F.R.G.

ABSTRACT. Titanocene and zirconocene-compounds in combination with methylaluminoxane as cocatalyst form extremely active soluble Ziegler-Natta-Catalysts. One mol chiral stereorigid ethylenebis(indenyl)zirconiumdichloride yields 43.000 kg per hour highly isotactic polypropylene. ^{13}C-nmr spectroscopically determined isotacticity lies above 99%. Working with non chiral zirconocenes as bis(cyclopentadienyl)zirconium compounds leads solely atactic polypropylene. Isotacticity is dependent on polymerization temperature.

With bis(neomenthylcyclopentadienyl)zirconiumdichloride propene can be polymerized to a stereo block polymer with long isotactic sequences of n =6. The molecular weight distribution $M_w/M_n = 2$ is narrow.

By separation of the chiral racemic ethylene bis(4.5.6.7 tetrahydroindenyl)zirconiumdichloride into the enantiomers an optically active catalyst component can be produced. Application of the S-enantiomer, displaying a value of specific rotation of 297°, together with methylaluminoxane, transforms prochiral monomeric propylene into chiral polypropylene. The hydrocarbon soluble amount of atactic polypropylene lies below 0.1 weight %. The isotactic sequences reaches 330.

1. INTRODUCTION

The combination of zirconocenes or titanocenes with methylaluminoxanes of the structure $(Al(CH_3)O)_{6-20}$ forms highly active Ziegler-Natta-Catalysts which are soluble in hydrocarbons. For the polymerization of ethylene a productivity of $200 \cdot 10^6$ g polyethylen per g zirconium can be reached. The concentration of the transition metal lies between 10^{-5} and 10^{-10} mol/l. So the zirconium concentration in the polymer is very low. There is still the need for a relatively high aluminoxane concentration of 10^{-3} mol/l, which serves to form the active complex and to bond poisons (1-6).

M. Fontanille and A. Guyot (eds.), Recent Advances in Mechanistic and Synthetic Aspects of Polymerization, 503–514.

Polymerizations of propene and butene with bis(cyclopentadienyl)-zirconium - or titanium compounds yield only atactic polypropylenes (Table 1). It is possible either to work in solutions or in liquid monomer. At low temperatures the mean molecular weight (M_η) of the atactic polypropylene reaches values of 590 000. With increasing polymerization temperatures it drops down decisively.

TABLE 1 Polymerization of Propene with Methylalumoxane 10^{-3} mol Al units as Cocatalyst and different Transition metal compounds

Catalyst	Toluene (ml)	Propene (ml)	T (°C)	Time (h)	Yield (g)	Activity (g PP/g M·h)	$\bar{M}_\eta$
$Cp_2Ti(CH_3)_2$	150	200	20	48	15,1	533	100 000
$Cp_2Ti(CH_3)_2$	-	250	10	76	18,0	430	93 000
$Cp_2Zr(CH_3)_2$	-	330	5	24	48,5	13 300	106 000
$Cp_2Zr(CH_3)_2$	5	340	5	24	120,0	15 700	106 000
$Cp_2Zr(CH_3)_2$	5	340	-10	12	32,0	8 400	191 000
$Cp_2Zr(CH_3)_2$	50	250	-30	50	202,0	1 300	84 000
$Cp_2Zr(CH_3)_2$	50	100	-70	58	8,0	75	590 000
Cp_2ZrCl_2	120	210	20	24	5,6	17 600	75 600
$Cp(CpMe_5)ZrCl_2$	120	210	20	24	2,6	8 600	180 000
$(CpMe_5)_2ZrCl_2$	120	210	20	24	0,7	2 300	147 000

^{13}C-NMR spectroscopic investigations of the polymers show that nearly ideally atactic samples are produced which are suitable for blending of elastomers (7).

By variation of the ligands at the zirconium atom asymmetric centres have been incorporated into the soluble catalyst, in order to synthesize isotactic polypropylene as well.

Although the molecule (pentamethylcyclopentadienyl)cyclopentadienylzirconiumalkylchloride (I) consists of four different ligands at the zirconiumatom, this compound in combination with methylaluminoxane yields only atactic polypropylene. With regard to the insertion reaction of the olefin the ligands change places too fast. The application of bis(cyclopentadienyl)zirconiumisobutylchloride (II), which contains a chiral carbon atom at one ligand, did not melt the target neither. The explanation here is that the chiral end of the growing polymer chain faded away by repeated insertion steps from the active centre of the catalyst, so that no effect is any longer observable.

2. ISOTACTIC POLYPROPYLENE

Isotactic polypropylene only can be synthesized by chiral catalysts. This is also possible with a homogeneous system. If the π-bonded cyclopentadienyl ligands of the zirconium compound are changed for ethylene bridged indenyl rings a chiral ethylene-bis(indenyl)zirconiumdichloride is formed.

The CH_2-CH_2-bridge serves for a very stereorigid conformation around the zirconium atom, so that fluctuation among the ligands is not possible. The two indenyl groups are arranged in a chiral array. Ethylene bis(tetrahydroindenyl)zirconiumdichloride appears in two isomeric forms (R and S form). The analogous titanium compound can exist in an additional mesomeric form which is not stable in the case of zirconium.

By application of the racemic mixture of ethylene bis(indenyl)zirconiumdichloride or ethylene bis(tetrahydroindenyl)zirconiumdichloride together with the cocatalyst methylaluminoxane in polymerizations of propene highly isotactic polypropylenes can be produced (Table 2) (8-9).

TABLE 2 Polymerization of 70 ml Propene or 1-Butene with $7\cdot10^{-7}$ mole/lit rac-Et(Ind)$_2$ZrCl$_2$(I) or $8,4\cdot10^{-6}$ mole/lit rac-Et(TH-Ind)$_2$ZrCl$_2$(II) and $1,6\cdot10^{-2}$ mole Al/l methylaluminoxane in 330 ml toluene

Monomer	Catalyst	T(°C)	Time (min)	Yield (g)	Activity (kg PP(PB)/mol Zr h)
Propene	I	15	620	3,6	1 500
Propene	I	21	500	31	16 500
Propene	I	35	120	26,6	43 000
Propene	II	-20	360	1,5	80
Propene	II	-10	270	4,5	300
Propene	II	0	255	12,5	880
Propene	II	8	180	13,0	1 300
Propene	II	15	170	26,7	2 900
Propene	II	20	120	31,3	4 750
Propene	II	60	90	38,7	7 700
Butene	II	-10	330	9,1	500
Butene	II	20	200	29,2	2 640

Polymerization activities of the bis(indenyl)zirconiumdichloride are distinctively greater than those of the corresponding hydrated compounds.

The activities referred to the concentration of the transition metal overcome those values which nowadays are achieved with carrier fixed heterogeneous catalysts. In a one hour polymerization period up to 43 000 kg polypropylene per mole zirconocene can be produced at a zirconium concentration of $5\cdot10^{-6}$ mol/l.

Aluminoxane concentrations in the range of $3\cdot10^{-2}$ mole Al/lit are quite high. For mean molecular weights M_w of 45 000 the growth time of one macromolecule is approx. 3.85 and the insertion time of a propene molecule is $3.5 \cdot 10^{-3}$s. These calculations are based on the assumption that every zirconium atom forms an active centre, which could be proved in the case of ethylene polymerizations (10).

But also the features of the polypropylenes produced with the soluble Ziegler-Natta-catalyst are remarkable (Table 3).

Molecular mass distribution M_w/M_n has typical values between 1.9 and 2.6 which are extraordinarily narrow. NMR-spectroscopic investigations show that the atactic amount is smaller than the toluene soluble amount; typical values lie below 0,2% which are by 10 smaller than for those polymers which are synthesized by technical heterogeneous catalysts. By variation of polymerization temperature between -10°C and 20°C molecular weights can be controlled in the range from 305 000 to 45 000.

TABLE 3 Features of Isotactic Polypropylene Produced with Ethylene-bis-(tetrahydroindenyl)zirconiumdichloride/methylaluminoxane

Polymerization Temperature	$\bar{M}_w$	M_w/M_n	Toluene Soluble (atactic) Amount
-10	305 000	2,6	0,25 %
0	144 000	2,4	0,2 %
15	62 000	2,0	0,7 %
20	45 000	1,8	1,0 %

The polymerization kinetics of forming polypropene with this soluble catalyst were studied in toluene. The resulting products were characterized by GPC, DSC, viscosimetry and ^{13}C-NMR-spectroscopy.

For the polymerization process, a laboratory scale bubble column was used (Figure 1). This reactor type was preferred because of the very high heat and mass transfer rates and the easy kinetic measurements (11).

The bubble column consists of an 2 l cylindrical glass tube, fitted with a perforated plate. It contains the solvent for the catalyst and the monomer . Propene is bubbled through the solution, polymerizates and precipitates. The propene gas is pumped by a compressor in the circuit. Consumed monomer is replaced by a pressure regulator. The dozed quantity of monomer is measured by a thermal mass-flow-controller. The plant is fitted with an efficient temperature and pressure (monomer concentration) regulation. Reachable constancy: $\Delta T < \pm 0,3$°C, $\Delta P < \pm 0,05$ bar.

Measurements of the dependence of the polymerization rate (R_p) upon the catalyst concentration (c_{Zr}) results in a linear dependence. The kinetics show usual first order related to the catalyst concentration (Figure 2).

The gradient of the straight line results in a value of 43 200 kg PP/mol Zr·h. In the assumption, that all catalyst molecules form active centers, the turnover time equals to $3.5 \cdot 10^{-3}$ sec. The weight average molecular weight is independent to the catalyst concentration and lies in the range of 47 000 g/mol.

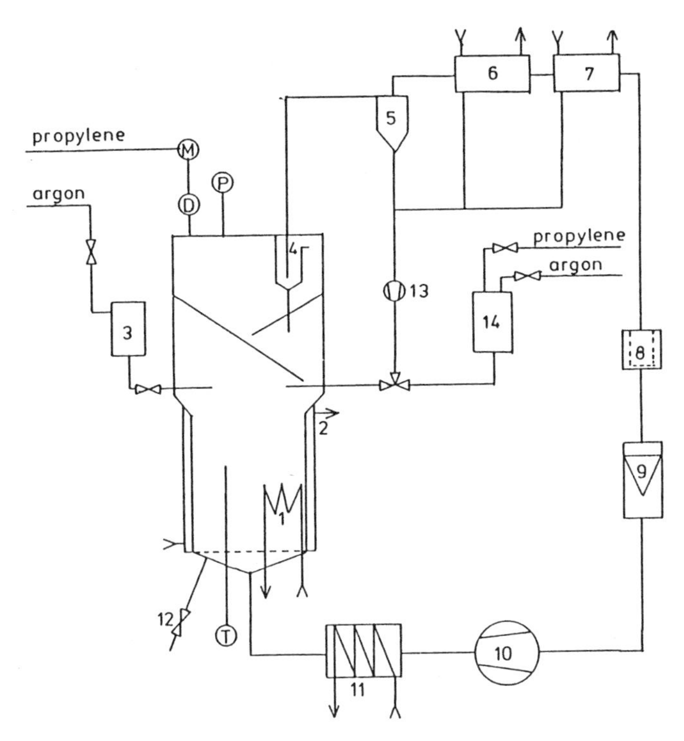

Figure 1. Flowsheet of the plant 1) cooling coil, 2) heating jacket, 3) dosing container for the catalyst, 4+5) cyclones, 6+7) condensers, 8) filger, 9) flowmeter, 10) compressor, 11) heat exchanger, 12) outlet valve, 13) solvent pump, 14) solvent container, P) pressure measurement, T) thermocouple

The measurement of dependence of the polymerisation rate upon the monomer concentration, also equals the linear dependence (Figure 3). At a lower concentration level than 2 mol/l signifies the influence of mass-transfer control. The average molecular weight increases with the growth of the monomer concentration.

The chiral zirconocene compound is able to polymerize besides propene also 1-butene to isotactic polybutylene. Furthermore it is possible to produce copolymers from ethylene and 4-methyl-1-pentene or iso-butylene.

Compared with the zirconium compound the activities of the analogous titanium catalyst are lower by a factor of 1 000. In addition to this the atactic amount is much greater and rises from 14% at -8°C to 94,4% at 27°C.

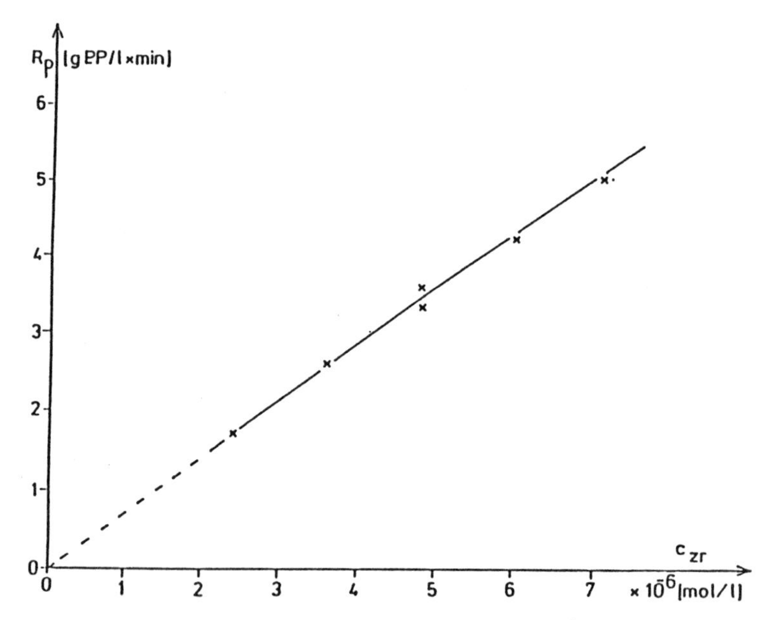

Figure 2. Dependence of polymerization rate (R_p) upon catalyst concentration (c_{Zr}), cocatalyst concentration c_{Al}=3,1·10^{-2} mol/l, propene concentration c_p=2,75 mol/l; polymerization temperature T_p=35°C

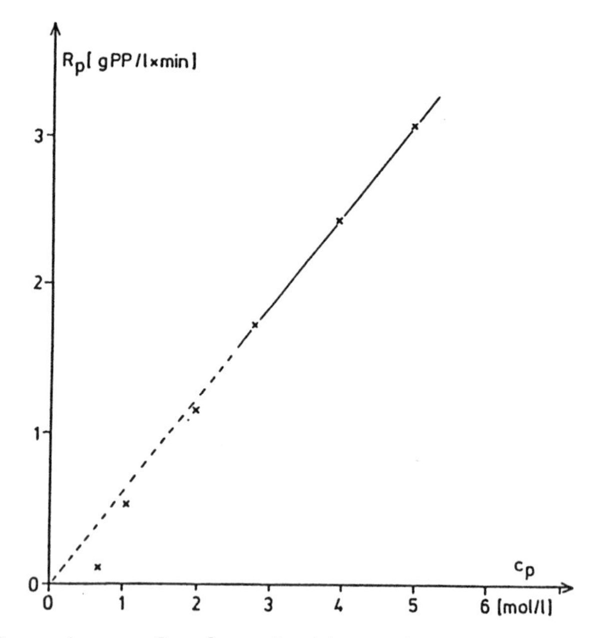

Figure 3. Dependence of polymerization rate upon propene concentration (c_p), c_{Zr}=2,4·10^{-6} mol/l, c_{Al}=3,1·10^{-2} mol/l, T_p=35°C

3. CHIRAL POLYPROPYLENE

With the support of optically active binaphthyl a racemate cleavage of a racemic mixture of ethylene bis(tetrahydroindenyl)zirconiumdichloride has been successfully performed (12, 13): This procedure yields a zirconocene compound whose optical activity reaches $(\alpha)^{RT}_{436}$ = 297°C. Table 4 contains polymerization results for (S)-ethylene bis(tetrahydroindenyl)zirconiumdichloride.

The activity is smaller than the activity for the racemic mixture. Obviously the cleavage of the binaphthyl-complex back into the dichloride has not been achieved completely. But the atactic amounts are even smaller. Several samples exhibit toluene soluble amounts below 0.1%. Working at higher polymerization temperatures (53°C) leads to a wax-like, low-molecular weight polypropylene (M_η = 6,900) which is toluene soluble.

TABLE 4 Propene Polymerization with (S)-ethylene-bis(tetrahydroindenyl) zirconiumdichloride and 300 mg Methylaluminoxane in 330 ml Toluene

Zr (mol/l)	Temperature (°C)	Propene (bar)	Activity (kg PP/mol Zr·h)	M_η	Solubility in Toluene	Atactic Amount(%)
1,8 10	74	6,0	930	1 000	100	18,3
2,5 10	53	5,5	3 300	6 900	100	0,2
1,1 10	37	4,9	326	27 400	-	0,1
1,2 10	15	2,5	178	42 300	-	0,1
6,3 10	11	4,0	147	59 500	-	0,05
1,3 10	0	2,3	10	135 600	-	0,05

Sequence lengths and isotacticities can be calculated from ^{13}C-nmr spectrum and is listed in Table 5.

TABLE 5 ^{13}C-NMR Spectroscopically Calculated Sequence Lengths and Isotacticities (I) of two Polypropylenes, Produced with Chiral Zirconocene/Methylaluminoxane Catalyst

Catalyst	Temp. (°C)	n_{iso}	n_{syn}	I
rac Et(TH-ind)$_2$ ZrCl$_2$	12	99	-	99
S-Enantiomere	11	330	-	99

Typical sequence lengths lie at 99 when the racemic catalyst is used and can be rised to 330 when the (S)-enantiomer of zirconocene is engaged. This is another way of saying that approximately 330 propene molecules are inserted into the polymer chain with the same configuration before one irregular insertion takes place. Accordingly the isotacticity values are very high with 99%. Under special condition the polypropylene can be optically active.

4. STEREO BLOCK POLYMERS

What happens when the chiral centre is more remote from the zirconium atom? In order to investigate this effect bis(neomenthylcyclopentadienyl)zirconiumdichloride has been synthesized and used as a catalyst component. Bis(neomenthylcyclopentadienyl)zirconiumdichloride consists of three chiral c-atoms (Figure 4)

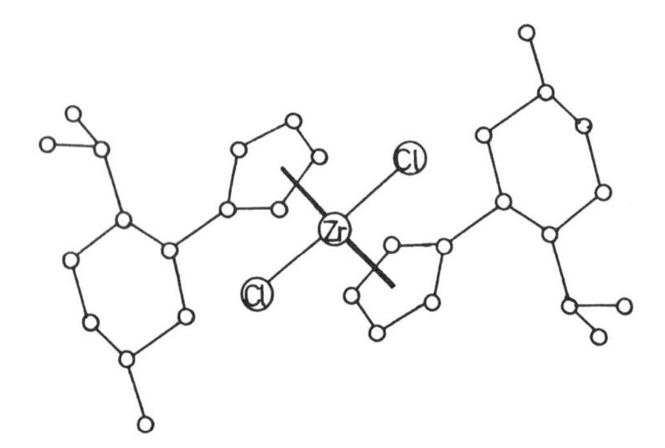

Figure 4. Structure of Bis(neomenthylcyclopentadienyl)zirconiumdichloride

Together with methylaluminoxane as cocatalyst it yields in polymerizations of propene stereo block polymers with isotactic sequences of n = 6. Table 6 displays the polymerization results (see next page).

At low temperatures relatively high mean molecular weights are obtained, reaching values of 640 000 at -60°C. A sequence analysis of polymers produced in this way has been carried out with the help of ^{13}C-NMR spectroscopy. In Figure 5 there the methyl proton section of the NMR-spectrum is depicted.

It becomes obvious that the isotactic sequences at 21 ppm raise with decreasing temperatures. Isotacticity indices and isotactic sequence lenghts of the neomenthyl-catalyst and some other bridged titanocenes and zirconocenes are shown in Table 7.

Dependence of isotacticity from the polymerization temperature is depicted in Figure 6.

Propene-bis(cyclopentadienyl)titaniumdichloride is extremely temperature dependent, whereas tetramethylethylene-bis(cyclopentadienyl) zirconiumdichloride has nearly no dependence.

From this it can be stated that these catalysts are able to produce stereoblock polymers in wide ranges, which afford narrow molecular

TABLE 6 Propene polymerization with bis(neomethylcyclopentadienyl) zirconiumdichloride/methylaluminoxane

T (°C)	Zr (mol)*10^5	P (bar)	Propene (mol)	t (h)	PP (g)	Activity (kg PP/mol Zr·h)	$\bar{M}\eta$ (g/mol)
70	0,36	8,0	0,95	4,0	7,2	530	720*
65	0,36	7,2	0,90	5,0	8,0	630	500*
60	0,36	6,6	0,95	5,0	50,0	3570	720*
50	0,36	5,8	1,0	5,0	46,4	3080	960*
40	0,36	4,8	0,93	5,0	38,0	2680	1220*
30	0,36	4,0	0,89	5,0	27,2	1950	
20	0,36	3,3	1,07	5,0	22,8	1300	2100*
20	0,36	3,3	1,07	5,0	19,3	1100	
20	1,45	4,0	1,52	4,1	72,8	881	1300*
0	1,45	2,0	1,43	19,8	77,3	198	16000
-20	1,45	-	1,00	21,3	26,7	86	82000
-40	1,45	-	1,43	65,5	42,5	31	360000
-60	1,45	-	1,43	70,5	3,0	2	643000

* cryoscopic determined mean molecular weight

TABLE 7 Determination of isotacticity indices and isotactic sequence length from ^{13}C-NMR-measurements of polypropylene, produced with different Zr and Ti compounds; bis(neomenthylcyclopentadienyl)-zirconiumdichloride$(NMCp)_2ZrCl_2$:tetramethylethylenbis(cyclopentadienyl)zirconiumdichloride ($Tm-Et-Cp_2ZrCl_2$), tetramethylethylenbis(cyclopentadienyl)titaniumdichloride($Tm-Et-Cp_2TiCl_2$), propene-bis(cyclopentadienyl)titaniumdichloride($Prop-Cp_2TiCl_2$).

	I.	n_{iso}	I.	n_{iso}	I.	n_{iso}	I.	n_{iso}	I.	n_{iso}
T (°C)	- 60°		- 40°		- 20°		0°		20°	
$(NMCp)_2ZrCl_2$	77,1	4,8	76,8	4,5	74,6	4,6	70,8	3,6	67,3	3,1
$Tm-EtCp_2ZrCl_2$	42,3	1,6	41,9	1,5	43,4	1,6	-		-	
$Tm-EtCp_2TiCl_2$	82,5	5,9	82,6	6,0	72,4	4,3	62,8	3,0	59,8	2,5
$Prop.-Cp_2TiCl_2$	83,4	6,3	83,1	6,1	69,5	3,6	57,7	3,1	-	

mass distributions of M_w/M_n = 2 as well. Sequence lengths can be regulated from 2 to 7 according to the temperature applied.

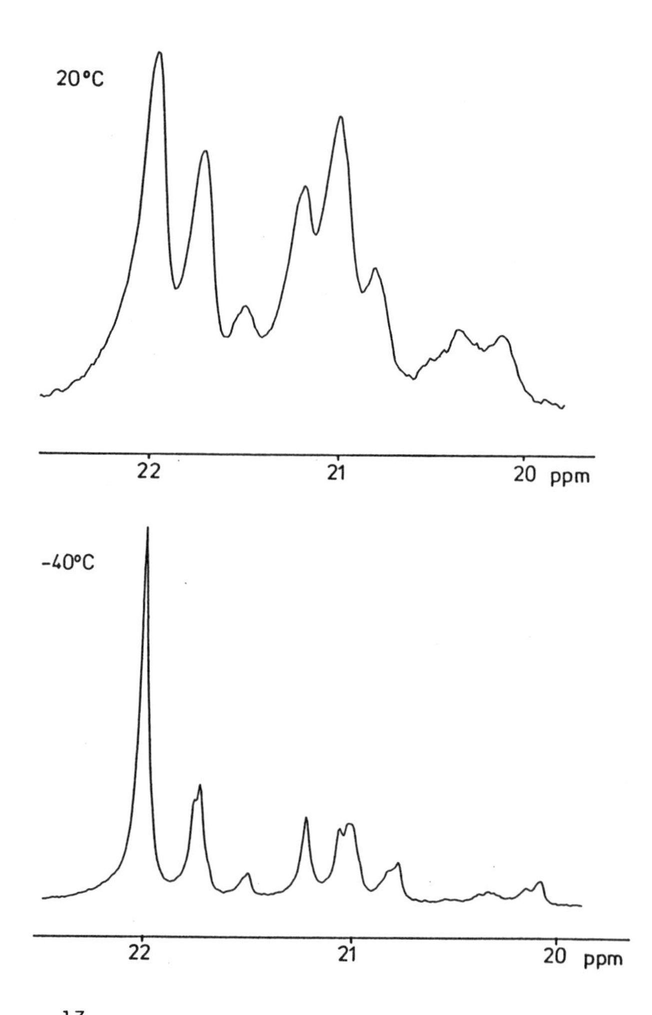

Figure 5. 75 MHz-^{13}C-NMR-spectrum of polypropylene, produced with bis-(neomenthylcyclopentadienyl)zirconiumdichloride at different temperatures

5. REFERENCES

1) H. Sinn, W.Kaminsky, H.-J. Vollmer, R. Woldt: Angew.Chem. 92 (1980) 396
2) H. Sinn, W. Kaminsky: Adv.Organomet.Chem. 18 (1980) 99
3) J.A. Ewen: J.Am.Chem.Soc. 106 (1984) 6355
4) E. Gianetti, G.H. Nicoletti, R. Mazzochi: J.Polymer Sci, Polym.Chem. Ed. 23 (1985) 2117
5) W. Kaminsky, H. Lüker: Makromol.Chem.Rapid Commun. 5 (1984) 225
6) W. Kaminsky, M. Miri: J.Polym.Sci., Polym.Chem.Ed. 23 (1985) 2151

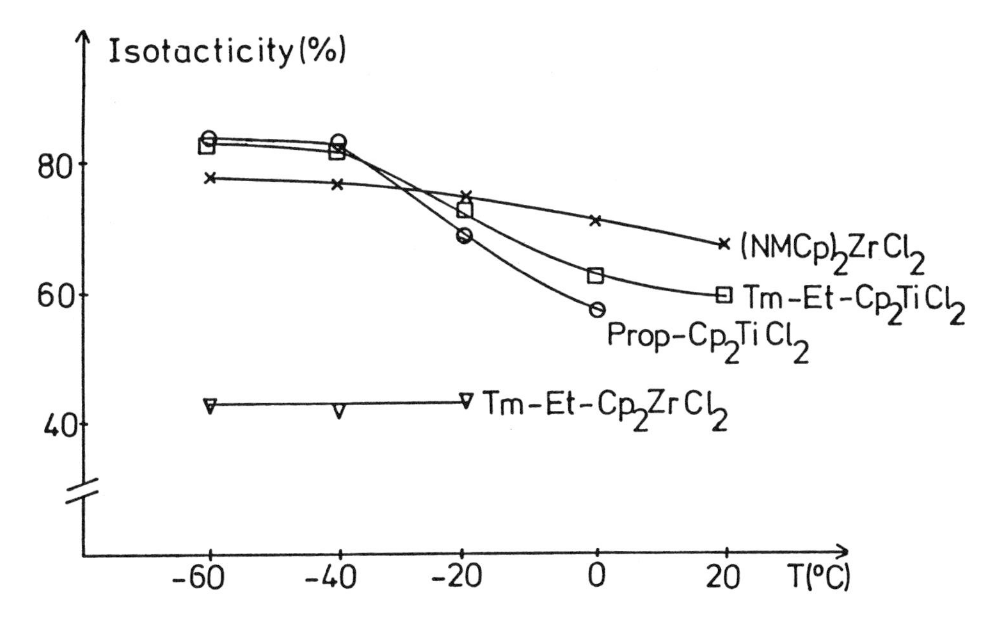

Figure 6. Depencdence of isotacticity from the polymerization temperature of polypropene synthesized with different catalysts

7) P. Pino, R. Mülhaupt: Angew.Chem. 92 (1980) 869
8) W. Kaminsky, K. Külper, H.H. Brintzinger, F.R.W.P. Wild: Angew. Chem. 97 (1985) 507
9) W. Kaminsky in: R.B. Seymour, T. Cheng: History of Polyolefins, Reidel Publ.Co., Dordrecht 1986, p. 257
10) W. Kaminsky, in (Hgs.) T. Keii, K. Soga: Catalytic Polymerization of Olefins, Elsevier-Kodansha, Tokyo 1986, p. 293
11) J. Dutschke, W. Kaminsky, H. Lüker in: Polymer Reaction Engineering (Hg.) K.H. Reichert and W. Geiseler, Hanser Publ., München 1983, p. 207
12) F.R.W.P. Wild, L. Zsolnai, G. Huttner, H.H. Brintzinger: J.Organomet.Chem. 232 (1982) 233
13) W. Kaminsky, K. Külper, S. Niedoba: Makromol.Chem.Macromol. Symp. 3 (1986) 377
14) E. Cesarroti, K. Kagan, R. Goddard, C. Krüger, J.Organomet.Chem. 162 (1978) 297

DISCUSSION

Measurements of the number of active centers with these homogeneous metallocene catalysts are in progress in Pr Tait and Pr Chien's laboratories. Because CO may be incorporated into the polymer, it is better to use the kinetic method. At 0° C about 30-50 % of the Ti or Zr atoms are active sites. Increasing the temperature increases the number of active sites up to 100 % at 50° C. The number decreases after 70° C ; the mesurements are possible using non isospecific catalytic systems which make the polymer to remain soluble.

EXAFS has been used to determine the first two coordination spheres of the active sites while working, precise indications on the mechanism can be obtained ; the presence of an oxygen atom of the aluminoxan in the second coordination sphere of the zirconium may be demonstrated. Among the other methods which can be used to study the homogeneous catalyst, UV and multinuclear NMR have been quoted ; however, in the last case, the method is sensitive to the dominating species which may not be the active species.

These metallocene catalysts are very sensitive to the temperature because of the flexibility of the ligands ; more atactic polymers are produced by increasing the temperature and β transfer is made possible. The flexibility of the ligands has been studied by dynamic methods. However, the bis methylene bridge between the indenyl groups inhibits their movement, so that no inversion of the helical sense were observed with the chiral catalysts. Evidence for agostic interaction in the catalytic cycle were obtained from EXAFS measurements of the soluble catalyst and the Xray structure of similar compounds.

The optically active polymers are only native polymers. Their melting point is about 10° C lower than that of the isotactic polymer prepared using heterogeneous catalysts, probably because they have lower molecular weight. It is not clear why they contain less atactic polymers.

The high amount of aluminoxan needed, as well as the very high dilution are factors that make the industrial applications of these catalysts difficult.

THE MIGRATORY CATALYST: α-OLEFIN POLYMERIZATION BY 2,ω-LINKAGE

Gerhard Fink
Max-Planck-Institut für Kohlenforschung
P.O. Box 10 13 53
D-4330 Mülheim a.d. Ruhr
Federal Republic of Germany

ABSTRACT. With the homogeneous catalyst system nickel(O)compound bis(trimethylsilyl)aminobis(trimethylsilylimino)phosphorane linear α-olefins and singly branched α-olefins can be polymerized. The structure of the poly-α-olefins is unusual. When a linear α-olefin is polymerized, the polymer contains only methyl branches, regularly spaced along the chain with a separation corresponding to the chain length of the monomer. Thus, in the polymer of a linear α-olefin with n CH_2 groups the distance between two methyl branches is (n+1) CH_2 groups. Analysis of the ^{13}C-NMR spectra of polymers of deuterated α-olefins revealed that the growing chain is bound to the next α-olefin via $C_\omega \rightarrow C^2$ linking; C^1 forms the latter methyl branch in the polymer. Taking all results into account, a reaction scheme was developed with which the origin of the special structure of the poly-α-olefin can be explained. The main points of the scheme are:

- The monomer can only insert into a CH_2-Ni bond at the end of the growing chain;
- the insertion is regioselective; only $C_\omega \rightarrow C^2$ coupling of the growing chain with the next monomer takes place;
- the nickel-catalyst complex 'migrates' along the polymer chain between two insertions. During this 'migration' transfer reactions to the monomer can occur, but not insertions.

The 'migration' can be explained in terms of an addition-elimination mechanism via alkylnickel/nickel hydride species with 1,2-hydride shift.

INTRODUCTION

Polymerization of ethene with the homogeneous catalyst system nickel(O)compound/bis(trimethylsilyl)aminobis(trimethylsilyl-

M. Fontanille and A. Guyot (eds.), Recent Advances in Mechanistic and Synthetic Aspects of Polymerization, 515–533.

imino)phosphane leads, according to Keim et al.[1] to short-chain, branched polymers. We have found that this system polymerizes α-olefins: surprisingly the structure of the products is consistent not with the usual 1,2-coupling of the monomers to give a comb-like branched product (Figure 1, above), but with a 2,ω-coupling (Figure 1, below).

RESULTS AND DISCUSSION

Examples of nickel(O)compounds[2]/aminobis(imino)phosphane[3] catalyst are shown in Figure 2. The catalyst components - preferably in equimolar ratio - are employed in situ in the pure liquid monomer or in aromatic solvents. The nature of the ligand of the Ni^{0} compound has no influence on the structure of the α-olefin polymer. This finding and the results of kinetic investigations suggest that the rate-determining step in the formation of the polymerization-active species is the release of the ligand from Ni^{0}.

Examples for the 2,ω-linkage and how it takes place formally are demonstrated in Figure 3. In this way, linear α-olefins and singly branched α-olefins can be polymerized, but not α-olefins with quaternary C-atoms in the chain (in other words: one hydrogen atom at every carbon atom is necessary for the latter mechanistic discussion) or olefins with vinylene or vinylidene groups.

The structure of the formed poly-α-olefins is unusual. The polymer (see again Figure 3, upper line) contains only methyl branches, regularly spaced along the chain with a separation corresponding to the chain length of the monomer. Thus, in the polymer of a linear α-olefin with n $-CH_2$ groups the distance between two methyl branches is (n+1) CH_2 groupe. Thus, their structure is well defined and can be predetermined by selection of the appropriate α-olefin. Thus, for example, the structure of the polymer synthesized from 1-pentene corresponds to that of a rigidly alternating copolymer of ethylene and propylene (see the third line in Figure 3).

The structure of the products was confirmed ^{13}C-NMR spectroscopically; this is demonstrated in Figure 4, for the example poly-2,9-(nonene-1). The assignment of the signals was carried out with the help of the increment rules of Lindemann and Adams[4]. All the signals to be expected for 2,ω-linked α-olefin polymers were found in the spectra in the corresponding intensity ratios.

Analysis of the ^{13}C-NMR spectra of polymers of deuterated α-olefins proofs that the growing chain is bound to the next α-olefin via $C_{\omega} \rightarrow C^{2}$ linking; C^{1} forms the latter methyl branch in the polymer. This is shown in Figure 5 in the case of a polymer from 1-hexene.

The gel permeation chromatographically investigated mol mass distribution in dependence on time and temperature, is

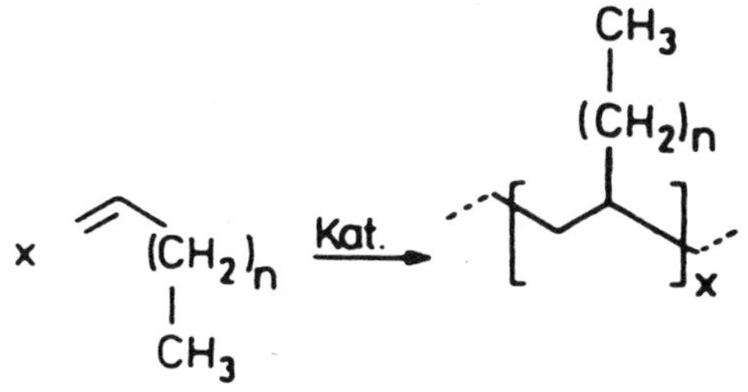

1.2 - linkage

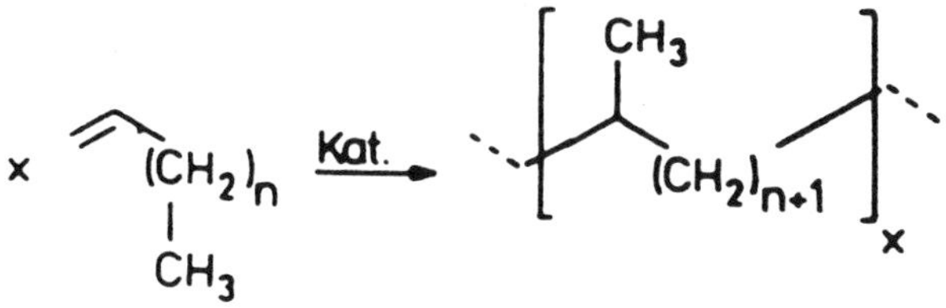

2.ω - linkage

Figure 1. Polymerization of linear α-olefins.
Above: 1,2-linkage of the monomers
Below: 2,ω-linkage of the monomers

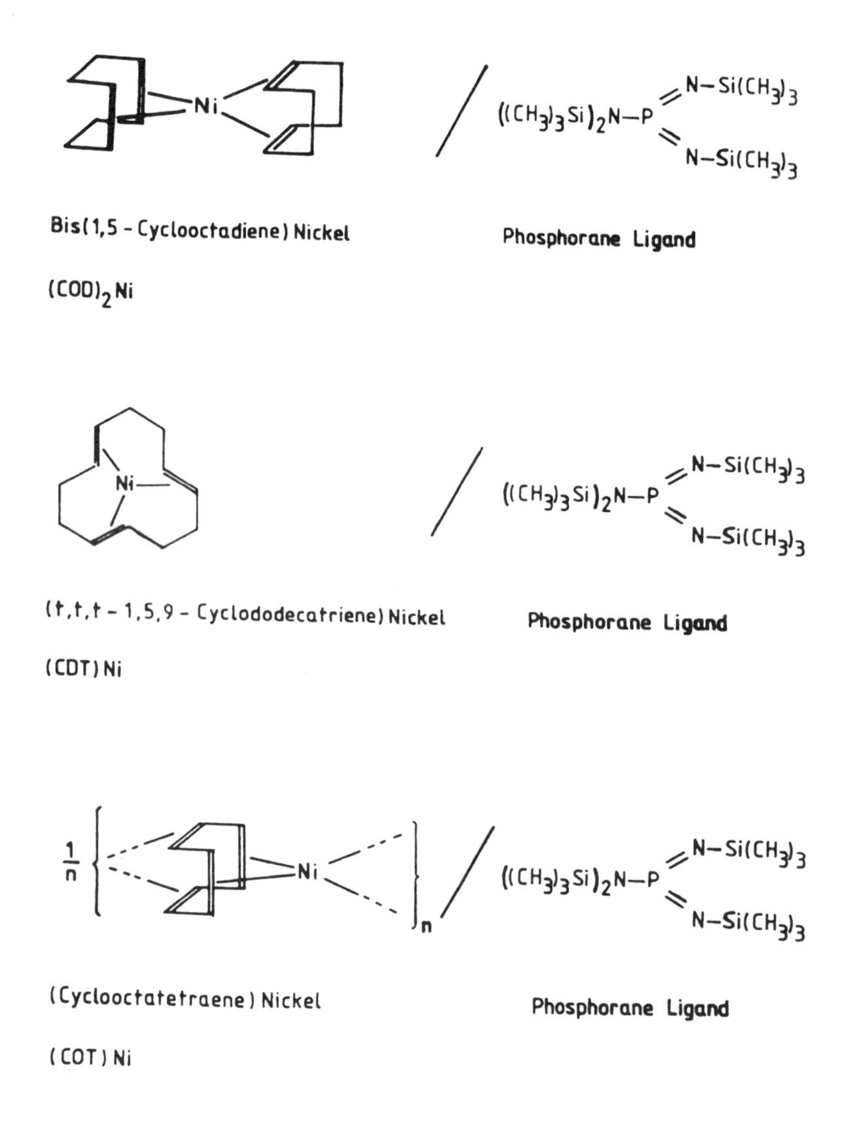

Figure 2. Ni(0)/Phosphorane catalysts.

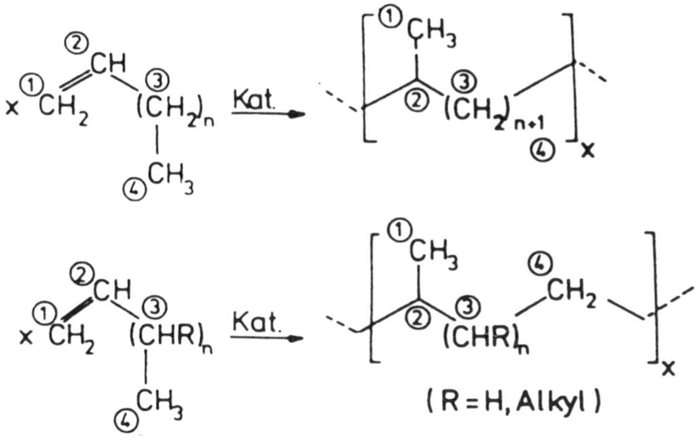

Example: Poly -(1 - penten)
Poly - 2.5 -(1 - penten)

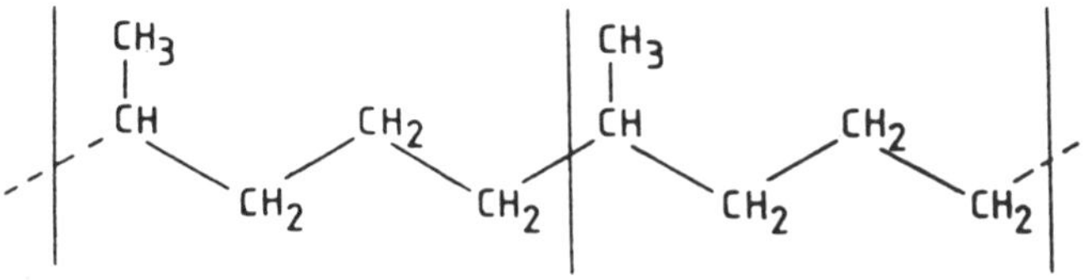

Strongly alternating ethene/propene copolymer

Example: Poly -(4 - methyl - 1 - penten)
Poly - 2.5 -(4 - methyl - 1 - penten)

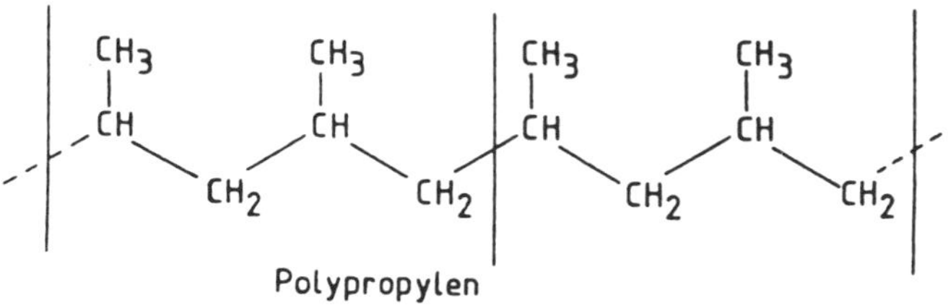

Polypropylen

Figure 3. Examples for the 2,ω-linkage.

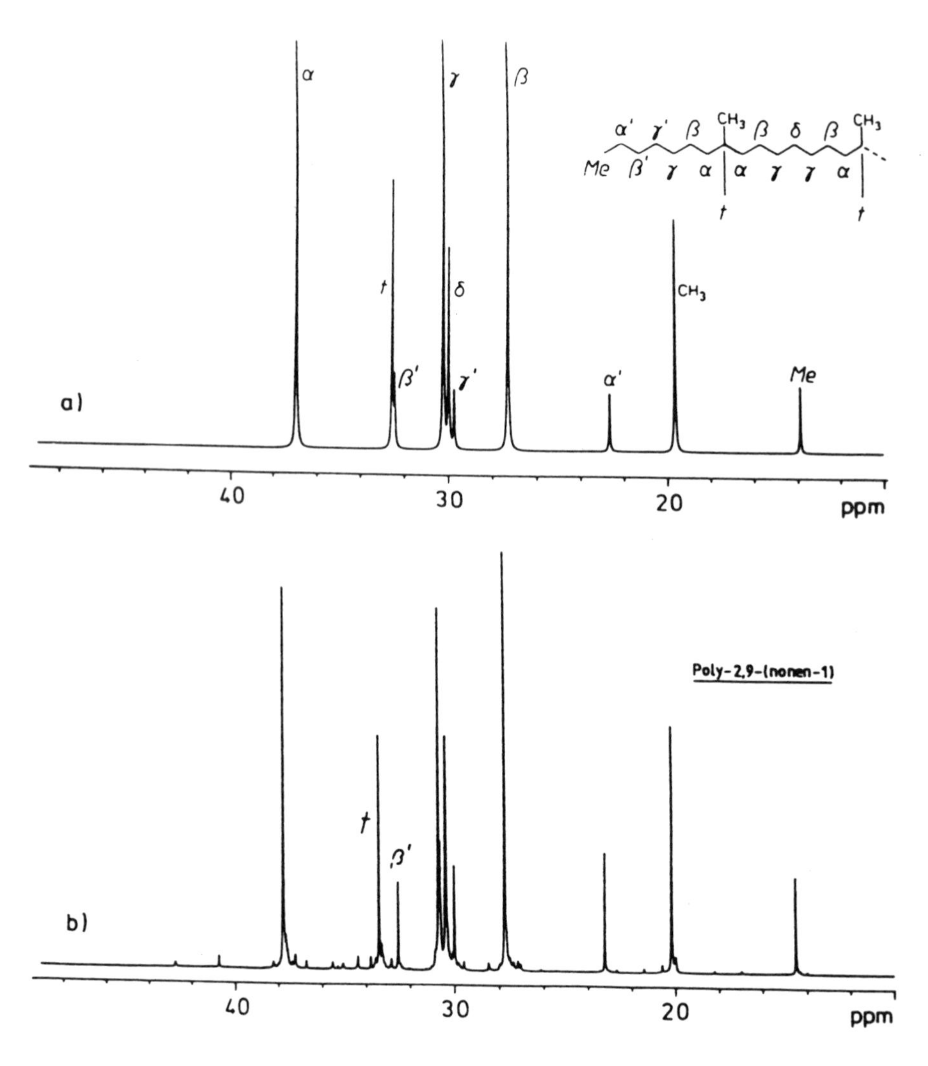

Figure 4. ^{13}C-NMR spectra.
a) Simulation (with end of the polymer chain)
b) Experimental spectrum of poly-2,9-(1-nonene)
(75.5 MHz, benzene-d_6, 303 K)

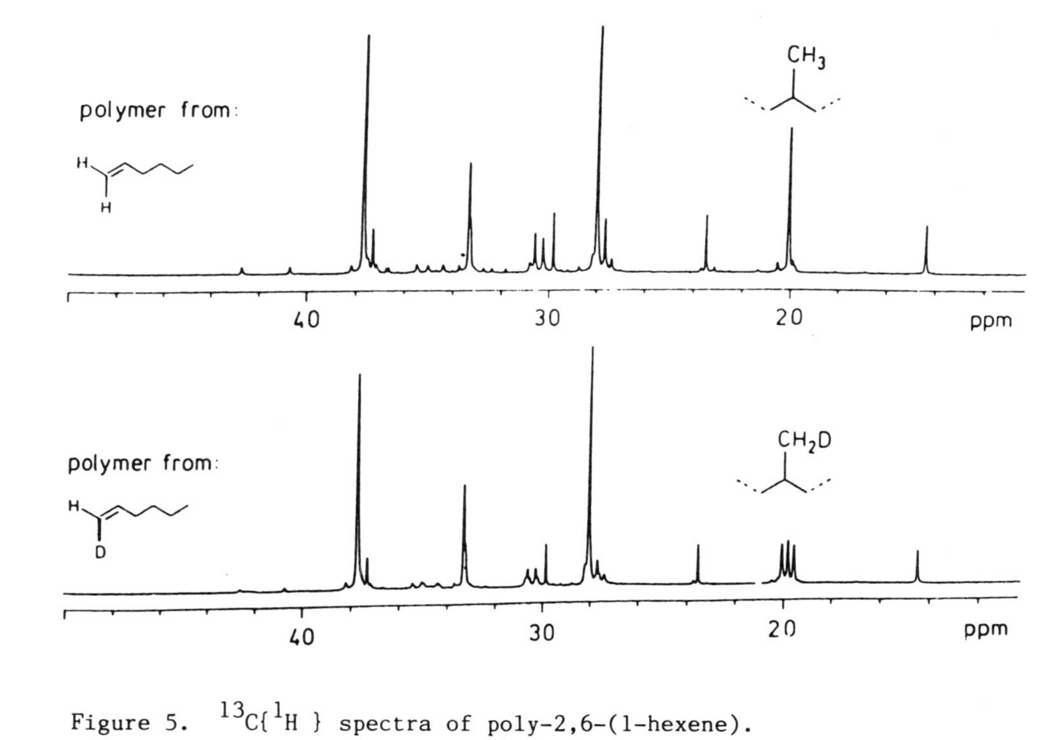

Figure 5. $^{13}C\{^{1}H\}$ spectra of poly-2,6-(1-hexene).

illustrated in Figure 6. At and below 273 K characteristic features of a living polymer are being observed: that is, the molecular weight increases with increasing time (Figure 6, upper part) or conversion, can be regulated by the ratio of monomer to catalyst-concentration, and the mol mass distribution remains narrow.

Above 273 K, for instances at 303 K, as becomes obvious in Figure 6 in the lower part, the molecular weight remains constant with increasing time and the mol mass distribution becomes broadend; at the same time the double bonding in the polymer (vinylene, vinyl, vinylidene groups) is increased. This finding indicates that now, at temperatures above ca. 273 K, transfer reactions to the monomer take place via β-H-elimination.

The fact that there is a living polymer, is confirmed with Figure 7. The time/conversion curve of the hexene-1 polymerization goes on after additon of new monomer.

Taking all the results into account, a reaction scheme was developed with which the origin of the special structure of the poly-α-olefins can be explained. This scheme is shown in Figure 8 and its main points are:

- The monomer can only insert into a CH_2-Ni bond at the end of the growing chain;
- the insertion is regioselective; only $C_\omega \rightarrow C^2$ coupling of the growing chain with the next monomer takes place;
- the nickel-catalyst complex 'migrates' along the polymer chain between two insertions. During this 'migration' transfer reactions to the monomer can occur, but not insertions.

These processes are represented schematically in Figure 8 for the polymerization of 1-butene and looking further, on the transfer reaction from left to right in the scheme, we should note: if the transfer reaction occurs immediately after insertion, then a vinylidine group will be formed. If the transfer reaction occurs during the migration of the nickel-catalyst complex along the polymer chain, a vinylene group will be formed and finally, if the transfer reaction occurs at the end of the growing chain, a vinyl group will be formed.

Taking again into account that one hydrogen atom at every carbon atom in the α-olefin is necessary for polymerization, the 'migration in itself' can be explained in terms of an addition-elimination mechanism via alkylnickel/nickel hydride species with 1,2-hydride shift. This is demonstrated in Figure 9. The intermediate Ni-hydrid could also function as a transferring species (as demonstrated in the box below in Figure 9), whereby the various types of detectable double bonds would be formed corresponding to the momentary position of the nickel-catalyst complex in the chain.

Figure 10 leads to another important detail: the degree of

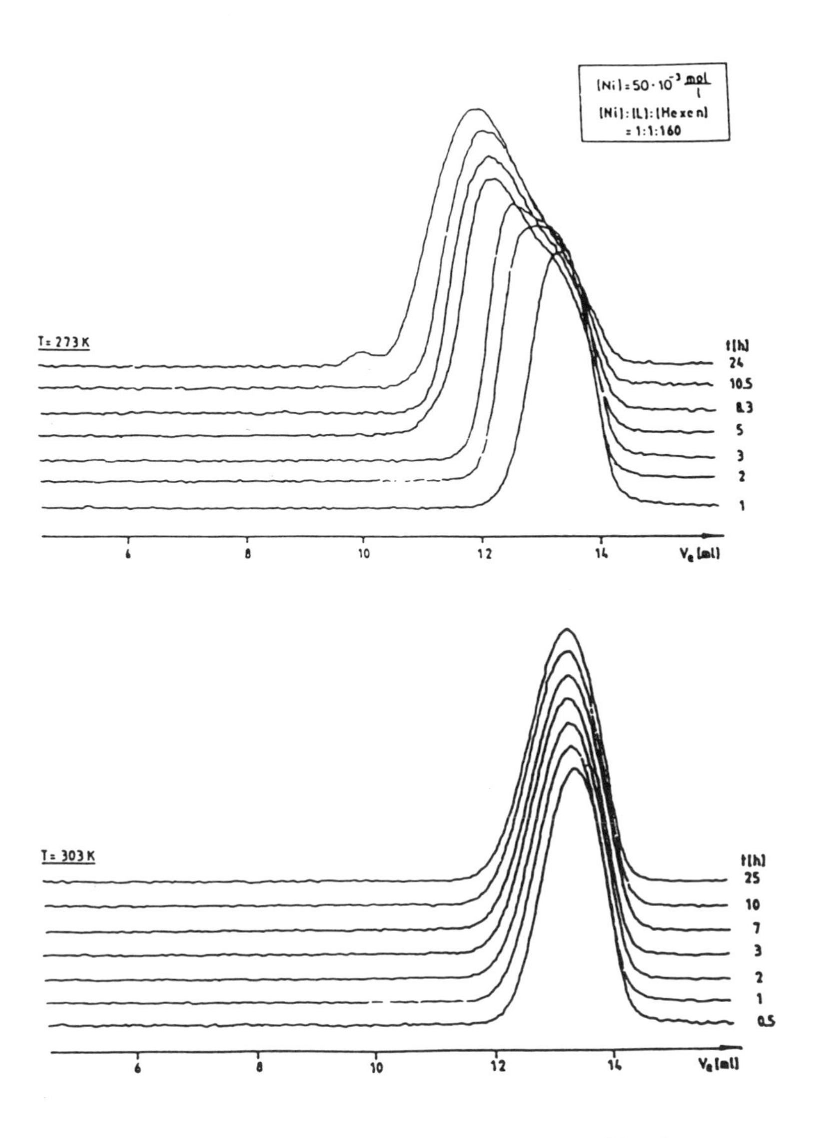

Figure 6. Dependence of the mol mass distribution on temperature and time in the case of hexene-1 polymerization.

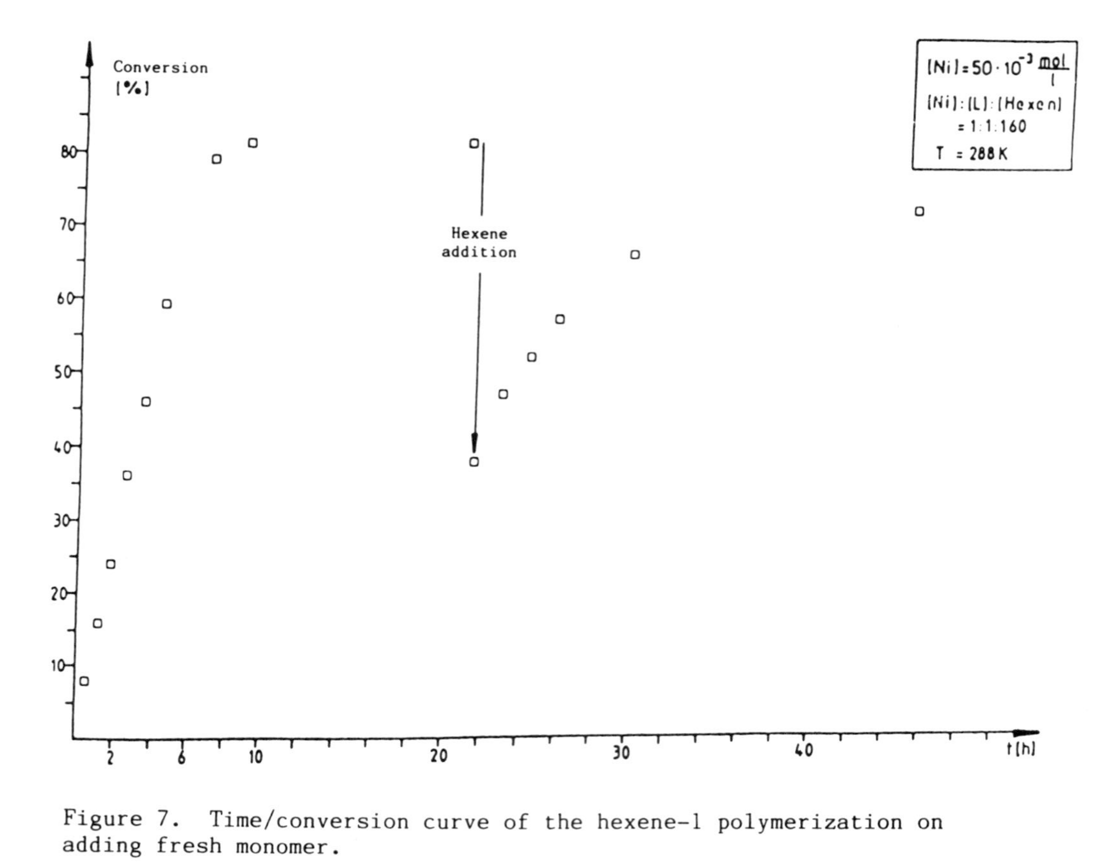

Figure 7. Time/conversion curve of the hexene-1 polymerization on adding fresh monomer.

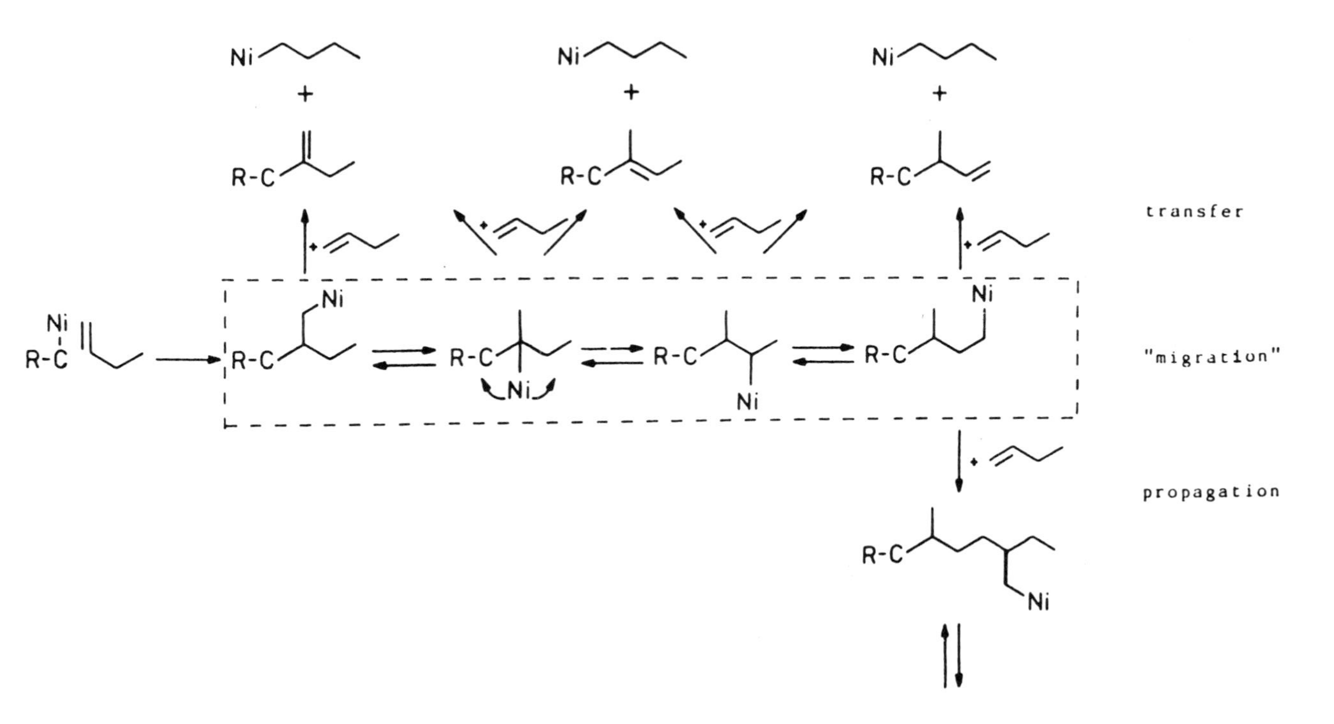

Figure 8. Migration mechanism for the formation of the different structural units found in the poly-α-olefins (here: polymerization of buten-1).

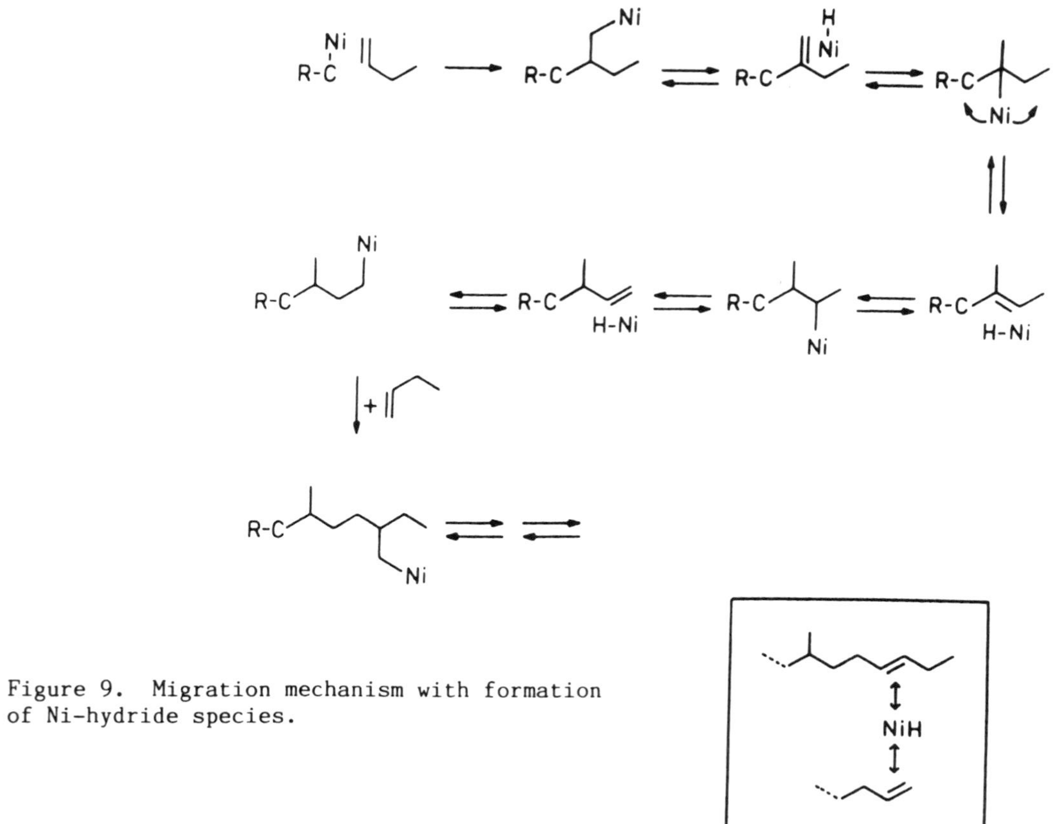

Figure 9. Migration mechanism with formation of Ni-hydride species.

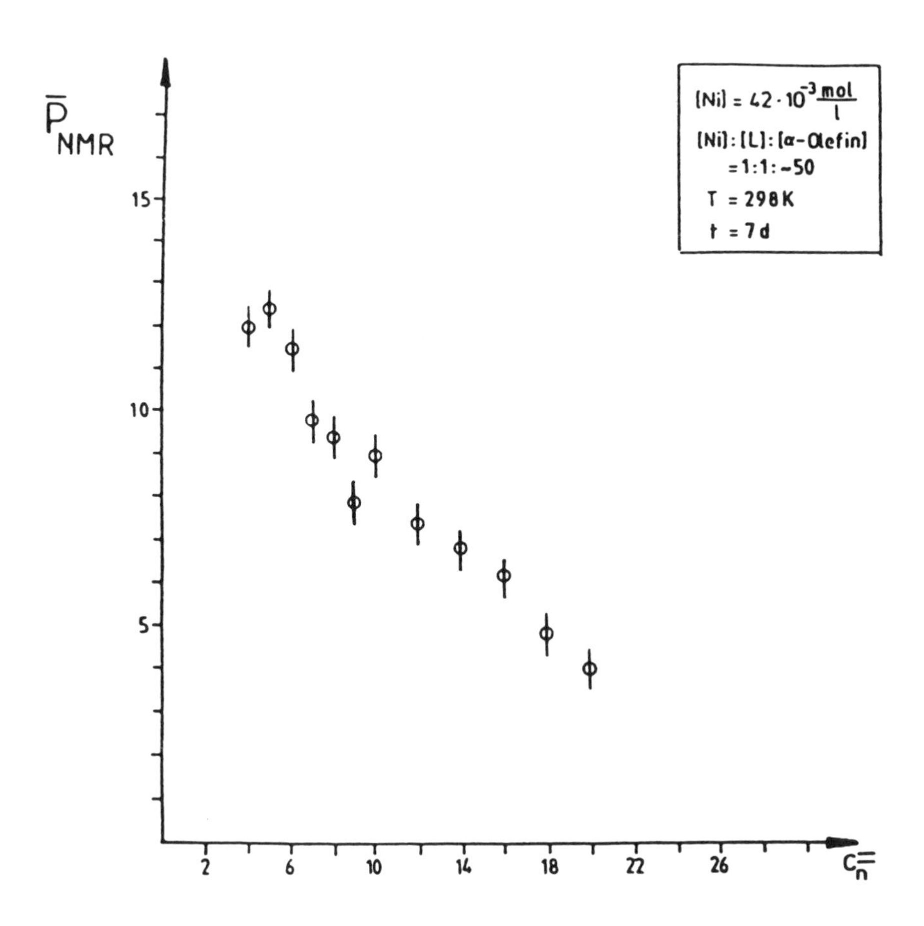

Figure 10. Degree of polymerization as a function of the chain length of the used α-olefin.
Polymerization conditions: $[(COD)_2] = 42x10^{-3}$ M in toluene, $[(COD)_2]$: [Phosphorane] : [α-olefin] = 1:1:50, T = 298 K, t = 7 d.

polymerization (determined from the ^{13}C-NMR spectra, i.e. from the ratio of the signal intensities of the sums of all C-atoms to those of the chain terminal C-atoms) is decreased with increasing chain length of the used α-olefin. The reason for it lies again in the migration mechanism. If each step of the migration is reversible, then statistically, the nickel-catalyst complex will not only move forwards to the 'correct' end of the chain, but also backwards in the direction of the beginning of the formed polymer chain. The ever increasingly longer chain being formed in the polymerization is structured in such a way that the migrating nickel-catalyst complex always finds a similar chemical environment after a certain chain length. This means, any possible influence of the chain end will progressively become smaller, so that there is an equal probability of migration in both directions. In other words, the nickel catalyst 'marks time' within the inner part of the polymer chain.

This behaviour is well confirmed on hand of the results of Figure 11, in which the amount of different double bonds, found in the polymer, is plotted again versus the C-number of the starting α-olefin. Whereas the amount of vinylidene bonds decreases and the amount of vinyl bonds remains constantly low with elongated α-olefin, the amount of vinylene bonds is drastically enhanced with the rising C-number of the used α-olefin. Please remember: the vinylene bond is formed, if the transfer reaction to the monomer takes place from a nickel catalyst position in the inner part of the polymer chain. Figure 12 reveals an other remarkable detail. In the ^{13}C-NMR spectrum of poly-2,5-(pentene-1), for an example, we can discern additional signals that arise from the difference between the beginning (as marked with A) and the end of the polymer chains: at the beginning there are not (n+1) but (n+3) CH_2 groups between the first two methyl branches. The reason for this particularity can be found in the chain start and the first two insertion steps. This is demonstrated in Figure 13. Assuming that we have a Ni-H-species as chain starting species, the insertion of the first α-olefin into the Ni-H bond leads to the first Ni-C-chain. Regioselective insertion, now of the second α-olefin into the Ni-C_1 bond, leads to the branched chain shown in the lowest line in Figure 13. And now, the Ni-catalyst complex migrates to the end of the longer chain. Only in this way is explainable the special structure at the beginning of the chains.

Finally, as a first conclusion, in Figure 14 is summarized our mechanistic insight up to now. Please realize: it is the same reaction coil in which either insertion or transfer or β-H-elimination and subsequent migration may proceed.

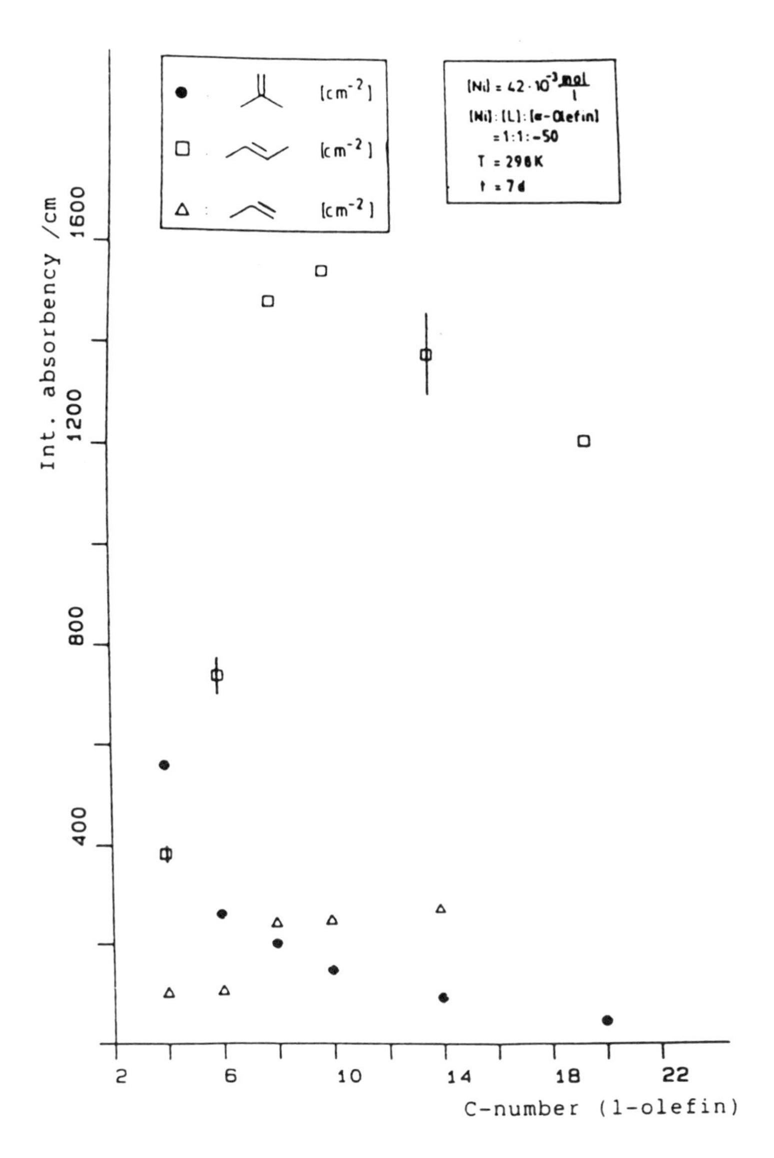

Figure 11. Integral absorbance per cm of the bands at 910 cm^{-1} (vinyl, △), 965 cm^{-1} (vinylene, □) and 890 cm^{-1} (vinylidene, ●) as a function of the polymerized α-olefins chain length. (Polymerization conditions: see Figure 10)

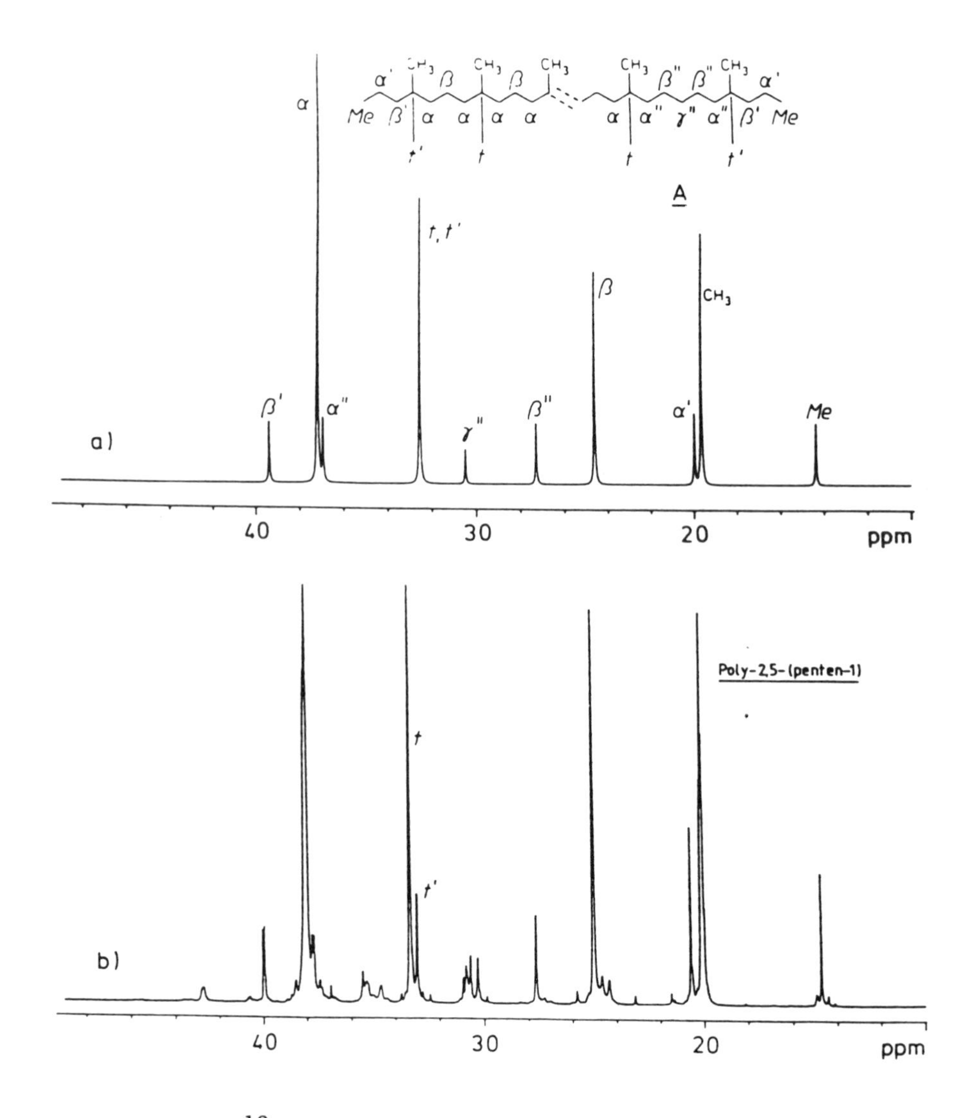

Figure 12. ^{13}C-NMR spectra.
a) Simulation (with end and beginning (A) of the polymer chain)
b) Experimental spectrum of poly-2,5-(1-pentene)

Chain start

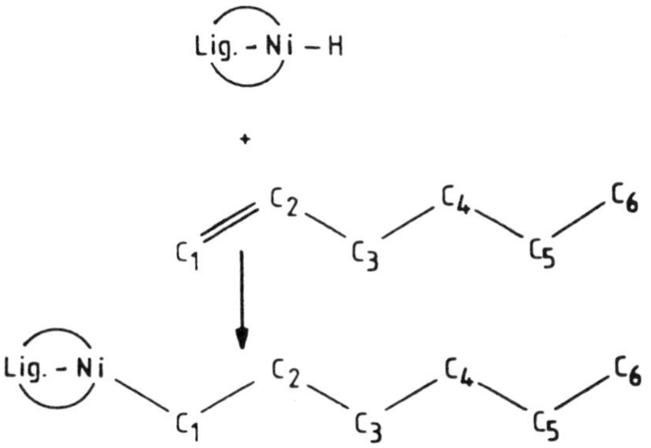

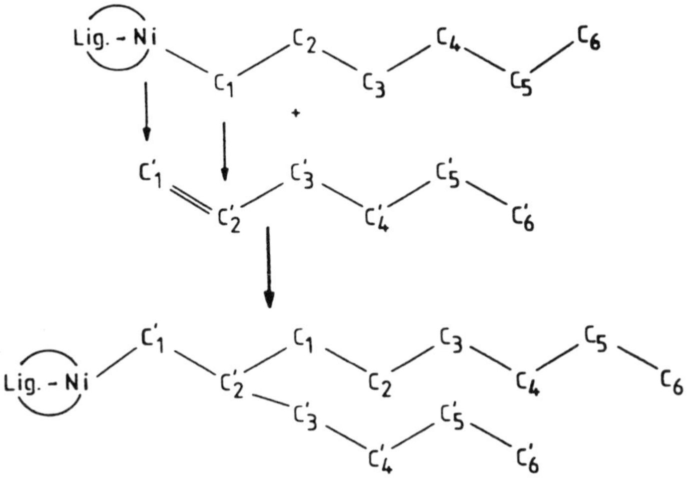

Migration of the Ni-catalyst complex to the end of the longer chain.

Figure 13. Chain start and first insertion steps.

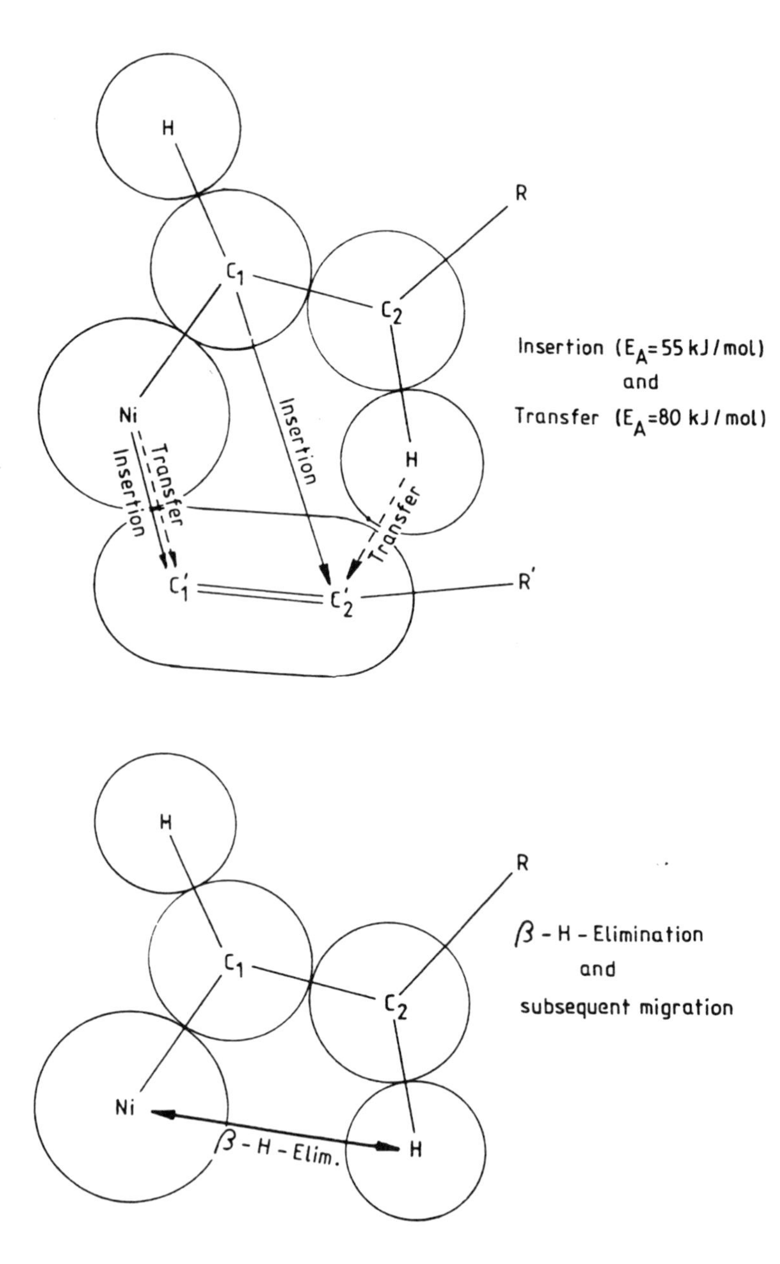

Figure 14. The same reaction coil in which either insertion or transfer or β-H-elimination may proceed.

ACKNOWLEDGEMENT

The author thanks the 'Fond der Chemischen Industrie' for support.

REFERENCES:

1) W. Keim, R. Appel, A. Storeck, C. Krüger and R. Goddard, Angew. Chem., 93 (1981) 91.
2) B. Bogdanovic, M. Kröner, G. Wilke, Justus Liebigs Ann. Chem. 699 (1961) 1.
3) O. J. Scherer, N. Kuhn, Chem. Ber., 107 (1974) 2123.
4) L. P. Lindemann, J. Q. Adams, Anal. Chem., 43 (1971) 1245.

DISCUSSION

The rate determining step is the migration of Ni along the monomers except in the first step, in which it is the insertion of the monomer to a Ni-H bond.

^{13}C NMR spectra show that the Ni catalyst choose to move to the longer end of the molecule.

Experiments are in progress using optically active olefins in order to get more information about polymerization mechanisms and stereochemistry.

A very large number of bidentate ligands : P-N, P-O, S-O... are possible to be used with these kinds of Nickel catalysts, especially in order to explore the possibilities of incorporation of polar monomers, or to work in the presence of water, for instance in emulsion or micellar systems.

NICKEL CATALYSTS FOR ETHYLENE HOMO- AND CO-POLYMERIZATION

Ulrich Klabunde and Steven D. Ittel

Contribution No. 4195 from the
Central Research & Development Department
Experimental Station
E. I. du Pont de Nemours & Company
Wilmington, DE 19898

To be submitted to the Proceedings of the International Symposium on Homogeneous Catalysis, Kobe, Japan, September 22, 1986, Journal of Molecular Catalysis

Summary

Nickel compounds based upon phosphorus-oxygen chelate ligands, $Ph_2PCR{=}CRO^-$, are effective catalysts for the homopolymerization of ethylene to high molecular weight polyethylene. They will also copolymerize ethylene with alpha-olefins, and more importantly, with polar monomers and carbon monoxide.

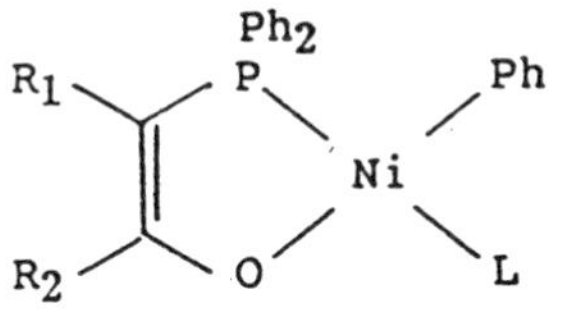

R_1, R_2, Ph_2 P, O, Ni, Ph, L

L = PR_3
Pyridine

When L is PPh_3 or other phosphine ligands, a cocatalyst is required to scavenge the phosphorus ligand. This cocatalyst can be $Ni(COD)_2$, $Rh(acac)(C_2H_4)_2$, or $Me_3N{=}O$. Alternatively, it is possible to prepare "lightly ligated" complexes with ligands like pyridine or "unligated" dimer complexes in which the oxygen

M. Fontanille and A. Guyot (eds.), Recent Advances in Mechanistic and Synthetic Aspects of Polymerization, 535–536.

atoms of the chelate ligands bridge two nickel centers. These species act as single component catalysts.

These catalysts are resistant to a variety of polar molecules such as nitriles, alcohols, and even water. They will copolymerize ethylene with a variety of polar comonomers under mild conditions to yield functionalized linear low density polyethylene (FLLDPE). Introduction of carbon monoxide to an active ethylene polymerization yields a rigorously alternating, 1:1 copolymer of ethylene and CO. If the complexes are exposed to CO before the ethylene polymerization is initiated, CO inserts into the Ni-Ph bond giving the benzoyl complex. The reaction continues, giving Ni(0) carbonyl complexes and the benzoic ester of the vinyl chelate ligand, $Ph_2PCH{=}C(Ph)OC(O)Ph$.

PENDING QUESTIONS IN COORDINATION POLYMERIZATION OF DIOLEFINS

Ph. TEYSSIE, P. HADJIANDREOU, M. JULEMONT and R. WARIN
Laboratory of Macromolecular Chemistry and Organic Catalysis
University of Liège - Sart-Tilman - 4000 LIEGE - BELGIUM

ABSTRACT : Recent advances in transition metal initiated polymerization of conjugated dienes are discussed, with emphasis on results obtained using simple η^3-allyl nickel complexes that are good models for Ziegler-Natta type catalysts. Besides a brief reassessment of the mechanism, three types of data will be envisioned, that have not been completely clarified or implemented as yet : 1) The possibility to master a "living" coordination polymerization of butadiene, and to use it for the synthesis of new (stereo)block copolymers; 2) The control of a highly regioselective 1.4 polymerization, with a stereoselectivity ranging from very pure cis to very pure trans, and depending not only on temperature and complex ligand field but also unexpectedly on initiator concentration; 3) The existence of a dynamic situation allowing to "code" the distribution of the cis and trans isomeric units in chains of equibinary composition, from a purely random placement to a highly stereoblock one with interesting bulk properties. As far as possible, these data will be related to those obtained with Ziegler-Natta systems.

1. INTRODUCTION

The understanding of coordination polymerization mechanisms of conjugated diolefins has remarkably progressed over the last two decades (1-2). Important fundamental advances came from the realization that in those polymerizations, chain growth is controlled by an η^3-η^1 allyl isomeric equilibrium (Scheme 1) due to the structure of the active transition metal (M_T) site itself (3).

M. Fontanille and A. Guyot (eds.), Recent Advances in Mechanistic and Synthetic Aspects of Polymerization, 537–545.

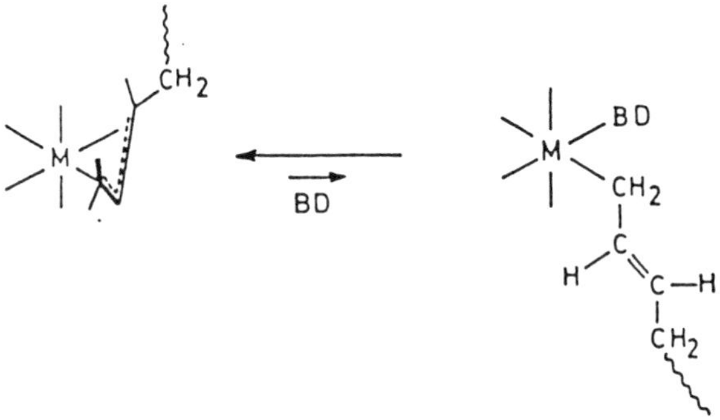

Scheme 1

Since η^3-allyl M_T complexes (also called π-allyl) are rather easy to synthesize, it was possible to isolate some particularly active ones, and to study their catalytic behaviour in a clear and systematic manner (4). These investigations have put in evidence the following characteristics (5-6) :

- propagation proceeds by a typical insertion process, the determining step being usually (besides polynuclear complexes dissociation) the conversion of the η^3 isomeric form into the η^1 one;
- polymerization rate and regio-stereo-chemistry are strongly determined by the ligand field around the M_T;
- initiation is relatively fast, and favored by a high monomer concentration;
- spontaneous chain termination occurs by typical β (or 5) elimination (hydride transfer).

Judging from the similarity of the processes and of the resulting products of high isomeric purity, there is little doubt that these conclusions can be extended to less defined Ziegler-Natta type systems. However, such model catalytic studies have also revealed three new facets of these diolefins coordination polymerizations, that are summarized hereafter and certainly deserve further in-depth investigation.

2. THE CONTROL OF THE "LIVING" CHARACTER

It had already been shown that steric blocking (apex) of M_T complex positions, prone to promote transfer eliminations, could convert oligomerization catalysts into polymerization ones (7). On these bases, different ligands were tested which proved to be quite efficient on maintaining the living character of the growing chains.

In particular, an equimolar amount of chloranil added to bis(trifluoroacetato-allylnickel), or $(ANiTFA)_2$, ensures a perfectly living process (8). That possibility is illustrated by the usual DP versus monomer-catalyst ratio linear relationship, and by polymerization resumption experiments. An intriguing fact is the relatively broad M.W. distribution of the resulting polydienes : at least 1.35 ($\bar{M}_w/\bar{M}_n$); it might however be explained by the interference of determinant dissociation-association phenomena in the catalytic species (including the η^3-η^1 isomerism).

Anyhow, that situation allowed the synthesis of new block copolymers with good efficiency (8); namely a distereoblock poly (cis-b-trans)butadiene obtained by adding an equimolar amount of a trans-directing phosphate ligand after the polymerization of a first crop of butadiene and then resuming polymerization; and a poly (1.4 butadiene-b-styrene) diblock, which upon hydrogenation yielded a poly(linear ethylene-b-styrene) diblock, an excellent emulsifier for polystyrene-high density polyethylene blends (9).

Most likely, these living systems should also provide an interesting approach to end-functionalization of polydienes with high isomeric purity. It would obviously be very worthwhile to also extend these possibilities to Ziegler-Natta catalysts.

3. THE CONTROL OF THE CIS-TRANS ISOMERISM

M_T complexes having a large orbitals extension (i.e. rare earth or uranium complexes (10)), or strongly electron-withdrawing ligands (11)(i.e.$(ANiTFA)_2$), promote in paraffinic solvents the formation of very high cis 1.4-polydienes (up to 99^+%), characterized by a relatively high degree of crystallinity and melting point (above 0°C).

Blocking coordination positions by strong electron-donating ligands (i.e. triphenylphosphite) charges the stereochemistry to a pure trans 1.4 one (12), at the expense of the polymerization rate though; this latter can be partially restored however by the addition of controlled amounts of electron-withdrawing additives such as chloranil. These stereochemical controls have been investigated in detail (in particular in the elegant experiments (3) of Porri), and ascribed to the fact that the cis isomer is produced by cisoïd bidentate coordination of the monomer on the same M_T yielding an intermediate anti-η^3 allyl complex; when ligands are added, only monodentate transoïd coordination of the monomer (more stable form) is allowed, producing the trans isomer through the syn-allyl isomer (another recently documented (14) but less likely possibility being a transoïd coordination on two μ-bonded metal atoms).

It has now been shown that the cis-isomer content could be increased significantly even in systems producing cis-trans mixtures simply by decreasing catalyst concentration (11). That seems to be a quite general phenomenon : i.e. $TiCl_4$ and $TiBr_4$ based Ziegler-Natta catalysts have so yielded (2) high cis products (over 90%), and the same behaviour has even been reported for lithium alkyl initiated polymerizations (15); the most obvious interpretation being that dilution dissociates catalyst dimers or aggregates, so liberating vicinal coordination vacancies favouring the cisoïd coordination of the monomer, thus the formation of cis 1.4 polymers.

4. THE CONTROL OF THE STATISTICS OF INSERTION : AN EXAMPLE OF A "CHRONOSELECTIVE CODING" PROCESS

When complexes such as $(ANiTFA)_2$ are used in solvents other than paraffins, i.e. benzene or chlorobenzenes they promote a peculiar mixed microstructure (16), usually close to 50% cis and 50% trans units (and therefore called "equilinary"); it does not represent a thermo-dynamic cis-trans content, and its reversible occurence depends on parameters typical of coordination equilibria involving a particular species, at least between certain boundaries. A high resolution H-NMR triads analysis (17) unexpectedly showed that isomeric units distribution in polybutadiene could go from purely random when prepared in benzene (Bernoullian statistics), through partially alternate in chlorobenzene (Markov 1st order), to stereosequential when using CH_2Cl_2. Changing the complex structure to $(ANiCl)_2$, again in CH_2Cl_2, led to a pure multistereoblock structure, poly(cis 1.4-b-trans 1.4)$_n$ butadiene with a block mean DP of 50 (11). Such modifications obviously have a profound influence on the resulting products bulk properties. The polymer with random units distribution is perfectly amorphous (T_g = - 105°C), while the stereoblock is an example of a "thermoplastic elastomer" (or at least a high green strength rubber)

obtained in one step from one monomer : it exhibits all of the thermal transitions of both pure cis and pure trans 1.4 polybutadienes.
Although a detailed understanding of such phenomena has not yet been acquired, they have been experimentally related to the dominant dynamic feature of these binuclear η^3-allyl complexes, i.e. the constant and fast opening and reforming of the M_T-counteranion bond, offering temporarily vicinal coordination vacancies on one side of the complex, which can then be viewed momentarily as an (allyl-M)$^{\delta+}$x (allyl-M-X)$^{\delta-}$ species (18) : this situation has been monitored in ^{1}H-NMR spectrometric investigations (6), which have also shown that the opening of the η^3 allyl structure into an η^1 active one was much slower. It can then be speculated (Scheme II) that if monomer insertion (in cisoïd or transoïd form depending on the complex side), during the short time of existence of the η^1 allyl structure, is faster than the M-X bond dissociation, a sequential placement will result; while a random stepwise insertion will occur if insertion is slower than the dissociation-reassociation process.

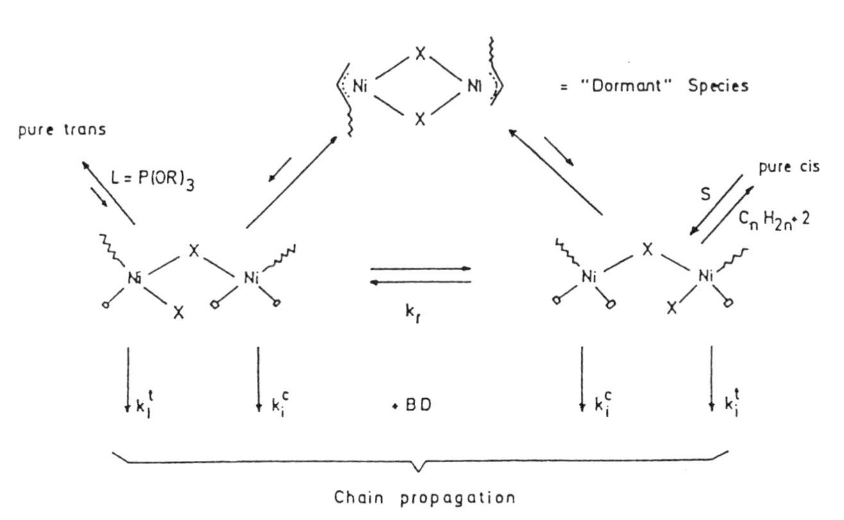

Scheme II

This hypothesis has received some support from copolymerizations of methylmethacrylate with isoprene (19) using an "equibinary cis 1.4 - 3.4" catalyst of this latter monomer polymerization; significantly, the MMA units replace only (when increasing their concentration) the 3.4 isoprene units and not the cis 1.4 ones.
It clearly appears that we have here a real (although simple) binary code, that can be used to modify the properties of 1.4 polydienes. It might also explain some intriguing features of 1.4 polybutadienes produced by Ziegler-Natta catalysts, and wherein the degree of crystallinity and the melting points (in the 70-90% cis range) can be significantly different for an identical microstructure (2), maybe due to the presence (or not) of long sequential placements of the isomeric units.

In conclusion, it can be said that despites our improved knowledge of diolefins coordination polymerization, there are still, even if at a more subtle molecular level, interesting new challenges in terms of both mechanistic analysis and synthesis of new products.
Moreover, such studies certainly can lead to a better understanding and a closer mastering of the action of the classical Ziegler-Natta systems.
The interested reader may find a detailed analysis of the mentioned experimental results in reference (2).

REFERENCES

1. The reader is referred to numerous recent reviews, e.g. :
   a) S. Horn, in "Transition Met. Catal. Polym.", MMI Press Symp. Ser. 4B, p. 527 sq. (1983)
   b) Ph. Teyssié et al. (and references therein) in "Preparation and Properties of Stereoregular Polymers", R. Lenz and F. Ciardelli Ed., D. Reidel Publ. Co., London, p. 131 sq (1980); and Ph. Teyssié and F. Dawans in "Stereo Rubbers", W.M. Saltman Ed., J. Wiley, New-York, p. 79 sq., (1977)
   c) L. Porri, in "Proc. Int. Symp. Macromol.", E.B. Mano Ed., Elsevier, Amsterdam, p. 77 sq. (1975); and in "Struct. Order Polym", F. Ciardelli and P. Giusti Ed., Pergamon, Oxford, p. 51 sq. (1881)
   d) J. Furukawa, Pure Appl. Chem., 42, (4) 495 (1975)

2. Ph. Teyssié, P. Hadjiandreou, M. Julémont and R. Warin, in "Transition Metal Catalyzed Polymerization", R. Quirk, Editor, Cambridge University Press, 1987.

3. a) G. Natta and G. Mazzanti, Tetrohedron 8, 86 (1960); and ref. 1b
   b) P. Cossee, in "Stereochemistry of Macromolecules", D. Ketley Ed., M. Dekker, New-York 1967, p. 145 sq.
   c) R.P. Hughes and J. Powell, J. Am. Chem. Soc., 94, 7723, (1972)
   d) G. Wilke et al., Angew. Chem., Int. Ed., 5, 151, (1966)
   e) F. Dawans and Ph. Teyssié, a review, Ind. Eng. Chem. Res. Dev., 10 261 (1971)
   f) B.A. Dolgoplosk et al., Izv. Akad. Nauk SSSR, Ser. Khim., 344 (1970); Vysokomol. Soedin., All, 1645 (1969)

4. a) J.P. Durand, F. Dawans and Ph. Teyssié, J. Polym. Sci., Al(8), 979 (1970)
   b) F. Dawans and Ph. Teyssié, J. Polym. Sci., B7, 111 (1969)

5. a) R. Warin, Ph. Teyssié, P. Bourdauducq and F. Dawans, J. POlym. Sci., B11, 177 (1975)
   b) J.M. Thomassin, E. Walckiers, R. Warin and Ph. Teyssié, J. Polym. Sci., Al, 13, 1147 (1975)

6. R. Warin, M. Julémont and Ph. Teyssié, J. Organomet. Chem., 185, 413 (1980); J. Molec. Cat., 7 523 (1980)

7. a) Y. Doi, S. Veki and T. Keii, Makromol. Chem., Rap. Commun., 3, 225 (1982)
   b) G. Wilke, in "Fund. Res. in Homog. Cat.", Vol.3, M. Tsutsui Ed., Plenum Publ. C° (1979)

8. P. Hadjiandreou, M. Julémont and Ph. Teyssié, Macromolecules, 17, 11 (1984); Makromol. Chem., in press.

9. R. Fayt, P. Hadjiandreou and Ph. Teyssié, J. Polym. Sci., Polym. Chem. Ed., 23, 337 (1985)

10. a) G. Lugli, W. Marconi, A. Mazzei, N. Paladino and U. Pedretti, Inorg. Chim. Acta, 3 253 (1969); G. Lugli, A. Mazzei and S. Poggio, Makromol. Chem., 175 2021 (1974)
    b) H.L. Hsich and H.C. Yeh, Rubber Chem. Technol., 58(1), 117 (1985)
    c) N.G. Marina, N.V. Duvakina, Yu.B. Monakov, U.M. Dzhemilev and S.R. Rafikov, Vysokomol. Soedin., A27 (6) 1203 (1985)
    d) Quyang Jun, Wang Fusong and Huang Baotong, in "Transition Metal Catal. Polym.", M.M.I. Press Symp. Ser. 4, p.265 (1983)

11. P. Hadjiandreou, PhD Thesis, University of Liège 1980

12. See e.g. J.P. Durand, F. Dawans and Ph. Teyssié, J. Polym. Sci., B5 785 (1967); ref. (6); and ref. (4) (this list)

13. See e.g. L. Porri et al. in ref. $1C_2$, p.51; Proc. IUPAC Int. Sym. Macrom., Amherst, p.239 (1982); Makromol. Chem. Rapid Com., 4, 485 (1983) and 5 679 (1984); "Transition Met. Catal. Polym.", M.M.I. Press Symp. Ser. 4B, 555 (1983).

14. See e.g. H. Yasuda, K. Tatsumi and A. Nakamura, Acc. Chem. Res., 18, 120 (1985)

15. M. Morton and J.R. Rupert, in "Initiation and Polymerization" A.C.S. Sym. Ser. 212, p.283 (1983)

16. Ph. Teyssié, F. Dawans and J.P. Durand, J. Polym. Sci., C22, 221 (1968); J.C. Maréchal, F. Dawans, and Ph. Teyssié, J. Polym. Sci., A1(8) 1993 (1970); see also ref. ($1b_1$)

17. E.R. Santee, V.D. Mochel, M. Morton, J. Polym. Sci., B11, 453 (1973)

18. J.J. Eisch, A.M. Piotrowski, S.K. Browstein, E.J. Gabe, and F.L. Lee, J. Am. Chem. Soc., 107, 7219 (1985)

19. F. Dawans and Ph. Teyssié, Makromol. Chem., 109, 68 (1967); Europ. Polym. J., 5 541 (1969)

DISCUSSION

A factor of 1000 in the overall polymerization rate may be observed by changing the electron with drawing character of the anion. The effect of the ligand is to help the dynamic opening of the active bond which is normally engaged in an association between two catalyst moieties, as shown by cryoscopic measurements.

The system is living because Mw increases with conversion. However, the polydispersity index being higher than 2, some termination must occur. The sites are distributed in between two forms ; a π-allyl form which is sleeping and a σ form which is active.

The blockiness may be roughly estimated from NMR. The formation of cis-trans stereoblock may be explained by the mechanism of Ni-X opening, but conversely to the random case, the 50-50 composition is not fixed and a range of cis-trans ratios can be obtained in the butadiene polymerization.

Application of a magnetic field has not been tried, but it might change the stereoregularity since the opening of the Ni-X bond depends on electronic and dielectric factors.

REPORT OF DISCUSSIONS AND RECOMMENDATIONS

Millions of tons of polyolefins are produced each year by coordination catalysts and also large amount of synthetic elastomers. However, except in the case of diene polymerization catalysts, our knowldege of the active sites remains very poor. Although most of the industrial catalysts are heterogeneous, very promising advances have been recently obtained using both bimetallic or monometallic homogeneous catalysts. The basic research concerning diene polymerization catalysts seems less active at the moment and should be revived again.

Industrial Catalysts

High activity heterogeneous catalysts for ethylene and propylene polymerization are completely adequate concerning :

1 - Activity
2 - Stereospecificity
3 - Morphology of the resulting polymer.

However, as shown by the long discussion related to these catalysts, a number of points concerning the mechanisms of their action, the exact roles of the components ($MgCl_2$ support, the Lewis bases, even the alkyl Al), the differences in the environment of the various kinds of active sites remain obscure.

A significant improvement would result from a better control of the uniformity of the catalytic sites which might lead to :
a) the synthesis of polymers with narrow MWD
b) the synthesis of copolymers with a uniform distribution of the monomers (LLDPE and EPR)

A better knowldege of the chemical and structural characteristics of the catalytic center is necessary to achieve this result. In this way, EXAFS and solid state NMR could be useful instruments provided that catalysts are studied with high efficiency (high proportion of active Titanium atoms).

M. Fontanille and A. Guyot (eds.), Recent Advances in Mechanistic and Synthetic Aspects of Polymerization, 547–551.

Homogeneous catalysts

Homogeneous catalysts are especially investigated in an academic way to look at the mechanism and side reactions. The solubility makes it easier to measure the number of active sites. All this helps to optimize also heterogeneous catalysts.

But in the last time homogeneous Ziegler-Natta catalysts also became more industrially interesting. Elastomers like ethylene-propylene copolymers, EPPM are synthetized with homogeneous vanadium, neodym or zirconium catalysts. There is an interest to supported soluble catalysts to handle them more easily or use build-up plants.

The zirconocene/aluminoxane system offers the possibility to produce only by changing the ligands pure atactic polymers, stereoblock isotactic or highly isotactic polyolefins with a high activity. For the first time it is possible to make optically active oligomers or polymers in the solid state from only prochiral -olefins. The stereoselectivity is higher than that with heterogeneous catalysts.

In the future, homogeneous catalysts will not only give a deeper view into the polymerization mechanism but will also be used to produce special copolymers with new properties. For example, copolymers like isotactic polypropylene/ethylene blocks can substitute PVC. Copolymers of longer chain -olefins (hexene, octene) together with ethylene or propylene are of industrial interest. One point to investigate this new polyolefin is the waste treatment and the fact that there will be no poison upon burning them.

Bridges : Homogeneous-heterogeneous systems

From the practical point of view, the extension of the progress in coordination chemistry might open the heterogeneous polymerization to new polymers and copolymers and reduce their sensitivity to poisoning.

From the basic point of view, it might be expected that the unsolved questions in heterogeneous

catalysis should be solved by studies in homogeneous medium :

- nature and distribution of active centers
- bimetallic or monometallic active centers
- oxidation state of the transition metal (role of the ligands)

However, the generalisation to heterogeneous catalysts of the knowledge gained about their homogeneous analogs should be made very cautiously, because the criteria of validation are not safely established.

New monometallic catalysts

A new family of catalysts, based on Nickel compounds with bidentate P-O, P-N, P-S chelated phosphine ligands have appeared recently, which possess very exciting properties : they can be handled in the presence of polar groups, they allow to copolymerize ethylene with functional monomers, generally unconjugated through the use of spacers ; they allow new kinds of polymerization mechanism with migration of the active site along the chain, giving regularly spaced methyl branches ; they allow to prepare with only ethylene monomers by "in situ branching" polyolefin with regularly spaced methyl and long branches, and by this way one gets both the control of crystallinity and flow properties ; finally, one gets the possibility of preparing block-copolymers between -olefins and ethylene. Some of these systems are living systems. However we are far from understanding the reasons for these very interesting properties. Obviously, very active research should be devoted to study the mechanism and the properties of these and other new monometallic catalysts.

Prospects in diolefin polymerization

From a practical point of view, the more interesting present trend is the use of rare earth catalysts that can provide for very high stereoregularity (i.e. cis 1.4) even in C_4 cuts (presence of olefins) and under rather agrêssive conditions.

In terms of products, two main points of interest appear now, that should be developed :

- the possibility of controlling living processes, leading to interesting new block copolymers and possibly end functionalized macromolecules ;
- partial control of multi-stereoblock situations, promoting new properties in polydienes based on classical monomers.

The soluble model catalyst for diene polymerization where most metal atoms are active, provide also unique opportunities for mechanistic and structural studies, the results of which can be extrapolated to Z-Ni type systems :

- spectroscopic (mainly multinuclei high resolution NMR) ; study of the structure and dynamics of the catalysts "in situ" ;
- fine study of the structural controls (from analysis of microstructure and statistical distribution of units in the chain, as related to catalyst structure) ;
- detailed kinetic description of the reactions involved.

All these points should be still further developed since the knowledge acquired will be very useful for the understanding of coordination polymerization in general.

RECOMMENDATIONS

1 - More active cooperation between scientists working in coordination chemistry, metathesis, and Ziegler-Natta catalyst fields.

2 - Study of Homogeneous catalysts as models for industrial catalysts through kinetics and spectroscopic measurements, to clarify the various roles of the different components.

3 - Establish criteria of validity for transposing the mechanism of model homogeneous catalysts to industrial catalysts. Establish simpler models of heterogeneous catalysts.

4 - Improve the methods of characterization of both the active sites in heterogeneous catalysts (EXSAFS, Solid NMR...) and the polymers (especially the very high molecular weight polymers).

5 - Anchor new homogeneous catalysts on suitable supports for industrial developments.

6 - Develop systems with uniform distribution fo active sites, and also systems with wide distribution of sites leading to broad but unimodal molecular weight distribution or copolymer composition.

7 - Develop the study of monometallic Ni/chelate ligand catalysts to reach new well defined polyolefin structures.

8 - Develop new catalysts (either mono- or bimetallic) for preparing functional copolymers of olefins.

10 - Study the new rare earth catalysts for diene polymerizations able to work with crude monomers.

11 - Develop new well controlled living systems to get stereosequences in diene and diene-olefin polymerization.

12 - Develop the use of the more advanced instruments (especially NMR) to understand the dynamic details of the polymerization mechanism.

INDEX

The number refers to the first page of related text

A

B

C

D

E

F

G

H

I

L

O

P

T

U

V

Z

LIST OF CONTRIBUTORS

M. AGLIETTO
University of Pisa - PISA (Italy)

T. AIDA
University of Tokyo - TOKYO (Japan)

E. ALBIZZATI
Montedison Research Center - NOVARA (Italy)

R.D. ALLEN
Virginia Polytechnic Institute - BLACSKSBURG,Va (USA)

D.G.H. BALLARD
ICI Laboratory - RUNCORN (Cheschire) (U.K.)

F. BANDERMANN
Universitat Gesamthorschule - ESSEN (F.R.G.)

P.C. BARBE
Natta Research Center - FERRARA (Italy)

S. BOILEAU
Collège de France - PARIS (France)

M. BUSCHERMOHLE
Universitat Hambourg (F.R.G.)

F. CANSELL
Université Paris XIII - VILLETANEUSE (France)

F. CIARDELLI
University of Pisa - PISA (Italy)

E. CHIELLINI
University of Pisa - PISA (Italy)

R.P.J. CORRIU
Université des Sciences et Techniques du Languedoc
MONTPELLIER (France)

A. DIGUA
Université Paris XIII - VILLETANEUSE (France)

W. DRENTH
Organisch Chemisch Laboratorium, Afd. Fyisch
Organisch Chemie - UTRECHT (The Netherlands)

M. FARINA
Universita deli Studi di MILANO (Italy)

R. FAUST
University of Akron - Institute of Polymer Science - AKRON, Ohio (USA)

J.F. FAUVARQUE
Laboratoire de Recherches de la CGE MARCOUSSIS (France)

R. FAYT
Université de LIEGE (Belgium)

A. FEHERVARI
University of Akron - Institute of Polymer Science - AKRON, Ohio (USA)

G. FINK
Max Planck Institut fur Kohlenforschung MULHEIM ad RUHR (F.R.G.)

M. FONTANILLE
Université de Bordeaux I - TALENCE (France)

R. FORTE
Institut de Chimie - Université de LIEGE (Belgique)

G. GALLI
University of Pisa - PISA (Italy)

U. GIANNINI
Montedison Research Center - NOVARA (Italy)

L. GILLIOM
California Institute of Technology - PASADENA, C
(USA)

G. GIUNCHI
Montedison Research Center - NOVARA (Italy)

R.H. GRUBBS
California Institute of Technology - PASADENA, C
(USA)

A. GUYOT
CNRS - Laboratoire des Matériaux Organiques
LYON (France)

P.A. HOLMES
ICI Laboratory - RUNCORN (Cheshire) - (U.K.)

E. HUMS
Universitat BAYREUTH (F.R.G.)

S. INOUE
University of Tokyo - TOKYO (Japan)

S. ITTEL
Experimental Station - EI Du Pont de Nemours - WILMINGTON, Delaware (USA)

K.J. IVIN
Queen's College - BELFAST (U.K.)

R. JEROME
Université de LIEGE (Belgium)

M. JULEMONT
Université de LIEGE (Belgium)

W. KAMINSKY
Universitat HAMBURG (F.R.G.)

J.P. KENNEDY
University of Akron - Institute of Polymer Science AKRON, Ohio (USA)

U. KLABUNDE
Experimental Station - EI Du Pont de Nemours - WILMINGTON, Delaware (USA)

R. KRAUSS
Universitat BAYREUTH (F.R.G.)

J. KRESS
Institut LE BEL - Université Louis Pasteur STRASBOURG (France)

T.E. LONG
Virginia Polytechnic Institute - BLACSKSBURG,Va (USA)

J.E. Mc GRATH
Virginia Polytechnic Institute - BLACSKSBURG,Va (USA)

O. MAHAMAT
Collège de France - PARIS (France)

M. MAJDOUB
Collège de France - PARIS (France)

F. MECHIN
Collège de France - PARIS (France)

M.K. MISHRA
University of Akron - Institute of Polymer Science
AKRON, Ohio (USA)

A.E. MULLER
Johannes Gutenberg Universitat - Institut fur
Physikalische Chemie - MAINZ (F.R.G.)

R.J.M. NOLTE
Organisch Chemisch Laboratorium, Afd. Fyisch
Organische Chemie - UTRECHT (The Netherlands)

N. OGATA
Sofia University - TOKYO (Japan)

A. OSBORN
Institut LE BEL - Université Louis Pasteur
STRASBOURG (France)

M.A. PETIT
Université Paris XIII - VILLETANEUSE (France)

S.L. REGEN
Lehigh University - BETHLEHEM, Pa (USA)

K.H. REICHERT
Technische Universitat BERLIN (F.R.G.)

J.J. ROONEY
Department of Chemistry - Queen's University
BELFAST (U.K.)

J.B. ROSE
University of Surrey - GUILFORD (U.K.)

G. RUGGERI
University of Pisa - PISA (Italy)

P.J. SENIOR
ICI Agricultural Division - BILLINGHAM (U.K.)

P. SIGWALT
Université Pierre et Marie Curie - PARIS (France)

B. SILLION
INSTITUT FRANCAIS DU PETROLE - LYON (France)

A. SIOVE
Université PARIS XIII - VILLETANEUSE (France)

H.D. SITZ
Universitat Gesamthorschule - ESSEN (F.R.G.)

H. SLEIMAN
Collège de France - PARIS (France)

D.Y. SOGAH
E.I. Du Pont de Nemours - WILMINGTON, Delaware (USA)

A. SOUM
Université de Bordeaux I - TALENCE (France)

R. SPITZ
CNRS - Laboratoire des Matériaux Organiques
LYON (France)

Ph. TEYSSIE
Université de Liège - LIEGE (Belgium)

K. WEISS
Universitat BAYREUTH (F.R.G.)

O.W. WEBSTER
E.I. Du Pont de Nemours - Central Research Dpt -
WILMINGTON, Delaware (USA)

G. WULFF
Universitat DUSSELDORF (F.R.G.)

R. WARI
Université de Liège - LIEGE (Belgium)